AI and Automation in Modern Manufacturing

Shalin Hai-Jew
https://orcid.org/0000-0002-8863-0175
Sedgwick County, USA

IGI Global
Scientific Publishing
Publishing Tomorrow's Research Today

Vice President of Editorial	Melissa Wagner
Director of Acquisitions	Mikaela Felty
Director of Book Development	Jocelynn Hessler
Production Manager	Mike Brehm
Cover Design	Jose Rosado

Published in the United States of America by
IGI Global Scientific Publishing
701 East Chocolate Avenue
Hershey, PA, 17033, USA
Tel: 717-533-8845 | Fax: 717-533-7115
Website: https://www.igi-global.com E-mail: cust@igi-global.com

Library of Congress Cataloging-in-Publication Data

Names: Hai-Jew, Shalin editor
Title: AI and automation in modern manufacturing / edited by Shalin Hai-Jew.
Description: Hershey, PA : IGI Global Scientific Publishing, [2026] | Includes bibliographical references and index. | Summary: "This book explores a range of relevant issues in the new manufacturing space. It presents insights on new manufacturing and focuses on techniques, technologies, human training, and more"-- Provided by publisher.
Identifiers: LCCN 2025038134 (print) | LCCN 2025038135 (ebook) | ISBN 9798337357560 h/c | ISBN 9798337357577 s/c | ISBN 9798337357584 ebook
Subjects: LCSH: Manufacturing processes--Automation | Artificial intelligence--Industrial applications | Robotics--Industrial applications | Machine learning--Industrial applications
Classification: LCC TS183 .A43 2026 (print) | LCC TS183 (ebook)
LC record available at https://lccn.loc.gov/2025038134
LC ebook record available at https://lccn.loc.gov/2025038135

British Cataloguing in Publication Data
A Cataloguing in Publication record for this book is available from the British Library.

Table of Contents

Detailed Table of Contents

Section 2
Smart Modern Manufacturing

This chapter examines the transformative role of artificial intelligence and automation in accelerating the transition toward net-zero factories and sustainable manufacturing ecosystems. It explores how intelligent systems, including digital twins, machine learning, predictive analytics, and cyber-physical production networks, enable real-time optimization of energy use, waste reduction, and carbon-efficient operations. The chapter further analyzes the integration of smart robotics, sensor-driven monitoring, and green supply chain automation to support low-emission, resource-efficient production cycles. Key challenges, emerging technologies, and future research opportunities are presented to highlight the strategic pathways required for achieving climate-aligned, resilient, and sustainable industrial environments.

This chapter examines the expanding role of drones in modern manufacturing by outlining their contributions to real-time inspection, environmental monitoring, and quality assurance. The chapter explains how drones equipped with optical, thermal, and gas-sensing technologies enhance operational visibility, detect anomalies, and strengthen regulatory compliance. It further discusses how drone-generated data supports digital twins, predictive maintenance, and smart-factory integration. Key considerations related to safety, workforce transformation, risk management, and implementation strategies are highlighted. Overall, the chapter demonstrates how drone-enabled systems are transforming manufacturing into a safer, more efficient, and data-driven industrial environment.

Preface

The inspiration for a book can seriously come from anywhere. For *AI and Automation in Modern Manufacturing,* this idea sprung from my work in grant-seeking at a community college (and one particular grant application in particular), my prior work with a major manufacturer of airplanes, a gap in the research literature about modern manufacturing, and a presidential push in the U.S. for reshoring various types of critical manufacturing (particularly those related to national security concerns). Another piece of the inspiration has come from living in an industrial town in Kansas in the central region of the U.S. and seeing a skyline of grain elevators, smokestacks, cranes, industrial vents, water towers, warehouse buildings, cranes, and scaffolding.

In academia, many may conflate DIY "making" in making spaces as a kind of small-scale and creative manufacturing. Using all sorts of stretching, that sure may seem so. But the reality of modern manufacturing is much larger, complex, difficult, and essential as compared to the lived-academic version.

In the real, AI and automation are not small-risk ventures. On-ground solutions have to be engineered with high precision. Major adaptations have to be made all along the way. They are tested in development and testing…in the virtual simulated spaces…in the small-scale real…under a range of conditions…long before anything goes into live production. No one is that careless or risk-taking to take AI, which is dynamic and high-dimensional and itself sufficiently black-box, and deploy it in expensive systems…without thorough work. Integrations magnify mystery and unknown unknowns (which are sometimes only discovered the hard way, at high cost).

Another inspiration came from several people I met who had spent years in heavy industry. One missed part of a finger. One had so battered his body and then gone to opioids for comfort that he had lost all of his teeth…as a middle-aged man. A former college student of mine had shared a story of the risks of pinhole breaks in pipes, with the intense steam emitted from those spaces invisible and capable of slicing through a passing person. The manufacturing space is a hard-driving and unforgiving one.

Then, there has been the common discussions of artificial intelligence, automation, and robotics taking on some of the high efficiencies and heavy lifting of industrial manufacturing. There are stories in mass media of dark factories, those that do not need lighting for long hours of the day because the machines essentially run themselves. Sensors monitor how the machines are running. Challenges may be troubleshot from a distance and handled remotely, in some cases.

If so, where are the pockets of expertise? What are the main research questions in the space? What are the optimizations in the world today? If most of the manufacturing for particular goods located in particular developing world spaces, is that where many of the research works would emerge?

FINDING EXPERTISE IN THE AI AND AUTOMATION SPACE

One of the challenges in this book, with work starting in 2025, was to find expertise, from those on the front lines. Would there be sufficient incentive for professionals in the field to share their insights in an academic research form? Would there be those willing to share inner workings and insights? Would there be sharing of benefits to be had? Failing that, perhaps there could be those involved in the research and development realm, and failing that, perhaps going into academia. In many cases, though, the expertise is in academia first, and then the innovations are infused out into the world.

The conundrum here is that many of the line staff may not fully understand what they are rolling out and deploying. And there is also sometimes a large gap between those working theoretical problems and challenges and the on-ground reality in which issues are being addressed. There is work in the conceptual and a lot of tweaking from the manufacturing floor.

Another element of finding expertise has to do with where large-scale and sophisticated manufacturing is happening in the world. While many in the West have a full-surround of available goods and services, the actual manufacturing work is done abroad. Some part of the supply chain is made (designed) in universities and R&D labs...but much of the doing in parts and in whole are done elsewhere. Multinational companies work their leadership hard to search the world for top-flight manufacturers to create particular goods, along a complex supply chain. They work through nation-states, cultures, political systems, economies, and personalities...to achieve the manufacturing with the optimized prices.

THE CALL FOR PROPOSALS TOPIC LIST

The call for proposals was necessarily broad: What hard problems were being solved with cutting-edge technologies in manufacturing? What sorts of smart manufacturing was going on? How were such endeavors made "green" and "environmentally sustainable"? How were human factory workers being trained and working within these more advanced factories? What skillsets did they need to have to thrive in such spaces? What was going on in R&D labs for manufacturing, and why?

- An overview of modern manufacturing in a particular discipline or space
- Hard problems in the manufacturing space
- Smart / intelligent manufacturing
- Green / sustainable manufacturing
- DIY manufacturing
- Artificial intelligence in modern manufacturing
- Automation in modern manufacturing
- Robotics in modern manufacturing
- Human factory workers and AI, automation, and robotics
- Modern factory spaces: factory floors, assembly lines, production lines, control rooms
- Dimensions in manufacturing quality
- Manufacturing work for people
- Necessary skillsets
- Manufacturing workforce development
- Skills-based learning
- Hands-on learning and practice
- Apprenticeships
- Manufacturing work for humans
- Interactions with AI
- Augmented reality (AR) interactions in the learning space
- Simulations (multisensory, immersive) in manufacturing education and training
- Enriched technology-based learning
- Opportunities for learning practice for manufacturing education and training
- The design of learning aids, aides de memoires in manufacturing education
- Professional development in manufacturing
- R&D labs in modern manufacturing
- Innovations in modern manufacturing
- Assessments in manufacturing certifications
- Supply chains

- Finding markets, market share
- Costing out manufactured products
- Setting up manufacturing processes and sequences, product lines
- Supply chains in manufacturing
- Coding various machines
- AI in manufacturing
- Automation in manufacturing
- Robotics in manufacturing
- Learning pipelines for workforce contexts

THE RESULTING DRAFTS

Over 110 some proposals were submitted, and some 88 proposals accepted. Several dozen drafts were submitted. Some dealt with topics that were not quite novel, like multi-sourcing of data to inform industrial processes, some about digital twinning of analog machinery, and the like. Quite a few used generative AI to create outlines and drafts, replete with hallucinated source citations. Many did not even bother to try for primary research. Many did not have the first-hand expertise that would enable insight while sifting through secondary and tertiary research. Many works did not make the cut.

The book is split off from the initial resulting manuscript, which had a broader focus. This collection is more succinct, with a focus on AI and automation in modern manufacturing. The collection is a stronger one with this focus. (Thanks to the editorial team at IGI Global for the inspired suggestion.)

A TABLE OF CONTENTS

The included chapters are original and insightful. They are organized into three sections:

1. Humans in Modern Technology-Augmented Manufacturing
2. Smart Modern Manufacturing
3. Applied Case Studies in Modern Manufacturing

SECTION 1: HUMANS IN MODERN TECHNOLOGY-AUGMENTED MANUFACTURING

Chapter 1: Manufacturing Workforce Development in Modern Ai and Automation Era

Senthil Kumar B, CMS Business School, Jain University, India; Elantheraiyan P, RMK Engineering College, India; and Vasudevan M., Hindustan Institute of Technology and Management, Chennai, India

The manufacturing industry is experiencing a significant transformation propelled by AI and automation, which is altering production methods, job functions, and labor frameworks. As routine tasks are automated, there is an increasing need for a workforce that possesses both technical and interpersonal skills. This transition requires interdisciplinary training, ongoing education, and collaboration between industry and educational institutions. While AI enhances efficiency, it also poses risks of job displacement and inequality if inclusive policies are not implemented. Workforce initiatives must focus on diversity, data literacy, and psychological support to facilitate a smooth transition. Customized strategies are essential on a global scale, as the effects of automation differ by region. By prioritizing human-centered automation, lifelong learning, and equity, we can cultivate a resilient workforce that is prepared for the future.

Chapter 2: Immersive Training for the AI-Enabled Factory: AR and Simulation-Based Workforce Development

Otmane Azeroual, University of Hagen, Germany

This chapter examines the potential of immersive training methods such as Augmented Reality (AR), Virtual Reality (VR), Mixed Reality (MR), and simulation-based approaches including Digital Twins in AI-enabled factories. It presents the theoretical foundations, pedagogical concepts, and practical applications, while discussing opportunities, challenges, and future developments. The chapter highlights how employees can be effectively prepared for complex, data-driven production processes and emphasizes the significance of immersive training for the competitiveness and innovation capacity of modern manufacturing companies.

SECTION 2: SMART MODERN MANUFACTURING

Chapter 3: Revolutionizing Production: AI and Automation Driving Industrial Excellence

Saranya V. S., SRM Institute of Science and Technology, India; Shobana J., SRM Institute of Science and Technology, India; Renugadevi Ramalingam, Saveetha Engineering College, India; L. Annie Isabella, Saveetha School of Engineering, Saveetha Institute of Medical and Technical Sciences, India; Priyanka N., Vellore Institute of Technology, India; Arthi K., SRM Institute of Science and Technology, India

Recent advancements in intelligent systems and industrial automation have accelerated the evolution of digital manufacturing ecosystems, contributing significantly to the realization of Industry 4.0. This chapter investigates the transformative role of Artificial Intelligence (AI)-driven intelligent automation, emphasizing the convergence of Robotic Process Automation (RPA) and AI technologies in optimizing complex industrial and enterprise operations. RPA systems are designed to execute repetitive, deterministic tasks based on structured logic, enabling improved operational throughput with minimal human intervention. When integrated with AI techniques such as machine learning algorithms, natural language processing (NLP), and neural network-based data modelling, these systems acquire cognitive capabilities, including data inference, anomaly detection, predictive analytics, and semantic classification.

Chapter 4: Artificial Intelligence in Manufacturing: A Conceptual Synthesis of Applications, Challenges, and Future Direction

Ardra Rianisa, Universitas Esa Unggul, Indonesia; Binastya Anggara Sekti, Universitas Esa Unggul, Indonesia

This chapter discusses the impact of Artificial Intelligence (AI) on manufacturing, focusing on applications like predictive maintenance, quality control, and automation. AI enhances operational efficiency by predicting equipment failures, reducing downtime, and improving product quality through automated defect detection. The concept of smart factories, powered by AI and the Internet of Things (IoT), optimizes workflows and adapts to market demands. However, challenges such as high implementation costs, integration with legacy systems, and data security concerns persist, especially for small and medium-sized enterprises. The chapter also covers the potential for AI to drive sustainability in manufacturing by optimizing energy consumption and reducing waste, and explores future trends like autonomous manufacturing systems and AI-powered supply chains

Chapter 5: Artificial Intelligence in Modern Manufacturing

Senthil Kumar B., CMS Business School, Jain University, India; Elantheraiyan P., RMK Engineering College, India; Thandayuthapani S., Kristu Jayanti University, India; Nirmal Raj R. J. T., SRM University, Sikkim, India; Loganatha Prasanna S., SRM Institute of Science and Technology, India; Vasudevan M., Hindustan Institute of Technology and Management, Chennai, India

The incorporation of Artificial Intelligence (AI) within the manufacturing sector signifies a crucial transition towards Industry 4.0, converting conventional systems into intelligent, automated, and data-centric operations. AI technologies—such as machine learning, computer vision, and the Internet of Things (IoT)—improve productivity, facilitate predictive maintenance, enhance quality control, and bolster supply chain resilience. Smart factories serve as a prime example of this transformation, allowing for real-time decision-making and promoting sustainability. Nevertheless, challenges such as cybersecurity threats, data privacy concerns, and workforce displacement remain, highlighting the necessity for ethical AI implementation and workforce reskilling. To fully harness the potential of AI, manufacturers must integrate it into their long-term strategies as a strategic enabler that aligns with human insight and innovation.

SECTION 3: APPLIED CASE STUDIES IN MODERN MANUFACTURING

Chapter 6: AI and Automation for Net-Zero Factories and Sustainable Manufacturing

Shalom Akhai, MMEC, Maharishi Markandeshwar University, Mullana, India; Mahapara Abbass, MMEC, Maharishi Markandeshwar University, Mullana, India

This chapter examines the transformative role of artificial intelligence and automation in accelerating the transition toward net-zero factories and sustainable manufacturing ecosystems. It explores how intelligent systems, including digital twins, machine learning, predictive analytics, and cyber-physical production networks, enable real-time optimization of energy use, waste reduction, and carbon-efficient operations. The chapter further analyzes the integration of smart robotics, sensor-driven monitoring, and green supply chain automation to support low-emission, resource-efficient production cycles. Key challenges, emerging technologies, and future research opportunities are presented to highlight the strategic pathways required for achieving climate-aligned, resilient, and sustainable industrial environments.

Chapter 7: Integrating Drones into Manufacturing: Real-Time Inspection, Environmental Monitoring, and Quality Assurance

Shalom Akhai, MMEC, Maharishi Markandeshwar University, Mullana, India; Mahapara Abbass, MMEC, Maharishi Markandeshwar University, Mullana, India

This chapter examines the expanding role of drones in modern manufacturing by outlining their contributions to real-time inspection, environmental monitoring, and quality assurance. The chapter explains how drones equipped with optical, thermal, and gas-sensing technologies enhance operational visibility, detect anomalies, and strengthen regulatory compliance. It further discusses how drone-generated data supports digital twins, predictive maintenance, and smart-factory integration. Key considerations related to safety, workforce transformation, risk management, and implementation strategies are highlighted. Overall, the chapter demonstrates how drone-enabled systems are transforming manufacturing into a safer, more efficient, and data-driven industrial environment.

Chapter 8: Strategic Navigation in the VUCA Manufacturing Landscape: AI, Automation, and Robotics Amidst Geopolitical Turbulence and Regulatory Transformation

Göran Roos, Adelaide University, Australia

Contemporary manufacturing organisations operate within volatile, uncertain, complex, and ambiguous (VUCA) environments characterised by rapid technological advancement, geopolitical instability, and regulatory transformation. This chapter examines strategic approaches enabling manufacturers to navigate these challenges through AI, automation, and robotics integration. Drawing upon dynamic capabilities theory, institutional theory, and strategic management literature, this analysis presents evidence-based frameworks for transforming operational volatility into competitive advantage. The chapter synthesises theoretical foundations with empirical evidence, demonstrating how smart manufacturing strategies enhance organisational resilience. Key contributions include a VUCA navigation framework, technology integration models, and guidance for managing institutional complexity. Analysis reveals successful VUCA navigation requires orchestration of technological capabilities, institutional adaptation strategies, and dynamic learning processes.

MUCH TO EXPLORE

This collection offers much to learn and to explore.

This book is presented with gratitude to the talented and hardworking contributors of this collection.

Shalin Hai-Jew
Sedgwick County, USA

Section 1
Humans in Modern Technology–Augmented Manufacturing

Chapter 1
Manufacturing Workforce Development in Modern AI and Automation Era

B. Senthil Kumar
http://orcid.org/0000-0002-1974-0645
CMS Business School, Jain University, India

P. Elantheraiyan
RMK Engineering College, India

M. Vasudevan
http://orcid.org/0009-0002-6672-7651
Hindustan Institute of Technology and Management, Chennai, India

ABSTRACT

The manufacturing industry is experiencing a significant transformation propelled by AI and automation, which is altering production methods, job functions, and labor frameworks. As routine tasks are automated, there is an increasing need for a workforce that possesses both technical and interpersonal skills. This transition requires interdisciplinary training, ongoing education, and collaboration between industry and educational institutions. While AI enhances efficiency, it also poses risks of job displacement and inequality if inclusive policies are not implemented. Workforce initiatives must focus on diversity, data literacy, and psychological support to facilitate a smooth transition. Customized strategies are essential on a global scale, as the effects of automation differ by region. By prioritizing human-centered automation, lifelong learning, and equity, we can cultivate a resilient workforce that is prepared for the future.

DOI: 10.4018/979-8-3373-5756-0.ch001

INTRODUCTION

The manufacturing sector is experiencing a swift transformation primarily fueled by advancements in artificial intelligence (AI) and automation technologies. These innovations are altering production processes, modifying job roles, and establishing new skill requirements for the workforce. Automation and AI-driven systems, including robotics, machine learning, and intelligent manufacturing platforms, significantly enhance operational efficiency, accuracy, and productivity. However, these advancements also raise concerns about job displacement, skills obsolescence, and inequalities within the workforce. Projections indicate that by 2030, automation may replace as much as 30% of manufacturing tasks worldwide, highlighting the urgent need for reskilling and upskilling initiatives (Manyika et al., 2017). This report explores various research studies that underscore the impact of AI and automation on workforce development, the evolving skill sets required by contemporary manufacturing, and effective strategies to address the related challenges.

TRANSFORMATION OF MANUFACTURING THROUGH AI AND AUTOMATION

Technological Advancements and Their Impact on Manufacturing

Automation technologies in the manufacturing sector encompass a range of solutions, from robotic process automation (RPA) to AI-driven predictive analytics. (Lee et al. 2018) highlights the role of cyber-physical systems in facilitating smart factories, where AI utilizes real-time data to enhance supply chains, maintenance, and quality assurance. AI algorithms are capable of forecasting equipment failures, scheduling maintenance, and identifying defects, thereby minimizing downtime and elevating product quality (Kusiak, 2018).

The incorporation of AI enhances decision-making by delivering detailed data insights, enabling managers to respond proactively to changes in operations (Zhong et al., 2017). Nevertheless, as automation takes over repetitive and manual tasks, there is a risk of workforce displacement. (Acemoglu and Restrepo, 2019) observe that while automation increases productivity, it simultaneously alters labor demand — diminishing routine jobs while amplifying the necessity for technical and analytical positions.

Workforce Skill Transformation

A significant theme in contemporary literature is the changing demand for skills. (Bessen, 2019) asserts that automation does not completely eradicate jobs but rather alters them. Workers are now expected to possess hybrid skills that integrate digital literacy, AI system management, and traditional manufacturing expertise. (Wamba-Taguimdje et al. 2020) emphasize that soft skills such as adaptability, problem-solving, and communication is essential for effective collaboration with AI systems.

Educational institutions must revise their curricula to meet this evolving landscape. Partnerships between industry and academia are crucial for developing interdisciplinary training that combines engineering, data science, and behavioral skills. Continuous learning frameworks are essential as AI tools and processes advance rapidly, necessitating ongoing employee development.

Addressing Job Displacement and Social Impacts

The uneven effects of automation across various regions and sectors pose significant social challenges. In advanced economies, skilled workers generally reap the benefits of automation, while low-skilled positions are at a greater risk of displacement, particularly in developing nations (UNIDO, 2019). To tackle this issue, governments and corporations need to emphasize retraining initiatives and establish robust social safety nets.

The implementation of inclusive workforce development strategies that concentrate on vulnerable populations, such as women and minorities, who are often underrepresented in AI-related manufacturing positions (Bessen et al. 2019) supports. Programs like Women in Manufacturing (WiM) and STEM outreach initiatives have demonstrated potential in improving diversity and closing the talent gap (Ely et al., 2011).

Strategies for Workforce Development

Effective strategies for workforce development focus on reskilling, upskilling, and promoting lifelong learning. Siemens and General Electric have taken the lead in establishing internal 'digital academies' that offer employees specialized training in AI and automation. Additionally, online platforms such as Coursera and Udacity provide scalable options for remote education.

It is crucial for industry, academia, and government to collaborate in order to ensure that training programs meet the demands of the labor market. Initiatives like apprenticeships, internships, and vocational training related to AI advancements enhance workforce preparedness and offer hands-on experience (Ameen, K., & Keizer,

A. 2023). Moreover, organizations that implement human-centered automation—where AI supports rather than replaces human workers—often experience greater job satisfaction and employee retention (Wilson & Daugherty, 2018).

Psychological and Organizational Considerations

Studies highlight the psychological effects of automation on employees, which include anxieties regarding job security and resistance to change (Frey & Osborne, 2017). Effective communication, engaging employees in the decision-making process, and incorporating change management strategies are essential for the seamless integration of AI technologies (Schiavone et al., 2020). Providing psychological support and fostering inclusive leadership can significantly improve resilience and adaptability within the workforce.

The transformation of the manufacturing industry through AI and automation requires a holistic approach to workforce development. It is vital to combine technical training with the development of soft skills, promote lifelong learning, and ensure inclusive participation to effectively navigate this transition. Policymakers, educators, and industry leaders must work together to devise adaptable and equitable strategies that equip workers for increasingly complex, AI-enhanced positions. By adopting these principles, the manufacturing sector can maintain its competitiveness while advancing social equity and workforce resilience.

EVOLVING SKILL REQUIREMENTS IN THE AGE OF AI AND AUTOMATION: BEYOND TECHNICAL TRAINING

As artificial intelligence (AI) and automation technologies become more deeply integrated into manufacturing and various other sectors, the nature of work and the skills required are undergoing a fundamental transformation. Scholars largely concur that conventional technical training alone does not adequately equip the workforce for the complexities presented by AI-enhanced environments. Rather, a comprehensive approach that merges digital literacy, technical proficiency, and soft skills—such as critical thinking, adaptability, and problem-solving—is crucial (Bessen, 2019; Wamba-Taguimdje et al., 2020). This section explores recent studies on the changing skill requirements and how interdisciplinary educational initiatives are influencing workforce development strategies to address the needs of contemporary technology-driven workplaces.

The Evolving Division of Labor: Automation versus Human Skills

Automation technologies, such as robotic process automation and AI-driven systems, are particularly adept at executing repetitive and routine tasks with accuracy and efficiency. This transition enables human workers to concentrate on positions that necessitate creativity, judgment, coordination, and intricate problem-solving (Marien, M, 2014). Autor (2015) underscores this division of labor by pointing out that while machines can automate predictable tasks, humans maintain an edge in areas requiring social intelligence and adaptive reasoning.

This changing dynamic carries substantial implications for workforce development. Technical skills, including the operation and maintenance of automated machinery, remain essential; however, cognitive and interpersonal skills that empower workers to interpret AI outputs, collaborate with machines, and innovate within complex production systems are equally critical (Deming, 2017). For instance, data analytics, decision-making, and troubleshooting have emerged as vital elements of numerous manufacturing roles as AI systems produce extensive amounts of operational data.

The Significance of Soft Skills in the Age of AI

Numerous studies highlight the increasing significance of soft skills alongside digital competencies. The World Economic Forum (2020) states that skills such as critical thinking, creativity, emotional intelligence, and adaptability will rank among the foremost workforce priorities by 2025. These competencies equip employees to navigate uncertain situations, devise innovative solutions, and respond adeptly to swift technological advancements.

Research conducted by Lee et al. (2021) demonstrates that adaptability and continuous learning are essential for employees collaborating with AI systems. Workers who can swiftly learn new technologies and dynamically adjust their roles make substantial contributions to organizational agility and innovation. Likewise, critical thinking skills are vital for evaluating AI-generated recommendations and making nuanced decisions that machines cannot entirely automate (Wilson & Daugherty, 2018).

Interdisciplinary Learning and Educational Collaborations

To meet the changing demands for skills, it is essential to transition from conventional, isolated educational approaches to interdisciplinary learning that integrates technical, analytical, and human-centered competencies. Educational institutions are required to establish stronger collaborations with industry to jointly develop curricula that reflect the realities of workplaces influenced by AI.

Programs that merge data science, engineering, and ethics equip students to responsibly manage AI systems and analyze data with contextual understanding (Bongomin et al., 2020). Vocational training and apprenticeship frameworks are increasingly incorporating modules on teamwork, communication, and problem-solving alongside technical subjects to cultivate well-rounded professionals who can effectively collaborate with AI (Ameen, K., & Keizer, A. 2023).

Moreover, lifelong learning structures supported by digital platforms like Coursera and edX allow workers to continuously enhance their skills as new technologies arise. Huang and Rust (2021) contend that integrating soft skills development into ongoing training programs creates a workforce that is resilient to disruptions and capable of leveraging the advantages of AI.

Workforce Development in AI-Driven Manufacturing

Numerous corporate examples illustrate effective practices in the integration of both technical and soft skills training. Siemens' Digital Academy offers employees extensive training that encompasses robotics operation, AI system management, and teamwork skills that are crucial for Industry 4.0 settings. In a similar vein, General Electric's Brilliant Learning platform provides tailored learning pathways that merge data analytics, cybersecurity, and leadership skills (GE, 2023).

Research conducted by Bessen et al. (2019) further suggests that organizations that invest in interdisciplinary workforce development experience greater productivity improvements and reduced turnover rates. In contrast, companies that concentrate exclusively on technical training often face challenges related to workforce adaptability and innovation.

Challenges and Policy Implications

Despite the increasing acknowledgment of the importance of soft skills, obstacles persist in the execution of comprehensive workforce development initiatives. Numerous educational systems are sluggish in adopting interdisciplinary curricula, and there remains a significant shortage of qualified trainers who can effectively deliver integrated technical and soft skills training (UNESCO, 2021).

Policymakers need to encourage collaboration between academic institutions and industry, enhance digital infrastructure for remote learning, and emphasize equitable access to upskilling opportunities to avert the expansion of skills gaps (Solomon, B. D., 2023). Furthermore, fostering diversity in AI-related sectors can enhance problem-solving and innovation by integrating varied perspectives into the design and implementation of AI systems (Ely et al., 2011).

The advancement of manufacturing and various sectors due to AI and automation necessitates a comprehensive workforce development strategy that transcends conventional technical training. By combining digital literacy with critical thinking, adaptability, and interpersonal skills, employees are better prepared to succeed alongside intelligent machines. It is essential for educational institutions, industry leaders, and policymakers to work together in creating interdisciplinary curricula and lifelong learning systems. This collaboration guarantees that the workforce stays flexible, innovative, and ready for the continuous technological changes that are defining the future of work.

LIFELONG LEARNING AND CONTINUOUS EDUCATION IN THE AGE OF AI AND AUTOMATION

The swift advancement of artificial intelligence (AI) and automation technologies is transforming the manufacturing industry and various sectors globally. Although these innovations have significantly enhanced productivity and operational accuracy, they have also posed considerable challenges concerning workforce skills and employability (Manyika et al., 2017). A crucial response to this issue is the focus on lifelong learning and continuous education as vital strategies to ensure workforce relevance and competitiveness in an AI-driven economy (Schwab, 2016). This report investigates recent studies and case analyses that delve into the importance of ongoing skill development, corporate digital academies, online learning platforms, and policy frameworks aimed at supporting workers during this transformative era.

The Importance of Lifelong Learning in a Rapidly Evolving Technological Environment

Artificial intelligence and automation technologies are progressing at a speed that surpasses the capacity of traditional educational systems to provide workers with current skills (Bessen, 2019). According to the World Economic Forum (2020), it is projected that by 2025, half of all employees will need to undergo reskilling due to the adoption of new technologies. Continuous learning has transitioned from being an option to becoming a necessity for workers to adjust to changing job roles, particularly in manufacturing, where automation increasingly performs routine tasks, allowing humans to focus on supervisory, analytical, and creative roles (Deming, 2017).

A study conducted by Noe et al. (2014) emphasizes that lifelong learning not only aids in the acquisition of technical skills but also enhances adaptability, critical thinking, and digital literacy—essential skills for effectively collaborating with AI systems. The notion of lifelong learning encompasses more than just formal education; it also includes informal and on-the-job training, highlighting the significance of ongoing knowledge renewal throughout an individual's professional journey.

Corporate Digital Academies: Addressing Internal Skill Gaps

Numerous large corporations acknowledge the critical need for continuous workforce development and have made substantial investments in internal digital academies aimed at equipping employees with current training in AI, robotics, data analytics, and related technologies (Cascio & Montealegre, 2016). Notable examples include Siemens' Digital Academy and General Electric's Brilliant Learning platform, which provide tailored learning pathways that merge technical skills with the development of soft skills.

These digital academies utilize blended learning methodologies that combine virtual classrooms, simulation environments, and project-based tasks, allowing employees to acquire hands-on experience in managing and utilizing AI-driven systems. Bessen et al. (2019) discovered that companies investing in such ongoing training experience higher levels of employee engagement, innovation, and operational efficiency, highlighting the significance of corporate-led lifelong learning initiatives.

Online Learning Platforms: Enhancing Access and Flexibility

The emergence of Massive Open Online Courses (MOOCs) and digital learning platforms like Coursera, edX, Udacity, and LinkedIn Learning has made education more accessible, enabling skill development for workers across various geographic and socioeconomic backgrounds. These platforms offer flexible and scalable opportunities for employees to learn new skills, update their existing knowledge, and even obtain certifications pertinent to AI and automation technologies.

Research conducted by Xu and Jaggars (2014) indicates that online learning boosts participation among working adults who may otherwise encounter obstacles to traditional classroom education due to time, financial, or geographical limitations. Furthermore, adaptive learning technologies enhance the learning experience by personalizing it, enabling individuals to concentrate on their most pressing needs and progress at their own pace.

Challenges to the Implementation of Lifelong Learning

Despite the encouraging advancements, various obstacles impede the successful execution of lifelong learning strategies. A major hindrance is the absence of comprehensive workforce development policies that unify the efforts of government, academia, and industry (Solomon, B. D., 2023). In the absence of coordinated initiatives, disparities in access to training may worsen existing social inequalities, leaving marginalized groups behind and contributing to labor market polarization (Autor, 2015).

Another significant challenge is the swift obsolescence of skills in the age of AI. As technology progresses, training programs require frequent updates, necessitating considerable investment in curriculum development and instructor proficiency (Bessen, 2019). Additionally, psychological factors such as employee resistance to change, anxiety over job loss, and digital apprehension can hinder participation in ongoing education programs (Frey & Osborne, 2017).

Policy Frameworks and Collaborative Efforts Among Stakeholders

To address these challenges, researchers and policymakers recommend multi-stakeholder strategies that align incentives across government, industry, and educational sectors. The European Union's Digital Skills and Jobs Coalition serves as a prime example of such collaboration, aiming to enhance digital skills throughout the workforce via partnerships that provide training, certification, and career guidance (European Commission, 2020).

In a similar vein, the U.S. Department of Labor's Workforce Innovation and Opportunity Act support funding for reskilling initiatives designed for emerging industries, fostering private-sector engagement and apprenticeship models (U.S. DOL, 2020). International entities like UNESCO stress the importance of incorporating digital skills training into lifelong learning policies to ensure equitable access and inclusivity (UNESCO, 2021).

The Influence of Lifelong Learning on Social Equity and Labor Market Outcomes

Empirical research indicates that ongoing education has a beneficial impact on employability, wage increases, and career advancement. For example, a study conducted by Klerk et al. (2020) revealed that individuals who participate in lifelong learning exhibit greater resilience to job displacement caused by automation and are more inclined to move into higher-skilled positions.

Furthermore, providing equitable access to lifelong learning opportunities can help bridge skills gaps that disproportionately impact women, minorities, and low-income workers (Ely et al., 2011). Initiatives focused on diversity and inclusion within training programs not only promote social equity but also boost organizational innovation by integrating diverse viewpoints in the development and implementation of AI technologies.

Exemplary Models and Insights Gained

The digital academies of Siemens and General Electric serve as exemplary models for integrating lifelong learning into corporate culture. Siemens has reported that more than 85% of participants show enhanced technical skills and greater confidence in AI-related positions after completing their digital academy programs.

In the educational sector, programs such as P-TECH, a partnership involving IBM, public schools, and community colleges, provide pathways for young individuals to acquire STEM skills that align with industry requirements, emphasizing the importance of continuous education from secondary to tertiary levels.

Nevertheless, challenges persist in expanding these initiatives within small and medium enterprises (SMEs) and ensuring access to lifelong learning in developing countries, where issues related to infrastructure and funding continue to be significant barriers (Ameen, K., & Keizer, A. 2023).

Continuous education and lifelong learning are essential strategies for ensuring workforce relevance in the face of rapid advancements in AI and automation. Corporate digital academies and online learning platforms offer scalable and flexible training opportunities; however, their effectiveness relies on well-coordinated policy frameworks and collaboration across multiple sectors. It is vital to tackle challenges associated with access, curriculum updates, and psychological barriers to avoid exacerbating social inequalities. When implemented effectively, lifelong learning not only improves individual employability but also fosters organizational innovation and promotes social equity within the changing manufacturing landscape.

PROMOTING DIVERSITY AND INCLUSION IN THE WORKFORCE IN THE ERA OF AI AND AUTOMATION

Artificial intelligence (AI) and automation are fundamentally reshaping the manufacturing workforce, necessitating a wide range of new technical and interpersonal skills. In conjunction with these technological advancements, critical discussions regarding workforce diversity and inclusion (D&I) are emerging. As the manufacturing sector increasingly integrates AI-driven systems, the promotion of diversity becomes not only a social necessity but also a strategic advantage that fosters innovation, enhances problem-solving capabilities, and reduces algorithmic biases (West et al., 2019). This report examines research on the challenges encountered by underrepresented groups in AI-enhanced manufacturing positions, assesses diversity-oriented initiatives, and investigates the significance of inclusive education and training in promoting equitable workforce development.

The Importance of Diversity in AI-Driven Manufacturing

Diversity within the workforce includes variations in gender, ethnicity, age, socioeconomic status, and cognitive viewpoints. Researchers contend that diversity is especially crucial in AI-driven settings, where the design, implementation, and management of algorithms have a profound effect on ethical and operational results (Buolamwini & Gebru, 2018). AI systems that are trained on non-representative datasets are at risk of perpetuating biases, which can worsen discrimination and social inequalities (Friedman & Nissenbaum, 1996).

West et al. (2019) emphasize that diverse teams enhance algorithm design by incorporating a range of perspectives that can uncover blind spots in AI development. Additionally, organizations that cultivate inclusive cultures demonstrate improved creativity, decision-making, and adaptability—traits that are essential for navigating the complexities brought about by AI and automation. Thus, promoting diversity transcends mere compliance; it serves as a catalyst for competitive advantage in the manufacturing sector.

Challenges Faced by Underrepresented Groups

In spite of the evident advantages, underrepresented groups—including women, racial and ethnic minorities, and individuals from low-income backgrounds—persistently encounter significant obstacles in securing AI-related manufacturing positions. The National Science Foundation (2020) reports that women represent less than 30% of the STEM workforce, with minority representation being disproportionately low in technical disciplines.

Structural challenges encompass gender stereotypes, a scarcity of role models, educational inequities, and workplace environments that may not foster diversity (Cheryan et al., 2017). The "leaky pipeline" phenomenon illustrates how women and minorities exit STEM pathways at various educational and career stages due to unwelcoming environments and insufficient support (Moss-Racusin et al., 2018). These challenges are intensified in AI-related fields, where swift technological advancements necessitate ongoing learning and adaptation, potentially worsening exclusion if training opportunities are not accessible or inclusive (West et al., 2019).

Initiatives to Promote Diversity and Inclusion in AI-Enhanced Manufacturing

Numerous initiatives and programs have been established to tackle diversity issues within the AI and manufacturing sectors. Women in Manufacturing (WiM) stands out as a key initiative aimed at boosting women's involvement through networking, mentorship, and advocacy (WiM, 2023). WiM provides training programs that focus on both technical and leadership competencies, empowering women to progress in AI-enhanced positions.

In a similar vein, STEM outreach initiatives aimed at underrepresented youth, including programs like Black Girls Code and Code2040, offer early exposure to coding and the fundamentals of AI, with the goal of broadening the talent pipeline (Scott et al., 2020). Collaborations between corporations and educational institutions have also resulted in scholarships, internships, and apprenticeships intended to reduce barriers for minorities and women pursuing careers in manufacturing that involve AI and automation (Google, 2022).

Research conducted by McKinsey & Company (2020) highlights that effective diversity and inclusion (D&I) strategies merge recruitment efforts with inclusive workplace practices, such as bias training, fair promotion policies, and adaptable work arrangements. Organizations that adopt these strategies tend to exhibit higher retention rates and enhanced innovation outcomes.

The Importance of Inclusive Training and Education

Providing equitable access to education and inclusive training programs is essential for enabling all demographic groups to engage fully in AI-enhanced manufacturing. Educational programs should be crafted to be culturally relevant, gender-sensitive, and accessible to learners from various backgrounds and learning styles (UNESCO, 2021).

Research indicates that interdisciplinary learning—integrating technical AI expertise with the development of soft skills—encourages wider participation and success. Engaging in hands-on, project-based learning within diverse teams promotes collaboration and aids in breaking down stereotypes by prioritizing competence over identity (et al., 2015).

Furthermore, if designed with inclusivity in mind, online platforms and digital academies can help close geographic and socioeconomic divides, providing flexible learning opportunities for underrepresented groups. Nevertheless, the issue of digital divides persists; therefore, investments in infrastructure and outreach are crucial to guarantee equitable digital access.

The Influence of Diversity on Innovation and Bias Reduction

Diverse teams engaged in AI manufacturing have shown improved problem-solving skills and stronger innovation results. Research conducted by Hewlett et al. (2013) indicates that diverse groups generate higher-quality ideas due to their varied perspectives and constructive disagreements.

From the perspective of AI ethics, diverse teams are more capable of recognizing potential biases in AI models and executing strategies to alleviate discriminatory results (Holstein et al., 2019). For instance, including women and minorities in data annotation and model validation processes aids in minimizing representational biases in AI systems utilized on the manufacturing floor (Buolamwini & Gebru, 2018). Furthermore, diversity within leadership shapes organizational priorities, promoting investments in responsible AI governance and strategies for inclusive workforce development (Ely et al., 2011).

Policy Suggestions for Advancing Diversity and Inclusion

Policy frameworks are vital in promoting diversity and inclusion (D&I) in AI-driven manufacturing. Government-led incentives, such as grants and tax credits for organizations that show a commitment to diverse hiring and inclusive training, have demonstrated beneficial outcomes (U.S. Department of Labor, 2020). Legislation that enforces equal opportunity, anti-discrimination laws, and transparency in hiring and compensation practices remains a critical foundation. Moreover, public-private partnerships that support scholarships, apprenticeships, and community outreach initiatives help to increase the participation of underrepresented groups (European Commission, 2020). Collaboration across sectors, including industry, academia, and civil society, is essential to track progress and exchange best practices, ensuring that D&I initiatives stay pertinent to the changing technological landscape (Ameen, K., & Keizer, A, 2023).

Promoting diversity and inclusion in AI-enhanced manufacturing presents a complex challenge that necessitates collaborative efforts across education, corporate practices, and policy. Research highlights the essential role that diverse teams play in driving innovation, reducing bias, and developing ethical AI. Initiatives like Women in Manufacturing and focused STEM outreach programs serve as effective models for increasing the participation of underrepresented groups. Education and training programs that promote inclusivity and tackle structural barriers are essential for preparing a diverse workforce with the skills required for the future of manufacturing. Ultimately, nurturing diversity is not merely an issue of fairness but a strategic necessity for maintaining competitiveness and fostering innovation in the AI-driven industrial environment.

REGIONAL AND SECTORAL VARIATIONS IN AUTOMATION'S IMPACT ON MANUFACTURING EMPLOYMENT

Artificial Intelligence and automation are transforming the manufacturing landscape worldwide; however, the repercussions differ markedly across regions and sectors. Advanced economies typically adopt automation in ways that enhance skilled labor and boost productivity, while developing nations experience a swift displacement of low-skilled workers, often without adequate compensation mechanisms. This disparity emphasizes the necessity for customized workforce development strategies.

Automation in Advanced Economies: Enhancing Skilled Labor

Research indicates that in developed economies, automation is more likely to enhance skilled labor rather than completely replace it. For example, the established manufacturing regions in China illustrate how automation can boost employment for high-skilled workers, even as low-skilled jobs diminish—leading to a complex "inverted U-shaped" employment impact that varies with regional AI development (Frontiers).

In Europe, while automation has heightened work intensity, it has had minimal adverse effects on job quality in terms of autonomy or working conditions (arXiv). Additionally, case studies such as Siemens' smart factories in Germany reveal that productivity can increase without a reduction in workforce numbers, attributed to reskilling initiatives and a focus on human-centered automation (TIME).

Developing Economies: Accelerated Displacement Without Safeguards

Developing nations encounter a unique challenge: a significant number of workers are employed in low-skilled, automation-susceptible jobs, coupled with insufficient retraining resources. For instance, the garment industry in Bangladesh faces the potential of displacing thousands of workers due to the lack of effective upskilling programs (Vogue Business, SpringerLink). The World Bank reports that in East Asia and the Pacific, robots have displaced around 1.4 million low-skilled workers while simultaneously generating approximately 2 million skilled positions—underscoring both the opportunities and disparities present (World Bank).

Other analyses suggest that automation poses a threat to as much as 77% of jobs in China, 69% in India, and potentially 85% in Ethiopia—representing a crisis-level threat for developing economies (Twn, Wikipedia). The notion of premature deindustrialization adds further complexity, as developing countries may lose manufacturing jobs without successfully transitioning to service-sector employment (Wikipedia).

Sectoral and Regional Disparities: A Deeper Dive

Sectoral Differences

The effects of automation differ significantly across various sectors. Routine manufacturing jobs are particularly at risk, whereas positions that are professional, technical, and require cognitive skills are on the rise (World Bank, arXiv). In certain developing areas, the increased use of robots has been associated with enhanced labor productivity—especially in regions where human capital and research and development are already robust (SpringerLink, World Bank, arXiv).

Regional Variability

Within individual countries, regions exhibit varying outcomes from automation. In China, certain areas—like the East Coast—demonstrate a favorable net impact on high- and middle-skilled employment, while others encounter employment polarization, especially among lower-skilled demographics (Frontiers, PMC). Smaller cities in the United States are more susceptible to automation due to a lack of occupational specialization—contrasting with larger urban areas where jobs that cannot be automated are more common (arXiv).

Compensation Mechanisms: Job Creation through Automation

A comprehensive literature review reveals that while automation does lead to job displacement, compensatory mechanisms—such as growth driven by demand and the emergence of new industries—frequently mitigate these losses (arXiv). Analysis from MIT further indicates that developing nations may experience a more significant impact on routine jobs due to automation compared to developed countries, yet they may still gain from productivity improvements if managed effectively (MIT Sloan Management Review).

Workforce Development: Strategies Centered on Humans

To effectively navigate this complex landscape, several focused strategies are necessary:

Human-Centered Automation

"Co-bots" facilitate collaboration between humans and machines, safeguarding jobs while enhancing efficiency. The UK is advocating for robotics apprenticeships to cultivate such skills (The Times). Concurrently, initiatives led by unions in Australia are advocating for AI training, equitable work hours, and inclusive design to ensure that technology elevates rather than diminishes the workforce (The Australian).

Apprenticeships and Technical Training

The UK government has allocated £275 million to support training and apprenticeships in AI and advanced manufacturing, with the goal of strengthening domestic talent pipelines (The Guardian). In Australia, regional technical facilities also provide accelerated programs in robotics, engineering, and emerging technology sectors (Vogue Business, Financial Times).

Scaling Across Regions

Local educational institutions and public universities are essential in equipping mid-tier regions. In the United States, rural universities utilize flexible credentialing to meet the demands of emerging skill requirements (Midland Daily News).

Policy Suggestions for Equitable Transitions

Differentiation by development stage: Advanced economies should concentrate on enhancing human roles, while developing nations need to focus on scalable reskilling and job support systems.

Emphasize inclusive automation: Implement human-centered models to maintain the quality of employment.

Invest in vocational education and apprenticeships: Collaboration between government and industry can bridge the skills gap.

Monitor and respond regionally: Data-driven, localized interventions enable customized workforce development.

Strengthening compensation mechanisms: Support for innovative-driven sectors can mitigate job losses and promote economic growth.

The impact of automation on manufacturing employment varies significantly, influenced by regional maturity, industrial structure, and human capital infrastructure. Advanced economies have the opportunity to leverage high-tech automation to enhance skilled labor, while developing economies face the risk of increased job polarization. Comprehensive and regionally tailored workforce development strategies, including apprenticeships, co-bot deployment, and policy-supported training investments—are vital to ensure equitable and resilient labor transitions.

THE GROWING SIGNIFICANCE OF DATA LITERACY AND DIGITAL COMPETENCIES IN MANUFACTURING

The integration of AI in manufacturing has significantly increased the amount, diversity, and speed of data produced throughout various processes—from predictive maintenance and quality control to supply chain management. Consequently, data literacy and digital competencies—such as cybersecurity, cloud computing, and software troubleshooting—are now imperative for frontline workers and supervisors (Wikipedia, "Digital literacy"; Coursera via WEF) (Wikipedia, World Economic Forum). Employees are required to convert raw data into actionable insights, thereby promoting a culture of continuous improvement across operations.

Understanding Data Literacy and Its Significance

Data literacy is defined as the capability to read, interpret, and effectively communicate data within a practical context (Wikipedia—Data Literacy) (Wikipedia). In the manufacturing sector, this skill is crucial as sensor-driven systems and IoT devices continuously provide complex data streams. Workers must transcend basic awareness to critically assess data outputs, align them with production goals, and facilitate AI-driven decision-making.

Essential Digital Skills for Contemporary Manufacturing

Data Analysis and Cloud Computing

According to Coursera's industry report, cloud computing and data analysis rank among the most in-demand skills across various sectors, including manufacturing (World Economic Forum). AWS has identified these skills—along with cybersecurity and machine learning—as vital by 2025, particularly in the post-pandemic era as workplaces become increasingly digitally integrated (The Hindu Business Line, World Economic Forum).

Cybersecurity and AI

The increasing need for cybersecurity experts is in line with the rise of cloud technology, with global skill shortages surpassing millions of positions (Finextra Research, OECD, ISC2). The ISC2 2024 report highlights that education and training in cybersecurity have become essential rather than optional, especially as AI is integrated into standard business practices (ISC2).

Applications in Manufacturing: Data-driven and Intelligent Maintenance

Predictive vs. Intelligent Maintenance

Artificial Intelligence and the Industrial Internet of Things facilitate predictive and intelligent maintenance strategies, utilizing deep learning, real-time sensor information, and mobile analytics for proactive system maintenance (arXiv).

Manufacturing IoT and Analytics

Smart manufacturing produces extensive IoT data that requires sophisticated analytics and Big Data capabilities to derive actionable operational insights (arXiv).

Cybersecurity in Digital Manufacturing

As Industry 4.0 systems incorporate sensors and networks, the importance of cybersecurity escalates to protect data and maintain operational integrity (arXiv).

Industry and Policy-backed Training Initiatives

Corporate and Educational Programs

- AWS & Gallup/Bucknell Research: Training in digital skills contributes to national GDP—$1.1 trillion in the U.S. and $6.3 trillion worldwide. Employees with advanced digital skills experience significantly greater salary increases, career advancement, and productivity (About Amazon).
- India-AWS Study: Manufacturing employees anticipate a need for cloud architecture design and software skills, with digital literacy deemed essential by 2025 (US Press Center).

- Nasscom-KPMG India Survey: 70% of companies emphasize the development of cloud, AI/ML, and related skills; digital skills have a lifespan of 2.5–5 years, necessitating agile training (FinTechBizNews).
- IBM SkillsBuild: Provides over 1,000 courses in AI, cloud, cybersecurity, data analytics, and soft skills—aimed at underrepresented communities (Wikipedia).

Government and Institutional Actions

- UK TechFirst Initiative: A £187M program aims to enhance digital and AI skills for individuals of all ages, focusing on inclusion, lifelong learning, and regional adaptability (TechRadar).
- University-Industry Collaboration: The University of Delhi's alliance with Google Cloud offers practical training in AI, cloud computing, cybersecurity, and data analytics (The Times of India).

Challenges and Workforce Implications

Despite the existence of training initiatives, skills gaps continue to be a challenge. Cloud computing and cybersecurity are identified as the primary unmet needs worldwide, with organizations facing difficulties in recruiting suitable talent (Verdict, Finextra Research). The rapid pace of technological advancement further shortens the relevance of essential skills, and a significant number of workers still do not have access to structured training—an issue underscored by the AI education crisis in Australia, prompting AWS to step in to enhance the skills of both students and parents (The Australian).

Recommendations for Workforce Development Strategy

- Invest in Core Digital Literacy: Training in data analysis, cloud computing, AI, and cybersecurity should be incorporated into both technical and frontline workforce development programs.
- Promote Lifelong Learning Models: Focus on accessible and inclusive upskilling through internal platforms, partnerships with technology providers, and flexible certification options.
- Align with Emerging Technologies: Equip workers for the demands of intelligent maintenance, IIoT, and cybersecurity through hands-on, scenario-based curricula.

- Leverage Public-Private Partnerships: Utilize funding and expertise from government entities, technology platforms, and educational institutions to enhance educational initiatives.
- Prioritize Inclusivity and Accessibility: Ensure that training programs reach diverse and underrepresented groups to address long-term inequality.

The transition of manufacturing towards data-driven operations amplifies the necessity for data literacy and sophisticated digital skills. Today's workforce is required to integrate mechanical knowledge with analytics, cybersecurity awareness, cloud expertise, and software proficiency. To effectively develop the workforce, it is essential to invest in education, ensure inclusive access, and promote lifelong learning, supported by collaboration between the public and private sectors. By fostering these skills, the manufacturing sector can develop a digitally resilient workforce that is capable of maintaining innovation and competitiveness in the era of Industry 4.0 and beyond.

ORGANIZATIONAL AND PSYCHOLOGICAL DIMENSIONS OF WORKFORCE DEVELOPMENT IN THE AGE OF AI AND AUTOMATION

Artificial Intelligence (AI) and automation are transforming manufacturing and other industries, altering job structures and workplace dynamics. In light of these transformations, workforce development must consider not only technical training but also the psychological and organizational effects on employees. Feelings of anxiety, job insecurity, resistance, and uncertainty regarding future career trajectories can impede successful adoption and integration. Research indicates that organizations tend to be more successful when they implement transparent communication, inclusive decision-making, and psychological support as integral components of their change management strategies.

Employee Perceptions, Anxiety, and Career Insecurity

Research indicates that employees frequently encounter increased career anxiety with the introduction of AI systems:

- A study conducted in Pakistan's manufacturing and services sectors identified a significant correlation between perceptions of AI and heightened career anxiety, with career insecurity serving as a crucial mediator in employee motivation and career decisions (Critical Social Sciences Review).

- A comprehensive review of research underscores how stress induced by AI manifests as job insecurity, heightened stress levels, diminished mental health, and burnout, particularly in positions that are highly susceptible to automation. Importantly, coaching leadership has the potential to alleviate these negative outcomes (PMC).

The transition of manufacturing towards data-driven operations has heightened the necessity for data literacy and advanced digital skills. Today's workforce must integrate mechanical knowledge with analytics, cybersecurity awareness, cloud expertise, and software proficiency. To effectively develop the workforce, it is essential to invest in education, ensure inclusive access, and promote lifelong learning, supported by collaboration between the public and private sectors. By fostering these skills, the manufacturing industry can develop a digitally resilient workforce that is capable of maintaining innovation and competitiveness in Industry 4.0 and beyond.

Mental Well-Being and Automation-Related Stress

The introduction of robot-assisted work and surveillance presents unique psychological challenges:

- Workers in environments dominated by robots, such as warehouses, have reported lower job satisfaction, a reduced sense of purpose, less autonomy, and heightened stress and feelings of isolation (Financial Times).
- AI tools may unintentionally increase workload and diminish job quality. For example, at Amazon, software engineers have indicated that AI assistance leads to tighter deadlines and fewer opportunities to acquire essential skills (New York Post).

Trust, Transparency, and Ethical Implementation

For employees to embrace AI, organizations must emphasize transparency and trust:

- The reliability of AI systems, particularly in human resources, relies significantly on ethical governance, fairness, transparency, and privacy. In the absence of these elements, the risk of backlash and perceptions of unfairness increases (SA Journal of Human Resource Management, PubMed).
- A systematic review indicates that successful AI integration necessitates clear communication, management support, training, ethical oversight, and reporting mechanisms (PubMed).

Employee Involvement and Change Management

Responsibility for automation is enhanced when employees are actively engaged:

- In studies on RAIA implementation, the importance of human decision-making and interpersonal skills remains significant. Employees who view AI as a tool for enhancement rather than a replacement generally experience greater security and satisfaction. Their confidence in acquiring new skills enhances their employability (SpringerLink).
- Surveys conducted among workers in nine countries indicated that employees who felt appreciated, motivated, and involved reported greater advantages from AI, including enhanced safety, autonomy, and compensation, even if they had reservations. Financial incentives further improved their perceptions of automation (arXiv).

Leadership Style, Technophilia, and Learning Culture

The willingness of employees to embrace AI is influenced by leadership and personal attitudes:

- Research published in Frontiers in Psychology revealed that paradoxical leadership, which effectively balances structure with flexibility, along with technophilia (a passion for technology), significantly promotes learning and knowledge sharing in workplaces that adopt AI (Frontiers).

Organizational Risks: Fatigue, Bias, and Alienation

Organizations must be vigilant against deeper challenges:

- Change fatigue, a form of burnout resulting from continuous organizational change, can diminish engagement and hinder transformation initiatives (Wikipedia).
- Employees may succumb to automation bias, excessively relying on AI suggestions even when they are flawed, necessitating training, calibration, and awareness (Wikipedia).
- Algorithm aversion—where employees distrust AI in critical decision-making due to perceived deficiencies in explainability or accountability—highlights the necessity for transparent and ethical AI design (Wikipedia).

Best Practices: Psychological Supports in Workforce Development

Combining insights from research, organizations can adopt best practices to build resilience and trust:

Transparent Communication & Inclusion

Transparent Communication and Inclusion are vital components in the management of change, particularly during major transitions such as the incorporation of artificial intelligence (AI) in the workplace. Engaging in open dialogue and establishing clear expectations can alleviate fear and uncertainty among employees. When individuals comprehend the reasons behind changes, what occur, and how they will be impacted, they are more inclined to feel secure and supported. Institutions like Arizona State University (ASU) exemplify this methodology by emphasizing inclusive decision-making, which incorporates diverse perspectives into the discussion and cultivates a sense of collective ownership. This approach not only enhances trust but also encourages greater commitment from stakeholders across all levels.

Furthermore, establishing accessible avenues for feedback and consultation underscores the importance of human input in an increasingly automated environment. For instance, the reversal of AI-related layoffs at the Commonwealth Bank illustrates how heeding employee feedback and recognizing their concerns can sway significant decisions. It demonstrates that organizations can, and ought to, modify their strategies based on ethical considerations and the well-being of employees. As noted by The Australian, such measures convey a powerful message that while AI may revolutionize work, it should not do so at the cost of transparency, fairness, and inclusion.

Psychological Support & Ethical Governance

Psychological Support & Ethical Governance are essential in guaranteeing that the implementation of AI and other technological advancements is both effective and conducted in a humane and responsible manner. During significant organizational changes, particularly those that involve automation or the redesign of jobs, employees frequently face feelings of uncertainty, anxiety, and even resistance. By merging change management with psychological counseling and well-defined ethical frameworks, organizations can more effectively support the mental health of their workforce. As highlighted in sources such as PMC and Wikipedia, this comprehensive approach fosters emotional resilience while reinforcing the organization's dedication to ethical practices and transparency in AI.

Furthermore, the establishment of dedicated ethics offices, the promotion of transparent reporting, and the facilitation of employee oversight serve as powerful mechanisms for cultivating long-term trust. These frameworks guarantee that AI technologies are implemented in a manner that is accountable and consistent with human values. For instance, ethics committees or governance bodies can evaluate AI use cases, monitor for bias or misuse, and ensure that employee perspectives are considered in the oversight process. As noted by PubMed, such initiatives not only enhance employee confidence but also position the organization as a responsible leader in the adoption of AI. Ultimately, ethical governance and psychological support are not secondary considerations—they are fundamental to achieving sustainable and trustworthy digital transformation.

Employee-Centric Design & Incentives

Employee-Centric Design and Incentives play a vital role in effectively incorporating AI into the workplace while preserving employee morale and engagement. Instead of enforcing top-down changes, organizations can empower their workforce by actively involving them in the transformation process. This can be achieved through the establishment of AI champions, internal innovation hubs, or collaborative environments where employees can experiment with AI tools, exchange insights, and help shape the application of these technologies. As highlighted by Business Insider, such participatory methods cultivate a sense of ownership and diminish resistance to change. When employees perceive themselves as co-creators of innovation—rather than mere passive recipients—they are more inclined to accept new systems and strive towards common objectives.

In addition to engagement strategies, financial incentives that align with AI adoption can further enhance employee attitudes. For example, connecting performance bonuses or rewards for upskilling to the successful integration of AI motivates workers to view these tools as opportunities instead of threats. Research from arXiv supports this notion, indicating that incentives—when carefully crafted—can alter perceptions, making employees more receptive to learning and adapting. Collectively, these strategies not only boost acceptance of AI but also guarantee that its implementation is inclusive, strategic, and centered around human needs.

Leadership & Learning Culture

Leadership and Learning Culture serve as essential pillars for effectively navigating technological changes, particularly in the realm of AI integration. Contemporary successful leaders are embracing a paradoxical leadership approach—one that harmonizes a clear structure and direction with the autonomy for employees to explore, adapt, and innovate. This approach acknowledges that while guidance is crucial, excessive control can hinder creativity and learning. Research published in Frontiers highlights that when such leadership is bolstered by strong training environments, organizations are more adequately prepared for technological advancements. Leaders who empower their teams while upholding strategic clarity foster conditions where employees feel both secure and motivated to progress alongside emerging technologies.

In addition to leadership style, a robust dedication to continuous skill development is vital for long-term adaptability. This commitment extends beyond mere technical training; it encompasses the cultivation of soft skills such as critical thinking, communication, and emotional intelligence, alongside AI literacy and ethical awareness. Investing in these domains equips employees not only to utilize AI tools proficiently but also to make responsible, informed decisions in a swiftly evolving landscape. Organizations that integrate learning into their culture—through workshops, mentorship, and ongoing development programs—nurture a workforce that is resilient, agile, and prepared for the future.

Guarding Against Fatigue & Bias

Safeguarding Against Fatigue & Bias is an essential component of overseeing AI-driven transformation within organizations. When changes are implemented too swiftly or without appropriate sequence, employees may feel overwhelmed, resulting in organizational fatigue and burnout. To mitigate this, it is vital to break down significant changes into manageable phases, prioritize initiatives, and consistently monitor employee well-being throughout the transition. As noted by Wikipedia, careful pacing is crucial for maintaining energy and morale, ensuring that teams can adjust without experiencing excessive pressure or disengagement.

Equally significant is the need to address cognitive biases that may emerge when engaging with AI. Two primary concerns are automation bias—the inclination to excessively depend on AI outputs without scrutiny—and algorithm aversion, where individuals reject AI support after witnessing a single error. By educating employees about these biases, they are empowered to engage with AI tools in a balanced and critical manner. Instead of unconditionally trusting or dismissing AI, they learn to utilize these tools as intelligent aids—assessing outputs, applying human judgment, and making more informed decisions. This approach not only improves the quality of decision-making but also fosters ethical and responsible utilization of AI in everyday workflows.

As manufacturing and organizations increasingly integrate AI, it is essential to prioritize psychological and organizational dynamics in workforce development. Failing to address employee anxiety, trust, autonomy, and well-being may lead to unsuccessful technological transitions. Implementing practices such as transparent communication, ethical implementation, participatory design, and supportive leadership fosters a resilient and adaptable workforce. This comprehensive approach protects both human dignity and organizational performance.

CONCLUSION: TOWARDS A COMPREHENSIVE WORKFORCE STRATEGY

Achieving a contemporary, resilient, and equitable manufacturing workforce necessitates a multi-faceted strategy that harmonizes technological advancements with social and organizational changes.

Strategic Workforce Planning

Strategic Workforce Planning (SWP) is a proactive, data-informed approach that organizations employ to guarantee they possess the appropriate personnel, equipped with the necessary skills, in the suitable positions—both presently and in the future. As technological innovations, especially in AI and automation, swiftly transform business environments, it is insufficient for companies to merely respond to talent shortages as they occur. Rather, they must actively foresee the skills and roles that will be required and plan accordingly. This is where SWP becomes crucial.

SWP utilizes workforce analytics, scenario planning, and business forecasting to synchronize talent strategies with the organization's long-term objectives. For instance, if a company anticipates the integration of more AI-driven tools over the next five years, SWP would pinpoint which roles are likely to evolve, which skills need enhancement, and where potential deficiencies may arise. It subsequently directs recruitment, reskilling, upskilling, and succession planning initiatives to fulfill those future requirements.

According to McKinsey & Company, organizations that successfully implement SWP can realize up to 300% greater revenue per employee compared to those that do not. This impressive statistic underscores the efficiency improvements and competitive edge that result from aligning human capital with changing business and technological demands. High-performing organizations leverage SWP not merely as a human resources tool but as a strategic function that connects talent management with business expansion. By doing so, they ensure that their workforce remains adaptable, pertinent, and capable of fostering innovation—rather than becoming an obstacle during periods of transition.

SWP enables companies to remain ahead of disruptions by regarding talent as a strategic asset, facilitating more informed decision-making and promoting sustainable performance in a swiftly changing world.

Closing Skill Gaps through Collaborative Training

According to the OECD, approximately 40–50% of manufacturing employers identify skill shortages as a significant barrier to the adoption of artificial intelligence (AI) in their operations. This observation highlights a crucial concern: although technology itself is becoming more accessible and affordable, the human ability to effectively implement and utilize AI is not keeping pace. Without a workforce that comprehends how to integrate, manage, and collaborate with AI systems, the full potential of these technologies remains untapped.

The skills gap impacts not only technical positions—such as data scientists, AI engineers, or machine operators—but also managerial and strategic roles. Leaders and decision-makers need to have a basic understanding of AI's capabilities, risks, and ethical considerations to facilitate effective and responsible implementation. Nevertheless, many organizations find it challenging to offer sufficient internal training or to recruit from a limited pool of qualified candidates.

To address this challenge, the OECD strongly recommends public-private training partnerships. This entails that governments, educational institutions, and private enterprises must collaborate to create and provide comprehensive training programs. These initiatives should focus on both job-specific technical skills (e.g., programming, data analytics, robotics maintenance) and managerial competencies (e.g., change management, AI strategy, ethical governance). By combining resources and coordinating efforts, such partnerships can swiftly enhance skill development across various industries.

Ultimately, tackling these workforce shortcomings is not merely an issue of economic competitiveness—it is also crucial for facilitating a fair and inclusive transition to AI. In the absence of adequate investment in talent development, there exists a danger of exacerbating inequality between companies that have the means to enhance their employees' skills and those that do not. Consequently, coordinated training initiatives serve a dual purpose: fostering AI adoption and enhancing overall social and economic resilience in response to technological advancements.

Responsible AI through Ethical Soft Law

The responsible integration of AI into the workforce extends beyond mere technical implementation, it necessitates a robust ethical foundation to steer the design, deployment, and governance of AI in practical environments. As noted by the OECD, numerous member nations are adopting soft law frameworks—such as ethical guidelines, voluntary principles, and industry codes of conduct—as a pragmatic and flexible approach to managing the risks and complexities linked to AI in the workplace.

In contrast to hard law (e.g., formal legislation or binding regulations), soft law offers non-binding yet impactful frameworks that influence behavior and decision-making without the constraints of legal enforcement. This adaptability is especially beneficial in the rapidly changing domain of AI, where regulatory frameworks frequently lag behind technological progress. Soft law enables governments, corporations, and institutions to react more swiftly to ethical issues, including algorithmic bias, surveillance, data privacy, and job displacement due to automation.

For the workforce, the application of ethical soft law is crucial to ensuring that AI is utilized in a manner that honors human rights, fosters fairness, and cultivates trust. For instance, ethical AI guidelines may stipulate that employees are made aware of when and how AI is employed to make decisions regarding their performance, hiring, or remuneration. They may also highlight the necessity of human oversight, transparency, and the right to challenge automated decisions.

Moreover, as AI transforms job roles and workplace interactions, these ethical frameworks play a vital role in safeguarding human dignity and agency. This encompasses protecting workers from exploitative surveillance practices, ensuring clarity in how AI systems impact labor, and advocating for inclusiveness in the design and deployment of AI.

The OECD advocates for this strategy as a means to harmonize innovation with social responsibility, acknowledging that trust in AI—particularly among employees—is an essential element for its effective implementation. By integrating ethics into AI governance via soft law, nations and organizations can facilitate not only technological advancement but also a more compassionate and fair transition to the future of work.

Collaborative Models between Industry and Government

Collaborative models between industry and government are increasingly acknowledged as vital for addressing the workforce effects of technological disruption, particularly the emergence of AI, automation, and the shift towards cleaner technologies. These collaborations leverage the strengths of both sectors: the technical innovation and operational capabilities of industry alongside the policy support, funding, and public accountability provided by government. Together, they can formulate coordinated strategies that not only tackle job displacement but also foster a more future-ready and inclusive workforce.

A notable example can be observed in the U.S. automotive sector, where the transition to electric vehicles (EVs) and battery technologies is reshaping traditional manufacturing positions. As reported by Reuters, the U.S. government and prominent automakers have initiated joint efforts that encompass multi-billion-dollar investments in retraining programs. These initiatives are designed to reskill workers from internal combustion engine manufacturing for new positions in EV production, battery assembly, and associated green technologies. This type of strategic collaboration not only saves existing jobs where feasible but also generates new employment opportunities in a swiftly evolving industry. It embodies a collective responsibility to ensure that workers are not overlooked during the green and digital transitions.

In a similar vein, China has experienced significant efficiency improvements across various industries due to the swift integration of automation technologies. Nevertheless, this progress has resulted in the displacement of a considerable number of workers, particularly among vulnerable or lower-skilled demographic segments. In response to this challenge, the government, in partnership with industry leaders, has launched training initiatives and social protection programs designed to facilitate the transition of these workers into new and more inclusive employment opportunities, as highlighted by the Financial Times. These initiatives emphasize not only technical competencies but also soft skills and adaptability, thereby assisting displaced workers in securing sustainable jobs in sectors such as services, logistics, and green energy.

Both instances exemplify a broader principle: that advancements in technology must be paired with inclusive workforce strategies; otherwise, the social repercussions may surpass the economic advantages. Through collaborative efforts, governments can influence policy frameworks and allocate resources, while industries can provide immediate insights and expand training within their operations.

Enhancing Human Labor through Automation

Automation, especially through the implementation of collaborative robots (often referred to as co-bots), is frequently misinterpreted as a direct threat to employment. Nevertheless, when executed with care, automation can enhance human labor instead of replacing it. Co-bots are engineered to collaborate with human workers, aiding in specific tasks rather than usurping entire job roles. This cooperative model of automation has the capacity to elevate job quality, alleviate physical and mental stress, and enhance overall employee satisfaction.

For instance, co-bots can manage repetitive, physically taxing, or dangerous tasks, allowing human employees to concentrate on higher-value, creative, or decision-making responsibilities. This transition not only boosts productivity but also empowers employees to cultivate and utilize more advanced skills—promoting a deeper sense of purpose and engagement in their positions.

A recent study conducted in Australia, as reported by The Australian, revealed that employees utilizing robotics in their daily tasks reported increased job satisfaction and were able to save up to 245 hours annually. This equates to more than six full working weeks—a considerable time gain that can be redirected towards more meaningful activities or utilized to enhance work-life balance. This illustrates that when automation is harmonized with human-centered design, it can yield improved results for both organizations and their employees.

Furthermore, by reducing the strain of repetitive or laborious tasks, automation can lead to decreased burnout levels and enhanced mental well-being in professional environments. It also creates avenues for reskilling and upskilling, as employees move into more strategic or technical positions in conjunction with automated systems.

Final Summary

An effective workforce development strategy necessitates a comprehensive and forward-looking approach that incorporates several essential components like innovative talent planning, targeted up skilling initiatives, ethical governance frameworks and collaborative policy frameworks.

When the manufacturing sector prioritizes human potential in its technological evolution, it cultivates a workforce that is inclusive, adaptable, and prepared for the future. This balanced strategy enhances industry competitiveness by harnessing innovation while ensuring that the advantages of progress are widely distributed throughout society, thereby fostering social equity. In the end, this cohesive approach lays the foundation for a resilient and prosperous manufacturing industry that can address the challenges of the digital era.

REFERENCES

Acemoglu, D., & Restrepo, P. (2019). Automation and new tasks: How technology displaces and reinstates labor. *The Journal of Economic Perspectives, 33*(2), 3–30. DOI: 10.1257/jep.33.2.3

Ameen, K., & Keizer, A. (2023). International Labour Organization. In *Encyclopedia of Human Resource Management, Second Edition.* https://doi.org/DOI: 10.4337/9781800378841.I.19

Autor, D. H. (2015). Why are there still so many jobs? The history and future of workplace automation. *The Journal of Economic Perspectives, 29*(3), 3–30. DOI: 10.1257/jep.29.3.3

Bessen, J. E. (2019). AI and jobs: The role of demand. *NBER Working Paper No. 24235.* https://www.nber.org/papers/w24235

Bessen, J. E., Goos, M., Salomons, A., & van den Berge, W. (2019). Automatic Reaction - What Happens to Workers at Firms that Automate? SSRN *Electronic Journal.* https://doi.org/DOI: 10.2139/ssrn.3328877

Bongomin, G. O. C., Ahiauzu, A., & Ntayi, J. M. (2020). Digital technology adoption and firm performance: Role of organizational capabilities. *Technological Forecasting and Social Change, 162*, 120330. DOI: 10.1016/j.techfore.2020.120330

Buolamwini, J., & Gebru, T. (2018, January). Gender shades: Intersectional accuracy disparities in commercial gender classification. In *Conference on fairness, accountability and transparency* (pp. 77-91). PMLR.

Cascio, W. F., & Montealegre, R. (2016). How Technology Is Changing Work and Organizations. In *Annual Review of Organizational Psychology and Organizational Behavior* (Vol. 3). https://doi.org/DOI: 10.1146/annurev-orgpsych-041015-062352

Cheryan, S., Ziegler, S. A., Montoya, A. K., & Jiang, L. (2017). Why are some STEM fields more gender balanced than others? *Psychological Bulletin, 143*(1), 1–35. DOI: 10.1037/bul0000052 PMID: 27732018

Deming, D. J. (2017). The growing importance of social skills in the labor market. *The Quarterly Journal of Economics, 132*(4), 1593–1640. DOI: 10.1093/qje/qjx022

Ely, R. J., Ibarra, H., & Kolb, D. M. (2011). Taking gender into account: Theory and design for women's leadership development programs. In *Academy of Management Learning and Education* (Vol. 10, Issue 3). https://doi.org/DOI: 10.5465/amle.2010.0046

European Commission. (2020). *The Digital Skills and Jobs Coalition*. Ec.Europa. Eu. https://digital-strategy.ec.europa.eu/en/policies/digital-skills-coalition

Frey, C. B., & Osborne, M. A. (2017). The future of employment: How susceptible are jobs to computerisation? *Technological Forecasting and Social Change, 114*, 254–280. DOI: 10.1016/j.techfore.2016.08.019

Friedman, B., & Nissenbaum, H. (1996). Bias in computer systems. [TOIS]. *ACM Transactions on Information Systems, 14*(3), 330–347. DOI: 10.1145/230538.230561

Holstein, K., Wortman Vaughan, J., Daumé, H., III, Dudík, M., & Wallach, H. (2019). Improving fairness in machine learning systems: What do industry practitioners need? In Proceedings of the 2019 CHI Conference on Human Factors in Computing Systems (CHI '19), Paper 600, 1–16. Association for Computing Machinery. https://doi.org/DOI: 10.1145/3290605.3300830

Huang, M. H., & Rust, R. T. (2021). Engaged to a Robot? The Role of AI in Service. *Journal of Service Research, 24*(1), 30–41. Advance online publication. DOI: 10.1177/1094670520902266

Klerk, N., van den Bosch, R., & Volberda, H. (2020). Lifelong learning and employee resilience: The moderating effect of learning goal orientation. *Human Resource Management Journal, 30*(3), 1–18. DOI: 10.1111/1748-8583.12312

Kusiak, A. (2018). Smart manufacturing. *International Journal of Production Research, 56*(1–2), 508–517. Advance online publication. DOI: 10.1080/00207543.2017.1351644

Lee, J., Bagheri, B., & Kao, H.-A. (2018). A cyber-physical systems architecture for industry 4.0-based manufacturing systems. *Manufacturing Letters, 3*, 18–23. DOI: 10.1016/j.mfglet.2014.12.001

Manyika, J., Chui, M., Miremadi, M., Bughin, J., George, K., Willmott, P., & Dewhurst, M. (2017). *A future that works: Automation, employment, and productivity*. McKinsey Global Institute. https://www.mckinsey.com/featured-insights/digital-disruption/harnessing-automation-for-a-future-that-works

Marien, M. (2014). Book Review: The Second Machine Age: Work, Progress, and Prosperity in a Time of Brilliant Technologies. *World Futures Review, 6*(2), 189–193. Advance online publication. DOI: 10.1177/1946756714541404

Moss-Racusin, C. A., Sanzari, C., Caluori, N., & Rabasco, H. (2018). Gender bias produces gender gaps in STEM engagement. *Sex Roles, 79*(11), 651–670. https://link.springer.com/article/10.1007/s11199-018-0902-z. DOI: 10.1007/s11199-018-0902-z

Noe, R. A., Clarke, A. D. M., & Klein, H. J. (2014). Learning in the twenty-first-century workplace. *Annual Review of Organizational Psychology and Organizational Behavior*, *1*(1), 245–275. DOI: 10.1146/annurev-orgpsych-031413-091321

Schiavone, F., Simoni, M., & Bresciani, S. (2020). Towards Industry 5.0: An analysis of the enabling technologies. *Technological Forecasting and Social Change*, *162*, 120346. DOI: 10.1016/j.techfore.2020.120346

Schwab, K. (2016). *The fourth industrial revolution.* World Economic Forum. https://www.weforum.org/about/the-fourth-industrial-revolution-by-klaus-schwab

Solomon, B. D. (2023). Organisation for Economic Co-operation and Development (OECD). In *Dictionary of Ecological Economics: Terms for the New Millennium.* https://doi.org/DOI: 10.4337/9781788974912.O.20

UNESCO. (2021). *Education and digital transformation.* United Nations Educational, Scientific and Cultural Organization. https://unesdoc.unesco.org/ark:/48223/pf0000374719

UNIDO. (2019). *Industrial development report 2020: Industrialization, trade, and development in the digital era.* United Nations Industrial Development Organization. https://www.unido.org/resources-publications-flagship-publications-industrial-development-report-series/industrial-development-report-2020

U.S. Department of Labor. (2020). *Workforce Innovation and Opportunity Act.* https://www.dol.gov/agencies/eta/wioa

Wamba-Taguimdje, S. L., Fosso Wamba, S., Kala Kamdjoug, J. R., & Tchatchouang Wanko, C. E. (2020). Influence of artificial intelligence (AI) on firm performance: The business value of AI-based transformation projects. *Business Process Management Journal*, *26*(7), 1893–1924. Advance online publication. DOI: 10.1108/BPMJ-10-2019-0411

West, S. M., Whittaker, M., & Crawford, K. (2019). Discriminating systems. *AI Now*, *2019*, 1–33.

Wilson, H. J., & Daugherty, P. R. (2018). Collaborative intelligence: Humans and AI are joining forces. *Harvard Business Review*, *96*(4), 114–123. https://hometownhealthonline.com/wp-content/uploads/2019/02/ai2-R1804J-PDF-ENG.pdf

World Economic Forum. (2020). The future of jobs report 2020. https://www.weforum.org/reports/the-future-of-jobs-report-2020

Xu, D., & Jaggars, S. S. (2014). Performance gaps between online and face-to-face courses: Differences across types of students and academic subject areas. *The Journal of Higher Education, 85*(5), 633–659. DOI: 10.1080/00221546.2014.11777343

Zhong, R. Y., Xu, C., Chen, C., & Huang, G. Q. (2017). Big Data Analytics for Physical Internet-based intelligent manufacturing shop floors. *International Journal of Production Research, 55*(9), 2610–2621. Advance online publication. DOI: 10.1080/00207543.2015.1086037

Chapter 2
Immersive Training for the AI-Enabled Factory:
AR, VR, MR, Simulation, and Digital Twin-Based Workforce Development

Otmane Azeroual
http://orcid.org/0000-0002-5225-389X
University of Hagen, Germany

ABSTRACT

This chapter examines the potential of immersive training methods such as Augmented Reality (AR), Virtual Reality (VR), Mixed Reality (MR), and simulation-based approaches including Digital Twins in AI-enabled factories. It presents the theoretical foundations, pedagogical concepts, and practical applications, while discussing opportunities, challenges, and future developments. The chapter highlights how employees can be effectively prepared for complex, data-driven production processes and emphasizes the significance of immersive training for the competitiveness and innovation capacity of modern manufacturing companies.

1. INTRODUCTION

Industrial manufacturing is currently undergoing a profound transformation driven by the increasing integration of Artificial Intelligence (AI), robotics, and automated processes. In so-called AI-enabled factories, intelligent systems not only perform routine tasks but also support employees in decision-making, process optimization, and real-time production control. This technological shift necessitates a redefinition of

DOI: 10.4018/979-8-3373-5756-0.ch002

workforce competencies, particularly in areas such as complex machinery operation, data analytics, human-robot interaction, and process supervision (Kumar, Kumar & Ramesh, 2024; Islam et al., 2025; Lodhi & Zeb, 2025). Employees are no longer expected to execute only manual or repetitive tasks; instead, they must collaborate with advanced systems, interpret data streams, and make informed decisions that affect production efficiency and safety.

Traditional training and education methods—including on-site workshops, printed manuals, and conventional e-learning programs—often reach their limits in this new industrial landscape. These methods rarely provide opportunities for employees to engage with production processes in a hands-on, interactive, and risk-free manner. As a result, a significant skills gap emerges, which can negatively impact both productivity and operational safety in manufacturing environments (Ras et al., 2017; Badri, Boudreau-Trudel & Souissi, 2018). Workers are challenged to develop not only technical skills but also cognitive and collaborative competencies required to navigate increasingly automated and data-driven production environments.

Immersive training methods—including Augmented Reality (AR), Virtual Reality (VR), Mixed Reality (MR), and simulation-based approaches—offer novel opportunities to address these challenges. AR refers to the overlay of digital information onto the physical environment in real time (Azuma, 1997), VR describes fully computer-generated, three-dimensional environments that enable a sense of presence (Steuer, 1992), and MR combines elements of both AR and VR along the reality–virtuality continuum, enabling real and virtual objects to co-exist and interact (Milgram & Kishino, 1994). These technologies allow employees to virtually experience production processes, practice complex procedures, and receive immediate feedback without interrupting actual production lines or compromising safety (Hancko, Majlingova & Kačíková, 2025; Zahid et al., 2025).

Through immersive environments, learners can experiment with different operational scenarios, explore potential hazards, and develop problem-solving and decision-making skills in a controlled yet realistic setting. This approach operationalizes experiential "learning by doing" (Kolb, 2014), allowing employees to safely practice real tasks, refine their skills, and gain situational awareness before applying them in actual production environments.

Crucially, immersive training does not stand apart from AI, robotics, and automation but is closely interwoven with them: AI-driven adaptive systems personalize learning pathways, robotics-integrated simulations prepare employees for human–robot collaboration, and automation-focused digital twins allow safe experimentation with production processes. In this way, immersive training becomes an integral enabler of workforce readiness for Industry 4.0 and 5.0 ecosystems.

The objective of this chapter is to explore the potential of immersive training methods in AI-enabled factories, combining theoretical foundations, pedagogical concepts, and practical examples. The discussion highlights both the opportunities and challenges associated with these technologies and outlines anticipated future developments in the field. In doing so, the chapter provides a comprehensive foundation for designing effective, technology-enhanced workforce development programs in modern manufacturing contexts.

The structure of the chapter follows a logical progression: it begins by presenting the theoretical background and contextual framework of AI-enabled factories, followed by an introduction to immersive technologies and relevant didactic concepts. Subsequently, detailed practical use cases are provided, illustrating how immersive training can be applied across different industrial sectors. Finally, the chapter examines the opportunities, challenges, and emerging trends, culminating in a summary of key insights and recommendations for practice.

This chapter is particularly relevant for:

- Academics and students in mechanical engineering, computer science, business administration, and Industry 4.0/5.0 programs seeking an in-depth understanding of immersive training applications.
- Human resources and corporate training departments in manufacturing companies aiming to prepare employees for AI-enabled systems.
- Researchers and developers working at the intersection of human factors, AI, and industrial manufacturing.
- Industry decision-makers responsible for planning innovation and digitalization projects while ensuring workforce qualification and operational readiness.

2. RELATED WORK

Immersive learning technologies such as VR, AR, and MR have attracted increasing attention in recent years as promising tools for workforce training and development. Early studies in higher education and technical training demonstrated that VR can significantly enhance learner engagement, spatial understanding, and procedural accuracy compared to traditional methods (Radianti et al., 2020; Salas, 2021). In industrial contexts, AR has been shown to improve on-the-job training by overlaying digital instructions onto physical work environments, thereby reducing error rates and training time (Palmarini et al., 2018; Falola et al., 2024).

Several systematic reviews confirm the effectiveness of immersive training in domains such as healthcare, engineering education, and safety-critical environments (Jensen & Konradsen, 2018; Hamilton et al., 2021). These reviews highlight that immersive methods foster experiential "learning by doing" and allow learners to practice high-risk procedures in safe, controlled environments. However, they also emphasize limitations such as high implementation costs, lack of standardized evaluation metrics, and the need for more longitudinal studies to assess long-term learning outcomes.

In the manufacturing sector, immersive technologies are increasingly linked with Industry 4.0 initiatives. Research has explored the use of AR for assembly tasks, remote collaboration, and equipment maintenance (Syberfeldt, Danielsson & Holm, 2017; Damiani et al., 2018). VR and MR environments have been piloted for operator training in digital twins of production lines, providing employees with opportunities to rehearse complex scenarios without disrupting actual operations (Pires et al., 2019). Despite these advances, the majority of studies remain focused on isolated use cases rather than integrated, enterprise-wide training strategies.

A second research stream addresses the convergence of immersive learning with Artificial Intelligence (AI). Adaptive learning systems powered by AI are capable of personalizing training content, dynamically adjusting task difficulty, and providing real-time feedback based on learner performance (Bratu, 2023; Häfneret al., 2024). While these approaches show promise, the integration of AI-driven personalization into immersive environments in real industrial settings is still in its infancy. The combination of immersive technologies, robotics, and process automation has only rarely been investigated, creating a gap between theoretical potential and practical implementation.

Table 1. Summary of prior studies on immersive training technologies.

Author(s) & Year	Technology Focus	Domain / Application	Key Findings	Limitations / Gaps Identified
Radianti et al. (2020)	VR	Higher education, technical training	Enhanced learner engagement, spatial understanding, and procedural accuracy.	General education focus; limited industrial validation.
Salas (2021)	VR	Procedural & technical learning	Demonstrated significant improvements in procedural accuracy.	Short-term focus; lacks industrial scale studies.
Palmarini et al. (2018)	AR	Industrial on-the-job training (assembly, maintenance)	Reduced error rates and training time through digital overlays.	Limited to case-specific pilot studies.
Falola et al. (2024)	AR	Workforce development in industry	Showed improved knowledge retention and task performance.	Implementation challenges; scalability unclear.

continued on following page

Table 1. Continued

Author(s) & Year	Technology Focus	Domain / Application	Key Findings	Limitations / Gaps Identified
Jensen & Konradsen (2018)	VR/AR/MR	Healthcare, engineering education	Systematic review: immersive tech supports experiential learning.	High costs and lack of standardized evaluation metrics.
Hamilton et al. (2021)	VR/AR/MR	Safety-critical environments	Confirmed benefits of practicing high-risk procedures safely.	Need for longitudinal studies to assess long-term outcomes.
Syberfeldt, Danielsson & Holm (2017)	AR	Manufacturing (assembly tasks, remote collaboration)	AR reduces complexity of assembly and supports remote expert collaboration.	Not enterprise-wide; case-based studies.
Damiani et al. (2018)	AR	Maintenance and industrial training	Demonstrated efficiency gains and fewer human errors.	Limited industrial domains tested.
Pires et al. (2019)	VR/MR + Digital Twins	Operator training in production lines	Enabled rehearsal of complex industrial scenarios without disruptions.	Pilot implementations only; not yet large-scale integrated.
Bratu (2023)	AI-driven adaptive learning	Personalized immersive training	Real-time adaptation of content and task difficulty.	Early-stage, limited industrial adoption.
Häfner et al. (2024)	AI + Immersive environments	Workforce skill training	Promising convergence of AI personalization with immersive learning.	Largely theoretical; practical implementation rare.

As the table illustrates, prior research confirms the benefits of immersive training technologies across multiple domains, but also reveals several recurring challenges. While VR has proven effective for engagement and procedural learning (Radianti et al., 2020; Salas, 2021), and AR has demonstrated measurable improvements in assembly and maintenance tasks (Palmarini et al., 2018; Falola et al., 2024), most studies remain confined to isolated, small-scale implementations. Systematic reviews (Jensen & Konradsen, 2018; Hamilton et al., 2021) further underline the lack of standardized evaluation frameworks, high costs, and insufficient longitudinal evidence. In manufacturing, research on AR and VR/MR combined with digital twins (Syberfeldt et al., 2017; Damiani et al., 2018; Pires et al., 2019) is promising but still fragmented and not yet enterprise-wide.

The integration of AI-driven adaptive learning into immersive environments (Bratu, 2023; Häfner et al., 2024) represents an emerging trend with significant potential to close this gap, yet its industrial application is still nascent. Overall, the literature demonstrates both the potential and the shortcomings of immersive technologies in workforce training. This chapter contributes to advancing the field by synthesizing these findings and proposing a holistic framework for systematically embedding immersive training within AI-enabled Industry 4.0 and 5.0 ecosystems.

Unlike prior studies, this chapter provides a holistic synthesis by combining the AI-enabled factory context with immersive training technologies and workforce development implications. This integrated perspective bridges the gap between isolated case studies and comprehensive, enterprise-wide workforce strategies, offering a novel contribution to both academic research and industrial practice.

3. THE CONTEXT: AI-ENABLED FACTORY

The AI-enabled factory represents a central element of modern manufacturing, embodying the convergence of traditional production processes with AI, robotics, Internet-of-Things (IoT) devices, and data-driven decision-making (Tyagi et al., 2024; Praveen & Mutya, 2024; Martínez-Mireles et al., 2025). Characterized by high levels of automation, self-optimizing machines, and the ability to analyze large datasets in real time, AI-enabled factories enable production processes to become more efficient, flexible, and resilient.

In such environments, human workers increasingly assume roles as supervisors, analysts, and coordinators of intelligent production systems. Their role shifts from simply operating machines to actively overseeing and managing complex processes, making decisions based on AI-generated recommendations. This transition requires a deep understanding of data analytics, machine learning, software architectures, and human-robot interaction (Bandi et al., 2025).

A key feature of AI-enabled factories is the capability to control and monitor autonomous systems in real time. Machine learning and AI algorithms enable predictive maintenance, detection of process deviations, and early identification of quality issues, supporting proactive and preventive production management. At the same time, the increasing complexity of manufacturing processes means that traditional training methods are no longer sufficient to adequately prepare employees (Cotton, 2001; Fitsilis, Tsoutsa & Gerogiannis, 2018).

Implementing an AI-enabled factory also introduces new challenges in terms of organization, workflow design, and safety requirements. Employees must not only operate new technologies but also understand the interfaces between humans, machines, and software systems. This necessitates targeted training and professional development programs that are practice-oriented, flexible, and adaptive, aligning with the demands of a dynamic, data-driven manufacturing environment (Bondin & Zammit, 2025).

Finally, the context of the AI-enabled factory is strongly shaped by global competitiveness. Companies that invest early in workforce qualification and innovative training methods are better positioned to respond rapidly to market changes, optimize production processes, and minimize operational risks. Immersive training approaches, such as AR, VR, and simulation-based methods, provide opportunities to efficiently and realistically prepare employees for their roles in AI-driven factories, thereby enhancing both individual competencies and overall production performance (MK et al., 2024; Deogaonkar, 2025).

Table 2 summarizes the core characteristics, challenges, and workforce implications of AI-enabled factories, providing a clear reference for the design of immersive training programs and strategic workforce development initiatives.

Table 2. Key Features and Implications of AI-Enabled Factories.

Aspect	Description	Implications for Workforce
Automation & Self-Optimizing Systems	Machines and processes adjust autonomously based on real-time data.	Employees transition from operating machines to supervising and analyzing autonomous systems.
Data-Driven Decision-Making	AI analyzes large datasets to support process optimization.	Employees are required to develop data literacy, analytical skills, and an understanding of AI algorithms.
Predictive Maintenance & Quality Control	Machine learning anticipates maintenance needs and detects anomalies.	Employees must proactively monitor systems and intervene when deviations occur, reducing downtime and defects.
Human-Robot Interaction	Collaboration with cobots and AI-assisted systems is integrated into production workflows.	Employees need to learn human-robot interaction principles, follow safety protocols, and coordinate effectively with autonomous systems.
Complexity & Flexibility	Adaptive processes respond to dynamic production demands.	Employees must engage in continuous learning and develop flexible skills to adapt to changing production requirements.
Training & Upskilling Needs	Traditional methods are insufficient for AI-driven environments.	Employees require immersive and simulation-based training to acquire practical skills safely and efficiently.
Competitiveness & Strategic Impact	Rapid adaptation to market changes is essential for maintaining competitive advantage.	Employees benefit from early adoption of advanced training methods, enhancing productivity, safety, and innovation within the organization.

These contextual challenges form the basis for exploring immersive technologies as workforce training solutions (Section 4).

4. IMMERSIVE TECHNOLOGIES IN INDUSTRIAL TRAINING

The increasing complexity of industrial production environments, driven by AI, robotics, and automated systems, requires innovative approaches to employee training. Traditional instructional methods, such as classroom lectures or on-the-job mentoring, are often insufficient to prepare workers for technology-intensive contexts where decision-making, real-time process monitoring, and human–robot collaboration are critical.

Immersive technologies—including Augmented Reality (AR), Virtual Reality (VR), Mixed Reality (MR), and simulation-based learning using digital twins—provide empirically supported approaches that enhance learning outcomes, reduce cognitive load, and enable safe, practice-based skill acquisition (Hazrat et al., 2023; Zhu et al., 2025). These technologies bridge the gap between theoretical knowledge and operational practice, allowing employees to acquire procedural, technical, and situational skills in realistic but controlled environments. By embedding feedback and guidance directly into the learning context, immersive training promotes accelerated skill acquisition and improves retention compared to conventional methods (Попова, 2024).

4.1 Augmented Reality (AR)

AR overlays digital information onto the physical environment, providing real-time, context-sensitive guidance during task execution. AR supports situated cognition by embedding instructions directly in the work context, reducing cognitive load and minimizing the need for memorization or frequent reference to manuals (Shelton & Hedley, 2004; Petersen, 2015; Buchner, Buntins & Kerres, 2022).

For example, AR can display step-by-step assembly instructions, highlight components to be installed, or indicate the correct tool usage, thereby reducing errors and onboarding time. Empirical studies in industrial assembly and maintenance tasks demonstrate error reductions of up to 25% and faster task completion compared to traditional methods (Yuan, Ong & Nee, 2008; Lavric et al., 2022). Additionally, AR enhances workplace safety by visualizing hazardous areas, alerting workers to procedural warnings, or simulating potential failure modes in real-time.

Beyond individual tasks, AR supports collaborative work. Multiple operators can share augmented views, coordinate complex assembly steps, and receive live guidance from remote experts, effectively extending the knowledge of senior personnel across distributed teams.

In practice, AR training solutions are implemented using platforms such as PTC Vuforia, Microsoft Dynamics 365 Guides, and Scope AR. These tools enable step-by-step guidance, component highlighting, and real-time feedback, supporting both individual and collaborative assembly or maintenance tasks.

4.2 Virtual Reality (VR)

VR provides fully immersive simulations, allowing employees to train in operational or hazardous scenarios without risk to themselves or production systems. VR supports the experiential learning cycle—comprising concrete experience, reflective observation, abstract conceptualization, and active experimentation (Kolb, 2014)—by enabling repeated practice in high-fidelity environments.

VR applications include emergency response training, robot interaction exercises, and complex machine operation scenarios. Repeated engagement in VR has been shown to improve procedural memory, situational awareness, and decision-making efficiency in industrial contexts (Colombo, Nazir & Manca, 2014; Conges et al., 2023). For example, chemical plants have implemented VR simulations for spill response and safety drills, allowing personnel to practice rare but critical emergency procedures in a safe, controlled manner.

VR also enables scenario-based assessment. Performance metrics such as task completion time, errors, and response strategies can be captured and analyzed to provide individualized feedback and guide adaptive learning paths.

Common VR platforms include Unity and Unreal Engine combined with headsets like Oculus Rift, Oculus Quest, or HTC Vive. These solutions facilitate fully immersive training simulations for emergency response, robotic operations, or complex machinery handling, providing measurable improvements in situational awareness and procedural accuracy.

4.3 Mixed Reality (MR)

MR integrates physical objects with virtual overlays, creating hybrid environments where employees can interact with real machinery augmented by digital information. MR leverages embodied cognition, allowing learners to integrate sensory-motor experience with conceptual knowledge, which is especially valuable for complex maintenance, calibration, and assembly tasks (Lehrman, 2025).

Applications include real-time maintenance support, where virtual instructions are superimposed on equipment, highlighting components, showing sensor data, or visualizing process flows. MR can also facilitate collaborative training, enabling multiple employees to interact simultaneously with hybrid environments and coordinate operations in ways that mirror actual production conditions.

MR applications typically rely on devices such as Microsoft HoloLens in combination with Dynamics 365 MR or Magic Leap Enterprise Suite. These systems overlay digital content on physical machinery, enabling real-time guidance, collaborative problem-solving, and hybrid operational training that mirrors real industrial conditions.

4.4 Simulation and Digital Twins

Digital twins are virtual replicas of machines, processes, or entire factories, enabling experimentation, process optimization, and learning-by-doing without operational risk (Asmat et al. 2025). They integrate real-time sensor data to reflect current operational states and allow predictive analysis, making them ideal for both training and process improvement.

By combining digital twins with AR, VR, or MR, organizations create cyber-physical learning ecosystems, where virtual and physical experiences reinforce each other. Employees can practice process adjustments in the digital twin, then apply learned strategies in the real environment with minimal risk. This integration supports continuous learning, rapid error correction, and iterative process improvement.

4.5 Integrated Overview

Figure 1 illustrates the synergistic relationships between immersive technologies in industrial training. AR provides context-aware guidance and cognitive support during real-world tasks, VR delivers immersive experiential learning for complex or hazardous scenarios, MR enables hybrid interaction with real equipment, and digital twins allow simulation, experimentation, and process optimization. Arrows in the diagram indicate data and feedback flows, showing how information from digital twins, AR, VR, and MR environments can be shared and integrated to maximize learning outcomes, support adaptive training, and improve operational performance.

While AR is best suited for real-time task guidance in live operational contexts, VR excels in preparing employees for rare or hazardous scenarios where mistakes would be costly in reality. MR provides the bridge between both, combining physical interaction with digital augmentation, which is particularly effective in collaborative and maintenance settings. Digital twins extend these capabilities by enabling predictive experimentation and optimization at the process level. Taken together, these technologies are not substitutes but complementary layers of an integrated industrial training ecosystem.

Figure 1. Overview of Immersive Technologies in Industrial Training (Author's own illustration).

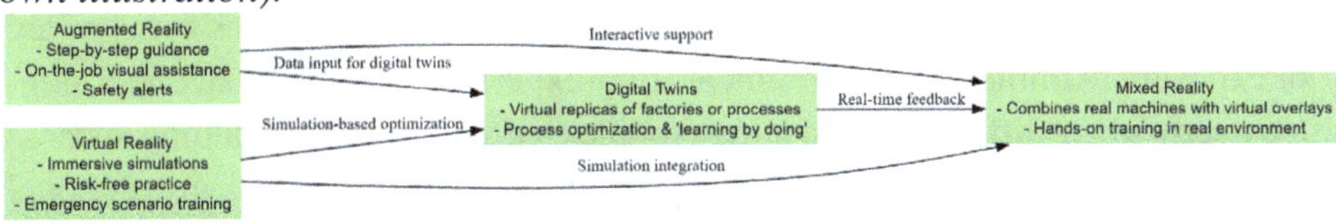

4.6 Challenges and Emerging Trends

The implementation of immersive technologies in industrial training is accompanied by a number of challenges that need to be addressed in order to achieve their full potential. At the same time, emerging technological and pedagogical trends are shaping the future of immersive learning in AI-enabled factories. This section provides an overview of both aspects, highlighting barriers to adoption and opportunities for innovation.

Technical and Infrastructure Challenges

Immersive technologies such as AR, VR, and MR demand substantial computational resources, high-quality graphics rendering, and low-latency connectivity to ensure smooth user experiences. In industrial settings, these requirements are often difficult to meet due to outdated IT infrastructure or limited access to high-performance servers. Moreover, while simulation-based learning benefits from digital twins, the integration of real-time sensor data into immersive environments remains a significant challenge. In particular, the lack of GPU capacity in many training institutions hinders large-scale deployment of realistic simulations.

Cost and Scalability

Despite falling hardware prices, immersive systems remain costly in terms of acquisition, maintenance, and content development. Developing high-quality training simulations requires interdisciplinary collaboration between engineers, educators, and software developers. This makes scaling solutions across different factories or industrial sectors difficult. Small- and medium-sized enterprises (SMEs) are particularly challenged, as they often lack the financial and organizational resources for large-scale adoption.

Human Factors and Usability

A critical barrier lies in human acceptance and usability. Employees with limited prior exposure to immersive environments may experience simulator sickness, cognitive overload, or difficulty navigating complex virtual systems. Training success thus depends not only on the technology itself but also on careful instructional design, ergonomic hardware solutions, and continuous user support. Research has also shown that intergenerational differences can impact the acceptance of immersive learning, with younger employees adapting more quickly than older colleagues (Hamilton et al., 2021).

Standardization and Evaluation

Currently, there is a lack of standardized frameworks for evaluating the effectiveness of immersive training. While many case studies report improved engagement and learning outcomes, systematic longitudinal studies are still rare. Without consistent evaluation metrics, it remains difficult to compare results across industries, validate training effectiveness, and justify investments.

Emerging Trends

At the same time, several promising trends are emerging that are likely to shape the next generation of immersive industrial training:

- *AI-Driven Personalization.* Advances in adaptive learning systems allow immersive training to be dynamically tailored to the learner's performance, skills, and cognitive load. AI-driven tutors can provide real-time feedback, optimize task difficulty, and recommend personalized learning paths.
- *Next-Generation Digital Twins.* Digital twins are evolving into real-time, multi-scale ecosystems that integrate machine learning, predictive analytics, and cross-factory connectivity. These next-generation twins enable continuous training and simulation, fostering both skill acquisition and process innovation.
- *Generative AI for Content Creation.* The emergence of generative AI tools accelerates the production of immersive training content, including 3D assets, scenarios, and simulations. This trend reduces costs and supports scalability by enabling rapid customization for different industrial contexts.
- *Integration with Human–Robot Collaboration.* As collaborative robots (cobots) become increasingly prevalent, immersive environments are being used to train employees in safe and efficient human–robot interaction. Mixed

Reality platforms, in particular, provide opportunities for simulating collaboration with autonomous systems in real-time.

- *Ethical and Social Dimensions.* With the rise of AI-enabled immersive training, issues of data privacy, fairness, and inclusiveness become increasingly important. Ensuring that training data is representative and unbiased, while protecting employee privacy, will be a central concern in future deployments.

In conclusion, the adoption of immersive training technologies in AI-enabled factories is influenced by both persistent challenges and exciting opportunities. Addressing infrastructure, cost, usability, and evaluation barriers is essential to enable sustainable implementation. At the same time, emerging trends such as AI-driven personalization, generative content creation, advanced digital twins, and collaborative robotics are expected to redefine the landscape of workforce training. By strategically aligning these developments with pedagogical goals and organizational needs, immersive training can evolve into a key enabler of workforce readiness for Industry 4.0 and 5.0.

The identified strengths and complementarities of immersive technologies lead directly to the discussion of didactic frameworks that can maximize their pedagogical impact (Section 5).

5. DIDACTIC CONCEPTS AND LEARNING APPROACHES

Effective training in technologically advanced industrial environments requires the integration of evidence-based pedagogical approaches. Immersive technologies—such as AR, VR, MR, and digital twins—do not operate in isolation but are closely intertwined with AI, robotics, and automated systems, enhancing their effectiveness in workforce development. They actively support practice-based learning by providing realistic, work-related scenarios directly connected to AI-driven processes and human–robot interactions. These technologies achieve their full potential only when embedded within didactically sound frameworks that enhance engagement, knowledge retention, and skill acquisition. The integration of AI-driven adaptive systems allows personalized learning pathways, robotics-integrated simulations prepare employees for human–robot collaboration, and automation-focused digital twins enable safe experimentation with production processes.

5.1 Experiential Learning

The Experiential Learning Theory (ELT) proposed by Kolb (2014) conceptualizes learning as a cyclical process comprising four stages: concrete experience, reflective observation, abstract conceptualization, and active experimentation. In industrial training, immersive technologies provide an optimal environment for experiential learning. VR simulations allow employees to experience high-risk scenarios, such as chemical spills or robotic malfunctions, without endangering themselves or the production process. AR overlays deliver immediate, context-sensitive feedback during assembly, maintenance, or calibration tasks. Digital twins facilitate experimentation with process parameters in a controlled virtual environment, allowing learners to analyze the effects of operational changes before implementing them in real life. This approach embodies the principle of "learning by doing," where employees can safely experiment with processes, receive immediate feedback, and iteratively improve their skills, closing the gap between theoretical knowledge and practical application. Empirical research demonstrates that immersive experiential learning enhances skill acquisition, reduces error rates, and improves the transfer of training to real-world tasks (Kolb, 2014; Fitrianto & Saif, 2024).

5.2 Adaptive Learning Platforms

Adaptive learning platforms leverage artificial intelligence to personalize training pathways based on individual learner performance and engagement data. These systems analyze metrics such as task completion times, error frequency, and interaction patterns, dynamically adjusting content difficulty, providing targeted feedback, and recommending remedial exercises (Song, Shin & Shin, 2024; Qadir et al., 2025). For example, AI can dynamically adapt VR modules for robotic assembly tasks based on real-time performance data, allowing learners to focus on the most challenging aspects while receiving immediate, tailored feedback for skill improvement. Research in educational technology shows that adaptive learning significantly improves knowledge retention and reduces the time required to achieve competency, particularly when combined with immersive experiential training (Johnson et al., 2021).

5.3 Gamification

Gamification refers to the application of game design elements in training environments to enhance motivation, engagement, and knowledge retention. Typical gamification elements include points, badges, leaderboards, scenario challenges, and immediate feedback loops (Kapp, 2012; Christopoulos & Mystakidis, 2023). In industrial training, gamification can be applied in various immersive contexts. Gamified tasks can be integrated with AR-guided instructions or digital twin simulations, where employees compete to optimize production processes. This strengthens procedural knowledge while encouraging active, practice-based problem solving. Empirical studies suggest that gamified learning promotes active participation, reinforces procedural knowledge, and increases intrinsic motivation, making learning experiences more engaging and effective (Tsay, Kofinas & Luo, 2018; Dehghanzadeh et al., 2024).

5.4 Collaborative Learning

Collaborative learning emphasizes social interaction and knowledge co-construction among learners. In industrial training, virtual platforms such as VR or MR enable trainees to collaborate in real time regardless of geographic location (Zaphiris & Ioannou, 2015; Van der Meer et al., 2023). MR environments, in particular, allow joint manipulation of real and virtual machine components, training teamwork, coordination, and collaborative decision-making in realistic scenarios. Research indicates that collaborative virtual training enhances cognitive understanding, supports retention, and mirrors the cooperative nature of real-world industrial operations (Singhaphandu & Pannakkong, 2024).

5.5 Integrated Didactic Model

Figure 2 illustrates the integration of immersive technologies with these didactic concepts. Experiential learning forms the core framework, providing the structured cycle of practice and reflection. Adaptive learning platforms personalize the experience, adjusting training content based on individual performance. Gamification enhances engagement and motivation, while collaborative learning promotes knowledge co-construction and social interaction. Immersive technologies act as the enabling layer, linking all four approaches into a cohesive, interactive, and scalable learning ecosystem. By integrating AI-driven adaptations, robotics-based simulations, and automation-focused digital twins, this model ensures that learning is both immersive and directly transferable to the demands of modern industrial environments.

This integrated model demonstrates that effective industrial training requires not only advanced technological tools but also pedagogically informed instructional design, ensuring that learning outcomes are maximized and directly transferable to operational performance.

Figure 2. Integrated Didactic Concepts in Immersive Industrial Training (Author's own illustration).

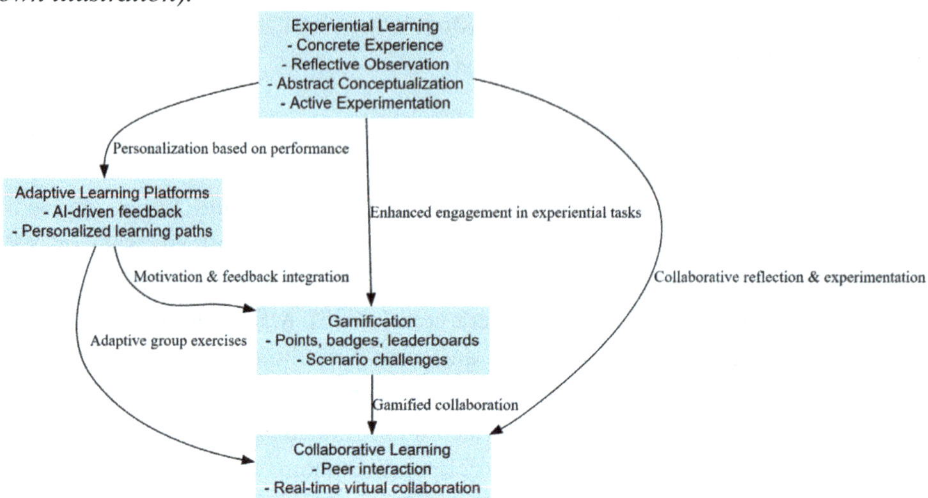

These didactic foundations support the practical use cases described in Section 6, where immersive technologies are applied to real industrial scenarios.

6. PRACTICAL EXAMPLES AND USE CASES OF IMMERSIVE TRAINING IN INDUSTRY

Immersive technologies have demonstrated transformative potential in industrial training, offering safe, scalable, and highly effective pathways for skill development. By leveraging AR, VR, and digital twins, organizations can provide employees with realistic, context-specific training while minimizing operational risks. In practice, the successful implementation of these immersive training technologies requires close alignment with existing production systems, workforce competencies, and organizational goals. Integration with AI-driven analytics allows real-time assessment of trainee performance, while robotics-integrated simulations prepare employees for direct interaction with automated systems. Digital twins enable safe experimentation with process adjustments before applying changes to live production lines. Early adoption in pilot projects has shown that phased implementation, combining AR, VR, and MR, along with structured feedback loops, ensures smoother technology adoption and higher learning outcomes. This section presents detailed examples from multiple sectors, illustrating how these technologies are applied to address real-world challenges in onboarding, safety, and maintenance training.

6.1 AR-based Onboarding in Automotive Assembly

Onboarding new employees in automotive production lines is traditionally both time-intensive and prone to errors due to complex assembly sequences, tight tolerances, and safety-critical components. Conventional methods, including manuals and in-person supervision, are limited in scalability and rarely provide real-time guidance. AR technology addresses these challenges by overlaying digital instructions directly onto the physical environment. Components are visually highlighted, assembly sequences are displayed step-by-step, and precise torque and fastening specifications are communicated through AR headsets. Supervisors can remotely monitor progress, and real-time feedback prevents errors before they occur, significantly improving both training efficiency and safety.

From a learning perspective, AR-based onboarding leverages the principles of situated cognition, embedding procedural knowledge directly within the work context (Padovano, Cardamone & Klaess, 2024). It also operationalizes the experiential learning cycle: trainees actively engage with tasks, observe outcomes, reflect on their actions, and adjust behaviors in real time. Empirical evidence supports the efficacy of this approach: a study conducted at a major European automotive manufacturer reported a 30–40% reduction in onboarding time and a 25% decrease in assembly errors compared to traditional training methods (Nee et al., 2012). Furthermore, AR-enabled training improved procedural knowledge retention by 20% over six months, demonstrating its long-term learning benefits. Furthermore, AR-based onboarding can be dynamically adapted using AI-driven adaptive learning platforms. Employees receive personalized instructions and task sequences depending on prior knowledge, real-time performance, and learning progress. This approach ensures that training is not only standardized but also tailored to individual needs, optimizing efficiency and reducing errors even further. Despite these advantages, challenges such as high initial investment in hardware and software, integration with legacy enterprise systems, and acceptance among employees with low digital affinity must be carefully managed.

6.2 VR Hazard Simulation in Chemical Manufacturing

Chemical plants present high-risk operational environments due to the handling of hazardous substances, high-temperature reactors, and complex emergency procedures. Traditional classroom-based safety training and low-fidelity drills often fail to adequately prepare employees for actual incidents. VR technology provides a controlled yet immersive solution by replicating plant layouts, chemical reactions, and potential hazard scenarios. Trainees navigate simulated emergencies, such as chemical leaks, equipment malfunctions, or fire outbreaks, with dynamic variables including pressure changes, alarm responses, and evacuation protocols.

VR enables repetitive, experiential learning, allowing employees to understand the consequences of decisions without exposure to physical risk. Immediate feedback on procedural errors reinforces knowledge and supports the development of situational awareness. In a German chemical plant, employees trained with VR responded 35% faster and committed 40% fewer errors in emergency drills than those trained conventionally (Srinivasan et al., 2022). VR training also improved confidence and decision-making capabilities, which are critical for effective operations in high-risk contexts. Limitations include the high cost and time investment for developing realistic VR simulations, the need for continuous updates to reflect regulatory changes, and occasional user acclimatization challenges such as simulator sickness. To enhance practical applicability, VR scenarios are increasingly linked with digital twin representations of the plant and integrated with automated feedback systems. Trainees' actions within the VR simulation can immediately adjust virtual process parameters, reflecting the potential consequences in the real environment. This integration ensures that learning is directly transferable to operational tasks and that employees develop confidence in decision-making under dynamic conditions.

6.3 Digital Twins in Aerospace Maintenance

Aircraft maintenance requires mastery of highly complex systems, where errors can lead to severe safety risks, financial losses, and operational delays. Traditional hands-on training using actual turbine engines is risky, resource-intensive, and limited in opportunities for repetitive practice. Digital twins offer a solution by creating high-fidelity virtual replicas of turbine engines, linked to real-time sensor data. Trainees can perform inspections, maintenance procedures, and fault diagnoses within a virtual environment that accurately reflects wear patterns, dynamic operational behavior, and system responses.

Digital twins facilitate learning by doing, providing immediate feedback and allowing trainees to experiment with different maintenance strategies safely. The integration of AR overlays enables hybrid training scenarios, where virtual guidance complements interaction with real components. A European aerospace manufacturer reported a 45% reduction in errors during actual turbine maintenance after implementing digital twin training. Additionally, technician proficiency was achieved 20% faster, contributing to reduced downtime and lower operational risks. Challenges include the high cost of developing and maintaining real-time data integration, sophisticated IT infrastructure requirements, and the need for continuous model validation. Moreover, the implementation of AR overlays in combination with digital twins facilitates hybrid training environments where theoretical instructions are seamlessly connected to physical operations. This approach allows employees to practice complex maintenance strategies iteratively, with AI systems providing adaptive recommendations based on performance metrics. Such integrated training ecosystems strengthen the workforce's readiness for AI- and robotics-enabled operations.

6.4 Sector-Specific Integrated Applications

Immersive technologies are increasingly applied in an integrated manner across industrial sectors. In automotive manufacturing, AR-assisted assembly lines, VR safety drills, and digital twin-based workflow optimization enable faster onboarding, safer operations, and predictive maintenance. In aerospace, VR simulations for emergency scenarios, AR-guided assembly, and digital twin maintenance training reduce operational errors and enhance compliance with stringent safety regulations. Within the chemical industry, VR-based hazard response training, AR-guided chemical handling, and digital twin process simulations improve safety, reduce material waste, and optimize workflow efficiency. In addition, the synergistic use of AR, VR, MR, and digital twins enables cross-sector learning and scenario sharing. Data collected from one application—such as a VR chemical spill simulation—can inform training strategies in another sector, enhancing adaptive learning and operational safety. This comprehensive approach ensures that immersive training is not isolated but fully embedded within the broader industrial ecosystem, bridging the gap between technological possibilities and practical workforce readiness.

Figure 3 illustrates the synergistic deployment of AR, VR, and digital twins across these sectors, showing how onboarding, hazard simulation, and process optimization are combined to maximize learning outcomes and operational efficiency. The figure emphasizes the interconnectedness of immersive technologies, highlighting how data and feedback from one modality can inform and enhance others, creating a comprehensive and adaptive industrial training ecosystem.

Figure 3. Immersive Training Use Cases in Industry (Author's own work).

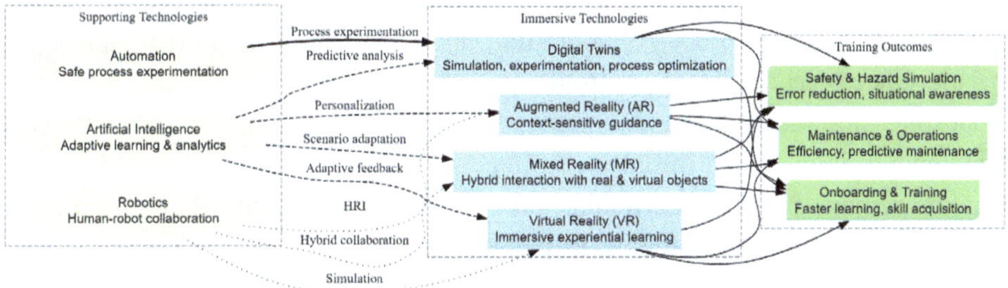

The practical examples and sector-specific use cases discussed here highlight both the benefits and the remaining challenges of immersive training, which are analyzed in Section 7.

7. OPPORTUNITIES AND CHALLENGES

The implementation of immersive technologies, including AR, VR, MR, and digital twins, in industrial training offers significant opportunities for enhancing workforce competence, productivity, and safety, while simultaneously presenting several challenges that must be addressed for effective and sustainable adoption. Immersive training solutions facilitate efficient skill development by allowing learners to engage in practical, controlled environments. For instance, AR-based guidance in manufacturing lines supports new employees during their onboarding by overlaying digital instructions directly onto the physical workspace. This immediate contextual assistance reduces errors and accelerates learning. Empirical studies indicate that AR-supported training can decrease onboarding time by up to 30–40% and reduce procedural errors in complex assembly tasks by approximately 25% (Yoo et al., 2024). Similarly, VR simulations enable employees to safely experience high-risk scenarios, such as chemical spills, machine failures, or emergency shutdown procedures, significantly improving safety and situational awareness (Yazdi, 2024). The repetitive, feedback-driven nature of immersive learning environments further accelerates skill acquisition and enhances knowledge retention, providing measurable benefits over traditional training methods (Kolb, 2014).

The practical implementation of these technologies requires careful consideration of platforms, hardware, and integration with operational systems. Key software and devices include AR guidance tools such as PTC Vuforia and Microsoft Dynamics 365 Guides, VR solutions like Strivr and ForgeFX, MR devices including HoloLens, and digital twin platforms provided by Siemens and Dassault Systèmes. These tools enable "learning by doing," offering real-time feedback while ensuring safety and minimizing operational disruptions. Integration with AI-driven adaptive systems allows personalized learning pathways, robotics-integrated simulations prepare employees for human-robot collaboration, and automation-focused digital twins provide safe experimentation environments.

Despite these advantages, companies face several challenges in implementing immersive training. High initial investments in hardware, software, and IT infrastructure remain a barrier, particularly for SMEs, requiring cost–benefit analyses and phased deployment strategies. Employee acceptance is critical; resistance may arise from unfamiliarity with technology, perceived complexity, or skepticism about effectiveness (Cham et al., 2024; Christopoulos et al., 2024). Structured change management, pilot testing, and hands-on demonstrations are essential to foster adoption and engagement. Furthermore, the lack of standardization in immersive training content and interoperability limits scalability, particularly for multinational organizations with diverse equipment and processes. Establishing industry-wide standards and best practices is essential to ensure effective and consistent deployment. Data privacy and security also play a central role, as immersive platforms collect extensive performance and behavioral data, including gaze tracking and operational metrics. Compliance with regulations such as GDPR is mandatory to mitigate legal risks and maintain employee trust.

In addition to operational benefits, immersive training contributes to sustainability by reducing material usage, energy consumption, and travel requirements for centralized training. Digital twins and VR simulations allow complex experiments and maintenance procedures to be conducted virtually, conserving resources while providing realistic learning experiences. Furthermore, immersive platforms support inclusive workforce development by enabling flexible, remote, and globally accessible learning environments.

Monitoring and evaluating the effectiveness of immersive training is crucial. Metrics such as error reduction, faster skill acquisition, improved safety compliance, and enhanced operational efficiency provide measurable evidence of success. Organizations should adopt iterative improvement cycles, continuously updating content and scenarios based on learner performance, technological advances, and operational changes. This ensures that training remains relevant, adaptive, and aligned with both workforce and organizational objectives.

In summary, immersive technologies in industrial training offer substantial potential to enhance efficiency, safety, cost-effectiveness, and accelerated skill development. However, their successful deployment depends on coordinated consideration of technological, organizational, and human factors, including infrastructure readiness, employee engagement, data security, and standardization. By strategically planning implementation, integrating AI, robotics, and automation, and leveraging software and hardware solutions effectively, organizations can maximize the benefits of immersive training while mitigating associated risks.

The challenges and opportunities identified here inform the future perspectives of immersive industrial training discussed in Section 8, highlighting emerging technologies, scalability, and strategic workforce considerations.

8. FUTURE PERSPECTIVES OF IMMERSIVE INDUSTRIAL TRAINING

The landscape of industrial training is undergoing a transformative shift driven by emerging technologies that complement existing immersive solutions. As industrial operations evolve toward Industry 4.0 and 5.0 paradigms, training moves beyond static procedures or classroom instruction, becoming a dynamic, adaptive, and globally accessible learning ecosystem. This ecosystem integrates immersive technologies with AI-driven personalization, robotics simulations, digital twins, and collaborative platforms, creating a unified framework for workforce development.

8.1 Generative AI for Personalized Training

Generative Artificial Intelligence (AI) has the potential to revolutionize industrial training by creating personalized, adaptive learning scenarios in real time (Rane et al., 2024). Unlike pre-programmed simulations, generative AI dynamically adjusts task complexity, simulates unforeseen scenarios, and provides context-specific feedback tailored to the learner's current skill level. For example, in automotive assembly, generative AI can detect repeated errors in part installation and generate additional training modules targeting the identified weaknesses, or create alternative scenarios to enhance problem-solving skills, fully integrated with AR/VR or digital twin environments. This approach aligns with adaptive learning theories, emphasizing individualized learning pathways to maximize knowledge retention and skill acquisition (Shanthi et al., 2025). Challenges include ensuring scenario accuracy, mitigating AI bias, and safeguarding operational data.

8.2 AR/VR-Supported Human-Robot Collaboration

The increasing presence of collaborative robots (cobots) in industrial environments necessitates immersive training focused on human-robot interaction (HRI). AR and VR technologies allow employees to practice coordinated workflows in simulated environments before real-world operations. AR interfaces can display robot motion trajectories, predicted movements, and safety zones, while VR recreates unexpected errors or hazardous scenarios safely. Integrating these simulations with real-time operational data and digital twins enhances training relevance and fidelity. Research demonstrates that immersive HRI training improves situational awareness, reduces human error, and enhances task performance (Roldán et al., 2017). Implementation challenges include sensor integration, latency reduction, and interface standardization.

8.3 Metaverse-Based Learning Environments

The concept of the metaverse offers a new paradigm for global, collaborative industrial training. By leveraging shared virtual spaces, companies can connect geographically distributed employees in a unified learning environment. Trainees from multiple locations can collaborate on simulated production processes, participate in joint problem-solving exercises, and exchange expertise across borders. Metaverse-based platforms also facilitate gamification and social learning, creating highly engaging and interactive training experiences that enhance motivation and knowledge retention (Damaševičius & Sidekerskienė, 2024).

For multinational corporations, metaverse environments provide a scalable solution to standardize training across facilities, ensuring consistency in skill development and operational procedures. Challenges in implementing metaverse-based training include network latency, interoperability of virtual environments, and cybersecurity, particularly when sensitive industrial processes are simulated in cloud-based platforms.

8.4 Scaling Immersive Training for SMEs

While large corporations have the resources to implement sophisticated immersive training infrastructures, small and medium-sized enterprises (SMEs) often face barriers related to cost, technical expertise, and infrastructure. Emerging trends, however, are making immersive training increasingly accessible to SMEs. Cloud-based VR/AR platforms, subscription models for digital twin simulations, and modular scenario libraries allow smaller organizations to deploy immersive learning without significant upfront investment.

This democratization of technology aligns with the broader objective of inclusive workforce development, enabling SMEs to maintain competitiveness in rapidly digitalizing industries. Furthermore, as immersive training becomes more affordable, SMEs can integrate flexible, hybrid learning models that combine virtual simulations with on-the-job mentoring, thereby optimizing both efficiency and safety. Despite these advances, careful planning is required to ensure integration with existing workflows, employee acceptance, and alignment with regulatory and safety standards.

8.5 Implications and Strategic Considerations

The convergence of generative AI, AR/VR-enabled human-robot collaboration, metaverse learning, and scalable platforms for SMEs represents a paradigm shift in industrial training. Companies that successfully adopt these technologies can achieve rapid, individualized skill development, enhance operational safety, and create globally consistent training programs. Strategic implementation requires not only technological readiness but also cultural adaptation, robust data governance, and alignment with organizational learning objectives.

Moreover, the integration of immersive technologies with AI-driven analytics and digital twins opens avenues for continuous learning and predictive skill management, where employee performance data informs ongoing adjustments to training content and intensity. This approach ensures that workforce competencies evolve in tandem with rapidly changing production technologies, preparing employees for increasingly complex roles in AI-enabled factories.

These future-oriented strategies provide a roadmap for integrating immersive training technologies into AI-enabled factories, supporting adaptive, scalable, and sustainable workforce development.

9. CONCLUSION

The integration of immersive training technologies into industrial environments represents a fundamental shift in workforce development, moving away from traditional, static forms of learning toward dynamic, flexible, and adaptive educational ecosystems. As industrial operations become increasingly complex due to automation, artificial intelligence, and collaborative robotics, conventional training methods are no longer sufficient to ensure safety, efficiency, and rapid competence acquisition. AR, VR, MR, and digital twin technologies offer immersive, hands-on learning opportunities that allow employees to practice complex tasks, respond to emergencies, and experiment with operational processes in controlled and risk-free environments. Empirical evidence demonstrates that such immersive approaches accelerate skill acquisition, improve knowledge retention, reduce procedural errors, and enhance situational awareness, thereby directly contributing to operational excellence (Kolb, 2014; Ahmed & Sutton, 2017).

The implementation of immersive training solutions also carries profound strategic implications. By leveraging technologies such as generative AI, industrial organizations can provide individualized learning paths that adapt in real-time to the learner's performance, fostering accelerated and targeted competence development. AR/VR-supported human-robot collaboration training ensures safe and efficient interaction with autonomous systems, mitigating risks associated with increasingly automated production lines. Moreover, metaverse-based learning environments enable global collaboration, standardization, and knowledge sharing, facilitating the integration of geographically distributed workforces into unified training programs. Finally, advances in cloud-based and modular platforms allow even small and medium-sized enterprises (SMEs) to adopt immersive training, democratizing access to cutting-edge educational technologies and enhancing competitiveness across industry sectors.

Despite these advantages, companies must carefully navigate several challenges to fully realize the benefits of immersive training. High initial investments, technological complexity, employee acceptance, data privacy, and the lack of standardization present significant barriers that require strategic planning, effective change management, and the development of industry-wide best practices. Additionally, the successful adoption of immersive training is not solely a technological question; it also requires organizational alignment, pedagogical expertise, and continuous evaluation to ensure that learning outcomes translate into measurable operational improvements.

In terms of sustainability, immersive technologies contribute to environmental and resource efficiency by reducing the consumption of physical materials, minimizing travel for centralized training, and lowering energy usage associated with conventional training methods. This aligns workforce development with corporate social responsibility and broader environmental objectives, offering a tangible contribution to sustainable industrial practices.

In conclusion, immersive training solutions are pivotal for the competitiveness of AI-enabled and smart factories, transforming rigid, linear training programs into adaptive, interactive, and scalable learning ecosystems. By integrating experiential learning, generative AI, collaborative robotics simulations, and metaverse-enabled training, organizations can enhance workforce competence, safety, and operational efficiency while simultaneously addressing sustainability and inclusivity goals. In the era of Industry 4.0 and 5.0, companies that strategically implement immersive training will be better positioned to navigate rapid technological change, foster continuous employee development, and maintain a competitive advantage in increasingly digitalized and automated production environments.

REFERENCES

Попова, Л. (2024). Immersive Technologies as a Modern Educational Strategy for Training Future Specialists. *Педагогічна освіта: Теорія і практика. Психологія. Педагогіка*, (42(2)), 51–58.

Ahmed, A., & Sutton, M. J. (2017). Gamification, serious games, simulations, and immersive learning environments in knowledge management initiatives. *World Journal of Science. Technology and Sustainable Development*, *14*(2/3), 78–83.

Asmat, H., Din, I. U., Almogren, A., & Khan, M. Y. (2025). Digital Twin With Soft Actor-Critic Reinforcement Learning for Transitioning From Industry 4.0 to 5.0. *IEEE Access : Practical Innovations, Open Solutions*, *13*, 40577–40593. DOI: 10.1109/ACCESS.2025.3546085

Azuma, R. T. (1997). A survey of augmented reality. *Presence (Cambridge, Mass.)*, *6*(4), 355–385. DOI: 10.1162/pres.1997.6.4.355

Badri, A., Boudreau-Trudel, B., & Souissi, A. S. (2018). Occupational health and safety in the industry 4.0 era: A cause for major concern? *Safety Science*, *109*, 403–411. DOI: 10.1016/j.ssci.2018.06.012

Bandi, A., Kongari, B., Naguru, R., Pasnoor, S., & Vilipala, S. V. (2025). The Rise of Agentic AI: A Review of Definitions, Frameworks, Architectures, Applications, Evaluation Metrics, and Challenges. *Future Internet*, *17*(9), 404. DOI: 10.3390/fi17090404

Bondin, A., & Zammit, J. P. (2025). Education 4.0 for Industry 4.0: A Mixed Reality Framework for Workforce Readiness in Manufacturing. *Multimodal Technologies and Interaction*, *9*(5), 43. DOI: 10.3390/mti9050043

Bratu, S. (2023). Generative Artificial Intelligence and Virtual Reality Immersive Training Tools, Algorithmic Monitoring and Cognitive Computing Systems, and Voice and Gesture Recognition Technologies in Virtual Immersive Workspaces. *Analysis and Metaphysics*, (22), 197–215.

Buchner, J., Buntins, K., & Kerres, M. (2022). The impact of augmented reality on cognitive load and performance: A systematic review. *Journal of Computer Assisted Learning*, *38*(1), 285–303. DOI: 10.1111/jcal.12617

Cham, T. H., Wei-Han Tan, G., Aw, E. C. X., Ooi, K. B., Jee, T. W., & Pek, C. K. (2024). Virtual reality in tourism: Adoption scepticism and resistance. *Tourism Review*, *79*(2), 337–354. DOI: 10.1108/TR-10-2022-0479

Christopoulos, A., & Mystakidis, S. (2023). Gamification in education. *Encyclopedia, 3*(4), 1223–1243. DOI: 10.3390/encyclopedia3040089

Christopoulos, A., Styliou, M., Ntalas, N., & Stylios, C. (2024). The impact of immersive virtual reality on knowledge acquisition and adolescent perceptions in cultural education. *Information (Basel), 15*(5), 261. DOI: 10.3390/info15050261

Colombo, S., Nazir, S., & Manca, D. (2014). Immersive virtual reality for training and decision making: Preliminary results of experiments performed with a plant simulator. *SPE Economics & Management, 6*(04), 165–172. DOI: 10.2118/164993-PA

Conges, A., Breard, L., Patruno, W., Ouro-Sao, A., Salatge, N., Fertier, A., Lauras, M., Graham, J., & Benaben, F. (2023). Situational awareness and decision-making in a crisis situation: A crisis management cell in virtual reality. *International Journal of Disaster Risk Reduction, 97*, 104002. DOI: 10.1016/j.ijdrr.2023.104002

Cotton, W. B. (2001). Transforming traditional training methods to meet the needs of a modern industry. *Journal of Petroleum Technology, 53*(02), 54–58. DOI: 10.2118/65351-JPT

Damaševičius, R., & Sidekerskienė, T. (2024). Virtual worlds for learning in metaverse: A narrative review. *Sustainability (Basel), 16*(5), 2032. DOI: 10.3390/su16052032

Damiani, L., Demartini, M., Guizzi, G., Revetria, R., & Tonelli, F. (2018). Augmented and virtual reality applications in industrial systems: A qualitative review towards the industry 4.0 era. *IFAC-PapersOnLine, 51*(11), 624–630. DOI: 10.1016/j.ifacol.2018.08.388

Dehghanzadeh, H., Farrokhnia, M., Dehghanzadeh, H., Taghipour, K., & Noroozi, O. (2024). Using gamification to support learning in K-12 education: A systematic literature review. *British Journal of Educational Technology, 55*(1), 34–70. DOI: 10.1111/bjet.13335

Deogaonkar, A. (2025). Immersion, realism and engagement: Exploring the impact of generative AI on virtual training performance across sectors. *European Journal of Training and Development.*

Falola, P. B., Adeniyi, A. E., Awotunde, J. B., Akinola, S. O., Olagunju, M., & Olanloye, O. D. (2024). The Role of Augmented Reality, Virtual Reality, and Mixed Reality in Industries 4.0 and 5.0. In *Computational Intelligence in Industry 4.0 and 5.0 Applications* (pp. 329-356). Auerbach Publications.

Fitrianto, I., & Saif, A. (2024). The role of virtual reality in enhancing Experiential Learning: a comparative study of traditional and immersive learning environments. *International Journal of Post Axial: Futuristic Teaching and Learning, 97-110.*

Fitsilis, P., Tsoutsa, P., & Gerogiannis, V. (2018). Industry 4.0: Required personnel competences. *Industry 4.0, 3*(3), 130-133.

Häfner, P., Bergmann, V., Häfner, V., Michels, F. L., Grethler, M., & Karande, A. (2024). An Adaptive Learning Environment for Industry 4.0 Competencies Based on a Learning Factory and Its Immersive Digital Twin. In *CSEDU (1)* (pp. 720-731).

Hamilton, D., McKechnie, J., Edgerton, E., & Wilson, C. (2021). Immersive virtual reality as a pedagogical tool in education: A systematic literature review of quantitative learning outcomes and experimental design. *Journal of Computers in Education, 8*(1), 1–32. DOI: 10.1007/s40692-020-00169-2

Hancko, D., Majlingova, A., & Kačíková, D. (2025). Integrating Virtual Reality, Augmented Reality, Mixed Reality, Extended Reality, and Simulation-Based Systems into Fire and Rescue Service Training: Current Practices and Future Directions. *Fire (Basel, Switzerland), 8*(6), 228. DOI: 10.3390/fire8060228

Hazrat, M. A., Hassan, N. M. S., Chowdhury, A. A., Rasul, M. G., & Taylor, B. A. (2023). Developing a skilled workforce for future industry demand: The potential of digital twin-based teaching and learning practices in engineering education. *Sustainability (Basel), 15*(23), 16433. DOI: 10.3390/su152316433

Islam, M. T., Sepanloo, K., Woo, S., Woo, S. H., & Son, Y. J. (2025). A review of the industry 4.0 to 5.0 transition: Exploring the intersection, challenges, and opportunities of technology and human–machine collaboration. *Machines (Basel), 13*(4), 267. DOI: 10.3390/machines13040267

Kapp, K. M. (2012). *The gamification of learning and instruction: game-based methods and strategies for training and education*. John Wiley & Sons.

Kolb, D. A. (2014). *Experiential learning: Experience as the source of learning and development*. FT Press.

Kumar, S. S., Kumar, S. R., & Ramesh, G. (2024). From industry 4.0 to 5.0: Enriching manufacturing excellence through Human–Robot interaction and technological empowerment. In *Intelligent systems and industrial internet of things for sustainable development* (pp. 24–51). Chapman and Hall/CRC. DOI: 10.1201/9781032642789-2

Lavric, T., Bricard, E., Preda, M., & Zaharia, T. (2022). A low-cost AR training system for manual assembly operations. *Computer Science and Information Systems, 19*(2), 1047–1073. DOI: 10.2298/CSIS211123013L

Lehrman, A. L. (2025). Embodied Learning Through Immersive Virtual Reality: Theoretical Perspectives for Art and Design Education. *Behavioral Sciences (Basel, Switzerland), 15*(7), 917. DOI: 10.3390/bs15070917 PMID: 40723701

Lodhi, S. K., & Zeb, S. (2025). AI-Driven Robotics and Automation: The Evolution of Human-Machine Collaboration. *Journal of Wood Science*, *4*(4), 422–437.

Martínez-Mireles, J. R., Rodríguez-Flores, J., García-Márquez, M. A., Austria-Cornejo, A., & Figueroa-Diaz, R. A. (2025). AI in Smart Manufacturing. In *Machine and Deep Learning Solutions for Achieving the Sustainable Development Goals* (pp. 463–494). IGI Global Scientific Publishing. DOI: 10.4018/979-8-3693-8161-8.ch023

Milgram, P., & Kishino, F. (1994). A taxonomy of mixed reality visual displays. *IEICE Transactions on Information and Systems*, *77*(12), 1321–1329.

MK. R. S., Poreddy, B. K. R., Sharmila, L., Rajesha, N., Anand, M. C. J., & Boopathi, S. (2024). Empowering Virtual Machine-Controlled and AI-Controlled Robots for Transportation in the Manufacturing Industry. In *Contemporary Solutions for Sustainable Transportation Practices* (pp. 436-466). IGI Global.

Padovano, A., Cardamone, M., & Klaess, J. (2024). Empowering Operator 5.0: Human-centric design of an augmented reality tool for a learning factory. *IFAC-PapersOnLine*, *58*(8), 180–185. DOI: 10.1016/j.ifacol.2024.08.117

Palmarini, R., Erkoyuncu, J. A., Roy, R., & Torabmostaedi, H. (2018). A systematic review of augmented reality applications in maintenance. *Robotics and Computer-integrated Manufacturing*, *49*, 215–228. DOI: 10.1016/j.rcim.2017.06.002

Petersen, N., & Stricker, D. (2015). Cognitive augmented reality. *Computers & Graphics*, *53*, 82–91. DOI: 10.1016/j.cag.2015.08.009

Pires, F., Cachada, A., Barbosa, J., Moreira, A. P., & Leitão, P. (2019). Digital twin in industry 4.0: Technologies, applications and challenges. In *2019 IEEE 17th international conference on industrial informatics (INDIN)* (Vol. 1, pp. 721-726). IEEE.

Praveen, R. V. S., & Mutya, B. (2024). *The Automated Future: AI-Driven Transformation in Manufacturing*. Addition Publishing House.

Qadir, H. M., Khan, R. A., Rasool, M., Sohaib, M., Shah, M. A., & Hasan, M. J. (2025). An adaptive feedback system for the improvement of learners. *Scientific Reports*, *15*(1), 17242. DOI: 10.1038/s41598-025-01429-w PMID: 40383780

Radianti, J., Majchrzak, T. A., Fromm, J., & Wohlgenannt, I. (2020). A systematic review of immersive virtual reality applications for higher education: Design elements, lessons learned, and research agenda. *Computers & Education*, *147*, 103778. DOI: 10.1016/j.compedu.2019.103778

Rane, J., Kaya, Ö., Mallick, S. K., & Rane, N. L. (2024). *Generative artificial intelligence in agriculture, education, and business*. Deep Science Publishing. DOI: 10.70593/978-81-981271-7-4

Ras, E., Wild, F., Stahl, C., & Baudet, A. (2017, June). Bridging the skills gap of workers in Industry 4.0 by human performance augmentation tools: Challenges and roadmap. In *Proceedings of the 10th international conference on PErvasive technologies related to assistive environments* (pp. 428-432). DOI: 10.1145/3056540.3076192

Roldán, J. J., Peña-Tapia, E., Martín-Barrio, A., Olivares-Méndez, M. A., Del Cerro, J., & Barrientos, A. (2017). Multi-robot interfaces and operator situational awareness: Study of the impact of immersion and prediction. *Sensors (Basel)*, *17*(8), 1720. DOI: 10.3390/s17081720 PMID: 28749407

Salas, E. (2021). Virtual Reality in Medical Education: A Systematic Review. *BMJ Quality & Safety*, *30*(4), 345–350.

Shanthi, D., Ashok, G., Biswal, C., Udharika, S., Varshini, S., & Sindhu, G. (2025). AI-driven adaptive IT training: A personalized learning framework for enhanced knowledge retention and engagement. *Metallurgical and Materials Engineering, 136-145.*

Shelton, B. E., & Hedley, N. R. (2004). Exploring a cognitive basis for learning spatial relationships with augmented reality. *Technology, Instruction. Cognition and Learning*, *1*(4), 323.

Singhaphandu, R., & Pannakkong, W. (2024). A review on enabling technologies of industrial virtual training systems. [IJKSS]. *International Journal of Knowledge and Systems Science*, *15*(1), 1–33. DOI: 10.4018/IJKSS.352515

Song, C., Shin, S. Y., & Shin, K. S. (2024). Implementing the dynamic feedback-driven learning optimization framework: A machine learning approach to personalize educational pathways. *Applied Sciences (Basel, Switzerland)*, *14*(2), 916. DOI: 10.3390/app14020916

Srinivasan, B., Iqbal, M. U., Shahab, M. A., & Srinivasan, R. (2022). Review of virtual reality (VR) applications to enhance chemical safety: From students to plant operators. *Journal of Chemical Health and Safety*, *29*(3), 246–262. DOI: 10.1021/acs.chas.2c00006

Steuer, J. (1992). Defining virtual reality: Dimensions determining telepresence. *Journal of Communication*, *42*(4), 73–93. DOI: 10.1111/j.1460-2466.1992.tb00812.x

Syberfeldt, A., Danielsson, O., & Gustavsson, P. (2017). Augmented reality smart glasses in the smart factory: Product evaluation guidelines and review of available products. *IEEE Access : Practical Innovations, Open Solutions*, *5*, 9118–9130. DOI: 10.1109/ACCESS.2017.2703952

Tsay, C. H. H., Kofinas, A., & Luo, J. (2018). Enhancing student learning experience with technology-mediated gamification: An empirical study. *Computers & Education*, *121*, 1–17. DOI: 10.1016/j.compedu.2018.01.009

Tyagi, A. K., Bhatt, P., Chidambaram, N., & Kumari, S. (2024). Artificial intelligence empowered smart manufacturing for modern society: a review. *Artificial Intelligence-Enabled Digital Twin for Smart Manufacturing*, 55-83.

Van der Meer, N., van der Werf, V., Brinkman, W. P., & Specht, M. (2023). Virtual reality and collaborative learning: A systematic literature review. *Frontiers in Virtual Reality*, *4*, 1159905. DOI: 10.3389/frvir.2023.1159905

Yazdi, M. (2024). Augmented reality (AR) and virtual reality (VR) in maintenance training. In *Advances in Computational Mathematics for Industrial System Reliability and Maintainability* (pp. 169–183). Springer Nature Switzerland. DOI: 10.1007/978-3-031-53514-7_10

Yoo, S., Reza, S., Tarashiyoun, H., Ajikumar, A., & Moghaddam, M. (2024). AI-Integrated AR as an Intelligent Companion for Industrial Workers: A Systematic Review. *IEEE Access : Practical Innovations, Open Solutions*, *12*, 191808–191827. DOI: 10.1109/ACCESS.2024.3516536

Yuan, M. L., Ong, S. K., & Nee, A. Y. C. (2008). Augmented reality for assembly guidance using a virtual interactive tool. *International Journal of Production Research*, *46*(7), 1745–1764. DOI: 10.1080/00207540600972935

Zahid, A., Ferraro, A., Petrillo, A., & De Felice, F. (2025). Exploring the Role of Digital Twin and Industrial Metaverse Technologies in Enhancing Occupational Health and Safety in Manufacturing. *Applied Sciences (Basel, Switzerland)*, *15*(15), 8268. DOI: 10.3390/app15158268

Zaphiris, P., & Ioannou, A. (2015). *Learning and Collaboration Technologies* (Vol. 9192). Springer. DOI: 10.1007/978-3-319-20609-7

Zhu, X., Lai, P. L., Li, X., Wang, Y., & Pei, X. (2025). A Systematic Review of Literature on the Use of Extended Reality in Logistics and Supply Chain Management Education: Evolution of Research Themes and System-Level Trends. *Systems*, *13*(7), 514. DOI: 10.3390/systems13070514

Section 2

Smart Modern Manufacturing

Chapter 3
Revolutionizing Production:
AI and Automation Driving Industrial Excellence

V. S. Saranya
http://orcid.org/0009-0001-6527-7022
SRM Institute of Science and Technology, India

J. Shobana
http://orcid.org/0000-0001-9754-2604
SRM Institute of Science and Technology, India

Renugadevi Ramalingam
http://orcid.org/0000-0001-7462-9205
Saveetha Engineering College, India

L. Annie Isabella
Saveetha School of Engineering, Saveetha Institute of Medical and Technical Sciences, India

N. Priyanka
http://orcid.org/0009-0004-2007-2973
Vellore Institute of Technology, India

K. Arthi
SRM Institute of Science and Technology, India

ABSTRACT

Recent advancements in intelligent systems and industrial automation have accelerated the evolution of digital manufacturing ecosystems, contributing significantly to the realization of Industry 4.0. This chapter investigates the transformative role of Artificial Intelligence (AI)-driven intelligent automation, emphasizing the convergence of Robotic Process Automation (RPA) and AI technologies in optimizing complex industrial and enterprise operations. RPA systems are designed to execute repetitive, deterministic tasks based on structured logic, enabling improved operational throughput with minimal human intervention. When integrated with AI techniques such as machine learning algorithms, natural language processing (NLP), and neural network-based data modelling, these systems acquire cognitive

DOI: 10.4018/979-8-3373-5756-0.ch003

capabilities, including data inference, anomaly detection, predictive analytics, and semantic classification. This review synthesizes recent advancements in AI-RPA integration, with particular focus on high-impact applications across key sectors like manufacturing, precision agriculture, healthcare diagnostics, financial services, and intelligent retail systems. The analysis highlights how intelligent automation enhances process efficiency, reduces operational costs, increases system scalability, and facilitates real-time decision-making. Furthermore, it addresses critical technical and ethical challenges, including interoperability issues, data governance, algorithmic bias, and cybersecurity vulnerabilities. By leveraging advanced AI models such as convolutional neural networks (CNNs), deep reinforcement learning, and knowledge-based reasoning RPA platforms are increasingly capable of supporting adaptive and autonomous operations. This convergence fosters the emergence of smart factories and cyber-physical production systems that are self-optimizing and resilient. The paper aims to provide a comprehensive technical perspective on the synergistic integration of RPA and AI, offering insights into its pivotal role in redefining intelligent automation within the framework of Industry 4.0.

1. INTRODUCTION

The global manufacturing landscape is undergoing a profound transformation driven by the convergence of Artificial Intelligence (AI), automation technologies, and the principles of Industry 4.0. These innovations are reshaping traditional industrial practices, giving rise to "smart factories"—digitally connected environments where intelligent systems autonomously optimize production processes, ensure quality, and adapt in real time. At the heart of this transition is the integration of Robotic Process Automation (RPA) with advanced AI techniques such as machine learning (ML), deep learning (DL), computer vision, and natural language processing (NLP), which collectively enable cognitive automation far beyond rule-based execution.

Automation in manufacturing has evolved from mechanical systems and programmable logic controllers (PLCs) to intelligent cyber-physical systems (CPS). These systems interact with humans and machines in real time, guided by sensor data, analytics, and decision-making algorithms. AI enhances automation by introducing data-driven intelligence into machines, allowing them to learn from historical patterns, adapt to new conditions, and make predictive and prescriptive decisions. This synergy improves productivity, efficiency, and system responsiveness while significantly reducing downtime, waste, and manual intervention.

One of the primary enablers of this transformation is the Internet of Things (IoT), which facilitates seamless data collection from machinery, tools, and products across the factory floor. The acquired data is processed by AI models to derive insights related to machine health, process efficiency, and product quality. Predictive maintenance, powered by AI, is a prominent application wherein algorithms analyze sensor data to detect equipment anomalies before failures occur—thus minimizing unplanned downtime and optimizing asset utilization.

Artificial intelligence (AI) and automation are redefining the foundations of modern industry, representing a technological shift as transformative as the Industrial Revolution itself, and their convergence is increasingly recognized as a driver of industrial excellence across productivity, quality, sustainability, and resilience. The combination of advanced data analytics, machine learning, robotics, and digital twin technologies is creating intelligent industrial ecosystems that sense, analyze, decide, and act in ways that were inconceivable a decade ago. These transformations are driven by several inflection points: the maturation of deep learning and reinforcement learning algorithms, the proliferation of industrial Internet of Things (IIoT) sensors and connected devices, the availability of high-performance computing and edge processing for real-time inference, and the deployment of digital twins that simulate complex production processes.

Together, these developments are enabling organizations to transition from reactive to predictive and prescriptive decision-making, thereby improving operational performance, reducing costs, and unlocking new opportunities for innovation. For example, predictive maintenance powered by hybrid AI models now allows manufacturers to estimate the remaining useful life of critical assets and schedule interventions before catastrophic breakdowns occur, avoiding costly downtime and enhancing equipment reliability (Ali Hamza et al., 2025). Similarly, AI-enabled computer vision systems in production lines are performing quality inspection at a level of accuracy and consistency surpassing human capabilities, identifying micro-defects in real time and minimizing rework and waste (Plathottam et al., 2023). The adoption of digital twins is allowing organizations to replicate machines, production lines, or entire factories in a virtual environment, enabling real-time optimization, scenario testing, and accelerated innovation cycles (Iranshahi et al., 2025).

Automation and robotics, once limited to repetitive and preprogrammed tasks, are evolving into adaptive, AI-driven agents capable of collaborative work alongside humans, with advances in perception, dexterous manipulation, and multi-robot coordination facilitating new applications in assembly, material handling, and logistics (Vespoli et al., 2025). Beyond efficiency and quality, AI and automation are proving critical to sustainability by optimizing energy consumption, reducing resource waste, and enabling circular manufacturing practices, with digital twin-enabled life-cycle analyses supporting carbon footprint reduction strategies (Schwark et al., 2023). These shifts are also enhancing industrial resilience, as organizations leverage AI-driven scenario planning, supply chain analytics, and predictive models to better withstand demand volatility, global disruptions, and resource constraints (IoT Analytics, 2024). Importantly, these benefits are not achieved through technology alone; effective integration requires human-machine collaboration, workforce reskilling, and governance frameworks that ensure ethical, safe, and transparent use of AI in industrial contexts (Lei et al., 2024). The challenges remain significant, ranging from data heterogeneity, legacy infrastructure, and model robustness under distributional shifts, to regulatory gaps around safety, explainability, and cyber-physical security (McKinsey & Company, 2023). The association of AI and Automation is given in figure 1.

Figure 1. Artificial Intelligence Vs Automation

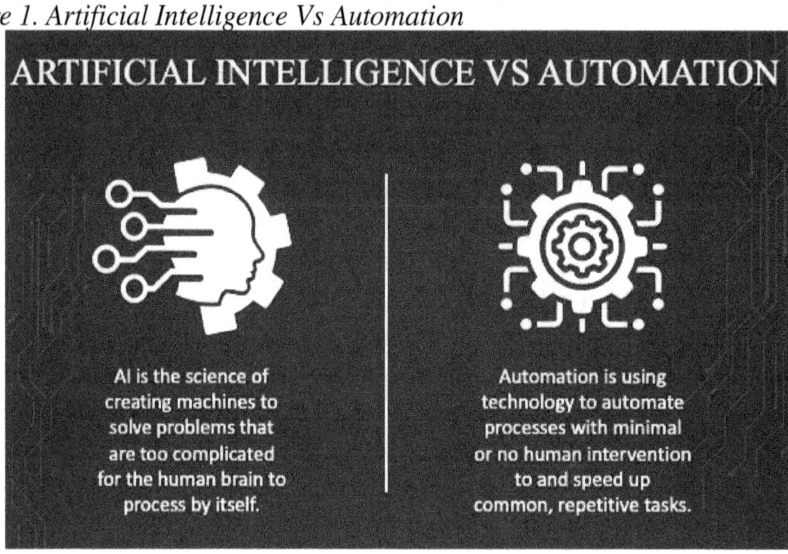

Organizations often adopt AI and automation in stages, beginning with localized pilot projects—such as defect detection on a single line or predictive analytics for a critical asset—before scaling to enterprise-wide platforms supported by standardized data lakes, MLOps practices, and cross-functional teams (Chowdhury et al., 2023). Emerging frontiers such as edge AI, explainable AI, generative foundation models, and hybrid physics-data digital twins are poised to further accelerate transformation, but their successful deployment depends on careful attention to organizational readiness, cultural adaptation, and policy alignment. As such, the convergence of AI and automation is not simply a matter of technology adoption but rather a socio-technical evolution requiring holistic strategies to maximize impact. Industrial excellence in the twenty-first century will therefore be defined not only by throughput and efficiency but also by an organization's ability to harness AI and automation responsibly, sustainably, and inclusively. This chapter situates AI and automation within this broader narrative, examining the technological underpinnings, value creation mechanisms, organizational challenges, and emerging research directions that collectively define the future of intelligent industrial ecosystems. Figure 2 represents the important operations involved in Business Process Automation when implemented with Artificial Intelligence. The integration of Artificial Intelligence (AI) into automation brings multiple transformative benefits across industries. One of the most significant outcomes is cost reduction, achieved through predictive maintenance, optimized resource allocation, and minimizing human errors. Alongside this, AI enables streamlined operations, where repetitive and time-consuming tasks are automated, leading to smoother workflows and faster turnaround times.

Another advantage is enhanced decision-making, as AI systems can analyze vast datasets in real time, detect patterns, and generate actionable insights for strategic planning. This directly supports enhanced efficiency, since processes are executed faster with fewer interruptions, maximizing productivity. Additionally, improved accuracy is realized through AI's ability to eliminate human bias and error, particularly in precision-driven fields like healthcare, engineering, and finance.

Figure 2. AI in Business Process Automation

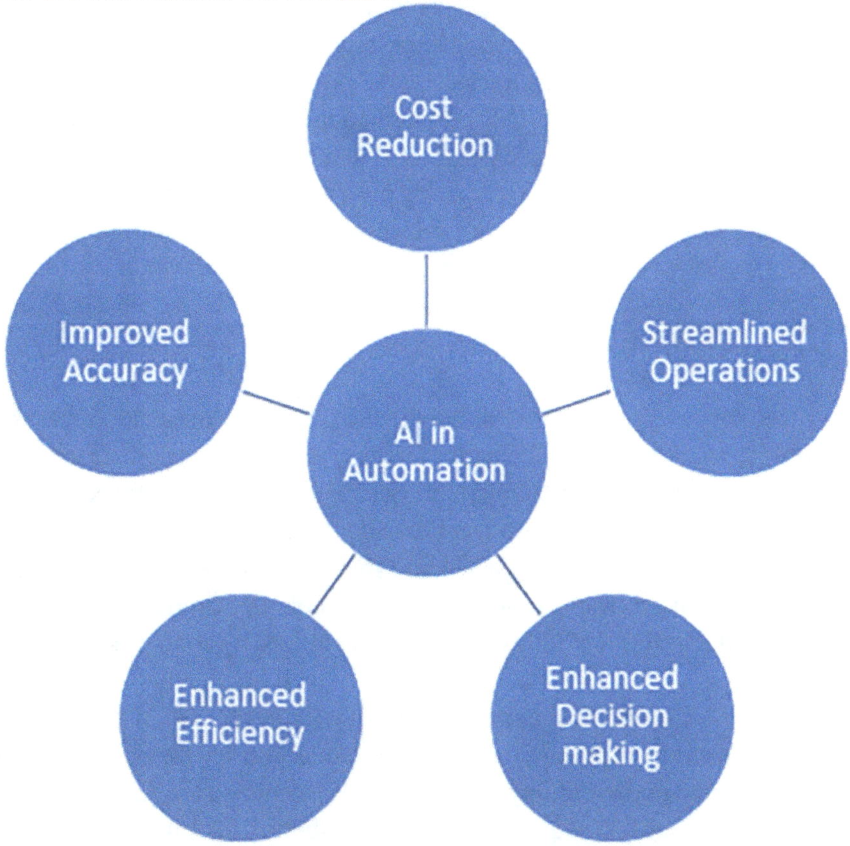

AI-powered computer vision systems are also revolutionizing quality control. Unlike conventional inspection methods, these systems employ convolutional neural networks (CNNs) and edge AI devices to detect surface defects, dimensional inconsistencies, or assembly errors in real time with sub-millimeter precision. Furthermore, AI algorithms are being deployed to dynamically adjust process parameters in manufacturing systems, ensuring consistent output quality under varying environmental or input conditions.

The integration of RPA and AI—often termed intelligent automation—extends the benefits of automation to administrative and knowledge-based functions within manufacturing enterprises. Tasks such as order processing, inventory management, procurement, and customer service can be automated using bots augmented with natural language understanding and ML-based decision-making. This enables end-to-end automation of both physical and digital processes, creating a unified and agile production ecosystem.

Despite its benefits, the adoption of AI and automation in manufacturing presents technical and strategic challenges. These include data quality and availability issues, system interoperability, workforce reskilling, cybersecurity threats, and ethical considerations related to transparency and accountability in automated decision-making. Overcoming these challenges requires a multidisciplinary approach involving industrial engineers, data scientists, IT professionals, and domain experts.

In summary, the integration of AI and automation in manufacturing is a key pillar of the digital industrial revolution. By enabling intelligent, autonomous, and adaptive operations, these technologies are redefining production paradigms and positioning manufacturing enterprises to be more competitive, resilient, and future-ready.

2. LITERATURE REVIEW

Over the past five years, artificial intelligence (AI) and automation have been increasingly recognized as transformative forces in the industrial domain, reshaping manufacturing systems, supply chains, and organizational strategies for competitiveness. AI enables machines to learn from data and improve their decision-making, while automation provides the physical and cyber-physical mechanisms to execute those decisions with precision. Together, these technologies contribute to the vision of "smart manufacturing" or "Industry 4.0," characterized by connected, intelligent, and adaptive systems that drive excellence in productivity, quality, sustainability, and resilience (McKinsey & Company, 2023; Plathottam et al., 2023). This literature review synthesizes recent research, classifying studies into key thematic domains to provide a structured understanding of how AI and automation are driving industrial excellence, while identifying current gaps and opportunities for future inquiry. Recent research highlights the transformative role of AI across sectors. Suresh et al. (2024) examine how integrating AI with UAVs can enhance precision and responsiveness in Industry 5.0 environments, enabling safer, more efficient drone operations. Meanwhile, Selvapriya's work (2025) investigates AI techniques to monitor and mitigate environmental toxins in civil engineering projects, demonstrating predictive capability for pollution control. Further, Selvapriya et al. (2025) survey broad AI applications in civil engineering—from structural health monitoring to smart city planning—emphasizing both opportunities and challenges such as data limitations and interpretability issues. Together, these works show AI's increasing centrality in sustainable, intelligent infrastructure.

Classification of Literature

a. AI in Predictive Maintenance

One of the most mature applications of AI in industrial contexts is predictive maintenance, where machine learning models analyze sensor data to predict failures before they occur. Ali Hamza et al. (2025) proposed a multi-stage framework combining anomaly detection and remaining useful life (RUL) prediction, demonstrating significant reductions in downtime in automotive plants. Similarly, Yang et al. (2023) integrated deep learning with physics-based models to increase model interpretability and reliability for critical assets. Other studies highlight the use of hybrid architectures that combine recurrent neural networks and Bayesian inference to manage uncertainty in predictions (Iranshahi et al., 2025). Despite successes, challenges remain in data availability and transferability across heterogeneous equipment, pointing to the need for federated learning and domain adaptation techniques (Ahn et al.,2023).

b. AI in Quality Control

AI-driven quality inspection, especially through computer vision, is another area with wide adoption. Plathottam et al. (2023) reviewed advancements in defect detection and classification, finding convolutional neural networks (CNNs) to outperform traditional methods in textile, electronics, and automotive industries. Recent research integrates vision with multimodal sensor fusion, such as vibration and acoustic data, to detect subtle process drifts (Hernandez et al., 2022). Vespoli et al. (2024) demonstrated how reinforcement learning could dynamically adjust inspection thresholds to balance false positives and false negatives in real-time production. However, implementation is often constrained by the lack of annotated defect datasets, prompting calls for synthetic data generation and transfer learning (Boikov et al.,2021).

c. Digital Twins and Simulation

Digital twins have become central to smart manufacturing, offering virtual representations of machines, lines, or systems that integrate real-time data for simulation and optimization. Iranshahi et al. (2025) presented a comprehensive review highlighting how digital twins enhance process design, predictive analytics, and scenario planning. Studies show that digital twins improve energy efficiency by simulating operational conditions and suggesting adjustments (Schwark et al., 2023). Moreover, when combined with AI, twins can autonomously update their parameters to reflect actual conditions, enabling closed-loop optimization (Besigomwe, 2025). Still, barriers include computational complexity, standardization issues, and cybersecurity risks associated with high levels of connectivity (IoT Analytics, 2024).

d. Robotics and Automation

The integration of AI into robotics has expanded industrial automation beyond repetitive tasks. Collaborative robots (cobots) now work alongside humans, leveraging vision, force sensing, and reinforcement learning for safety and adaptability (Lei et al., 2024). Vespoli et al. (2025) demonstrated coordinated multi-robot systems for logistics that increased throughput by 30% in large-scale warehouses. Advances in dexterous manipulation and tactile sensing also expand robotics' role in assembly and finishing processes (Rahman et al., 2025). However, the literature points out ongoing challenges in robot interoperability, safety standards, and cost-effectiveness in small and medium enterprises (SMEs) (Chowdhury et al., 2023).

e. Edge AI and Industrial IoT

Edge AI, which enables computation near the data source, is increasingly studied for latency-sensitive and bandwidth-limited industrial environments. Areo (2025) surveyed edge AI applications and highlighted their role in enabling real-time anomaly detection and adaptive process control. Edge deployment has been particularly effective in sectors such as chemical processing and energy, where milliseconds matter for safety (Alam et al., 2022). Studies also show that edge AI enhances cybersecurity by reducing dependence on centralized systems (Zhukabayeva et al.,2025). Nevertheless, resource constraints, model compression challenges, and lack of unified frameworks remain hurdles for broader adoption.

f. Human–AI Collaboration and Workforce Transformation

AI and automation not only impact processes but also reshape workforce roles and organizational strategies. Ghosh et al. (2023) emphasized the importance of human-AI collaboration, where AI augments rather than replaces human judgment in decision-making. Empirical studies reveal that workforce acceptance depends heavily on trust, transparency, and explainability of AI systems (Kristof et al., 2024). Upskilling programs have become essential to enable workers to interpret AI insights and operate advanced automation systems. However, resistance to change and ethical concerns about surveillance remain barriers, highlighting the need for participatory design and inclusive implementation strategies (McKinsey & Company, 2023).

g. Sustainability and Resilience in Industry

A growing body of literature emphasizes the role of AI and automation in promoting sustainable manufacturing and resilient supply chains. Maria et al. (2023) demonstrated how AI-enabled digital twins reduced energy consumption in machining operations. Ahn et al. (2023) highlighted AI's role in enabling circular economy practices through optimized material recovery and recycling. IoT Analytics (2024) reported that companies adopting AI-driven scenario planning were significantly more resilient to supply disruptions during global crises. However, literature also raises concerns about the environmental footprint of AI itself, particularly in training large models, pointing to the need for energy-efficient AI strategies (Rahman et al., 2025).

h. Challenges and Limitations

Despite significant progress, the literature identifies persistent challenges. Data quality and heterogeneity are common issues, with many industrial datasets lacking labeling or suffering from sensor drift (Boikov et al.,2021). Cybersecurity remains a major concern as connectivity increases attack surfaces (Zhukabayeva et al.,2025). Scalability of AI solutions beyond pilot projects remains limited due to integration costs and lack of standardized frameworks (Chowdhury et al., 2023). Furthermore, there is an ongoing tension between performance and interpretability, as black-box AI models may deliver high accuracy but reduce trust and regulatory compliance (Kristof et al., 2024).

Synthesis and Gaps in Literature

Synthesizing across domains, it is evident that AI and automation have progressed from experimental pilots to mainstream industrial deployment, yet adoption remains uneven across sectors and firm sizes. Predictive maintenance and quality control are the most mature, while digital twins, edge AI, and collaborative robotics represent emerging but less standardized fields. There is consensus on the value of AI-human collaboration, but empirical studies on long-term workforce impacts are limited. Sustainability is an emerging priority, though research is still fragmented, often focusing on isolated case studies rather than systemic analysis. Finally, standardization, explainability, and energy-efficient AI represent important gaps that must be addressed to achieve industrial excellence at scale.

The reviewed literature confirms that AI and automation are not incremental improvements but foundational enablers of industrial excellence, reshaping productivity, quality, sustainability, and resilience. Yet, their impact depends on overcoming persistent challenges related to data, interoperability, workforce transformation, and governance. The classification presented here—spanning predictive maintenance, quality control, digital twins, robotics, edge AI, human-AI collaboration, sustainability, and challenges—offers a structured lens to understand the state of the field. Future research must focus on integrating these domains into holistic, standardized frameworks while addressing ethical, social, and environmental dimensions to ensure that AI and automation deliver excellence responsibly and inclusively.

3. INTEGRATION OF ARTIFICIAL INTELLIGENCE (AI) WITH RPA

In the current era of digital transformation, Artificial Intelligence (AI) algorithms and Machine Learning (ML) techniques are being widely adopted across nearly every sector. ML plays a pivotal role in enabling machines to process and interpret large volumes of data efficiently. With the application of AI, it is now possible to explore complex datasets, extract valuable insights, analyze trends, and make data-driven predictions. These capabilities have significantly amplified the potential and effectiveness of Robotic Process Automation (RPA), making it more intelligent and context-aware than ever before.

There are numerous real-world scenarios where outcomes are probabilistic rather than deterministic—meaning results cannot be guaranteed with absolute certainty. For example, estimating the value of a property, assessing the likelihood of loan default, or predicting inventory demand all involve multiple dynamic variables. Consider property valuation: factors such as the age of the building, geographic location, and market trends must be analyzed in combination. Traditionally, such complex multivariable analysis required human expertise. Today, AI can perform these evaluations by leveraging sophisticated predictive models and statistical learning techniques.

Similarly, in recruitment, identifying suitable resumes for job openings is often a time-consuming and cognitively intensive task for human recruiters. AI-powered bots can now automate this process by parsing resumes and matching candidates' skills and experience to job requirements with high accuracy. This dramatically reduces the manual workload and enhances the precision of candidate selection.

While conventional RPA systems are effective in automating rule-based tasks, they are typically limited to structured data and yield consistent, predefined outputs. However, they lack the cognitive ability to interpret unstructured data such as text, images, and videos. When AI is integrated with RPA, the resulting system can understand patterns in unstructured content—enabling tasks such as image recognition, sentiment analysis, and document classification. This integration vastly expands the application range of RPA from routine automation to intelligent decision-making.

To overcome the limitations of standalone technologies and unlock their full potential, it is essential to combine RPA and AI. This convergence leads to what is termed *Intelligent Automation (IA),* a synergistic framework where RPA handles process automation while AI adds cognitive intelligence. As noted in, the integration of RPA, AI, and soft computing techniques under the umbrella of IA is driving a new era of automation that is adaptive, flexible, and capable of handling complex, data-intensive operations across industries.

Intelligent Process Automation (IPA) integrates Robotic Process Automation (RPA), Artificial Intelligence (AI), and other advanced technologies into a unified, pre-configured software environment designed to execute both routine, rule-based processes and cognitively demanding tasks with minimal human intervention. A key feature of IPA is its ability to incorporate cognitive technologies, including Natural Language Processing (NLP), enabling it to interpret and act on unstructured inputs such as text and speech. Compared to traditional RPA systems, IPA requires significantly less manual oversight, as it can adapt and make context-aware decisions. However, effective deployment of IPA still depends on careful calibration, including the selection and fine-tuning of appropriate machine learning models to achieve optimal performance.

IPA has found widespread application across both internal (back-office) and external (front-office) functions in sectors such as banking and finance. In recent years, a growing body of research has explored the practical use cases, implementation challenges, and transformative potential of IPA. These investigations include detailed case studies focusing on automated process discovery, intelligent data transformation, and real-time auditing. Moreover, IPA has shown considerable promise in enhancing business process management and improving operational efficiency and productivity.

Additionally, various tools have emerged to streamline the integration of AI capabilities into RPA platforms, further simplifying the development and deployment of IPA solutions. The next section of this paper will delve into these tools and platforms, highlighting their roles in enabling scalable and intelligent automation frameworks.

Implementation and impact of AI-driven Robotic Process Automation (RPA) across a range of key sectors:

Finance and Banking: AI-integrated RPA is transforming financial services by automating processes such as loan approvals, fraud detection through automated credit assessments, and enhancing customer support experiences. This section also explores critical concerns surrounding data security, regulatory compliance, and the challenges of integrating AI with traditional banking infrastructure.

Auditing: In the auditing domain, AI-based RPA significantly enhances both the accuracy and efficiency of financial audits, minimizing the potential for human error and enabling continuous financial monitoring. This discussion includes practical use cases and addresses challenges such as managing large data volumes and ensuring transparency and accountability in AI-driven decision-making processes.

Healthcare and Medical Services: The healthcare industry has seen substantial benefits from the use of AI-enabled RPA in automating patient data management, streamlining administrative operations, and supporting diagnostic workflows. However, this section also highlights pressing issues such as patient data privacy, security concerns, and the ethical implications of deploying AI in clinical settings.

Recruitment and Human Resources: AI-powered RPA is widely adopted in recruitment processes to automate tasks such as resume filtering, candidate ranking, and scheduling interviews. While these systems increase efficiency, they also raise important concerns about algorithmic bias and the opacity of AI-driven decisions, which are addressed in detail.

Manufacturing: In manufacturing, AI-based RPA contributes to optimizing production workflows, managing supply chains, and improving quality assurance processes. This part of the paper discusses the technical and operational challenges of integrating AI into existing manufacturing systems, particularly in complex industrial environments.

Back-Office Operations: A real-world case study of AI-RPA implementation in back-office functions is presented, demonstrating how automation can be deployed without disrupting existing IT infrastructures. This section further examines the technical difficulties associated with deploying AI-driven RPA in environments characterized by large-scale, high-volume data processing.

In addition to sector-specific insights, the paper explores broader challenges associated with AI-RPA adoption, including technological complexity, integration with legacy systems, data privacy and security, and ethical concerns regarding transparency and fairness in automated decision-making. Successfully addressing these issues requires robust governance structures, proactive data management strategies, and a sustained commitment to innovation and organizational learning.

4. TOOLS AND PLATFORMS FOR INTELLIGENT PROCESS AUTOMATION

The successful deployment of IPA relies on a combination of RPA frameworks, AI toolkits, machine learning libraries, and integration platforms. These tools enable the automation of both structured and unstructured tasks while embedding cognitive capabilities into workflows. Below are some widely used tools and platforms in IPA:

a. RPA Platforms

These tools form the backbone of process automation by handling rule-based and repetitive tasks.

- **UiPath**: One of the most popular RPA platforms, UiPath offers built-in AI integration, task mining, and a drag-and-drop interface for workflow creation.
- **Automation Anywhere**: Provides an intelligent automation platform with features like Bot Insight (analytics), IQ Bot (for document understanding), and integration with AI/ML models.
- **Blue Prism**: Offers scalable RPA solutions with AI connectors and a Digital Exchange (DX) for integrating third-party AI capabilities.
- **Microsoft Power Automate**: A cloud-based platform that allows users to automate workflows and integrate AI Builder for tasks like form processing and object detection.

Robotic Process Automation (RPA) platforms are the foundation of Intelligent Process Automation (IPA) as they focus on automating routine, repetitive, and rule-based tasks across industries. These tools act as the execution layer that mimics human actions on digital systems such as data entry, report generation, and system integrations. Among the most notable platforms, *UiPath* stands out as one of the most widely adopted RPA solutions, featuring a user-friendly drag-and-drop interface, task mining, and AI-assisted workflow design. It supports seamless integration with artificial intelligence and machine learning models, making it an essential component in modern automation ecosystems. *Automation Anywhere* provides an advanced automation suite with unique components such as *Bot Insight*, which delivers real-time analytics on bot performance, and *IQ Bot*, designed for intelligent document understanding. Its AI/ML integration makes it highly adaptable in complex business environments. *Blue Prism*, another market leader, emphasizes scalability and security while offering AI connectors and a *Digital Exchange (DX)* marketplace that allows enterprises to leverage third-party AI capabilities. Finally, *Microsoft Power Automate*, part of the Microsoft ecosystem, is a cloud-based solution that integrates with Office 365 and Dynamics, offering low-code workflow creation and access to *AI Builder* for tasks like form recognition, object detection, and sentiment analysis. Together, these RPA platforms form the backbone of IPA by enabling enterprises to automate at scale while incorporating cognitive functions.

b. AI and ML Libraries

These tools empower IPA systems with cognitive capabilities like prediction, classification, NLP, and decision-making.

- **TensorFlow**: An open-source deep learning framework developed by Google, used for building and deploying machine learning models.
- **PyTorch**: A flexible ML library developed by Facebook, widely used for research and production in AI-driven automation.
- **Scikit-learn**: A robust library for classical machine learning algorithms (e.g., decision trees, clustering, regression) often used in business process analytics.
- **Hugging Face Transformers**: Provides state-of-the-art pre-trained NLP models that can be integrated into IPA systems for sentiment analysis, summarization, and more.

Artificial Intelligence (AI) and Machine Learning (ML) libraries provide the "cognitive intelligence" that extends the capabilities of traditional RPA. They allow IPA systems to move beyond simple rule execution into prediction, classification, decision-making, and pattern recognition. *TensorFlow*, an open-source deep learning framework developed by Google, supports both training and deployment of machine learning models across various environments, including edge and cloud. It is widely applied in predictive analytics, computer vision, and NLP tasks in automation contexts. Similarly, *PyTorch*, developed by Facebook, has become the preferred choice in research and production due to its flexibility, dynamic computation graphs, and ease of debugging. It is extensively used in deep learning-driven automation tasks such as fraud detection, anomaly detection, and process optimization. For more classical

ML methods, *Scikit-learn* offers algorithms for clustering, classification, regression, and data preprocessing, which are particularly useful in business process analytics and optimization. Another significant tool, *Hugging Face Transformers*, provides state-of-the-art pre-trained NLP models like BERT, GPT, and RoBERTa, which are easily integrated into IPA systems to handle tasks such as sentiment analysis, summarization, entity recognition, and chat automation. These libraries make IPA systems more adaptive, data-driven, and capable of intelligent decision-making.

c. Natural Language Processing (NLP) Tools

Used for interpreting and automating tasks involving unstructured textual data.

- **SpaCy**: A fast and efficient NLP library for text processing, entity recognition, and dependency parsing.
- **NLTK (Natural Language Toolkit)**: A toolkit for text analysis, sentiment extraction, and tokenization in academic and research contexts.
- **Google Cloud Natural Language API**: A cloud service offering real-time NLP capabilities like sentiment analysis and content classification.
- **AWS Comprehend**: Amazon's NLP service that integrates easily with IPA workflows to extract insights from text.

NLP tools play a crucial role in IPA by enabling machines to interpret, understand, and automate processes involving unstructured textual data. This is particularly important in industries where vast amounts of customer feedback, emails, documents, and reports need to be processed. *SpaCy* is a high-performance NLP library known for its speed and accuracy in tasks such as tokenization, dependency parsing, and named entity recognition. It is often used in business automation for extracting structured information from unstructured text. *NLTK (Natural Language Toolkit)*, widely used in academia, provides a wide range of text processing libraries for sentiment analysis, stemming, lemmatization, and classification, making it suitable for research and prototyping IPA solutions. On the cloud service side, *Google Cloud Natural Language API* provides ready-to-use NLP functionalities such as entity analysis, sentiment analysis, and content classification that can be integrated directly into IPA workflows without extensive model training. Similarly, *AWS Comprehend* allows organizations to extract insights from textual data by detecting entities, key phrases, sentiment, and topics. Its integration with AWS services makes it suitable for large-scale enterprise automation projects. With these tools, IPA systems gain the ability to understand and process natural language, making customer interaction, document processing, and knowledge management more efficient.

d. Process Mining and Discovery Tools

These tools help identify automation opportunities and analyze workflows.

- **Celonis**: A leading process mining platform that visualizes business processes and pinpoints inefficiencies.
- **UiPath Process Mining**: Integrated within the UiPath platform, it helps in discovering, monitoring, and optimizing automated processes.
- **Minit**: Offers AI-powered process discovery and performance analytics to guide automation strategies.

Process mining and discovery tools are essential for identifying automation opportunities and understanding how business processes operate in reality compared to documented workflows. By analyzing digital footprints left in enterprise systems, these tools reveal inefficiencies, bottlenecks, and hidden variations in processes. *Celonis*, a market leader in this field, uses event logs from IT systems to create visual process maps, allowing organizations to pinpoint inefficiencies and prioritize automation opportunities. Its AI-driven insights help businesses not only to automate but also to optimize their workflows. *UiPath Process Mining* integrates directly within the UiPath ecosystem, providing organizations with end-to-end visibility of processes, enabling continuous monitoring and fine-tuning of automation strategies. *Minit* is another powerful tool that combines process discovery with AI-based analytics to uncover deviations and suggest data-driven improvements. By using process mining tools, enterprises ensure that automation is strategically targeted at high-value processes, thereby maximizing return on investment in IPA initiatives.

e. Intelligent Document Processing (IDP) Tools

Essential for processing semi-structured and unstructured documents.

- **ABBYY FlexiCapture**: An AI-based OCR and document processing solution.
- **Kofax Transformation**: Enables extraction, classification, and validation of data from various documents.
- **IQ Bot (by Automation Anywhere)**: Specifically designed for intelligent document processing within RPA workflows.

Intelligent Document Processing (IDP) tools are vital for handling semi-structured and unstructured documents such as invoices, contracts, medical forms, and legal papers. These tools use a combination of optical character recognition (OCR), natural language processing, and machine learning to extract, classify, and validate data. *ABBYY FlexiCapture* is a leading IDP platform that leverages AI-based OCR technology to process complex documents with high accuracy, making it suitable for industries like finance, insurance, and government. *Kofax Transformation* provides capabilities to extract and classify data from a wide range of document types while ensuring data validation and compliance with regulatory standards. *IQ Bot*, developed by Automation Anywhere, is designed specifically for intelligent document processing within RPA workflows. It learns continuously from human inputs, improving accuracy over time in handling invoices, receipts, and other business-critical documents. These tools allow organizations to automate previously manual and error-prone document processing tasks, thereby reducing costs, improving compliance, and accelerating turnaround times.

f. Integration and Orchestration Platforms

These tools ensure smooth communication between RPA bots, AI models, and enterprise systems.

- **Apache Airflow**: A workflow orchestration platform to schedule and monitor automation tasks.
- **Zapier / Integromat**: For low-code/no-code automation between web applications.
- **Kubernetes**: Used for deploying and scaling ML models and microservices in IPA systems.

While RPA, AI, and IDP tools provide specialized functionalities, integration and orchestration platforms ensure seamless coordination among them. These platforms connect disparate systems, manage workflows, and enable scalability of automation across an enterprise. *Apache Airflow* is a popular open-source orchestration platform that allows users to schedule, monitor, and manage workflows in a programmatic way. It is especially useful for data-driven automation pipelines. *Zapier* and *Integromat* (now known as Make) provide low-code/no-code environments for connecting various web applications, making them suitable for quick automation solutions in SMEs and startups. For large-scale enterprise deployments, *Kubernetes* plays a critical role in orchestrating containerized applications, including ML models and microservices that power IPA systems. Kubernetes ensures scalability, reliability, and resilience of automation environments by handling load balancing, scaling, and fault tolerance. Together, these platforms bridge the gap between different automation technologies, enabling organizations to build a cohesive and intelligent automation ecosystem.Top of Form

Intelligent Process Automation (IPA) has emerged as one of the most transformative technological advancements driving digital transformation across industries in the last decade. Building upon the foundations of Robotic Process Automation (RPA), IPA combines the precision of automation with the adaptability of artificial intelligence (AI), machine learning (ML), natural language processing (NLP), computer vision, and advanced analytics. Its purpose extends beyond automating repetitive and rule-based tasks; it strives to infuse intelligence into organizational workflows, enabling decision-making, optimization, and adaptability. Unlike conventional RPA, which functions primarily as a software bot mimicking human keystrokes and actions, IPA introduces cognitive and contextual understanding to business processes. This integration allows enterprises to automate not only routine processes but also semi-structured and unstructured workflows involving natural language, documents, and images. As global competition intensifies and organizations navigate challenges such as workforce shortages, rising customer expectations, and the growing need for resilience, IPA platforms and tools have become indispensable for achieving operational excellence.

The landscape of IPA tools and platforms is broad and dynamic, comprising commercial enterprise-grade solutions, open-source alternatives, AI-driven cloud-native systems, and industry-specific applications. Tools like UiPath, Automation Anywhere, and Blue Prism have evolved from RPA vendors into holistic IPA ecosystems, embedding AI engines and process mining capabilities within their suites. At the same time, enterprise technology leaders such as IBM, Microsoft, Appian, and Pega have positioned themselves as strategic providers of IPA by integrating automation with cloud services, business process management (BPM), and AI frameworks. Open-source solutions such as Robot Framework and low-code platforms have further democratized automation, enabling smaller organizations and developers to experiment with IPA without heavy upfront investments. Moreover, industry-specific applications are being designed for healthcare, banking, insurance, and manufacturing, where compliance, domain-specific data models, and specialized workflows necessitate tailored solutions.

One of the defining characteristics of IPA platforms is their reliance on RPA as the execution backbone while layering intelligence for contextual understanding. RPA bots remain the foundation, executing routine workflows, but IPA introduces AI services for natural language understanding, predictive analytics, and decision-making. For instance, NLP engines embedded in IPA platforms allow automated bots to read, interpret, and classify emails, customer complaints, or regulatory documents with high accuracy. Similarly, ML algorithms integrated into IPA platforms can continuously learn from historical data to improve invoice processing, fraud detection, or predictive maintenance in manufacturing. Computer vision further expands the scope of IPA by enabling automation of workflows involving scanned documents, forms, or images, thus broadening applicability in sectors like logistics and healthcare where visual data is abundant. Process mining tools embedded within IPA platforms serve as diagnostic instruments, identifying inefficiencies, bottlenecks, and automation opportunities across the enterprise. The convergence of these technologies illustrates the hybrid nature of IPA, where rule-based automation is augmented with intelligent analytics to produce adaptive systems.

The commercial enterprise IPA market is dominated by a few flagship platforms. UiPath has positioned itself as a leading player by offering an end-to-end automation ecosystem that integrates RPA with AI Fabric, Document Understanding, and Process Mining modules. Its marketplace, AI Center, and automation hubs are designed to provide enterprises with scalability, collaboration, and lifecycle management of automation projects. Automation Anywhere has pursued a similar trajectory with its Automation 360 platform, emphasizing cloud-native design, bot store marketplaces, and built-in cognitive services. Blue Prism, an early pioneer in RPA, has redefined itself as an IPA provider by embedding AI and ML capabilities through its Decipher and Interact products, while also integrating with third-party AI services. Microsoft Power Automate, leveraging its tight integration with Office 365, Dynamics, and Azure Cognitive Services, has gained traction among enterprises seeking seamless IPA adoption without the need for standalone platforms. Its low-code environment lowers the barrier to entry, enabling business users to design workflows while leveraging Microsoft's cloud ecosystem for scalability.

IBM Cloud Pak for Automation exemplifies a cognitive-first approach to IPA, combining workflow automation with AI-powered decision-making, content services, and robotic automation. By building on Watson AI and hybrid cloud infrastructure, IBM provides enterprises with advanced capabilities for scaling automation across multiple environments. Appian, with its roots in business process management, has evolved into a low-code automation suite that unifies RPA, AI, and case management, enabling organizations to orchestrate end-to-end business processes. Similarly, Pega Systems has integrated its decision management and AI capabilities with automation, focusing on real-time customer engagement and adaptive decision-making. Collectively, these platforms represent the mainstream enterprise-grade IPA landscape, catering to global organizations with complex, large-scale automation needs.

Alongside commercial platforms, open-source tools have gained prominence for offering cost-effective experimentation and flexibility. Robot Framework, one of the most widely used open-source RPA frameworks, provides extensibility for building IPA workflows when combined with AI libraries in Python, TensorFlow, or PyTorch. Tools such as TagUI, OpenRPA, and Taskt further provide community-driven alternatives for small and medium enterprises seeking lightweight automation solutions. Open-source IPA tools also play a critical role in academic and research contexts, where flexibility and customizability outweigh enterprise support. While they lack the extensive governance, compliance, and lifecycle management features of enterprise platforms, open-source solutions offer agility and innovation, complementing the broader IPA ecosystem.

Cloud-native IPA platforms represent another emerging trend. As organizations increasingly migrate workloads to the cloud, automation platforms are designed with elasticity, scalability, and AI integration at their core. Providers like Microsoft, AWS, and Google Cloud are embedding automation features into their ecosystems, enabling developers to integrate AI models, APIs, and process orchestration directly within cloud-native workflows. The advantage of cloud-native IPA lies in the seamless integration of automation with AI services such as speech recognition, natural language understanding, and image classification, all accessible through APIs. These platforms also facilitate real-time analytics, monitoring, and global scalability, enabling enterprises to rapidly deploy automation at scale without the need for heavy on-premise infrastructure.

Industry-specific IPA platforms further demonstrate the maturity of the ecosystem. In banking and financial services, automation platforms are designed with built-in compliance frameworks to handle anti-money laundering (AML), know-your-customer (KYC), and fraud detection workflows. Healthcare IPA platforms emphasize compliance with HIPAA and GDPR, while offering document understanding capabilities for processing electronic health records (EHRs) and insurance claims. Manufacturing-focused IPA platforms integrate with industrial IoT, predictive maintenance systems, and quality assurance processes. The tailoring of IPA to specific domains reflects the recognition that generic automation tools must be adapted to meet regulatory, technical, and operational nuances of different industries.

Comparative studies of IPA platforms reveal significant differences in scalability, AI integration, governance features, cost models, and ecosystem maturity. Enterprise platforms like UiPath and Automation Anywhere offer extensive AI integration and governance features but require substantial investment. Microsoft Power Automate, though less feature-rich in terms of advanced AI, provides affordability and accessibility within Microsoft's ecosystem. Open-source tools offer unmatched flexibility but lack enterprise-grade support. Cloud-native IPA platforms stand out for scalability and elasticity but may pose challenges in data residency and regulatory compliance. Industry-specific platforms strike a balance by offering tailored compliance features but may lack cross-domain applicability. Enterprises must therefore evaluate IPA platforms not merely on technical features but on alignment with strategic goals, regulatory environment, and organizational readiness.

Despite significant progress, challenges persist in IPA adoption. Integration with legacy systems remains a primary obstacle, as many enterprises operate on decades-old infrastructure not designed for automation. Governance and compliance frameworks are still evolving, with organizations grappling with issues of data privacy, auditability, and ethical AI. Scalability also remains elusive; while pilot projects often succeed, enterprise-wide rollouts encounter resistance, skill shortages, and escalating costs. Additionally, the rapid evolution of generative AI introduces new opportunities and risks. Integrating large language models (LLMs) into IPA platforms promises to enhance natural language understanding, customer service automation, and decision support. Yet, challenges related to model transparency, bias, and hallucination complicate their adoption. As such, the future of IPA will likely involve hybrid systems combining deterministic RPA bots, AI-powered decision-making, and human oversight.

In conclusion, tools and platforms for Intelligent Process Automation represent the vanguard of enterprise digital transformation, bridging the gap between traditional automation and cognitive intelligence. From commercial giants like UiPath and Automation Anywhere to open-source alternatives and industry-specific solutions, the IPA ecosystem is diverse and rapidly evolving. The convergence of RPA, AI, process mining, and cloud-native services has created platforms capable of not only executing tasks but also understanding, predicting, and adapting. Organizations must navigate this complex landscape by carefully selecting platforms that align with their strategic objectives, technical environment, and regulatory requirements. As IPA continues to evolve, its integration with generative AI, sustainability imperatives, and human-centric governance will shape its trajectory. Ultimately, the success of IPA lies not only in its technological sophistication but in its capacity to augment human intelligence, optimize workflows, and deliver value across industries.

5. FUTURE DIRECTIONS IN AI-DRIVEN INDUSTRIAL TRANSFORMATION

Artificial intelligence (AI) and automation are reshaping industries at an unprecedented pace, with transformative effects already visible in manufacturing, logistics, healthcare, automotive, and energy sectors. Yet, as powerful as current deployments are, the trajectory of technological advancement suggests that we are still in the early stages of industrial transformation. The next decade promises not only incremental efficiency gains but also fundamental reconfigurations of how industries design, produce, distribute, and sustain value. Future directions in AI-driven industrial transformation can be understood through three interconnected lenses: the evolution of technologies, the restructuring of organizational and workforce models, and the integration of sustainability and ethics into industrial ecosystems. Together, these forces will shape the contours of "Industry 5.0," where collaboration between human ingenuity and machine intelligence will redefine the meaning of industrial excellence.

At the technological frontier, several breakthroughs are expected to drive the next wave of transformation. Generative AI, powered by large language models (LLMs) and multimodal systems, is emerging as a powerful enabler of design, simulation, and decision support in production environments. Unlike earlier machine learning models limited to narrow tasks, generative AI can create synthetic data, propose design alternatives, optimize supply chain strategies, and even generate executable code for automation systems (Dwivedi et al., 2023). In manufacturing, this translates into faster product development cycles, customized production lines, and the ability to simulate failure scenarios before actual implementation. Coupled with digital twins—virtual replicas of machines, processes, and entire factories—generative AI will enable real-time experimentation, predictive diagnostics, and optimization without disrupting physical systems (Iranshahi et al., 2025). As a result, industries can expect not only operational resilience but also accelerated innovation pipelines.

The rise of edge AI and distributed intelligence represents another significant direction. With billions of connected devices generating massive data streams, reliance on centralized cloud computing introduces latency, bandwidth, and security concerns. Edge AI addresses this challenge by embedding intelligence directly into machines, robots, and sensors, allowing for local data processing and real-time decision-making . For industrial transformation, this means that autonomous robots, collaborative cobots, and smart sensors will operate with minimal human oversight, making decentralized production a reality. Edge intelligence also improves resilience in remote or critical environments such as oil rigs, mining operations, or disaster zones where connectivity is unreliable. Over the long term, we can expect industrial networks to function as federated ecosystems where localized AI systems collaborate with centralized platforms, balancing efficiency with autonomy.

Interoperability and integration will also become central to future AI-driven industrial systems. Current deployments often suffer from vendor lock-in and fragmented standards, making it difficult to scale automation across complex supply chains. The next phase of industrial transformation is likely to prioritize open standards, cross-platform compatibility, and AI-as-a-service models. This shift will allow enterprises to flexibly combine best-of-breed solutions while maintaining data governance and security. Cloud-native IPA (Intelligent Process Automation) platforms, already gaining momentum, will mature into hybrid ecosystems where AI services—such as computer vision, natural language understanding, and predictive analytics—can be seamlessly integrated into business workflows (Chowdhury et al., 2023). The modularization of AI will democratize access to advanced capabilities, enabling even small and medium enterprises to benefit from sophisticated automation without heavy infrastructure investments.

In parallel, advances in robotics will extend the reach of automation beyond structured environments. Collaborative robots, or cobots, designed to safely work alongside humans, are evolving with enhanced perception, dexterity, and learning capabilities. Future cobots will be context-aware, adaptive to unstructured environments, and capable of transferring knowledge across tasks through reinforcement learning (Vespoli et al., 2025). In logistics, autonomous mobile robots will manage dynamic warehouse operations with minimal human input, while in healthcare, surgical and care-assistive robots will augment human staff in precision-critical tasks. Importantly, the convergence of robotics with AI will not only expand functional capabilities but also reshape human–machine collaboration models. Rather than replacing workers, robots will increasingly function as intelligent assistants, enhancing safety, productivity, and ergonomics.

The workforce dimension of AI-driven industrial transformation cannot be overlooked. Future directions will involve not only technological augmentation but also a paradigm shift in skill requirements, work design, and organizational culture. As routine and repetitive tasks are further automated, human roles will migrate toward complex problem-solving, creativity, and oversight. This will necessitate large-scale reskilling and upskilling initiatives, with emphasis on data literacy, systems thinking, and human–AI collaboration skills (Lei et al., 2024). Moreover, organizational structures will evolve from hierarchical models to adaptive, cross-functional teams empowered by real-time AI insights. Companies that invest in continuous learning cultures, flexible work environments, and inclusive strategies for human–machine collaboration will be better positioned to harness the full potential of AI.

Another defining aspect of the future lies in embedding sustainability and circular economy principles into AI-driven industrial systems. Current discourse often highlights AI's potential to optimize energy consumption, reduce waste, and enable predictive maintenance, but the future requires going further by addressing the environmental costs of AI itself. Training large AI models and maintaining digital twins demand substantial energy resources, raising concerns about the carbon footprint of industrial AI (Schwark et al., 2023). Future research and practice will therefore focus on energy-efficient algorithms, green AI, and integration of renewable energy sources into AI infrastructures. Additionally, AI will play a pivotal role in enabling closed-loop manufacturing systems, where products are designed for disassembly, recycling, and reuse. By aligning AI with sustainability imperatives, industries can simultaneously pursue economic growth and environmental stewardship.

From a governance and regulatory perspective, the future of AI-driven industrial transformation will emphasize trust, transparency, and accountability. The opacity of black-box AI models poses significant risks when decisions affect safety, quality, and compliance. Explainable AI (XAI) will thus become a core requirement, enabling stakeholders to understand and audit AI-driven decisions (Kristof et al.,

2024). Regulatory frameworks are already evolving in regions like the European Union, where the AI Act introduces risk-based categorization of AI systems. In the industrial domain, future compliance will likely demand standardized audits, certification processes, and ethical governance boards. Enterprises that proactively embed governance and transparency into their AI strategies will be more resilient against regulatory shifts and reputational risks.

Cybersecurity will also be an enduring concern. As AI-driven automation increases connectivity across machines, cloud platforms, and supply chains, the attack surface for malicious actors expands. Future directions will involve embedding AI-driven cybersecurity within industrial systems to detect anomalies, predict threats, and respond autonomously (Zhukabayeva et al.,2025). Moreover, advances in quantum computing pose both opportunities and risks: while quantum algorithms may accelerate optimization tasks in supply chains, they may also undermine traditional encryption protocols. Preparing for this duality will require industries to invest in quantum-resistant cybersecurity strategies alongside next-generation AI deployments.

Looking forward, the convergence of AI with other emerging technologies will further accelerate industrial transformation. Integration with blockchain promises tamper-proof supply chain tracking, while 5G and beyond-5G networks will enable ultra-low latency communication between machines and AI systems. Augmented reality (AR) and virtual reality (VR), powered by AI, will enhance training, remote maintenance, and collaborative design, bringing new levels of immersion into industrial workflows. Quantum AI, though still nascent, holds potential for solving complex optimization problems in logistics, energy distribution, and material science (Dwivedi et al., 2023). The interplay of these technologies will not only expand AI's capabilities but also create entirely new industrial paradigms characterized by hyper-connectivity, intelligence, and resilience.

Finally, the broader vision of industrial transformation is moving toward Industry 5.0, where human-centricity, sustainability, and resilience complement technological innovation. Unlike Industry 4.0, which focused primarily on digitalization and automation, Industry 5.0 emphasizes symbiotic collaboration between humans and machines. In this paradigm, AI and automation are not ends in themselves but enablers of inclusive, sustainable, and adaptive industrial systems. The future will therefore be defined not only by technological breakthroughs but by the ability of industries to align AI adoption with human values, ethical governance, and environmental responsibility.

In summary, the future of AI-driven industrial transformation will be characterized by generative and multimodal AI, edge intelligence, interoperable ecosystems, advanced robotics, and integration with emerging technologies such as blockchain and quantum computing. Equally important will be the parallel evolution of workforce models, governance frameworks, and sustainability imperatives. Success in this future will not be measured solely by efficiency or productivity gains but by the ability to create intelligent, ethical, and sustainable industrial ecosystems. Organizations that adopt a forward-looking, holistic approach—balancing technological ambition with human-centric values—will lead the way in shaping the next era of industrial excellence.

6. CHALLENGES IN AI AND AUTOMATION FOR DRIVING INDUSTRIAL EXCELLENCE

Artificial intelligence (AI) and automation are widely acknowledged as catalysts of industrial transformation, enabling organizations to achieve unprecedented gains in productivity, quality, and resilience. Yet, the deployment of these technologies is not without significant challenges. The complexities of integrating intelligent algorithms with legacy infrastructure, ensuring robust data pipelines, addressing ethical and workforce concerns, and safeguarding industrial systems against cyberattacks collectively create obstacles that can undermine the promise of industrial excellence. Over the past five years, scholars and practitioners have examined these barriers in detail, often emphasizing that technological readiness must be complemented by organizational, regulatory, and socio-technical adaptation (McKinsey & Company, 2023; Kristof et al., 2024). This section provides a comprehensive literature-driven review of challenges, classifying them into seven categories: data-related, technical and integration, workforce and organizational, ethical and social, sustainability and environmental, regulatory and governance, and cybersecurity. Each category is discussed with reference to recent research and case studies to highlight not only the barriers but also potential pathways forward.

Classification of Challenges

a. Data-Related Challenges

The success of AI-driven industrial systems hinges on data quality, availability, and accessibility. However, industries often face fragmented, incomplete, or noisy datasets. Many legacy machines were not originally designed for IoT integration, creating difficulties in retrofitting them with sensors to generate reliable data (Boikov et al.,2021). Furthermore, industrial datasets frequently suffer from class imbalance—rare failure events are underrepresented, making it difficult for models to learn robust predictive patterns (Ali Hamza et al., 2025). Data heterogeneity across equipment, sites, and vendors also hampers the transferability of AI models, often requiring costly customization (Ahn et al.,2023). Moreover, data privacy and ownership remain contentious, particularly in multi-stakeholder supply chains where proprietary data is sensitive. As a result, federated learning and domain adaptation methods are being explored to mitigate these challenges, but their adoption remains nascent and resource-intensive (Iranshahi et al., 2025).

b. Technical and Integration Challenges

Beyond data, industries encounter technical barriers in deploying AI and automation. Integrating AI models into existing manufacturing execution systems (MES) or enterprise resource planning (ERP) frameworks is often complex, requiring middleware and substantial reengineering (Chowdhury et al., 2023). Edge AI promises real-time decision-making, yet model compression, hardware limitations, and energy constraints create performance bottlenecks. Digital twins, while powerful, demand high computational capacity and standardized frameworks for interoperability, which are currently lacking (Iranshahi et al., 2025). Robotics integration is also hindered by interoperability between vendors, safety certification requirements, and difficulties in achieving flexible automation in dynamic environments (Vespoli et al., 2025). As a result, many organizations remain trapped at the pilot-project stage, unable to scale AI solutions across the enterprise (IoT Analytics, 2024).

c. Workforce and Organizational Challenges

While technology adoption is central, workforce readiness and organizational culture play equally critical roles. Automation and AI disrupt job roles, often eliminating routine tasks while creating demand for higher-level analytical and technical skills (Lei et al., 2024). Resistance to change is common, especially when employees perceive AI as a threat to job security. Upskilling and reskilling programs are vital, but implementation is uneven, with small and medium enterprises (SMEs) struggling to invest in training (Kristof et al., 2024). Additionally, there is often a lack of cross-functional collaboration between domain experts, data scientists, and operators, leading to a gap between AI model development and practical deployment. Organizational inertia, siloed decision-making, and limited leadership commitment further slow transformation (McKinsey & Company, 2023).

d. Ethical and Social Challenges

AI and automation raise significant ethical questions around transparency, fairness, and accountability. Black-box AI models make decisions that directly affect product quality, safety, and workforce operations, yet they often lack explainability, undermining trust and regulatory compliance (Kristof et al., 2024). Worker surveillance through AI-driven monitoring tools raises privacy concerns, potentially eroding morale (Lei et al., 2024). Social inequality is another risk, as automation disproportionately affects low-skilled labor, potentially widening economic divides. Additionally, reliance on AI for decision-making raises accountability concerns: who is responsible when an AI-driven industrial process fails? Addressing these ethical issues requires not only technical solutions, such as explainable AI, but also organizational policies that emphasize human oversight and transparency.

e. Sustainability and Environmental Challenges

While AI and automation are often celebrated for enabling sustainable practices, they also present environmental challenges. The energy consumption associated with training large AI models and running digital twins can offset sustainability gains, especially if powered by non-renewable energy (Schwark et al., 2023). Manufacturing automation often leads to increased electronic waste from sensors, robotics, and embedded systems that have limited lifespans (Rahman et al., 2025). Furthermore, sustainability strategies driven by AI remain siloed, focusing on energy optimization in specific processes rather than addressing systemic resource efficiency across the value chain (Ahn et al.,2023). Balancing the promise of AI-driven sustainability with its ecological footprint remains a pressing challenge.

f. Regulatory and Governance Challenges

Industrial AI and automation operate in a regulatory grey zone, with many jurisdictions lacking clear guidelines on model validation, safety certification, and liability. For instance, explainability and transparency requirements differ across industries and countries, creating compliance complexities for multinational manufacturers (Chowdhury et al., 2023). Safety standards for collaborative robots and AI-driven decision systems are evolving but remain fragmented (Vespoli et al., 2025). Governance frameworks within organizations are also underdeveloped, with few firms establishing robust AI ethics boards or model governance protocols (McKinsey & Company, 2023). Without standardized benchmarks, organizations risk deploying systems that are difficult to audit or regulate, exposing them to operational and reputational risks.

g. Cybersecurity Challenges

The convergence of AI, automation, and IIoT has significantly expanded the cyber-attack surface of industrial systems. Connected sensors, robots, and cloud platforms create multiple entry points for malicious actors. Cybersecurity threats range from data poisoning attacks on AI models to ransomware targeting industrial control systems (Zhukabayeva et al.,2025). Edge AI systems, while reducing reliance on centralized servers, introduce vulnerabilities at endpoints that may lack robust protection . Moreover, the integration of legacy systems with modern AI-driven platforms creates weak links, as older systems often lack the necessary security protocols (Alam et al., 2022). Addressing these threats requires not only technical safeguards but also comprehensive cyber-resilience strategies, including workforce training and regulatory compliance.

Synthesis

Across the classification, several cross-cutting themes emerge. First, the challenges are not purely technical; they are socio-technical, involving the interplay of data, infrastructure, workforce, and governance. Second, scalability remains the central obstacle—many AI and automation solutions succeed in pilots but fail to achieve enterprise-wide adoption due to integration, cost, or governance barriers. Third, while sustainability and resilience are often cited as benefits, the literature also underscores the paradox of increased environmental and cybersecurity risks. Finally, the workforce dimension remains critical, with human factors shaping both adoption success and ethical implications. Addressing these challenges requires an integrated strategy combining technological innovation with governance frameworks, workforce empowerment, and cross-industry collaboration.

7. CONCLUSION

Artificial intelligence (AI) and automation are reshaping production systems by driving efficiency, flexibility, and resilience across industries. Beyond streamlining routine processes, these technologies empower organizations with predictive insights, intelligent decision-making, and adaptive workflows that enhance competitiveness in dynamic markets. The convergence of robotics, digital twins, machine learning, and intelligent process automation demonstrates how technology serves as a strategic enabler of industrial excellence. However, realizing this potential requires addressing persistent challenges in data quality, system integration, workforce readiness, ethics, and sustainability. Success depends on balancing technological adoption with human oversight, reskilling initiatives, and governance frameworks that ensure transparency and accountability. The true revolution lies not in replacing human workers but in enabling collaborative ecosystems where AI complements human creativity and judgment. Moving forward, industries that adopt responsible and inclusive approaches to AI and automation will set the benchmark for sustainable industrial transformation and long-term excellence.

REFERENCES

Ahn, J., Lee, Y., Kim, N., Park, C., & Jeong, J. (2023). Federated Learning for Predictive Maintenance and Anomaly Detection Using Time Series Data Distribution Shifts in Manufacturing Processes. *Sensors (Basel)*, *23*(17), 7331. DOI: 10.3390/s23177331 PMID: 37687787

Areo, G. (2025). *Edge AI for Real-Time Industrial Automation: Architecture*. Challenges, and Case Studies.

Besigomwe, K. (2025). Closed-Loop Manufacturing with AI-Enabled Digital Twin Systems. *Cognizance Journal of Multidisciplinary Studies.*, *5*(1), 18–38. DOI: 10.47760/cognizance.2025.v05i01.002

Boikov, A., Payor, V., Savelev, R., & Kolesnikov, A. (2021). Synthetic Data Generation for Steel Defect Detection and Classification Using Deep Learning. *Symmetry*, *13*(7), 1176. DOI: 10.3390/sym13071176

Chowdhury, S., Dey, P., Joel-Edgar, S., Bhattacharya, S., Rodriguez-Espindola, O., Abadie, A., & Truong, L. (2023). Unlocking the value of artificial intelligence in human resource management through AI capability framework. *Human Resource Management Review*, *33*(1), 100899. DOI: 10.1016/j.hrmr.2022.100899

Dwivedi, Y. K., Hughes, L., Baabdullah, A. M., Ribeiro-Navarrete, S., Giannakis, M., Al-Debei, M., & Wamba, S. F. (2023). Metaverse beyond the hype: Multidisciplinary perspectives on emerging challenges, opportunities, and agenda for research, practice and policy. *International Journal of Information Management*, *71*, 102642. DOI: 10.1016/j.ijinfomgt.2023.102642

Hamza, A., Ali, Z., Dudley, S., Saleem, K., Uneeb, M., & Christofides, N. (2025), A multi-stage review framework for AI-driven predictive maintenance and fault diagnosis in photovoltaic systems, Applied Energy,Volume 393,126108,ISSN 0306-2619, DOI: 10.1016/j.apenergy.2025.126108

IoT Analytics. (2024). *Industrial AI market: 10 insights on how AI is transforming manufacturing*. IoT Analytics GmbH. https://iot-analytics.com

Iranshahi, K., Brun, J., Arnold, T., Sergi, T., & Müller, U. (2025). Digital Twins: Recent Advances and Future Directions in Engineering Fields. *Intelligent Systems with Applications*, *26*, 200516. DOI: 10.1016/j.iswa.2025.200516

Lei, Z., Shi, J., Luo, Z., Cheng, M., & Wan, J. (2024). Intelligent manufacturing from the perspective of Industry 5.0: Application review and prospects. *IEEE Access : Practical Innovations, Open Solutions*, *12*, 167436–167451.

McKinsey & Company. (2023). *The state of AI in 2023: Generative AI's breakout year*. McKinsey & Company. https://www.mckinsey.com

Plathottam, S. J., Rzonca, A., Lakhnori, R., & Iloeje, C. O. (2023). A review of artificial intelligence applications in manufacturing operations. *Journal of Advanced Manufacturing and Processing*, *5*(3), e10159. DOI: 10.1002/amp2.10159

Rahman, M. A., Shahrior, M. F., Iqbal, K., & Abushaiba, A. A. (2025). Enabling Intelligent Industrial Automation: A Review of Machine Learning Applications with Digital Twin and Edge AI Integration. *Automation*, *6*(3), 37. DOI: 10.3390/automation6030037

Schwark, F., Dawel, L., & Pehlken, A. (2023). Sustainability Digital Twin: A tool for the manufacturing industry. *Procedia CIRP*, *116*, 143–148.

Vespoli, S., Mattera, G., Marchesano, M. G., Nele, L., & Guizzi, G. (2025),Adaptive manufacturing control with Deep Reinforcement Learning for dynamic WIP management in industry 4.0, Computers & Industrial Engineering, Volume 202, 2025,110966, ISSN 0360-8352, doi:DOI: 10.1016/j.cie.2025.110966

Zhukabayeva, T., Zholshiyeva, L., Karabayev, N., Khan, S., & Alnazzawi, N. (2025). Cybersecurity Solutions for Industrial Internet of Things–Edge Computing Integration: Challenges, Threats, and Future Directions. *Sensors (Basel)*, *25*(1), 213. DOI: 10.3390/s25010213 PMID: 39797003

Chapter 4
Artificial Intelligence in Manufacturing:
A Conceptual Synthesis of Applications, Challenges, and Future Directions

Ardra Rianisa
http://orcid.org/0009-0005-0304-1195
Universitas Esa Unggul, Indonesia

Binastya Anggara Sekti
http://orcid.org/0000-0001-5489-4888
Universitas Esa Unggul, Indonesia

ABSTRACT

This chapter discusses the impact of Artificial Intelligence (AI) on manufacturing, focusing on applications like predictive maintenance, quality control, and automation. AI enhances operational efficiency by predicting equipment failures, reducing downtime, and improving product quality through automated defect detection. The concept of smart factories, powered by AI and the Internet of Things (IoT), optimizes workflows and adapts to market demands. However, challenges such as high implementation costs, integration with legacy systems, and data security concerns persist, especially for small and medium-sized enterprises. The chapter also covers the potential for AI to drive sustainability in manufacturing by optimizing energy consumption and reducing waste, and explores future trends like autonomous manufacturing systems and AI-powered supply chains

DOI: 10.4018/979-8-3373-5756-0.ch004

1. INTRODUCTION

The manufacturing industry has undergone a substantial transformation with the increasing integration of artificial intelligence (AI), driving advancements in productivity, operational efficiency, and sustainability. AI technologies such as machine learning, robotics, predictive analytics, and computer vision have reshaped industrial operations by enabling manufacturers to optimize processes in real time and make more informed decisions (de Luis García, 2024; Schultheis et al., 2024; Sekti, 2025a). These capabilities facilitate the transition toward data-driven, flexible, and intelligent manufacturing systems.

A core feature of AI's contribution to manufacturing lies in predictive maintenance, which enables manufacturers to anticipate equipment failures and schedule interventions before disruptions occur, reducing downtime and extending asset lifespan (Fan et al., 2025). AI-driven quality inspection further enhances product consistency through automated defect detection, reducing human error and improving reliability (Simeth et al., 2025). Together, these applications establish AI as a foundation for modern industrial operations and support the development of smart factories.

Smart factories an essential component of Industry 4.0 leverage cyber-physical systems, IoT devices, and interconnected data infrastructures to support autonomous or semi autonomous decision making (Gao et al., 2025). Within these environments, AI systems monitor production, detect anomalies, optimize scheduling, and enhance supply chain visibility, resulting in greater operational flexibility and adaptability in dynamic market conditions (de Luis García, 2024).

Despite these advancements, AI implementation in manufacturing remains challenging, particularly for small and medium sized enterprises (SMEs). Barriers include high upfront investment costs, legacy system integration constraints, cybersecurity risks, and workforce skill gaps (Hassnain et al., 2025; Schultheis et al., 2024). Moreover, the dependence of AI technologies on data introduces privacy and security concerns related to the protection of intellectual property and operational data (Fan et al., 2025). These challenges demonstrate that AI adoption is not solely a technological endeavor but also an organizational, financial, and governance issue requiring strategic planning and workforce adaptation.

This chapter synthesizes current developments in AI driven manufacturing by examining key applications, performance outcomes, technological and organizational challenges, and emerging trends. To complement the literature-based analysis, an illustrative case study from the Indonesian manufacturing sector is included to highlight practical considerations and staged deployment strategies..

2. OBJECTIVES

This chapter aims to:

1. Examine the role of artificial intelligence in enhancing operational efficiency within modern manufacturing environments.
2. Identify and analyze major AI applications in predictive maintenance, quality control, supply chain optimization, and automation.
3. Discuss challenges associated with AI deployment in manufacturing, including cost, system integration, cybersecurity, and workforce adaptation.
4. Explore emerging trends shaping AI enabled manufacturing systems and their potential implications for future industrial practices.
5. Identify strategic considerations and research gaps relevant to the sustainable adoption of AI in manufacturing ecosystems.

3. THE ROLE OF AI IN MODERN MANUFACTURING

Artificial intelligence plays a central role in the transformation of modern manufacturing by enabling data driven decision-making, real time performance optimization, and increased automation. AI technologies such as machine learning, deep learning, computer vision, and robotics enhance operational efficiency by improving predictive capabilities, resource utilization, and production flexibility (Gao et al., 2025). These capabilities allow manufacturers to transition from reactive to proactive operational models, particularly in areas such as maintenance, quality control, and scheduling.

AI integration has also contributed to the emergence of smart factory architectures that combine cyber physical systems and IoT enabled data infrastructures to support continuous monitoring of equipment, energy consumption, and production processes (de Luis García, 2024). Smart factories leverage AI to automate repetitive functions, detect anomalies, and dynamically adjust operating parameters, thereby increasing responsiveness to fluctuating market and production conditions (Hassnain et al., 2025). These developments align with broader Industry 4.0 objectives, which emphasize automation, interoperability, and real-time optimization at scale.

From a process standpoint, AI has enhanced several core manufacturing functions, including production scheduling, inventory management, and supply chain forecasting. AI driven forecasting models can predict demand fluctuations, optimize inventory levels, and adjust logistics decisions in real time, enabling manufacturers to reduce waste and improve coordination across the value chain (Colaw et al., 2025; Gao et al., 2025). Furthermore, computer vision-based inspection systems support quality assurance by detecting defects during production, improving product consistency, and reducing rework, scrap, and customer returns ((Bernárdez et al., 2025).

AI also contributes to sustainability by optimizing energy usage and material consumption. AI based control systems can identify energi efficient machine configurations, minimize resource inefficiencies, and support compliance with green manufacturing practices (Hassnain et al., 2025). These functions reinforce the strategic importance of AI for improving competitiveness in increasingly dynamic and environmentally constrained industrial environments.

Overall, AI's role in manufacturing extends beyond technical enhancements to include organizational and strategic implications. Successful AI deployment requires alignment between technological systems, workforce competencies, and operational governance. Manufacturers that incorporate AI into their production environments therefore position themselves for higher efficiency, greater responsiveness, and long term sustainability (Fan et al., 2025).

4. APPLICATIONS OF AI IN MANUFACTURING

AI technologies have been applied across multiple manufacturing functions to enhance operational performance, reduce downtime, and improve product quality. Four major application areas examined in the literature include predictive maintenance, quality inspection, supply chain optimization, and automation.

4.1. Predictive Maintenance

Predictive maintenance is one of the most established applications of AI in manufacturing. In contrast to reactive or schedule based maintenance strategies, AI enabled predictive maintenance leverages sensor and operational data to forecast equipment failures and plan maintenance activities proactively. By identifying degradation trends and anomaly patterns, AI systems allow maintenance teams to intervene before failures occur, reducing unplanned downtime and extending asset lifespan (Fan et al., 2025; Simeth et al., 2025).

Predictive maintenance systems also support more efficient resource allocation by minimizing emergency repairs and avoiding unnecessary maintenance procedures. These improvements yield operational and cost benefits that are especially relevant for aset intensive sectors such as automotive, aerospace, and large scale discrete manufacturing (Alfaqiyah et al., 2025). Over time, the accumulation of historical machine data enhances model accuracy and supports more refined scheduling decisions tailored to equipment usage profiles.

4.1.1 Key Performance Metrics

The performance of AI based predictive maintenance systems is commonly evaluated using operational indicators such as reductions in unplanned downtime, increases in mean time between failures (MTBF), and decreases in maintenance cost per asset. Empirical studies demonstrate performance improvements; however, outcomes vary depending on data availability, sensor reliability, and model calibration quality (Fan et al., 2025; Simeth et al., 2025).

4.2. Quality Control and Inspection

AI has significantly transformed quality control processes through the use of computer vision and machine learning for defect detection. Unlike manual inspection, which is labor intensive and prone to inconsistency, AI based inspection systems operate continuously and maintain uniform accuracy across production cycles. Computer vision models can detect defects related to shape, alignment, surface texture, and color, thereby improving product consistency and reducing scrap and rework (Bernárdez et al., 2025).

An important advantage of AI based inspection systems is their capacity to trigger corrective actions during production. By detecting anomalies at earlier stages, manufacturers can prevent defective products from reaching later processing stages or final customers, contributing to lower waste and improved customer satisfaction. Additionally, AI enables compliance with industry quality standards by providing traceability and documentation for inspection outcomes.

4.2.1 Key Metrics for AI-Based Quality Control

Key evaluation metrics include defect detection accuracy, false positive rates, defect rate reduction, and inspection throughput. While AI-driven inspection systems consistently outperform manual inspection, performance remains sensitive to variations in lighting conditions, product design changes, and training data coverage, requiring periodic model retraining for sustained reliability (Bernárdez et al., 2025; Colaw et al., 2025).

4.3. Supply Chain Optimization

AI also contributes to supply chain visibility and coordination by improving forecasting accuracy, inventory management, and logistics planning. Demand forecasting models utilize historical data and market indicators to predict consumption patterns, allowing manufacturers to align production output with demand and avoid overproduction or stockouts (Bubeník et al., 2025). AI enabled inventory optimization further reduces holding costs and improves responsiveness to volatile market conditions.

In logistics operations, AI algorithms identify optimal transportation routes and scheduling decisions to minimize delivery times and operational costs, improving customer service levels. As global supply chains become more interconnected and complex, AI driven optimization is increasingly recognized as a strategic capability for enhancing resilience and agility in manufacturing networks (Gao et al., 2025).

4.3.1 Key Metrics for AI Driven Supply Chain Optimization

Common indicators include inventory turnover, order fulfillment lead time, forecast accuracy, and stockout frequency. The effectiveness of AI systems in supply chain contexts depends heavily on data integration across suppliers, manufacturers, and distributors, as incomplete or fragmented data can limit system performance and adoption (Bubeník et al., 2025; Gao et al., 2025).

4.4. Automation and Robotics

Automation and robotics are perhaps the most well-known applications of AI in AI powered automation and robotics enable the execution of complex production tasks traditionally performed by human workers. Compared to rule-based or pre programmed automation, AI-enabled robots can learn from operational data and adapt to variations in production requirements, enabling more flexible manufacturing environments. AI supports applications such as assembly, material handling, packaging, and inspection, contributing to higher throughput and lower operational errors (Colaw et al., 2025).

Robotics integration also reduces dependency on manual labor for repetitive and ergonomically challenging tasks, improving workplace safety. As robotic systems become more intelligent, collaborative robots (cobots) are increasingly deployed to work alongside human operators rather than replace them. This trend reflects a shift toward hybrid human AI production environments characterized by shared decision making and complementary task allocation (Fan et al., 2025).

5. BENEFITS OF AI IN MANUFACTURING

5.1. Increased Efficiency

AI enhances production efficiency by automating repetitive tasks, optimizing scheduling decisions, and enabling real time performance monitoring. Automation reduces bottlenecks, minimizes idle time, and increases throughput by allowing production lines to operate continuously without performance degradation (Colaw et al., 2025). AI driven scheduling systems also improve machine utilization and production flow by aligning operational parameters with demand and resource availability, supporting leaner and more responsive manufacturing environments.

From an operational metrics perspective, efficiency gains are typically reflected in reductions in cycle time, improvements in machine utilization rates, and increases in throughput. However, efficiency improvements vary across firms depending on digital maturity and data infrastructure, indicating that system integration and data readiness are critical determinants of performance (Schultheis et al., 2024).

5.2. Cost Reduction

Cost reduction is frequently cited as a major economic benefit of AI adoption. Automation reduces labor costs associated with manual and repetitive tasks, while predictive maintenance reduces emergency repair expenditure and extends equipment lifespan (Simeth et al., 2025; Sun et al., 2025). Cost reductions may also arise from improved production planning and minimized scrap rates, contributing to more efficient resource utilization.

Economic assessments generally evaluate AI's financial impact using metrics such as maintenance cost per asset, cost of downtime, and operational expenditure trends. However, realizing cost benefits often requires a long term perspective, as initial investments in hardware, software, training, and integration may delay return on investment, particularly for small and medium-sized enterprises (Qu & Kim, 2025).

5.3. Improved Product Quality

AI enhances product quality through continuous monitoring and automated inspection, reducing variability associated with manual inspection processes. Computer vision based systems improve defect detection accuracy and reduce rework by identifying quality deviations earlier in the production process (Bernárdez et al.,

2025). These improvements lead to lower defect rates, reduced customer returns, and greater consistency across production batches.

AI based quality systems also support compliance with industry standards by providing traceability and detailed documentation of inspection outcomes, which is valuable in regulated manufacturing sectors such as automotive and electronics (Colaw et al., 2025). Sustained quality improvements depend on model maintenance and retraining, especially when product specifications or environmental conditions change.

5.4. Sustainability

AI contributes to sustainability objectives by optimizing energy consumption, reducing material waste, and improving resource efficiency. AI systems can identify energi efficient machine configurations and detect wasteful operating conditions, supporting reductions in energy usage and associated emissions (Hassnain et al., 2025). Waste reduction occurs through better demand forecasting, more precise material planning, and improvements in production accuracy, which collectively reduce scrap and excess inventory.

While existing studies emphasize short term operational sustainability outcomes, less attention has been given to lifecycle environmental impacts or the energy consumption of AI infrastructure itself. This gap highlights the need for more comprehensive assessments of AI-enabled sustainability outcomes in future research (Colaw et al., 2025).

6. CHALLENGES OF AI IN MANUFACTURING

6.1. High Implementation Cost

One of the primary challenges of adopting AI in manufacturing is the high implementation cost. Integrating AI systems into existing manufacturing setups often requires significant investment. This includes the cost of purchasing AI software, upgrading hardware, and training employees to use the new systems effectively. For small and medium sized businesses, the upfront costs can be prohibitive, making it difficult to compete with larger companies that have the financial resources to invest in AI. Moreover, the complexity of implementing AI technologies, such as machine learning or robotics, requires skilled professionals, which further drives up the cost.

As a result, while AI can lead to long term savings and increased productivity, many companies struggle to bear the initial financial burden (Ryalat et al., 2025).

Additionally, the integration of AI into existing systems may require significant changes to the infrastructure. Legacy systems, which have been in place for years, might not be compatible with AI driven technologies, and significant time and money may be needed to ensure that everything works seamlessly. The transition from traditional systems to AI powered ones can cause temporary disruptions in production, which further adds to the overall cost and complexity. This can be a particularly difficult hurdle for smaller manufacturers who lack the resources to make such large-scale investments (Simeth et al., 2025).

6.2. Integration with Legacy Systems

Another significant challenge is the integration of AI with legacy systems. Many manufacturing companies still rely on older equipment and software that may not be compatible with new AI driven technologies. This creates a barrier to AI adoption, as companies need to either upgrade or completely replace their existing systems to take full advantage of AI's capabilities. The process of integrating AI with legacy systems is not only costly but also time consuming. It often requires significant modifications to the entire production line, including hardware upgrades, software updates, and new data management systems. The risk of disrupting existing operations during this transition is high, which can lead to delays in production and financial losses in the short term (Sama et al., 2025).

Moreover, the integration process is not always smooth. There is often a need to develop custom solutions or find ways to make older systems work with the newer AI technologies. This can add further complexity to the implementation process. Additionally, not all AI solutions are designed to be easily adaptable to different types of manufacturing environments, meaning companies might need to invest in tailored AI solutions that can address their specific needs. As a result, the challenge of integrating AI with legacy systems remains a key barrier for many manufacturers looking to adopt AI technologies (Qu & Kim, 2025).

6.3. Data Security and Privacy Concerns

Data security and privacy are major concerns when implementing AI in manufacturing. AI systems rely heavily on data to make decisions, which means manufacturers need to collect vast amounts of information from machines, sensors, and even employees. This data is often sensitive, and there is a risk that it could be accessed by unauthorized individuals or hacked, leading to significant breaches of privacy or intellectual property. For instance, AI systems might store proprietary designs or sensitive information about production processes that could be exploited if security measures are not properly implemented. Ensuring that data is securely stored and transmitted is a critical challenge for manufacturers looking to adopt AI technologies (Essaid & Ju, 2025).

Moreover, AI driven manufacturing systems often need to be connected to the internet or other networks, which makes them more vulnerable to cyberattacks. The increasing use of IoT (Internet of Things) devices in smart factories adds another layer of complexity to data security. These devices collect and transmit real time data, and any breach in security can have far-reaching consequences. Manufacturers must invest in strong cybersecurity measures, such as encryption, firewalls, and secure data storage solutions, to protect their AI systems and the data they rely on. Without proper security, the risks of using AI in manufacturing can outweigh the benefits (Ren et al., 2025).

6.4. Workforce Displacement

AI also poses a significant challenge in terms of workforce displacement. As AI and automation take over more repetitive and manual tasks, there is a growing concern about job losses. Workers who have been performing these tasks may find themselves replaced by machines or AI systems, leading to unemployment in certain sectors. While AI can create new job opportunities in areas such as software development, AI system maintenance, and data analysis, it also requires workers to have different skills. Many employees in manufacturing today may not have the technical skills necessary to work with AI-driven systems, creating a skills gap in the workforce (Schultheis et al., 2024).

This shift in the workforce also requires companies to consider how they will handle employee transitions. Reskilling programs are essential for helping workers adapt to the changes brought about by AI, but these programs require investment and time. Some companies might be reluctant to invest in reskilling, especially when faced with the upfront costs of AI implementation. This creates a dilemma, as manufacturers need to balance the cost of retraining their workforce with the need to stay competitive and adopt AI technologies that can improve their operations (Sama et al., 2025).

6.5. Deployment Arhitecture and Operational Challenges of AI in Manufacturing

Artificial intelligence deployment in manufacturing environments is strongly influenced by architectural choices, which in turn shape the nature and severity of implementation challenges. One commonly discussed approach in the literature is centralized cloud-based deployment, where data from production systems are transmitted to cloud platforms for processing and model inference. This approach enables scalable analytics and centralized model management, but it may introduce latency issues and increase exposure to cybersecurity risks. Such limitations are particularly critical for time-sensitive manufacturing operations that require real-time or near-real-time decision making (Essaid & Ju, 2025; Schultheis et al., 2024).

Edge based deployment has emerged as an alternative strategy that addresses latency constraints by processing data closer to production equipment. In manufacturing contexts such as predictive maintenance and automated quality inspection, edge based artificial intelligence enables faster response times and greater operational autonomy. However, edge deployment presents its own challenges. These challenges include limited computational resources, difficulties in updating and maintaining models across multiple devices, and increased system complexity as the number of deployed edge nodes grows (Fan et al., 2025). These constraints illustrate that architectural decisions involve trade-offs rather than universally optimal solutions.

Recent studies increasingly highlight hybrid edge cloud architectures as a pragmatic approach for industrial artificial intelligence systems. In such configurations, time critical inference tasks are executed at the edge, while model training, optimization, and long-term data analysis are conducted in the cloud. Although this approach balances responsiveness with analytical capability, it introduces integration challenges related to data synchronization, system interoperability, and operational governance. These issues are particularly pronounced in manufacturing environments that rely on legacy equipment and heterogeneous information systems (Gao et al., 2025; Schultheis et al., 2024).

Beyond technical considerations, deployment architecture also shapes organizational and workforce related challenges. The operation of distributed artificial intelligence systems requires new competencies in data governance, system monitoring, and model lifecycle management. Manufacturing organizations that lack these capabilities may struggle to sustain artificial intelligence performance over time, even when initial deployment is technically successful. This observation reinforces prior findings that workforce reskilling and close collaboration between operational technology and information technology teams are critical enablers of effective artificial intelligence adoption (Schultheis et al., 2024; Sekti, 2025a).

Overall, the literature suggests that many challenges associated with artificial intelligence adoption in manufacturing are not purely technological but are closely connected to deployment architecture and operational governance. Addressing issues such as system integration, cybersecurity, scalability, and workforce readiness requires coordinated architectural planning and organizational alignment. These findings indicate that successful artificial intelligence adoption in manufacturing depends not only on algorithmic performance but also on strategic decisions regarding system design and long-term operational management (Qu & Kim, 2025).

7. CASE STUDY: AI ADOPTION AT PT INDO SURYA KENCANA

PT Indo Surya Kencana is an electronics manufacturing company located in Pasar Kemis, Tangerang, Indonesia, that has begun adopting artificial intelligence as part of its digital transformation strategy. Although artificial intelligence was initially introduced through an AI assisted customer interaction system, the implementation reflects a broader organizational shift toward data driven decision making within the manufacturing environment rather than customer facing services.

Artificial intelligence deployment at PT Indo Surya Kencana has enabled systematic collection and utilization of production related data generated from equipment usage, process monitoring, and operational records. These capabilities provide a foundation for extending artificial intelligence applications to core manufacturing functions such as predictive maintenance and process optimization. Prior studies suggest that such incremental adoption pathways are common in small and medium sized manufacturing enterprises, where artificial intelligence is first deployed in low risk domains before being scaled to production critical processes (Qu & Kim, 2025; Schultheis et al., 2024).

From an operational performance perspective, artificial intelligence based analytics have enhanced manufacturing visibility by enabling the identification of recurring operational issues, emerging process bottlenecks, and maintenance related risks. Although predictive maintenance has not yet been fully implemented across all assets, the existing artificial intelligence infrastructure constitutes a technical and organizational foundation for future deployment. This staged implementation approach aligns with findings that phased integration reduces implementation risk and facilitates workforce adaptation (Fan et al., 2025; Simeth et al., 2025).

The case also highlights key challenges associated with artificial intelligence adoption in manufacturing contexts. Integration with legacy production equipment remains a significant constraint, as many machines were not designed for continuous data exchange with artificial intelligence platforms. Furthermore, workforce readiness and data governance capabilities materially affect the effectiveness of artificial intelligence supported operational decision making. These issues are consistent with broader empirical evidence identifying system interoperability and skill gaps as major barriers to artificial intelligence enabled manufacturing transformation (Schultheis et al., 2024; Sekti, 2025b).

Overall, the case of PT Indo Surya Kencana demonstrates how artificial intelligence adoption in manufacturing can evolve from peripheral applications toward operationally meaningful functions. Rather than representing an isolated customer service innovation, the implementation supports a transition toward data centric manufacturing practices characterized by increased operational visibility, improved reliability, and long term scalability. This evidence reinforces the argument that successful artificial intelligence adoption in manufacturing requires not only technical capability but also organizational readiness and phased deployment strategies aligned with production objectives (Gao et al., 2025; Qu & Kim, 2025).

8. ADDRESSING THE CHALLENGES OF AI IN MANUFACTURING

Addressing the challenges associated with AI adoption in manufacturing requires coordinated strategies that encompass technological, economic, and organizational dimensions. Literature highlights several approaches that enable firms to mitigate implementation barriers and achieve sustainable deployment of AI enabled systems

8.1. Reducing Implementation Costs

High upfront investment represents a major barrier to AI adoption, particularly for small and medium sized enterprises. One effective strategy involves phased or modular deployment in which AI solutions are introduced incrementally in low-risk operational areas before scaling to core production processes simeth(Simeth et al., 2025). This approach enables organizations to evaluate performance outcomes, build technical capabilities, and justify further investment through demonstrated returns. Cloud based AI services may also reduce capital expenditure by shifting computational resources away from on premises infrastructure toward subscription-based platforms. Cost benefit analysis and long term return on investment assessments support more informed decision making regarding AI investment (Colaw et al., 2025).

8.2. Improving System Integration

System integration challenges often arise due to incompatibility between AI solutions and legacy manufacturing equipment. Modular and interoperable architectures facilitate gradual integration with existing systems, reducing disruption and minimizing replacement costs(Essaid & Ju, 2025). Hybrid deployment strategies that combine edge processing for time-sensitive tasks with cloud based analytics for complex computation also help balance latency, scalability, and processing requirements (Qu & Kim, 2025). Standards for data interoperability and communication protocols contribute to effective long term integration in heterogeneous production environments. System integration challenges often arise due to incompatibility between AI solutions and legacy manufacturing equipment. Modular and interoperable architectures facilitate gradual integration with existing systems, reducing disruption and minimizing replacement costs (Essaid & Ju, 2025). Hybrid deployment strategies that combine edge processing for time sensitive tasks with cloud-based analytics for complex computation also help balance latency, scalability, and processing requirements (Qu & Kim, 2025). Standards for data interoperability and communication protocols contribute to effective long-term integration in heterogeneous production environments.

8.3. Workforce Reskilling

Workforce reskilling is essential for addressing concerns regarding job displacement and skill mismatch. Training programs that equip employees with competencies in AI supervision, data interpretation, and technical troubleshooting are widely recommended in the literature (Fan et al., 2025). Rather than fully replacing human labor, AI tends to shift labor demand toward higher-value activities such as decision-making, coordination, and oversight (Schultheis et al., 2024). Reskilling initiatives support organizational acceptance of AI technologies and reduce resistance to adoption, contributing to more sustainable, human centered implementation strategies.

8.4. Data Protection Strategies

Data driven manufacturing environments elevate cybersecurity and privacy concerns due to increasing data exchange across machines, processes, and supply chain partners. Effective data protection requires the implementation of encryption, secure storage, access control mechanisms, and adherence to legal and regulatory frameworks governing industrial data (Ren et al., 2025). Continuous monitoring and incident response systems strengthen cybersecurity resilience in AI enabled industrial settings (Essaid & Ju, 2025). Data governance frameworks establish policies for data ownership, usage, and sharing, which are critical in interconnected production ecosystems where intellectual property risks are significant.

Figure 1. Data Protection in AI-Driven Manufacturing

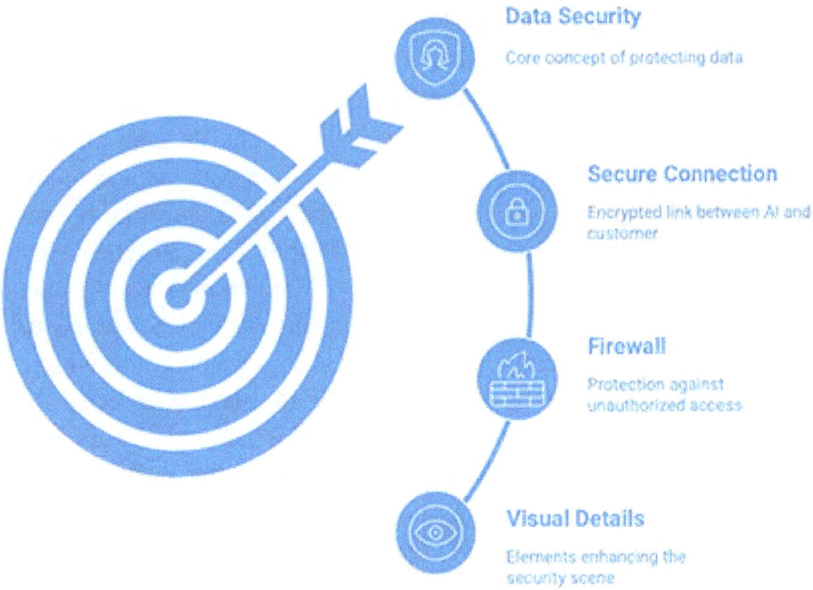

8.5. Economic Evaluation and Business Justification of AI Adoption in Manufacturing

Economic evaluation frameworks determine the feasibility of AI implementation by quantifying expected operational and financial benefits relative to capital and operating expenditures. Return on investment and payback period analyses are frequently used to assess the value of AI in manufacturing contexts (Gao et al., 2025). Cost structure analysis further distinguishes between short-term expenditure increases and long term operational gains, supporting strategic investment decisions under budget constraints. Risk adjusted evaluation methods consider uncertainties related to data quality, integration complexity, and cybersecurity threats, allowing firms to balance innovation with risk exposure (Schultheis et al., 2024; Sekti, 2025a). These strategies support strategic alignment between AI investments and organizational capabilities, improving the likelihood of successful and sustainable adoption.

9. FUTURE TRENDS IN AI-DRIVEN MANUFACTURING

Future developments in artificial intelligence driven manufacturing indicate a shift toward higher levels of autonomy, integration, and data centric decision making. Current studies suggest that manufacturing systems are evolving from isolated artificial intelligence applications toward coordinated environments in which data from machines, processes, and supply chains are continuously analyzed to support operational and strategic decisions. This evolution aligns with broader Industry 4.0 objectives that emphasize flexibility, responsiveness, and real time optimization across manufacturing operations (Gao et al., 2025; Lindberg, 2025).

One prominent trend is the increasing adoption of autonomous manufacturing systems. Artificial intelligence is expected to enable production systems to adapt operating parameters dynamically in response to changing conditions with reduced human intervention. While existing research highlights the technical feasibility of such systems, empirical evidence regarding long term reliability, accountability, and governance remains limited. This gap indicates the need for further investigation into how autonomous decision making can be aligned with safety requirements, operational responsibility, and regulatory compliance in manufacturing environments (Ren et al., 2025; Schultheis et al., 2024).

Another emerging trend involves the integration of artificial intelligence with sustainability objectives. Future manufacturing systems are increasingly expected to leverage artificial intelligence to support energy efficiency, waste reduction, and resource optimization. Although prior studies report short term efficiency gains, comprehensive evaluations of lifecycle environmental impact remain scarce. This suggests an opportunity for future research to assess the broader sustainability implications of artificial intelligence adoption, including the environmental costs associated with data processing and computational infrastructure (Hassnain et al., 2025; Sun et al., 2025).

Human artificial intelligence collaboration is also anticipated to play a growing role in manufacturing. Rather than fully replacing human labor, artificial intelligence systems are increasingly positioned as decision support tools that assist operators in monitoring processes, diagnosing anomalies, and making informed interventions. However, the effectiveness of such collaboration depends on workforce skills, trust in artificial intelligence outputs, and organizational readiness. Existing literature provides limited longitudinal evidence on human artificial intelligence interaction in manufacturing, highlighting a need for further research on training models, user acceptance, and organizational adaptation (Schultheis et al., 2024; Sekti, 2025a).

Overall, future trends in artificial intelligence driven manufacturing point toward more integrated and intelligent production environments. At the same time, unresolved challenges related to governance, sustainability evaluation, and workforce integration persist. Addressing these challenges requires interdisciplinary research that combines technological innovation with economic analysis, organizational studies, and policy perspectives. Such efforts are essential to ensure that artificial intelligence adoption in manufacturing delivers sustainable value while supporting inclusive and responsible industrial development (Gao et al., 2025; Qu & Kim, 2025).

10. AI IN MANUFACTURING: GLOBAL IMPACT

Artificial intelligence (AI) has emerged as a strategic driver of global industrial transformation by enabling increases in efficiency, precision, and responsiveness within modern manufacturing systems. Nations and industries that successfully integrate AI into their production ecosystems benefit from optimized workflows, reduced operational errors, and enhanced product quality. These advancements are accelerating structural shifts in global competition as countries reassess their industrial policies and reposition themselves within international value chains.

The role of AI in manufacturing is increasingly aligned with global sustainability objectives. Intelligent production systems contribute to reductions in waste, energy usage, and emissions, mirroring approaches in sustainable construction where intelligent methods support the reprocessing of waste materials into valuable resources (Gunasekaran et al., 2024). This demonstrates that AI not only strengthens economic performance but also advances ecological goals, making it relevant for countries pursuing green industrial development.

Beyond shop-floor productivity, AI contributes to the resilience of global supply networks by enhancing forecasting, coordination, and operational continuity. Similar efficiency effects have been observed in agriculture, where AI enabled decision systems support resource optimization and generate economic benefits through intelligent planning (Gunasekaran, Boopathi, Suresh, Suresh, Vanitha, et al., 2024). In manufacturing, these capabilities are essential for navigating geopolitical disruptions, fluctuating demand patterns, and logistical uncertainties within interconnected international supply chains.

AI also supports environmental safety and regulatory compliance through improved process monitoring. Studies in civil and environmental engineering reveal that intelligent systems can reduce exposure to harmful materials, promote safer operational practices, and mitigate environmental toxins (Selvapriya et al., 2025). When translated into industrial manufacturing contexts, such capabilities allow firms to meet increasingly stringent environmental standards demanded by global markets and international regulatory bodies.

The ongoing evolution toward smart and autonomous factories highlights the broadening scope of AI beyond traditional automation. In engineering sectors, AI is already being applied to explore novel forms of process innovation and advanced material manipulation (Suresh, Velmurugan, et al., 2024). Manufacturing industries are adopting similar pathways, developing intelligent production cells and sensor rich environments capable of reconfiguring operations in real time to respond to shifting priorities and market conditions.

These changes are closely associated with the transition toward Industry 5.0, which emphasizes collaborative interaction between humans and intelligent machines. Applications such as AI-enhanced drones have demonstrated value in industrial inspection, surveying, and maintenance, showcasing new forms of labor augmentation rather than labor substitution (Suresh, Natarajan, et al., 2024). This reinforces the notion that AI can expand human capability within industrial systems rather than simply replace it.

In parallel, additive manufacturing processes are leveraging AI to create environmentally sustainable production pathways through more efficient material usage and reduced processing waste (Suresh, Natarajan, et al., 2024). These capabilities position AI as a catalyst for innovation in manufacturing materials science and reinforce its role in shaping the future of global sustainable production systems. Countries investing in such technologies are likely to gain long-term competitive advantages in emerging clean-technology markets.

Human capital development also influences the global adoption of AI in manufacturing. Virtual reality-based training has emerged as a tool for skill development in engineering, helping to accelerate the preparation of technical workforces for highly digital production environments (Suresh, 2024). Countries able to integrate such educational innovations may experience smoother industrial transitions and reduced labor market frictions during technological transformation.

Despite its potential, the global diffusion of AI technologies is uneven. Countries with advanced digital infrastructure, research capacity, and investment capabilities are moving faster than those without comparable resources. This disparity risks widening global technological divides and may create new asymmetries in industrial power. Policymakers and international institutions are therefore increasingly recognizing the need for coordinated strategies to ensure inclusive technological progress.

In summary, AI is becoming a foundational component of modern manufacturing systems worldwide, not only enhancing operational performance and industrial competitiveness but also reshaping environmental, educational, and economic dimensions of industrial development. The integration of AI across multiple engineering and production domains demonstrates its growing influence on the structure of global value chains. As adoption continues to expand, AI is expected to play a defining role in shaping the future of manufacturing toward more intelligent, sustainable, and human-centered industrial ecosystems.

11. CONCLUSION

Artificial intelligence is fundamentally reshaping the landscape of modern manufacturing by enabling a shift from reactive and manually coordinated operations toward environments characterized by data driven decision-making, predictive capabilities, and increasing levels of autonomous functionality. Through the integration of machine learning, robotics, computer vision, and advanced analytics, manufacturers are increasingly able to optimize workflows, anticipate disruptions, and respond to market dynamics with greater efficiency and flexibility. These technological advancements support improvements in production throughput, product quality, and resource utilization, while simultaneously contributing to broader sustainability objectives.

The influence of artificial intelligence on manufacturing extends beyond performance enhancement at the production floor level. The emergence of smart factory concepts illustrates the potential for interconnected systems to coordinate seamlessly through real time data exchange and autonomous control. As manufacturing environments become more complex and volatile, artificial intelligence provides tools for aligning production activities with supply chain conditions, customer demand, and environmental constraints. In doing so, artificial intelligence serves as a critical enabler for achieving operational resilience, organizational agility, and long-term competitiveness within increasingly globalized markets. This global interconnectedness underscores the strategic significance of artificial intelligence in shaping industrial value chains, regional production hubs, and economic development trajectories.

Despite these advancements, the diffusion of artificial intelligence in manufacturing is accompanied by substantial structural and organizational challenges. Implementation often involves high capital expenditure requirements, modifications to legacy infrastructures, and the development of new workforce competencies. These challenges demonstrate that the successful adoption of artificial intelligence is not solely a matter of technological capability but also depends on organizational readiness, economic feasibility, and strategic planning. Firms that approach implementation as a phased transformation rather than a discrete technological upgrade tend to achieve more sustainable outcomes, as such approaches allow for iterative learning, risk mitigation, and alignment between technological capacity and operational strategy.

Another important dimension concerns the governance of artificial intelligence systems. As data becomes central to manufacturing operations, issues related to cybersecurity, data interoperability, intellectual property, and ethical use emerge as strategic considerations. Effective data governance frameworks are increasingly required to ensure that artificial intelligence deployments align with organizational priorities and regulatory contexts while maintaining trust and safeguarding proprietary knowledge. Governance also encompasses the establishment of standards and protocols that facilitate interoperability across machines, platforms, and production networks, enabling artificial intelligence to function at scale within complex industrial ecosystems.

Looking toward the future, manufacturing systems are expected to evolve toward higher levels of autonomy, integration, and human-machine collaboration. Autonomous production cells, intelligent supply networks, and digital twins represent examples of emerging paradigms that build upon artificial intelligence capabilities to support more dynamic, predictive, and adaptive industrial operations. While the technical feasibility of such systems continues to progress rapidly, questions regarding long-term reliability, governance, labor implications, and environmental impact remain areas of ongoing investigation and represent important opportunities for further research. These areas of inquiry are likely to shape not only how artificial intelligence is implemented, but also how future industrial systems are regulated and measured.

At a broader societal level, the transformative potential of artificial intelligence in manufacturing extends into labor markets, education systems, environmental policy, and national development strategies. The reconfiguration of skills, job roles, and human machine interfaces requires coordinated efforts across industry, academia, and government to ensure that the workforce is prepared to operate within increasingly digital and automated production environments. In parallel, national policy frameworks must address the implications of technological concentration, supply chain dependencies, and disparities in digital infrastructure that influence global competitiveness and industrial sovereignty.

In conclusion, artificial intelligence exhibits significant transformative potential across manufacturing environments. It enables process improvements, enhances decision making, and supports strategic objectives related to sustainability, resilience, and competitiveness. However, realizing these benefits requires more than the implementation of algorithms or automation technologies. It demands coordinated investments in infrastructure, workforce development, organizational capability, research, and system governance. As both industrial practice and academic scholarship continue to advance, artificial intelligence is positioned to become a cornerstone of future manufacturing systems, supporting more intelligent, resilient, and sustainable industrial ecosystems. The trajectory of these developments suggests that artificial intelligence will not merely enhance existing manufacturing paradigms, but will contribute to the emergence of new models of industrial operation that redefine the role of technology, labor, and innovation in the global economic system.

REFERENCES

Alfaqiyah, E., Alzubi, A., Aljuhmani, H. Y., & Öz, T. (2025). How Industry 4.0 Technologies Enhance Supply Chain Resilience: The Interplay of Agility, Adaptability, and Customer Integration in Manufacturing Firms. *Sustainability (Basel), 17*(17), 7922. Advance online publication. DOI: 10.3390/su17177922

Bernárdez, J. M., Boo, J., Díaz, J. I., & Medina, R. (2025). Interdepartmental Optimization in Steel Manufacturing: An Artificial Intelligence Approach for Enhancing Decision-Making and Quality Control. *Applied System Innovation, 8*(3), 63. Advance online publication. DOI: 10.3390/asi8030063

Bubeník, P., Rakyta, M., Buzalka, M., Biňasová, V., & Kovaríková, Z. (2025). Optimization of Business Processes Using Artificial Intelligence. *Electronics (Basel), 14*(11), 2105. Advance online publication. DOI: 10.3390/electronics14112105

Colaw, C. L., Madison, G., Tseng, B., Griser, G. M., Truelson, G., Gallo, A., & Hurmuzlu, Y. (2025). Methodology for Enablement of Human Digital Twins for Quality Assurance in the Aerospace Manufacturing Domain. *Sensors (Basel), 25*(11), 3362. Advance online publication. DOI: 10.3390/s25113362 PMID: 40968904

de Luis García, E. (2024). COMBATING CHILD SEXUAL ABUSE ON THE INTERNET: CONSIDERATIONS IN THE LIGHT OF THE PROPOSAL FOR A REGULATION. *Actualidad Juridica Iberoamericana, 21*, 104–129. DOI: 10.13039/501100011033

Essaid, M., & Ju, H. (2025). Blockchain Solutions for Enhancing Security and Privacy in Industrial IoT. *Applied Sciences (Basel, Switzerland), 15*(12), 6835. Advance online publication. DOI: 10.3390/app15126835

Fan, H., Liu, C., Janvisloo, N. E., Bian, S., Fuh, J. Y. H., Lu, W. F., & Li, B. (2025). MaViLa: Unlocking new potentials in smart manufacturing through vision language models. *Journal of Manufacturing Systems, 80*, 258–271. DOI: 10.1016/j.jmsy.2025.02.017

Gao, Y., Liu, Y., & Wu, W. (2025). How Does Artificial Intelligence Capability Affect Product Innovation in Manufacturing Enterprises? Evidence from China. *Systems, 13*(6), 480. Advance online publication. DOI: 10.3390/systems13060480

Gunasekaran, K., Boopathi, S., Suresh, S., Suresh, P., & Gokulakannan, S. (2024). Sustainable building construction practices for transforming waste into wealth through vernacular architecture. In *Innovations in Energy Efficient Construction Through Sustainable Materials* (pp. 387–413). DOI: 10.4018/979-8-3693-3398-3.ch014

Gunasekaran, K., Boopathi, S., Suresh, S., Suresh, P., Vanitha, N. S., & Radhika, K. (2024). Eco-friendly smart agricultural process using artificial intelligence: Economic benefits. In *Reshaping Environmental Science Through Machine Learning and IoT* (pp. 160–175). DOI: 10.4018/979-8-3693-2351-9.ch009

Hassnain, M., Lee, S. M. W., & Azhar, M. R. (2025). Soft sensing of biological oxygen demand in industrial wastewater using machine learning models. *Journal of Water Process Engineering*, *78*, 108699. Advance online publication. DOI: 10.1016/j.jwpe.2025.108699

Lindberg, D. G. (2025). Harnessing AI for smart manufacturing: Insights from Industry 4.0. *Discover Artificial Intelligence*, *5*(1), 111. Advance online publication. DOI: 10.1007/s44163-025-00363-0

Qu, C., & Kim, E. (2025). Artificial-Intelligence-Enabled Innovation Ecosystems: A Novel Triple-Layer Framework for Micro, Small, and Medium-Sized Enterprises in the Chinese Apparel-Manufacturing Industry. *Sustainability (Basel)*, *17*(11), 5019. Advance online publication. DOI: 10.3390/su17115019

Ren, Y., Liu, Y., Ji, T., & Xu, X. (2025). *AI Agents and Agentic AI-Navigating a Plethora of Concepts for Future Manufacturing*. DOI: 10.1016/j.jmsy.2025.08.017

Ryalat, M., Almtireen, N., Al-refai, G., Elmoaqet, H., & Rawashdeh, N. (2025). Research and Education in Robotics: A Comprehensive Review, Trends, Challenges, and Future Directions. *Journal of Sensor and Actuator Networks*, *14*(4), 76. DOI: 10.3390/jsan14040076

Sama, M., Sata, A., Joshi, G., & Mohan, D. G. (2025). Smart Manufacturing With Industrial Internet of Things: Advances in TIG Welding for SS304 Stainless Steel. *Engineering Reports*, *7*(7), e70257. Advance online publication. DOI: 10.1002/eng2.70257

Schultheis, A., Alt, B., Bast, S., Guldner, A., Jilg, D., Katic, D., Mundorf, J., Schlagenhauf, T., Weber, S., Bergmann, R., Bergweiler, S., Creutz, L., Dartmann, G., Malburg, L., Naumann, S., Rezapour, M., & Ruskowski, M. (2024). EASY: Energy-Efficient Analysis and Control Processes in the Dynamic Edge-Cloud Continuum for Industrial Manufacturing. *KI -. Kunstliche Intelligenz*. Advance online publication. DOI: 10.1007/s13218-024-00868-3

Sekti, B. A. (2025a). Adversarial machine learning in industrial cybersecurity: Challenges and solutions. In *AI-Enhanced Cybersecurity for Industrial Automation*. DOI: 10.4018/979-8-3373-3241-3.ch018

Sekti, B. A. (2025b). Machine learning and artificial intelligence (AI) in manufacturing. In *Integrating Digital Innovation and Integrated Frameworks in Manufacturing*. DOI: 10.4018/979-8-3373-1082-4.ch011

Selvapriya, R., Gopinath, M., Sampathkumar, P., Velmurugan, D., & Suresh, S. (2025). AI reduces environmental toxins in civil engineering. In *AI Methods for Environmental Protection and Resource Conservation* (pp. 83–106). DOI: 10.4018/979-8-3373-3246-8.ch004

Simeth, A., Kumar, A. A., & Plapper, P. (2025). Flexible and robust detection for assembly automation with YOLOv5: A case study on HMLV manufacturing line. *Journal of Intelligent Manufacturing*, *36*(5), 3447–3463. DOI: 10.1007/s10845-024-02411-5

Sun, Y., Zhang, M., Chang, J., & Wang, C. (2025). Unpacking Boundary-Spanning Search and Green Innovation for Sustainability: The Role of AI Capabilities in the Chinese Manufacturing Industry. *Sustainability (Basel)*, *17*(14), 6439. Advance online publication. DOI: 10.3390/su17146439

Suresh, S. (2024). Virtual reality in engineering: Innovations and applications. In *Creating Immersive Learning Experiences Through Virtual Reality (VR)* (pp. 287–310). DOI: 10.4018/979-8-3693-6407-9.ch013

Suresh, S., Natarajan, E., Sudhagar, S., Boopathi, S., & Elayaraja, R. (2024). Additive manufacturing for a sustainable production: Materials, processes, and impact. In *Futuristic Technology for Sustainable Manufacturing* (pp. 1–16). DOI: 10.4018/979-8-3693-1862-1.ch001

Suresh, S., Velmurugan, D., Priya, E. L. D., & Selvapriya, R. (2024). AI for enhanced drone capabilities in industry 5.0. In *Drone Applications for Industry 5.0* (pp. 75–91). DOI: 10.4018/979-8-3693-2093-8.ch006

Chapter 5
Artificial Intelligence in Modern Manufacturing

B. Senthil Kumar
http://orcid.org/0000-0002-1974-0645
CMS Business School, Jain University, India

P. Elantheraiyan
RMK Engineering College, India

S. Thandayuthapani
http://orcid.org/0000-0001-9559-745X
Kristu Jayanti University, India

R. J. T. Nirmal Raj
SRM University, Sikkim, India

S. Loganatha Prasanna
http://orcid.org/0000-0002-1696-1303
SRM Institute of Science and Technology, India

M. Vasudevan
http://orcid.org/0009-0002-6672-7651
Hindustan Institute of Technology and Management, Chennai, India

ABSTRACT

The incorporation of Artificial Intelligence (AI) within the manufacturing sector signifies a crucial transition towards Industry 4.0, converting conventional systems into intelligent, automated, and data-centric operations. AI technologies—such as machine learning, computer vision, and the Internet of Things (IoT)—improve productivity, facilitate predictive maintenance, enhance quality control, and bolster supply chain resilience. Smart factories serve as a prime example of this transformation, allowing for real-time decision-making and promoting sustainability. Nevertheless, challenges such as cybersecurity threats, data privacy concerns, and workforce displacement remain, highlighting the necessity for ethical AI implementation and workforce reskilling. To fully harness the potential of AI, manufacturers must integrate it into their long-term strategies as a strategic enabler that aligns with human insight and innovation.

DOI: 10.4018/979-8-3373-5756-0.ch005

INTRODUCTION

The manufacturing sector is experiencing a significant transformation propelled by the swift incorporation of Artificial Intelligence (AI) technologies. This change, frequently referred to as Industry 4.0, is marked by the merging of cyber-physical systems, the Internet of Things (IoT), cloud computing, and AI, resulting in intelligent and interconnected production environments. The capability of AI to analyze extensive datasets, automate intricate tasks, and provide predictive insights is revolutionizing the methods by which goods are produced, managed, and delivered.

The implementation of Artificial Intelligence (AI) within the manufacturing sector has led to substantial economic and operational shifts, altering how industries manage production, cost control, and customer interaction. AI technologies, including machine learning, computer vision, and robotics, improve productivity, lower operational expenses, and facilitate customized manufacturing processes that swiftly adapt to market needs.

AI-driven automation significantly enhances manufacturing productivity by streamlining workflows, reducing human error, and allowing for continuous operations. Intelligent systems are capable of monitoring production lines in real-time, adjusting parameters to sustain optimal performance, and identifying anomalies at an early stage. Predictive maintenance, powered by AI analytics, aids in anticipating equipment failures before they happen, thus averting costly unplanned downtimes and prolonging the lifespan of machinery.

AI improves operational efficiency by automating repetitive tasks such as assembly, inspection, and packaging. Robots that are equipped with AI can function with greater precision and speed than human labor, thereby enhancing output quality and throughput. Beyond physical automation, AI supports intelligent scheduling and resource allocation, ensuring the optimal utilization of materials, labor, and energy. This approach minimizes waste and operational costs, resulting in significant financial savings.

The economic advantages of artificial intelligence go beyond mere cost savings. AI empowers manufacturers to optimize supply chains by utilizing real-time data analysis and demand forecasting, which helps to minimize inventory holding expenses and mitigate the risk of overproduction. This just-in-time strategy decreases the capital invested in raw materials and finished products, thereby enhancing cash flow and profitability.

In addition to cost efficiency and increased productivity, AI transforms the manufacturing landscape by facilitating mass customization, marking a transition from conventional mass production to the creation of personalized products on a large scale. AI technologies process extensive consumer data, encompassing preferences, behaviors, and feedback, to dynamically modify production parameters and product designs. This functionality enables manufacturers to provide customized solutions swiftly without compromising on efficiency.

By incorporating AI into flexible manufacturing systems, organizations can swiftly adapt to fluctuating market demands and consumer trends. AI-enhanced production lines can transition between different product variants with minimal downtime, fostering agile manufacturing cycles and shortening lead times. This adaptability allows businesses to sustain a competitive edge in markets where customer expectations are rapidly changing.

AI-driven quality control systems that utilize computer vision are capable of identifying product defects with remarkable precision, ensuring that tailored products adhere to rigorous quality standards. Algorithms that continuously enhance inspection processes over time, thereby improving reliability and customer satisfaction. The integration of customization with unwavering quality bolsters brand reputation and cultivates customer loyalty.

The operational enhancements facilitated by AI result in significant economic benefits. Lower costs and heightened productivity improve profit margins, allowing manufacturers to reinvest in innovation and growth. The ability to customize products creates new revenue opportunities by addressing niche markets and increasing product differentiation. Nevertheless, these advantages necessitate careful strategic planning and investment in digital infrastructure, data management, and workforce development. Organizations must find a balance between automation and human expertise to maximize the effectiveness of AI integration. Successful manufacturers perceive AI not just as a technological upgrade but as a strategic catalyst for business transformation.

AI's influence in contemporary manufacturing goes far beyond mere automation; it serves as a vital engine of economic efficiency and market responsiveness. Improvements in productivity and reductions in costs through AI facilitate leaner, more agile operations. At the same time, AI enables manufacturers to provide customized products that satisfy the increasingly complex demands of consumers. As industries progressively adopt AI, those that strategically leverage its capabilities will secure sustainable competitive advantages, thereby shaping the future of manufacturing.

HISTORICAL CONTEXT OF AI IN MANUFACTURING

Automation has historically been a fundamental element of industrial advancement, tracing its origins to the early 20th century. Over the years, it has progressed from basic mechanized operations to advanced intelligent systems that possess self-learning capabilities and can make autonomous decisions. This evolution has been hastened by the incorporation of Artificial Intelligence (AI), which signifies a transition from conventional rule-based frameworks to intelligent automation. The present wave of innovation—commonly known as Industry 4.0—is transforming manufacturing by integrating intelligence throughout the production ecosystem. This report delves into this evolution, focusing particularly on the technological milestones, enabling factors, and the pivotal role of AI in shaping the contemporary manufacturing environment.

The origins of manufacturing automation can be linked to the introduction of mechanized assembly lines, which were popularized by Henry Ford in the early 1900s. These systems transformed mass production by standardizing and optimizing labor-intensive tasks. However, the automation of that period was predominantly mechanical and inflexible, necessitating considerable human supervision and re-configuration for any alterations in processes.

By the mid-to-late 20th century, the emergence of Computer Numerical Control machines and industrial robotics represented the next significant advancement in automation. These systems were capable of executing predefined tasks with exceptional accuracy and consistency, particularly in machining, welding, and assembly. Nonetheless, they remained rule-based and deterministic, demonstrating high efficiency in controlled settings but lacking adaptability. These technologies were unable to handle real-time variability or learn from data, rendering them inadequate for dynamic production environments.

The advent of Artificial Intelligence in the 21st century has transformed the landscape from deterministic automation to intelligent automation. Initially, the application of AI in manufacturing was restricted to expert systems and rule-based algorithms focused on quality control and scheduling. However, significant advancements in computational power, sensor technologies, and data storage have since enhanced AI capabilities, enabling real-time learning, decision-making, and autonomous control.

The 21st century marked a significant paradigm shift with the integration of AI into manufacturing processes. At first, AI's involvement was limited to expert systems and decision support tools that utilized hardcoded logic to assist in scheduling, fault diagnosis, and basic quality control (Crespo Márquez, 2007). Although these early implementations were beneficial, they were limited by their inability to adapt or react to unexpected situations.

A pivotal transformation took place with the emergence of machine learning (ML) and deep learning (DL). Unlike traditional rule-based systems, these methodologies can identify patterns within extensive datasets and enhance their performance over time. The rapid increase in computational power, driven by cloud computing and GPUs, along with advancements in sensor technologies and a dramatic rise in data storage capacity, has allowed AI to be utilized in real-time manufacturing settings (Lee, Bagheri, & Kao, 2015). Currently, computer vision algorithms surpass human capabilities in detecting product defects, and reinforcement learning methods are being employed to autonomously optimize robotic functions (Wuest et al., 2016).

The notion of Industry 4.0, which was first proposed in Germany around 2011–2012, characterizes this new industrial era where AI and the Internet of Things (IoT) serve as the foundation for manufacturing operations (Kagermann, Wahlster, & Helbig, 2013). In smart factories, AI technologies enable self-optimization, self-configuration, and self-diagnosis.

These capabilities signify a shift away from conventional automation, in which machines simply followed set instructions. Modern intelligent automation systems evaluate raw data streams produced by interconnected machines, sensors, and enterprise systems, converting them into actionable insights for immediate decision-making (Lu, 2017). A prime illustration is predictive maintenance, where AI algorithms analyze sensor data to accurately predict equipment failures, thereby minimizing downtime and prolonging asset lifespan (Jardine, Lin, & Banjevic, 2006). Another example is the digital twin concept, which creates virtual replicas of physical systems that utilize AI to simulate, forecast, and enhance performance across various scenarios (Tao et al., 2018).

The transition from mechanized assembly lines to AI-driven smart factories signifies more than mere technological advancement; it represents a fundamental rethinking of manufacturing processes. Intelligent automation, supported by AI and real-time data, facilitates unprecedented levels of flexibility, efficiency, and responsiveness. Nevertheless, this transformation also introduces challenges concerning workforce adaptation, system integration, and the ethical deployment of AI. As the manufacturing sector progresses deeper into the digital age, those who view AI as a strategic asset rather than merely a technical enhancement will be optimally positioned to spearhead the forthcoming industrial revolution.

AI TECHNOLOGIES TRANSFORMING MANUFACTURING

Machine Learning (ML)

Machine learning algorithms examine both historical and real-time data to identify patterns, generate predictions, and enhance decision-making. Machine Learning (ML), a crucial component of Artificial Intelligence (AI), significantly impacts contemporary manufacturing by allowing systems to learn from data, recognize patterns, and make decisions with minimal human involvement. In contrast to conventional rule-based automation, ML enables manufacturing processes to be adaptive and predictive, thereby elevating efficiency, quality, and innovation to unprecedented levels. A key application of ML in manufacturing is predictive maintenance. Sensors integrated into machinery consistently gather operational data such as temperature, vibration, and pressure. ML algorithms scrutinize this data to identify anomalies and forecast equipment failures before they happen. This forward-thinking strategy decreases unplanned downtime, prolongs machinery lifespan, and substantially reduces maintenance expenses (Lee, Bagheri, & Jin, 2016).

Conventional inspection techniques frequently depend on manual visual assessments that are labor-intensive and susceptible to human error. ML-driven computer vision systems can evaluate products on the production line in real time, identifying minute defects with greater accuracy and speed. These systems continuously enhance their performance as they are exposed to increasing amounts of data, ensuring consistent product quality and minimizing waste (Zhang et al., 2019). Furthermore, ML improves supply chain optimization by predicting demand patterns and managing inventory more efficiently. Sophisticated ML models analyze historical sales data, external influences such as market trends, and even social media sentiment to anticipate demand fluctuations. This ability assists manufacturers in reducing inventory costs while guaranteeing timely product availability (Choi et al., 2018).

Machine Learning (ML) enhances process optimization through adaptive control systems. By examining production data, ML models can pinpoint inefficiencies and recommend real-time adjustments to improve resource utilization, decrease energy consumption, and boost throughput (Lu, 2017). This advancement contributes to sustainable manufacturing practices and aligns with environmental objectives. The integration of ML in manufacturing also fosters mass customization, allowing factories to create personalized products without compromising efficiency. By evaluating consumer preferences alongside production constraints, ML algorithms can dynamically modify production parameters, facilitating flexible and scalable customization (Wuest et.al., 2016).

However, the implementation of ML presents challenges, including the necessity for extensive, high-quality datasets, compatibility with existing systems, and ensuring workforce preparedness. Organizations are required to invest in data infrastructure, cultivate specialized skills, and promote a culture of continuous learning to fully harness the capabilities of ML. Machine Learning serves as a fundamental element of AI's influence on contemporary manufacturing. Its capacity to extract insights from intricate data sets, forecast outcomes, and automate decision-making enables manufacturers to attain enhanced productivity, superior quality, and increased adaptability in a progressively competitive market.

Computer Vision

AI-powered computer vision systems examine products for defects, assess quality parameters, and oversee assembly processes.

These systems outperform human inspectors in both speed and precision, facilitating early defect detection and minimizing waste.

Computer Vision (CV), an essential sector of Artificial Intelligence (AI), is transforming contemporary manufacturing by allowing machines to interpret and analyze visual data with human-like accuracy and rapidity. By utilizing sophisticated algorithms, cameras, and sensors, computer vision systems improve various aspects of manufacturing processes, including quality control, safety, inventory management, and predictive maintenance.

Conventional quality control techniques depend on manual visual inspections, which are frequently labor-intensive, inconsistent, and prone to human error. Computer vision systems that incorporate deep learning models can evaluate products on production lines in real time, identifying defects such as cracks, scratches, misalignments, or contaminations with remarkable accuracy (Zhang et al., 2019). This leads to enhanced product quality, decreased waste, and increased throughput, as defects are identified early before they advance through the manufacturing cycle. Additionally, computer vision significantly enhances process monitoring and control. By continuously scrutinizing video feeds and images, these systems can oversee assembly processes, detect anomalies, and ensure compliance with production standards. In the realm of electronics manufacturing, CV algorithms confirm the correct placement of components and the quality of soldering, thereby reducing errors that could result in expensive recalls (Wang et al., 2020a).

AI-driven vision systems are capable of identifying unsafe behaviors among workers or hazardous conditions, such as being too close to dangerous machinery or neglecting to wear protective gear, and can initiate real-time alerts to avert accidents (Choi et al., 2021). This proactive strategy promotes a safer workplace and reduces downtime caused by workplace injuries. By merging computer vision with robotics and automated guided vehicles, manufacturing facilities can accurately monitor materials and finished products, enhance storage efficiency, and streamline the order fulfillment process (Wywicka & Murgalska, 2017). This improved visibility boosts supply chain responsiveness and lowers operational expenses.

Computer vision systems play a vital role in predictive maintenance by visually assessing the condition of equipment, including surface deterioration or leaks, and complementing sensor data to forecast potential failures. This multimodal strategy improves maintenance precision and helps prevent unforeseen breakdowns. Ongoing advancements in deep learning, edge computing, and high-resolution imaging are broadening the capabilities and accessibility of computer vision in the manufacturing sector. Nevertheless, challenges such as the necessity for extensive labeled datasets, integration complexities, and high computational requirements persist. Tackling these issues necessitates investments in infrastructure and skilled personnel who can develop and sustain computer vision solutions. Computer vision is a pivotal AI technology in contemporary manufacturing, revolutionizing visual inspection, safety monitoring, and operational efficiency. By automating and refining visual tasks, computer vision plays a significant role in achieving smart factories, which support higher quality standards, cost reductions, and safer work environments.

Natural Language Processing (NLP)

Natural Language Processing (NLP), a vital subfield of Artificial Intelligence (AI), empowers machines to comprehend, interpret, and respond to human language. In contemporary manufacturing, NLP is emerging as a formidable tool that connects complex industrial systems with human operators, thereby enhancing communication, automation, and decision-making processes throughout the production environment. One of the most practical applications of NLP in manufacturing is through voice-assisted human-machine interfaces. By utilizing NLP-driven virtual assistants, workers can control machinery, inquire about system status, and receive notifications using spoken or typed natural language commands. This diminishes reliance on conventional control panels and facilitates more intuitive, hands-free interactions—especially beneficial in high-speed, high-risk industrial environments (Lee et al., 2020).

NLP also supports real-time monitoring and problem resolution by extracting actionable insights from unstructured textual data, such as operator notes, maintenance logs, and quality reports. Through sentiment analysis and keyword extraction, manufacturers can identify patterns, such as recurring complaints or frequent equipment failures, which lead to quicker root cause analysis and ongoing improvement (Chatterjee et al., 2021).

Another significant application of NLP is in automated document processing and compliance management. Manufacturing organizations handle a substantial volume of technical documentation, including product specifications, safety protocols, and regulatory standards. NLP tools can automatically read, summarize, and cross-reference these documents to ensure adherence to regulations and reduce human oversight (Ghosh et al., 2022). This is particularly crucial in heavily regulated industries such as pharmaceuticals, aerospace, and food manufacturing.

In the realm of supply chain and customer relationship management, Natural Language Processing improves communication by examining supplier emails, customer feedback, and market data. It can prioritize support tickets, generate suggestions for responses, and translate technical terminology across different languages, thereby enhancing responsiveness and minimizing miscommunication in global operations.

NLP plays a significant role in predictive analytics by analyzing internal communication channels such as emails and chat logs to identify early indicators of supply chain disruptions, workforce issues, or quality concerns. When integrated with machine learning, NLP systems can anticipate emerging challenges and propose corrective measures before they escalate.

Despite its advantages, the application of NLP in manufacturing faces obstacles, including the use of domain-specific language, the presence of noisy or ambiguous data, and the challenge of integrating with legacy systems. Nevertheless, with advancements in large language models and industrial datasets, NLP applications are becoming more precise and customized to meet manufacturing requirements. Natural Language Processing is revolutionizing contemporary manufacturing by facilitating more natural, efficient, and intelligent interactions between humans and machines. As factories evolve to become smarter and more interconnected, NLP is set to assume an increasingly crucial role in automating knowledge work, improving safety, and optimizing complex processes in the digital manufacturing age.

Robotics and Automation

The fusion of Artificial Intelligence (AI) with robotics and automation is transforming the modern manufacturing landscape. Traditional automation, which is characterized by fixed, pre-programmed machines executing repetitive tasks, has progressed into intelligent automation, where robots not only perform tasks but also learn, adapt, and make decisions based on real-time data. This transformation is pivotal to the rise of smart factories and the wider Industry 4.0 revolution.

One of the most significant contributions of AI-powered robotics lies in flexible manufacturing. In contrast to traditional robotic systems that necessitate extensive reprogramming for new tasks, AI-driven robots can swiftly adapt to alterations in product design, batch size, or workflow by employing machine learning and computer vision. Cobots operate safely alongside human workers, learning from human actions and modifying their behavior accordingly. This facilitates a combination of precision and adaptability, which is essential for mass customization and agile production.

Intelligent robots are capable of detecting defects, adjusting tool paths, or rerouting operations in response to equipment malfunctions or material inconsistencies. These abilities minimize human intervention, decrease operational errors, and enhance overall production uptime (Wang et al., 2020b). In high-speed assembly lines, robotic arms guided by AI-powered vision systems achieve inspection and assembly rates that far exceed human capabilities, all while ensuring consistent quality. AI-enabled mobile robots traverse production floors, retrieve components, and manage inventory with real-time responsiveness. These robots utilize reinforcement learning to optimize their routes and enhance task efficiency over time, thereby improving supply chain agility and lowering labor costs.

AI-supported robotics is crucial in dangerous or high-precision environments, including chemical manufacturing, microelectronics, and cleanroom operations. By automating processes in these areas, manufacturers can mitigate health risks and enhance process consistency. The implementation of AI-based robotics faces challenges such as significant initial costs, complex system integration, and the need for workforce reskilling. Nevertheless, cloud robotics platforms and modular AI solutions are making these technologies more attainable for small and medium enterprises (SMEs), thus creating a more equitable landscape. Robotics and automation powered by AI are essential drivers of contemporary manufacturing efficiency, flexibility, and innovation. By merging mechanical accuracy with cognitive capabilities, these systems are ushering in a new era of industrial productivity and revolutionizing the production, transportation, and maintenance of goods.

KEY APPLICATIONS OF AI IN MANUFACTURING

Predictive maintenance employs AI algorithms to anticipate equipment failures before they occur, allowing for timely and cost-effective interventions. AI-driven predictive maintenance systems gather and analyze real-time data from sensors integrated into machinery, such as temperature, vibration, pressure, and acoustic signals. Machine learning models interpret this data to identify anomalies and estimate the remaining useful life (RUL) of components. This approach minimizes unexpected downtime, prolongs equipment lifespan, and lowers maintenance expenses. Predictive maintenance also enhances safety by averting catastrophic equipment failures and ensures product quality by keeping machine performance at optimal levels. As factories increasingly embrace digital technologies, predictive maintenance will remain a fundamental aspect of smart manufacturing, boosting operational reliability, efficiency, and competitiveness.

Quality control represents a vital application of Artificial Intelligence (AI) within the manufacturing sector, focused on ensuring that products adhere to established standards and fulfill customer expectations. Historically, quality inspection depended on manual evaluations or inflexible rule-based systems, which were frequently time-consuming, inconsistent, and susceptible to human error. AI revolutionizes this procedure by incorporating machine learning and computer vision technologies that can identify defects with enhanced speed, precision, and consistency. AI-driven quality control systems assess images or sensor data in real time to detect even minor defects such as surface cracks, misalignments, or absent components. These systems progressively improve by learning from data, thereby becoming increasingly dependable. Deloitte (2021) indicated that organizations utilizing AI-based visual inspection tools achieved up to 90% accuracy in defect detection and notable reductions in scrap and rework expenses. Beyond defect detection, AI assists manufacturers in monitoring process variations and optimizing quality at every stage of production, resulting in fewer recalls, greater customer satisfaction, and enhanced brand trust. As AI continues to advance, it emerges as an essential tool for realizing real-time, automated, and scalable quality assurance.

Supply chain optimization represents a crucial domain in which Artificial Intelligence (AI) facilitates substantial enhancements in manufacturing efficiency, agility, and cost reduction. Conventional supply chains frequently experience issues such as delayed forecasts, poor inventory management, and sluggish responses to market fluctuations. AI systems are capable of processing extensive amounts of both structured and unstructured data, including sales history, weather conditions, market conditions, and even social media interactions, to produce precise demand forecasts and optimize inventory levels. This capability enables manufacturers to minimize overstock situations, prevent stock shortages, and enhance just-in-time (JIT) production processes.

Smart factories are central to the Industry 4.0 revolution, a period characterized by the incorporation of digital technologies into manufacturing. In these advanced environments, artificial intelligence (AI) facilitates seamless interconnection and communication among machines, systems, and humans via the Internet of Things (IoT), cloud computing, and real-time analytics. AI is pivotal in promoting self-optimizing processes, autonomous decision-making, and predictive analytics. Smart factories employ AI algorithms to dynamically adjust production parameters based on sensor data, customer demand, or fluctuations in supply. This leads to enhanced efficiency, minimized waste, and highly adaptable production workflows. Real-time monitoring allows for immediate identification of process deviations or equipment malfunctions, coordinating corrective measures without the need for human intervention. This degree of automation and insight significantly enhances productivity, decreases downtime, and guarantees consistent product quality. Smart factories signify a transition from conventional manufacturing to digitally interconnected, intelligent production ecosystems, where AI serves as the operational brain, fostering agility, customization, and sustainable growth.

ECONOMIC AND OPERATIONAL IMPACTS OF AI IN MANUFACTURING

The integration of Artificial Intelligence (AI) in manufacturing has resulted in significant economic and operational transformations, altering how industries manage production, costs, and customer interactions. AI technologies, including machine learning, computer vision, and robotics, improve productivity, lower operational expenses, and facilitate customized manufacturing processes that swiftly adapt to market demands. This report examines these effects, emphasizing how AI enhances efficiency while promoting customization and agility in competitive markets.

AI-driven automation greatly enhances productivity in manufacturing by refining workflows, reducing human error, and facilitating continuous operations. Advanced systems are capable of monitoring production lines in real-time, making adjustments to maintain peak performance and identify anomalies promptly. AI analytics used in predictive maintenance can anticipate equipment failures before they happen, thereby avoiding expensive unplanned downtimes and prolonging the lifespan of machinery. Research shows that manufacturers utilizing AI-based predictive maintenance have achieved a 70% decrease in breakdowns and a 30% reduction in maintenance costs.

AI improves operational efficiency by automating repetitive tasks such as assembly, inspection, and packaging. Robots powered by AI can function with greater accuracy and speed than human labor, enhancing output quality and throughput. In addition to physical automation, AI supports intelligent scheduling and resource allocation, ensuring the optimal use of materials, labor, and energy. This minimizes waste and operational expenses, leading to significant financial savings.

The economic advantages of AI go beyond mere cost reductions. AI allows manufacturers to optimize supply chains through real-time data analysis and demand forecasting, which reduces inventory holding costs and mitigates the risk of overproduction. This just-in-time strategy decreases the capital tied up in raw materials and finished products, thereby improving cash flow and profitability.

In addition to cost savings and increased productivity, AI transforms manufacturing by enabling mass customization, moving away from traditional mass production towards personalized products at scale. AI systems evaluate extensive consumer data, including preferences, behaviors, and feedback, to dynamically modify production parameters and product designs. This ability enables manufacturers to provide customized solutions quickly without compromising efficiency.

By merging AI with adaptable manufacturing systems, organizations can swiftly react to fluctuating market demands and consumer preferences. For instance, AI-enhanced production lines can transition between different product variants with minimal downtime, facilitating agile manufacturing processes and shortening lead times. This agility allows businesses to sustain a competitive edge in markets where customer expectations are rapidly changing.

Additionally, AI-driven quality control systems that utilize computer vision can identify product defects with remarkable precision, ensuring that tailored products adhere to rigorous quality standards. Continuous learning algorithms enhance inspection capabilities over time, boosting reliability and customer satisfaction. The synergy of customization and consistent quality fortifies brand reputation and cultivates customer loyalty.

Operational enhancements brought about by AI result in significant economic benefits. Lower costs and heightened productivity improve profit margins, allowing manufacturers to reinvest in innovation and growth. Customization features create new revenue opportunities by addressing niche markets and enhancing product differentiation. Nevertheless, these advantages necessitate careful planning and investment in digital infrastructure, data management, and workforce development. Organizations must strike a balance between automation and human expertise to maximize AI integration. Successful manufacturers perceive AI not just as a technological upgrade but as a strategic facilitator of business transformation.

AI's function in contemporary manufacturing goes far beyond mere automation; it serves as a vital catalyst for economic efficiency and market responsiveness. The enhancements in productivity and reductions in costs facilitated by AI lead to more streamlined and agile operations. At the same time, AI enables manufacturers to provide tailored products that cater to the increasingly complex preferences of consumers. As sectors progressively adopt AI, those that strategically leverage its capabilities will secure lasting competitive advantages, thereby influencing the future direction of manufacturing.

WORKFORCE IMPLICATIONS OF AI IN MANUFACTURING

• Job Displacement vs. Job Evolution

The incorporation of AI into manufacturing has ignited significant discussions regarding its effects on the workforce. On one side, AI-driven automation is taking over repetitive, low-skilled tasks that were once carried out by humans. For example, robotic arms and AI-enhanced machines can execute assembly, packing, and quality assurance tasks more swiftly and accurately than human workers, which raises fears of job loss. Nevertheless, a more nuanced perspective indicates that AI is not solely about the complete elimination of jobs but rather about their transformation. New positions are emerging that necessitate technical expertise, including AI specialists, data scientists, robotics engineers, and automation technicians. A report from the

World Economic Forum (2023) suggests that while 85 million jobs may be lost to AI worldwide by 2027, 97 million new positions could be generated, highlighting a net positive transition towards job evolution.

• Upskilling and Reskilling

To effectively manage this transition, initiatives for upskilling and reskilling are essential. Organizations need to invest in training programs that provide employees with digital literacy, skills in AI management, and knowledge of advanced manufacturing. Nations such as Germany and Singapore have initiated national programs aimed at preparing the workforce for 'Industry 4.0,' highlighting the importance of collaboration among government, academia, and industry. Manufacturers implementing AI should foster a culture of continuous learning, ensuring that workers receive ongoing training that aligns with new technologies. This strategy not only safeguards jobs but also boosts employee engagement, productivity, and innovation capabilities.

CYBERSECURITY AND DATA PRIVACY CHALLENGES

The rise of AI and IoT devices in manufacturing brings increased cybersecurity threats. Smart factories rely on real-time data exchanges among machines, cloud platforms, and human operators, which broadens the attack surface for cybercriminals. Incidents of ransomware attacks, data breaches, and cyber threats to supply chains have escalated in recent years. A report from Cybersecurity Ventures in 2024 indicates that manufacturing ranks among the top five sectors targeted by cyberattacks. A compromised AI system could trigger cascading failures, disrupt production processes, or result in the theft of intellectual property. The ransomware attack on Colonial Pipeline in 2021 underscored how cyber threats to critical infrastructure can cripple entire supply chains.

Robust cybersecurity strategies are essential. Manufacturers are required to adopt multi-layered security protocols, which include network segmentation, AI-driven threat detection, and secure access controls. Regular penetration testing and employee training in cybersecurity are also vital components. Ethical governance of AI guarantees transparent decision-making processes and compliance with privacy regulations such as GDPR. AI models must be designed to be auditable and accountable to mitigate biases or unintended consequences that could jeopardize system integrity or human safety. As Artificial Intelligence becomes more integrated into contemporary manufacturing, the industry is confronted with increasing concerns regarding cybersecurity and data privacy. AI systems depend significantly on vast amounts of data gathered through interconnected devices, sensors, and cloud-based platforms. Although this connectivity fosters efficiency and intelligence, it simultaneously broadens the attack surface, rendering smart manufacturing systems more susceptible to cyber threats.

One of the main risks pertains to data breaches and unauthorized access to sensitive production data, customer details, and machine learning models. A compromised AI system could lead to the theft of intellectual property, sabotage of production processes, or erroneous decision-making. Attackers might take advantage of adversarial machine learning by supplying manipulated inputs to AI models, thereby altering results, such as incorrectly identifying product defects or interfering with predictive maintenance algorithms. This could result in production delays, quality issues, or safety risks. Adversarial attacks in industrial environments present distinct challenges due to the intricacies and potential physical repercussions of AI failures. Data privacy also becomes a significant issue, particularly as manufacturers gather personal information from employees, partners, and consumers via AI-driven systems. Adhering to regulations such as GDPR, CCPA, and industry-specific guidelines complicate matters further. Manufacturers are required to guarantee data anonymization, secure storage, and responsible data governance to fulfill these legal and ethical responsibilities (Zhou et al., 2021).

Numerous manufacturers continue to utilize outdated systems that are not equipped for AI integration or resilient against cybersecurity threats. These deficiencies can result in inadequately secured endpoints and insufficient encryption, rendering the entire network vulnerable to ransomware and various other attacks. To fortify these systems, it is essential to implement coordinated efforts in the convergence of IT and OT (Information Technology–Operational Technology), establish robust cybersecurity protocols, and maintain continuous monitoring. Incorporating security protocols during the design phase of systems, rather than treating them as an afterthought which aligns with the principles of secure AI development. Organizations are progressively embracing zero-trust architectures, AI-specific firewalls, and blockchain technology to bolster security and ensure data integrity within industrial settings. While AI presents significant advantages in contemporary manufacturing, a strong emphasis on cybersecurity and data privacy must support it. The creation of secure AI ecosystems necessitates not only technical safeguards but also governance frameworks, employee training, and interdepartmental collaboration. As AI becomes increasingly integral to production processes, safeguarding it is vital for operational resilience, regulatory compliance, and the long-term success of businesses.

ENVIRONMENTAL AND SUSTAINABILITY CONSIDERATIONS

Artificial Intelligence (AI) is increasingly acknowledged as a crucial facilitator of energy efficiency and waste minimization in contemporary manufacturing, aligning industrial practices with sustainability objectives. By utilizing real-time data, predictive analytics, and advanced control systems, AI enables manufacturers to reduce resource usage, enhance production processes, and lessen environmental effects. AI enhances energy efficiency through sophisticated energy management systems. These systems gather data from sensors integrated into machinery, utilities, and production lines to evaluate energy consumption trends. AI algorithms subsequently pinpoint inefficiencies—such as machinery operating during non-productive times or energy-heavy processes occurring at peak hours—and suggest or automate necessary modifications. (Wuest et al. 2016) indicate that AI-driven optimization can decrease energy usage in manufacturing plants by as much as 20%, particularly when integrated with IoT and machine learning frameworks.

AI also promotes waste reduction via predictive quality control and process enhancement. Conventional manufacturing frequently produces considerable waste due to overproduction, defects, or suboptimal resource utilization. AI-enabled systems, including computer vision and machine learning, identify quality problems early in the production cycle, preventing defective items from progressing further along the production line. This anticipatory defect identification leads to reduced material scrap rates and enhanced yield (Zhang et al., 2019).

In additive manufacturing and closed-loop production systems, artificial intelligence facilitates real-time modifications to material feed rates, temperature, and speed, ensuring that only the necessary amount of energy and material is utilized. These advanced systems also aid in the adoption of circular economy models, where waste materials are identified, recovered, and reused more effectively. AI plays a crucial role in scheduling and resource planning, guaranteeing that machines function under optimal conditions. For example, dynamic production scheduling powered by AI assists in balancing workloads, minimizing idle times, and preventing unnecessary energy surges. Additionally, The combination of AI with life cycle assessment (LCA) tools enables manufacturers to evaluate the environmental impact of products and processes with greater precision. This fosters decision-making that emphasizes sustainability alongside productivity and cost-effectiveness.

It encompasses the necessity for high-quality data, integration with legacy systems, and the energy footprint of AI computing infrastructure itself. Nevertheless, advancements in edge computing, green AI practices, and energy-efficient algorithms are contributing to the resolution of these issues (Strubell et al., 2020). AI acts as a formidable instrument in promoting energy efficiency and waste reduction in manufacturing. Its capacity to optimize processes, anticipate inefficiencies, and propel sustainability-focused innovation is assisting industries in transitioning towards greener, more responsible production systems, which is essential for achieving environmental objectives and enhancing long-term operational resilience.

AI FOR ENERGY EFFICIENCY AND WASTE REDUCTION

Artificial Intelligence (AI) is increasingly becoming essential in enhancing energy efficiency and minimizing waste in contemporary manufacturing. As manufacturers encounter mounting pressure to achieve environmental sustainability objectives, lower carbon emissions, and operate more cost-effectively, AI technologies provide robust solutions to optimize resource utilization, reduce waste, and establish more streamlined production systems. Central to AI-driven energy efficiency is the application of real-time data analytics and intelligent monitoring. Sensors integrated into manufacturing machinery gather data regarding energy usage, temperature, pressure, and load levels. AI algorithms analyze this information to identify inefficiencies and suggest corrective measures. According to Wuest et al. (2016), AI-enabled optimization can reduce energy consumption in industrial systems by as much as 20%, directly contributing to both environmental and economic benefits.

AI predictive energy management enables facilities to anticipate energy requirements based on production schedules, weather patterns, and historical data. This results in more precise planning and load balancing, which diminishes peak energy demand and prevents expensive overuse. For instance, the integration of AI with HVAC and lighting systems can dynamically modify energy consumption based on occupancy or production needs, enhancing operational efficiency without compromising performance (Lu, 2017). In terms of waste reduction, AI is particularly adept at pinpointing sources of process inefficiencies and quality concerns. Machine learning models can scrutinize production data to identify early indicators of defects, equipment misalignment, or process irregularities, facilitating prompt intervention before waste is produced. AI-enhanced computer vision systems are extensively utilized for quality inspection, detecting defects in real time and minimizing the necessity for post-production rework (Zhang et al., 2019).

AI enhances the efficient use of materials by examining consumption trends and production results to optimize input resources. In the realm of additive manufacturing, AI assists in refining deposition techniques to reduce material excess and enhance the precision of parts. Furthermore, by promoting closed-loop production and circular economy frameworks, AI allows manufacturers to more effectively reuse by-products or recycle scrap, thereby further decreasing material waste. Accurate AI predictions rely heavily on high-quality data, yet traditional manufacturing systems may not possess the necessary infrastructure for smooth data collection and integration. Additionally, the energy consumption associated with AI, particularly in deep learning and extensive computations, requires careful management. Strubell et al. (2020) highlight the significance of adopting "green AI" practices, advocating for the creation of energy-efficient AI models to guarantee net positive environmental impacts. AI presents a groundbreaking opportunity for manufacturers to improve energy efficiency and reduce waste throughout the production lifecycle. When implemented thoughtfully, AI not only fosters sustainability but also enhances profitability, ensures regulatory compliance, and bolsters brand reputation, thereby establishing smart manufacturing as a fundamental component of a low-carbon industrial future.

REAL-TIME MONITORING AND SMART GRIDS

The incorporation of Artificial Intelligence (AI) into real-time monitoring systems and smart grids is profoundly altering contemporary manufacturing by improving energy efficiency, operational transparency, and production flexibility. These innovations constitute the foundation of Industry 4.0, allowing manufacturers to establish highly adaptable and sustainable production settings. Real-time monitoring, driven by AI and the Industrial Internet of Things (IIoT), enables manufacturers to consistently monitor machine health, production conditions, environmental factors, and energy usage throughout their facilities. AI analyzes this continuous data flow to detect irregularities, forecast equipment malfunctions, and enhance production processes. As noted by Wang et al. (2020c), such systems have the potential to decrease unplanned downtime by as much as 40% and enhance overall equipment effectiveness (OEE) by facilitating predictive maintenance and prompt action during process anomalies.

Real-time AI analytics offer manufacturers detailed insights into energy consumption patterns, enabling them to eliminate inefficiencies, balance loads, and ensure optimal machine performance. In conjunction with real-time monitoring, the integration of smart grids—intelligent energy distribution systems that utilize AI to manage electricity flow among producers, consumers, and storage systems—is essential. In a manufacturing environment, smart grids empower plants to dynamically modify energy usage based on grid conditions, time-of-use pricing, and the availability of renewable energy sources. AI algorithms predict energy demand, optimize grid load balancing, and coordinate the utilization of on-site renewable energy sources such as solar or wind. Smart grids also support demand response programs, allowing manufacturers to decrease or shift their energy consumption during peak hours in exchange for financial incentives. This approach not only reduces operational costs but also contributes to overall grid stability and carbon reduction objectives (Fang et al., 2012).

The integration of AI with smart grids facilitates microgrid management, enabling factories to function independently from the main grid during outages, thereby enhancing resilience. However, the implementation of real-time monitoring and smart grid technologies presents challenges, including the necessity for significant infrastructure investment, cybersecurity risks associated with interconnected systems, and the complexities involved in integrating legacy machinery. Nevertheless, advancements in edge computing, cloud AI platforms, and secure communication protocols are rendering these systems increasingly accessible, even for small and medium-sized enterprises (SMEs). AI-driven real-time monitoring and smart grids are transforming the manufacturing industry, providing greater control over operations, lowering energy expenses, and fostering environmental sustainability. As manufacturers continue their journey towards digital transformation, these technologies will be crucial in developing agile, efficient, and future-ready production systems.

STRATEGIC IMPLEMENTATION OF AI IN MANUFACTURING

Manufacturers are progressively incorporating AI with robotics and the Industrial Internet of Things (IIoT) to improve operational efficiency. Autonomous mobile robots (AMRs) are transforming logistics within factories by autonomously transporting materials and products, thereby minimizing the necessity for fixed infrastructure and facilitating dynamic production layouts. This adaptability is especially advantageous in responding to varying demand and labor shortages. The strategic application of Artificial Intelligence (AI) in manufacturing focuses on converting traditional processes into AI-driven automation and intelligent systems. AI empowers machines to learn from data, adjust to changes, and make real-time decisions, thereby supporting dynamic production environments.

AI-driven automation enables manufacturers to transcend mere task execution and move towards the intelligent orchestration of entire workflows. Systems that utilize machine learning, computer vision, and predictive analytics can autonomously enhance production schedules, identify and rectify quality issues, and react to equipment failures. AI-enabled robotic systems can modify assembly tasks in real-time based on sensor feedback, significantly boosting efficiency and minimizing downtime (Caggiano et al., 2022). Smart systems also merge AI with IoT and edge computing to create self-regulating ecosystems where data circulates among connected devices, facilitating real-time process optimization. These systems promote mass customization, predictive maintenance, and just-in-time manufacturing, aligning operational performance with changing customer demands. The strategic implementation of AI necessitates more than just the adoption of technology; it requires organizational alignment, robust data infrastructure, and workforce upskilling. As noted by Lu (2017), successful adoption hinges on a long-term vision, cross-functional collaboration, and mechanisms for continuous improvement.

Cloud-Based Design and Manufacturing (CBDM) signifies a pivotal transition in the application of AI within the manufacturing sector, facilitating collaborative, data-driven innovation via the cloud. CBDM utilizes AI-enhanced tools and platforms that are hosted on cloud infrastructure to optimize product design, simulation, and production workflows across teams and systems that are geographically dispersed. The integration of AI in CBDM improves product development by automating design iterations, refining geometry in accordance with performance objectives, and forecasting manufacturing viability. Machine learning algorithms are capable of examining user inputs, historical data, and performance metrics to produce intelligent design suggestions and minimize development timelines. As stated by Wu et al. (2013), CBDM promotes accelerated innovation, reduced capital expenditure, and enhanced agility by leveraging shared resources and scalable computing capabilities. CBDM facilitates real-time collaboration, allowing designers, engineers, and suppliers to simultaneously access and modify digital twins or CAD models. AI also aids in on-demand production planning, where intelligent scheduling algorithms adjust to fluctuations in supply chain dynamics or customer requests. Notwithstanding its advantages, the strategic implementation of CBDM necessitates secure data sharing, a robust network infrastructure, and alignment with pre-existing digital ecosystems. As highlighted by Lu (2017), the effective integration of AI and cloud functionalities requires both technical preparedness and a progressive digital strategy.

WORKFORCE EVOLUTION IN AI-ENHANCED MANUFACTURING

The emergence of AI-enhanced manufacturing is significantly transforming the industrial workforce, not merely by removing traditional jobs, but by fostering the creation of new, high-skill roles. As automation assumes control of repetitive and manual tasks, there has been a notable increase in the demand for positions centered around AI development, management, and integration. New roles such as data scientists, machine learning engineers, robotics technicians, and AI ethicists have become essential components of contemporary factory environments. According to the World Economic Forum (2020), while automation is projected to displace 85 million jobs by 2025, it is also anticipated to generate 97 million new positions across various sectors, many of which are based in data analytics, programming, and human-machine collaboration. In the manufacturing sector, this transition has given rise to smart factory specialists and digital twin engineers, who oversee AI-integrated production systems and virtual simulations for real-time optimization.

Research conducted by Frank et al. (2019) emphasizes that the successful evolution of the workforce relies not only on technological adaptation but also on comprehensive upskilling and reskilling programs. As AI becomes increasingly central to operations, employers are required to invest in training initiatives that prepare workers with hybrid skills, merging mechanical knowledge with digital proficiency.

The incorporation of Artificial Intelligence (AI) into the manufacturing sector has generated an urgent requirement for upskilling and reskilling programs aimed at equipping the workforce for increasingly intricate, data-centric environments. As conventional roles are enhanced or supplanted by AI-driven systems, employees must develop new skills in digital technologies, data analytics, robotics, and human-machine interaction to maintain their relevance and competitiveness. Upskilling programs concentrate on improving the capabilities of current workers, such as training a machine operator to analyze AI-generated performance data or utilize predictive maintenance tools. In contrast, reskilling entails shifting workers into completely new positions, such as robotics technicians, AI system monitors, or industrial data analysts. These initiatives are crucial for alleviating job displacement and promoting inclusive growth within the industry. According to the McKinsey Global Institute (2020), nearly 50% of manufacturing workers will require reskilling by 2025 to meet the challenges posed by Industry 4.0. Additionally, Frank et al. (2019) highlight that organizations that invest in structured training and lifelong learning initiatives are more likely to successfully adopt AI technologies. Partnerships among industry, government, and educational institutions, along with apprenticeships, online certifications, and AI bootcamps, have demonstrated effectiveness in expediting workforce transformation. Nevertheless, the success of these efforts hinges on aligning training materials with practical industrial applications and nurturing a culture of ongoing learning.

SUSTAINABILITY AND AI IN MANUFACTURING

The integration of Artificial Intelligence (AI) in the manufacturing sector is hastening the shift towards circular economy practices, which emphasize the reduction, reuse, and recycling of materials to lessen waste and environmental impact. Instead of adhering to a conventional linear "take-make-dispose" approach, manufacturers are utilizing AI to establish closed-loop systems that promote sustainability and resource efficiency. AI facilitates real-time monitoring and data analysis that assist in identifying excessive material consumption, forecasting product longevity, and enhancing end-of-life recovery methods. Machine learning algorithms can predict when components are likely to fail and recommend refurbishing or recycling instead of replacement. This approach supports the extension of product life and material recovery, which are two fundamental tenets of the circular economy (Geissdoerfer et al., 2017). AI-driven design tools promote eco-design, enabling engineers to develop products that are simpler to disassemble, recycle, or remanufacture. Predictive analytics can aid in demand forecasting and inventory management, thereby minimizing overproduction and the waste that accompanies it. AI not only fosters sustainable manufacturing but also paves the way for innovative business models such as product-as-a-service and remanufacturing networks (Bressanelli et al. 2020)

Artificial Intelligence (AI) serves as a transformative element that enhances energy efficiency in contemporary manufacturing, thereby directly contributing to global sustainability objectives. By merging AI with intelligent sensors and control systems, manufacturers are able to monitor, analyze, and optimize energy consumption in real time, which leads to a significant reduction in operational waste and carbon emissions. AI algorithms are capable of identifying patterns in energy usage, forecasting peak demand, and automatically adjusting equipment settings for optimal performance. This facilitates dynamic energy management, allowing processes to be refined to minimize idle time, prevent energy-intensive operation cycles, and ensure that machines operate only when necessary. As indicated by Wuest et al. (2016), such AI-driven strategies can lead to a reduction in energy consumption of up to 20% in industrial environments. In smart factories, AI collaborates with the Industrial Internet of Things (IIoT) to facilitate real-time load balancing and the incorporation of renewable energy sources, such as solar or wind power. These functionalities assist manufacturers in transitioning towards cleaner energy profiles while preserving production efficiency. Furthermore, Edge AI and cloud-based platforms offer comprehensive analytics dashboards, enabling operators to monitor energy KPIs and make well-informed sustainability choices. As Lu (2017) points out, realizing significant energy savings through AI necessitates not only the appropriate technological infrastructure but also strategic planning and skilled personnel capable of interpreting AI outputs and aligning them with environmental goals.

EMERGING TECHNOLOGIES SHAPING THE FUTURE

Quantum Machine Learning (QML) is swiftly becoming a significant force at the convergence of AI and manufacturing, allowing manufacturers to transcend the limitations of traditional machine learning in scenarios involving high-dimensional, small sample, and complex data. Techniques in QML, such as quantum neural networks, clustering, regression, and support-vector methods, are increasingly being utilized in smart manufacturing applications, particularly in areas like quality control, predictive maintenance, process optimization, and materials innovation. Furthermore, it highlights the leadership of nations such as the U.S. and China in the progression of these technologies (Taylor & Francis Online). In the realm of defect detection, hybrid classical–quantum architectures have shown a remarkable capacity to identify complex manufacturing anomalies that classical AI alone cannot detect, providing tangible advantages in yield and cost reduction.

Beyond the semiconductor industry, the applicability of QML extends to manufacturing tasks that require significant optimization. There is a strong global interest: a SAS survey conducted in May 2025 revealed that more than 60% of business leaders across various sectors are actively investigating quantum AI, with manufacturing identified as a crucial area—although prevalent challenges include high costs, a lack of expertise, and ambiguous use cases (thequantuminsider.com). Concurrently, industry analyses forecast that hybrid quantum-classical pilots in logistics and production scheduling could provide a competitive edge within 3 to 5 years as hardware advances beyond current constraints.

At the software and systems level, contemporary frameworks are paving the way for simulation, optimization, and design activities within the manufacturing sector. A multi-layered ecosystem where QML enhances engineering simulations across materials design, production processes, and service-oriented manufacturing analytics (arXiv). In addition to these structural insights, the forecast for supervised QML anticipates a decade of development (2025–2035), illustrating how variational circuits, quantum neural networks, and kernel methods may evolve from specialized enterprise research into widely used industrial tools, while also acknowledging existing challenges such as noise, barren plateaus, and the absence of formal proofs of quantum advantage (arXiv).

Despite the evident potential, expert reviewers advise caution against overstating QML's capabilities. Commentary from practitioners on Reddit indicates that numerous QML proposals do not demonstrate provable speedup and are effective only when the data is quantum-native, suggesting that broader applications in manufacturing remain speculative for the time being. Nevertheless, practical hybrid implementations—such as defect detection systems, material simulations, and chip resistance modeling—offer initial evidence of the benefits QML can provide when integrated thoughtfully into production settings.

QML is establishing a crucial role in AI-driven modern manufacturing by facilitating enhanced pattern recognition, more intelligent optimization, and small data learning that traditional methods find challenging. Although adoption is still in its infancy, successful pilot projects and increasing research funding imply that hybrid QML platforms will develop into essential tools for the next generation of smart factories, quality control systems, and advanced materials engineering.

The development of AI-driven chips is transforming contemporary manufacturing by facilitating the creation of smarter, faster, and more energy-efficient chips that support AI workloads across various sectors—from intelligent factories and robotics to self-driving vehicles. Central to this evolution are sophisticated AI-powered EDA tools like Synopsys.ai and Cadence's AI-enhanced workflow suite, which incorporate reinforcement learning and generative AI throughout the design, verification, testing, and manufacturing processes to significantly enhance power performance area (PPA) metrics and throughput. Synopsys's DSO.ai has successfully enabled over 100 commercial tape-outs by autonomously exploring design space to shorten development cycles and boost efficiency, yielding tangible benefits such as up to a 25% reduction in power consumption, a smaller die size, and threefold productivity increases for major clients including STMicroelectronics, Samsung, and SK Hynix (Infosys, Forbes). Its auxiliary tools, VSO.ai and TSO.ai, expedite verification and test planning—where TSO.ai alone decreases test patterns by 20 to 30%, thereby reducing both test runtime and silicon validation expenses (Forbes). In addition, the Synopsys.ai Copilot provides generative AI workflows for real-time RTL and test bench creation, decision support throughout design phases, and cross-domain analytics that connect design choices to manufacturing results, thereby facilitating data-driven root-cause analysis and enhancements in yield (Synopsys).

Independent research and industry evaluations support these advancements: Deloitte reports that AI-enhanced design produces outcomes nearly on par with those of expert engineers, achieving results in hours instead of weeks, alongside significant enhancements in energy efficiency—MIT recorded a 2.3× improvement, Cadence increased the performance of a 5 nm mobile chip by 14% while decreasing power consumption by 3%, and NVIDIA applied reinforcement learning to reduce circuit sizes by 25% while maintaining comparable overall performance (deloitte.com). Infosys highlights that generative AI speeds up intellectual property reuse, automates floor planning, and optimizes error log analysis—allowing engineers to focus on innovation with fewer manual interruptions and shortening time-to-market and verification cycles.

In the field of manufacturing, the development of AI-enabled chips works in harmony with AI-powered production lines, allowing for swift adjustments to new process nodes and a more cohesive integration of design and fabrication workflows. Chips that are optimized for AI tasks (such as AI accelerators, custom NPUs, and photonic chips) assist manufacturing systems in processing sensor data more rapidly, enhancing predictive maintenance, and facilitating dynamic scheduling, thereby providing smart factories with both advanced hardware and cognitive capabilities at the same time. Companies such as AMD and OpenAI are even exploring the creation of custom AI chips (like NPUs and accelerators) to satisfy internal requirements, thereby emphasizing the strategic importance of AI-driven design within manufacturing ecosystems (pcgamer.com, businessinsider.com).

AI-driven chip development is not only advancing next-generation AI hardware but is also transforming contemporary manufacturing by producing optimized chips specifically designed for industrial intelligence. Through practical tape-outs, quantifiable PPA and yield improvements, and the development of AI-native frameworks, this technology is paving the way for a future where chip design and manufacturing evolve in closer alignment, propelling smarter factories, accelerating innovation, and enabling more efficient large-scale deployment of AI.

CONCLUSION

In summary, artificial intelligence (AI) has emerged as a transformative element in contemporary manufacturing, building upon decades of advancements in automation to usher in a new age of intelligent, responsive, and sustainable production systems. Historically grounded in rule-based frameworks and early robotics, the integration of AI has progressed remarkably with the advent of machine learning, computer vision, and generative models, facilitating unprecedented levels of automation, precision, and adaptability.

AI technologies—such as predictive analytics, reinforcement learning, digital twins, and generative design—are revolutionizing factory operations, ranging from real-time monitoring to self-optimizing production lines and AI-driven chip development. These innovations have resulted in significant applications in predictive maintenance, quality control, supply chain optimization, energy efficiency, and waste reduction, while also fostering the development of smart grids and interconnected infrastructure.

From an economic perspective, the adoption of AI provides measurable benefits—reducing costs, minimizing downtime, and improving scalability—yet it necessitates careful integration to navigate capital investment and data complexity. Operationally, AI enhances agility and resilience within global supply chains, particularly in volatile market conditions. Nevertheless, the implications for the workforce are crucial: while AI augments roles through human-machine collaboration, it also necessitates reskilling and changes in labor dynamics. At the same time, issues surrounding cybersecurity and data privacy must be tackled, especially as interconnected systems become increasingly susceptible to attacks and data exploitation. AI's contribution to sustainability is also significant, providing intelligent systems that lower energy consumption, optimize material use, and reduce carbon emissions.

From real-time factory oversight to AI-augmented grid management, these technologies contribute to worldwide sustainability objectives. Strategic execution is crucial—effective adoption relies not only on technical implementation but also on visionary leadership, interdisciplinary cooperation, and a flexible workforce. As AI progresses, new technologies such as quantum machine learning, neuromorphic computing, and edge AI will define the forthcoming era of manufacturing innovation. In conclusion, AI in contemporary manufacturing transcends being merely a collection of tools—it represents a fundamental transformation, facilitating smarter factories, more sustainable processes, and a stronger industrial future.

REFERENCES

Bressanelli, G., Adrodegari, F., Perona, M., & Saccani, N. (2020). Exploring how usage-focused business models enable circular economy through digital technologies. *Sustainability*, *12*(6), 1–20. DOI: 10.3390/su12062365

Caggiano, A., Grant, R., Peng, C., Li, Z., & Simeone, A. (2022). Manufacturing Process Impacts on Occupational Health: A Machine Learning Framework. *Procedia CIRP*, *112*, 561–566. Advance online publication. DOI: 10.1016/j.procir.2022.09.100

Chatterjee, S., Rana, N. P., Dwivedi, Y. K., & Baabdullah, A. M. (2021). Understanding AI adoption in manufacturing and production firms using an integrated TAM-TOE model. *Technological Forecasting and Social Change*, *170*, 120880. Advance online publication. DOI: 10.1016/j.techfore.2021.120880

Choi, E., Johnson, D. A., Moon, K., & Oah, S. (2018). Effects of Positive and Negative Feedback Sequence on Work Performance and Emotional Responses. *Journal of Organizational Behavior Management*, *38*(2–3), 97–115. Advance online publication. DOI: 10.1080/01608061.2017.1423151

Choi, E., Kim, C., & Lee, K. C. (2021). Consumer Decision-Making Creativity and Its Relation to Exploitation–Exploration Activities: Eye-Tracking Approach. *Frontiers in Psychology*, *11*, 557292. Advance online publication. DOI: 10.3389/fpsyg.2020.557292 PMID: 33505328

Crespo Márquez, A. (2007). *The maintenance management framework: Models and methods for complex systems maintenance*. Springer., DOI: 10.1007/978-1-84628-821-0

Fang, X., Brown, D. S., Florence, C. S., & Mercy, J. A. (2012). The economic burden of child maltreatment in the United States and implications for prevention. *Child Abuse & Neglect*, *36*(2), 156–165. Advance online publication. DOI: 10.1016/j.chiabu.2011.10.006 PMID: 22300910

Frank, A. G., Dalenogare, L. S., & Ayala, N. F. (2019). Industry 4.0 technologies: Implementation patterns in manufacturing companies. *International Journal of Production Economics*, *210*, 15–26. Advance online publication. DOI: 10.1016/j.ijpe.2019.01.004

Geissdoerfer, M., Savaget, P., Bocken, N. M. P., & Hultink, E. J. (2017). The Circular Economy – A new sustainability paradigm? In Journal of Cleaner Production (Vol. 143). DOI: 10.1016/j.jclepro.2016.12.048

Ghosh, S., Hughes, M., Hodgkinson, I., & Hughes, P. (2022). Digital transformation of industrial businesses: A dynamic capability approach. *Technovation, 113*, 102414. Advance online publication. DOI: 10.1016/j.technovation.2021.102414

Kagermann, H., Wahlster, W., & Helbig, J. (Eds.). (2013). Securing the future of German manufacturing industry: Recommendations for implementing the strategic initiative INDUSTRIE 4.0. Final report of the Industrie 4.0 Working Group. acatech – National Academy of Science and Engineering. https://www.acatech.de/wp-content/uploads/2018/03/Final_report__Industrie_4.0_accessible.pdf

Lee, J., Bagheri, B., & Jin, C. (2016). Introduction to cyber manufacturing. *Manufacturing Letters*, *8*, 11–15. Advance online publication. DOI: 10.1016/j.mfglet.2016.05.002

Lee, J., Bagheri, B., & Kao, H. A. (2015). A Cyber-Physical Systems architecture for Industry 4.0-based manufacturing systems. *Manufacturing Letters*, *3*, 18–23. Advance online publication. DOI: 10.1016/j.mfglet.2014.12.001

Lin, D., Banjevic, D., & Jardine, A. K. S. (2006). Using principal components in a proportional hazards model with applications in condition-based maintenance. *The Journal of the Operational Research Society*, *57*(8), 910–919. Advance online publication. DOI: 10.1057/palgrave.jors.2602058

Lu, Y. (2017). Industry 4.0: A survey on technologies, applications and open research issues. In Journal of Industrial Information Integration (Vol. 6). DOI: 10.1016/j.jii.2017.04.005

Strubell, E., Ganesh, A., & McCallum, A. (2020). Energy and policy considerations for modern deep learning research. AAAI 2020 - 34th AAAI Conference on Artificial Intelligence. DOI: 10.1609/aaai.v34i09.7123

Tao, F., Cheng, J., Qi, Q., Zhang, M., Zhang, H., & Sui, F. (2018). Digital twin-driven product design, manufacturing and service with big data. *International Journal of Advanced Manufacturing Technology*, *94*(9-12), 3563–3576. DOI: 10.1007/s00170-017-0233-1

Wang, C., Pan, R., Wan, X., Tan, Y., Xu, L., McIntyre, R. S., Choo, F. N., Tran, B., Ho, R., Sharma, V. K., & Ho, C. (2020a). A longitudinal study on the mental health of general population during the COVID-19 epidemic in China. *Brain, Behavior, and Immunity*, *87*, 40–48. Advance online publication. DOI: 10.1016/j.bbi.2020.04.028 PMID: 32298802

Wang, L., Wang, X. V., & Bi, Z. M. (2020b). Smart manufacturing and real-time monitoring: AI-enabled adaptive production control. *Journal of Manufacturing Systems*, *57*, 457–467. DOI: 10.1016/j.jmsy.2020.10.006

Wang, S., Wan, J., Zhang, D., Li, D., & Zhang, C. (2020c). Towards smart factory for Industry 4.0: A self-organized multi-agent system with big data-based feedback and coordination. *Computer Networks*, *101*, 158–168. DOI: 10.1016/j.comnet.2015.12.017

Wu, D., Rosen, D. W., Wang, L., & Schaefer, D. (2013). Cloud-based design and manufacturing: A new paradigm in digital manufacturing and design innovation. *Computer Aided Design*, *45*(2), 104–118. DOI: 10.1016/j.cad.2012.08.006

Wuest, T., Weimer, D., Irgens, C., & Thoben, K. D. (2016). Machine learning in manufacturing: Advantages, challenges, and applications. *Production & Manufacturing Research*, *4*(1), 23–45. Advance online publication. DOI: 10.1080/21693277.2016.1192517

Zhang, Y., Xu, C., Yu, S., & Xie, S. (2019). Machine vision-based inspection for industrial quality control: A review. *Journal of Manufacturing Systems*, *51*, 1–11. DOI: 10.1016/j.jmsy.2019.08.001

Zhou, T., Dallasanta, K., Nazarenko, L., Schmidt, G. A., & Jin, Z. (2021). The impact of increasing stratospheric radiative damping on the quasi-biennial oscillation period. *Atmospheric Chemistry and Physics*, *21*(9), 7395–7407. Advance online publication. DOI: 10.5194/acp-21-7395-2021

Section 3
Applied Case Studies in Modern Manufacturing

Chapter 6

AI and Automation for Net-Zero Factories and Sustainable Manufacturing

Shalom Akhai
http://orcid.org/0000-0002-7533-457X
MMEC, Maharishi Markandeshwar University, Mullana, India

Mahapara Abbass
http://orcid.org/0000-0002-0838-2598
MMEC, Maharishi Markandeshwar University, Mullana, India

ABSTRACT

This chapter examines the transformative role of artificial intelligence and automation in accelerating the transition toward net-zero factories and sustainable manufacturing ecosystems. It explores how intelligent systems, including digital twins, machine learning, predictive analytics, and cyber-physical production networks, enable real-time optimization of energy use, waste reduction, and carbon-efficient operations. The chapter further analyzes the integration of smart robotics, sensor-driven monitoring, and green supply chain automation to support low-emission, resource-efficient production cycles. Key challenges, emerging technologies, and future research opportunities are presented to highlight the strategic pathways required for achieving climate-aligned, resilient, and sustainable industrial environments.

DOI: 10.4018/979-8-3373-5756-0.ch006

CHAPTER HIGHLIGHTS

- *Establishes the conceptual foundations of net-zero manufacturing,* emphasizing circular production, life cycle thinking, and digital-driven carbon accountability.
- *Explains how AI technologies - machine learning, deep learning, NLP, and reinforcement learning - enhance energy efficiency, process stability, and predictive decision-making* across modern factories.
- *Demonstrates the role of automation systems*, including robotics, PLC/SCADA architectures, collaborative robots, and additive manufacturing, in minimizing waste and improving resource productivity.
- *Highlights the transformative power of digital twins, IIoT sensors, and data analytics*, enabling real-time monitoring, predictive maintenance, simulation-based optimization, and carbon-footprint reduction.
- *Presents emerging trends, global case studies, and an actionable roadmap* showing how industries can transition toward intelligent, climate-aligned, and resilient net-zero manufacturing ecosystems.

GRAPHICAL ABSTRACT

Figure 1. Graphical Abstract.

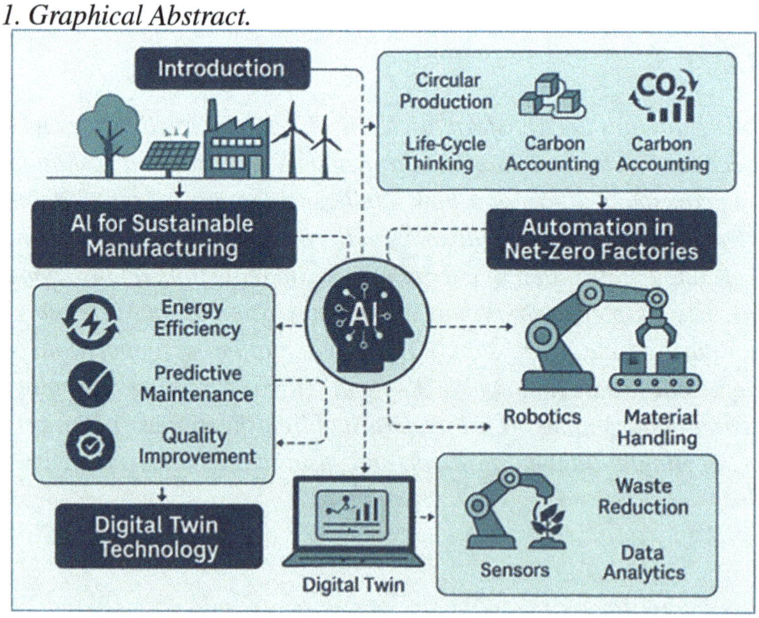

ABBREVIATIONS

AI: Artificial Intelligence
AM: Additive Manufacturing
AMR: Autonomous Mobile Robot
AGV: Automated Guided Vehicle
AR: Augmented Reality
AS/RS: Automated Storage and Retrieval System
BIM: Building Information Modelling
CPS: Cyber-Physical Systems
CPPS: Cyber-Physical Production Systems
DCS: Distributed Control System
DL: Deep Learning
DT: Digital Twin
EMS: Energy Management System
ERP: Enterprise Resource Planning
ESG: Environmental, Social, and Governance
GHG: Greenhouse Gas
HRC: Human–Robot Collaboration
HVAC: Heating, Ventilation, and Air Conditioning
IIoT: Industrial Internet of Things
IoT: Internet of Things
IP: Intellectual Property
IT: Information Technology
KPI: Key Performance Indicator
LCA: Life-Cycle Assessment
LCI: Life-Cycle Inventory
ML: Machine Learning
MLOps: Machine Learning Operations
NLP: Natural Language Processing
O&M: Operation and Maintenance
OPC UA: Open Platform Communications Unified Architecture
PDCA: Plan–Do–Check–Act
PLC: Programmable Logic Controller
QA: Quality Assurance
QC: Quality Control
RL: Reinforcement Learning
SCADA: Supervisory Control and Data Acquisition
SME: Small and Medium-Sized Enterprise
TRL: Technology Readiness Level
WMS: Warehouse Management System

1. INTRODUCTION

Global industries are entering a decisive period where environmental responsibility and operational competitiveness must advance together. Growing pressures from climate change, resource depletion, and international regulatory frameworks are compelling manufacturers to rethink long-established production practices (Akhai 2023; Akhai 2024; Abbas et al. 2025). In this context, artificial intelligence and automation have emerged as pivotal enablers of the transition toward net-zero factories—industrial systems capable of minimizing carbon emissions, maximizing resource efficiency, and maintaining high levels of productivity (Alghieth, 2025; Almusharraf, 2025; Kumar & Shahin, 2025; Mayer et al., 2025; Miele et al., 2024; Setyadi et al., 2025). This introductory section outlines the larger sustainability context, the shortcomings of conventional manufacturing, and the technological shifts that position AI and automation as strategic tools for building resilient, climate-aligned production ecosystems.

1.1 Background: Global Climate Commitments and Industry Decarbonization

Over the past decade, global climate agreements such as the Paris Agreement, IPCC directives, and national net-zero pledges have intensified the call for drastic emission reductions across all sectors (Bataille, 2019; Bataille et al., 2018; Elmegaard et al., 2021; Sharma et al., 2023). Manufacturing, being one of the most energy-intensive pillars of the global economy, faces specific expectations to accelerate decarbonization (Bataille et al., 2018, 2021). Countries are designing long-term climate strategies that require factories to integrate renewable energy, reduce waste, and adopt low-carbon technologies. International standards, carbon pricing mechanisms, ESG reporting norms, and corporate sustainability commitments are strengthening the shift toward greener production models (Ghayal, 2025; Glenk et al., 2024; Samuel, 2025; Tsai & Lin, 2024). This broad governance landscape creates both a responsibility and an opportunity: industries must not only comply but also innovate, ensuring that environmental sustainability and industrial vitality progress hand in hand.

1.2 Role of Manufacturing in Global Emissions and Sustainability Challenges

Manufacturing contributes a significant portion of global greenhouse gas emissions through energy-intensive operations, fossil-fuel dependence, raw material extraction, and waste-driven processes (Filho et al., 2025; Jayawardane et al., 2023;

Panagiotopoulou et al., 2021; Román & Kabir, 2024). Traditional production systems still rely heavily on linear consumption patterns—take, make, and dispose—which generate resource strain and long-term environmental degradation (Badhoutiya et al., 2023; Barteková et al., 2022; Bharadwaj et al., 2023; Dennison et al., 2024; Dixit et al., 2023; Mihai & Minea, 2021). Challenges such as unpredictable energy demand, inefficient equipment usage, excess material consumption, and poor monitoring of process emissions make sustainability difficult to integrate into daily operations. Additionally, complex supply chains spread across continents often lack transparency, contributing to hidden emissions and inefficiencies. These concerns highlight the urgent need for intelligent, data-driven frameworks capable of restructuring manufacturing into a more circular, efficient, and low-carbon domain.

1.3 Need for AI-Driven Transformation Towards Net-Zero Factories

The transition to net-zero factories requires a level of precision, adaptability, and predictive capability that conventional management approaches cannot consistently achieve (Colledani et al., 2014; May et al., 2024; Mo et al., 2023; Ren et al., 2025). AI and automation fill this gap by offering real-time analytics, optimized energy allocation, predictive maintenance, and carbon-aware decision-making (Abadi et al., 2024; Alghieth, 2025; Kronberger et al., 2020; Vater et al., 2019). Machine learning models can forecast energy demand, identify inefficiencies, and adjust process parameters before waste occurs (Abadi et al., 2024; Zou et al., 2017). Digital twins create virtual replicas of equipment and systems to simulate and refine operational conditions, leading to measurable energy savings and emission reductions (Ba et al., 2025; Gutmann et al., 2023; C. Li et al., 2024; H. Li et al., 2023; Resman et al., 2025; Vance et al., 2025; Xia et al., 2023). Robotics and intelligent automation enhance process consistency, reduce scrap, and improve material recovery (Xia & Ahmad, 2025). These features make AI-enabled systems indispensable for designing factories that are not only technologically advanced but also environmentally restorative.

1.4 Evolution of Automation in Industrial Systems

Automation in manufacturing has evolved through multiple phases—from mechanical automation to electronics-based control systems, to today's interconnected digital environments (Morel et al., 2019; Murugesan, 2006; Akhai & Balu 2022; Akhai & Khang 2024). Early automation primarily focused on improving speed and minimizing human error (Kumar et al., 2024; Wadhwa et al., 2025; Barzegaran, 2021). Later generations introduced programmable logic controllers, sensor-based monitoring, and supervisory control systems, which improved process reliability but lacked advanced decision-making capabilities (Singh et al., 2024; Singh et al., 2025; Brusso, 2018; Hansen, 2022). The emergence of the Industrial Internet of Things, cloud platforms, and data-centric software expanded the possibilities for deeper integration (Kaur et al., 2022; Pal & Joshi, 2023; Sharma et al., 2022). AI now represents the next stage of this evolution, enabling machines to learn from operational patterns, adapt autonomously, and collaborate seamlessly with human operators (Pal & Joshi, 2023; Sharma et al., 2022). This shift marks a fundamental departure from rigid, rule-based automation toward intelligent, self-optimizing industrial ecosystems (Höse et al., 2023).

1.5 Problem Statement and Research Motivation

Despite global momentum toward sustainability, many manufacturers struggle to operationalize net-zero strategies. Key barriers include fragmented data systems, lack of real-time visibility into energy use, limited predictive capabilities, and dependence on conventional control architectures (Akella et al., 2023; Hasan et al., 2024; Sharma et al., 2022). Renewable energy integration remains inconsistent, waste minimization is often reactive rather than proactive, and supply chains lack the automation required for carbon-efficient coordination (Bosona&Gebresenbet, 2023; Kaur et al., 2022; Santos et al., 2021; Sinha & Dhanalakshmi, 2021). These gaps motivate the need for a comprehensive review of how AI, automation, sensor networks, and digital twins can collectively address systemic inefficiencies. Understanding these technologies through the lens of sustainability provides a pathway for industries to move beyond incremental improvements toward transformative environmental performance.

This chapter examines how AI, automation, digital twins, and sensor-driven intelligence enable net-zero manufacturing by improving energy efficiency, reducing emissions, and transforming industrial operations as reflected in *Figure 1*. It also outlines real-world applications, key barriers, emerging trends, and strategic pathways that support sustainability-driven industrial transformation.

Figure 2. Architecture of an AI-Enabled Net-Zero Factory

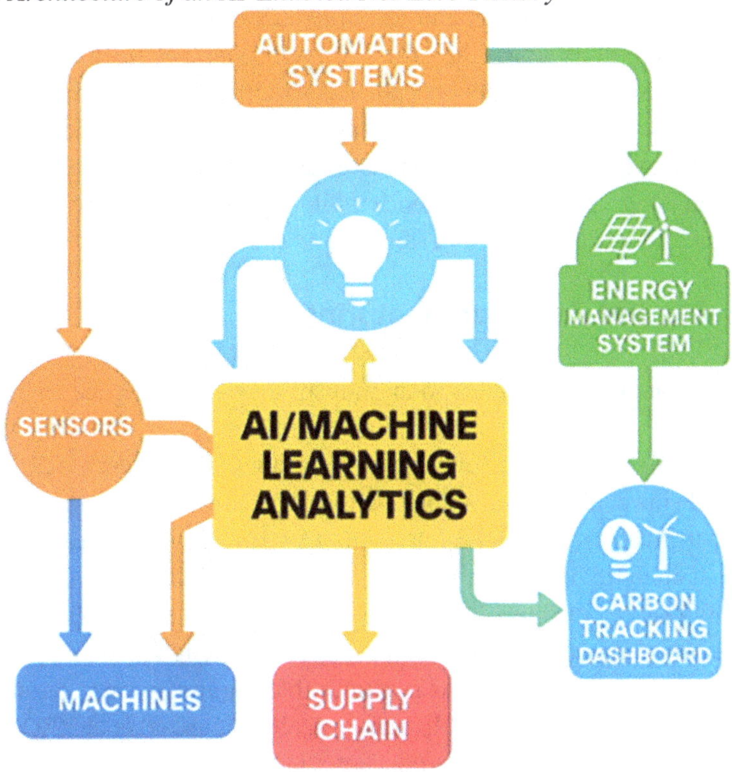

Figure 2 illustrates the overall architecture of an AI-enabled net-zero factory, depicting the interaction between sensors, analytics platforms, automation systems, digital twins, and energy management modules for sustainable operations.

2. CONCEPTUAL FOUNDATIONS OF NET-ZERO FACTORIES

Net-zero factories are manufacturing facilities engineered to reach a state of carbon balance in which operational greenhouse-gas emissions are reduced to the lowest technically feasible level and any residual emissions are compensated through verified mitigation or removal strategies. Such factories rely on digitalization, automation, renewable energy integration, circular-resource management, and advanced analytics to maintain consistent alignment with international net-zero commitments (Filho et al., 2025; Hanenkamp & Zipse, 2023).

Building net-zero factories requires more than integrating cleaner technologies; it demands an entire shift in how industries understand materials, energy, carbon flows, and long-term environmental responsibility. Modern manufacturing systems must evolve into intelligent, adaptive, and resource-efficient environments that operate with minimal emissions across their full life cycle (Motia et al., 2024; Motia et al., 2025). This section introduces the conceptual underpinnings of net-zero manufacturing by defining its scope, examining circular production principles, and exploring the analytical frameworks necessary to evaluate carbon impact. It further explains how digitalization supports climate resilience while highlighting the structural barriers that limit traditional sustainability efforts.

2.1 Defining Net-Zero Manufacturing

Net-zero manufacturing refers to an industrial system designed to balance the greenhouse gases it produces with equivalent reductions or removals, ultimately resulting in no net increase in atmospheric emissions (Bataille et al., 2021; Gebler et al., 2020; Prinz et al., 2023). Achieving this state goes beyond improving energy efficiency; it involves optimizing materials use, redesigning processes, and embedding sustainability into every operational decision (Filho et al., 2025; Akhai & Singh 2013; Akhai & Wadhwa 2024; Siddique, 2024; Triebe et al., 2023). A net-zero factory is characterized by low-carbon energy sources, high resource productivity, intelligent automation, and circular practices that minimize waste throughout the value chain (Alevizos et al., 2023; Hasan et al., 2024; Imdahl et al., 2021; Kerdlap et al., 2019; Zhu et al., 2023). This concept aligns manufacturing with global climate commitments while preserving competitiveness and operational resilience in a rapidly changing economic landscape (Rissman et al., 2020).

2.2 Key Principles of Sustainable and Circular Production

Sustainable and circular production forms the foundation upon which net-zero strategies are built (Mayer et al., 2025; Sartal et al., 2014). At its core, circularity emphasizes the continuous use, recovery, and regeneration of materials rather than the traditional linear consumption model (Filho et al., 2025; Triebe et al., 2023). Within manufacturing, this involves designing products for longer lifespans, enabling easy disassembly, and recovering valuable components at end-of-life. Sustainable production complements this by promoting efficient resource use, reducing emissions, and protecting ecological systems (Alevizos et al., 2023; Abbass & Akhai 2025; Siddique, 2024). Together, these principles reshape manufacturing by encouraging renewable inputs, minimizing waste streams, improving durability, and maximizing the value extracted from every resource (Filho et al., 2025). They also support the transition to cleaner and more stable supply networks by balancing environmental responsibility with industrial productivity.

2.3 Life-Cycle Thinking in Net-Zero Manufacturing

Life-cycle thinking provides a holistic approach for evaluating environmental impacts from raw material extraction to product disposal (Barbhuiya & Das, 2023; Blasco &Lambán, 2025; Dip, 2024; Hellweg et al., 2023; Indicated, 2010; Kleinekorte et al., 2020; Leonzio, 2023; Mendonça & Vieira, 2022; Turner et al., 2022; Valdivia et al., 2021). Instead of focusing solely on factory-level emissions, life-cycle assessment examines upstream processes such as mining, transportation, and component fabrication, as well as downstream aspects including distribution, use-phase efficiency, and end-of-life handling (D'Eusanio& Petti, 2024; Meinrenken et al., 2020; Patouillard et al., 2017). By adopting this perspective, manufacturers gain deeper insights into hidden emissions, embedded energy, and long-term resource burdens (Obydenkova et al., 2021; Zhao et al., 2018). Life-cycle thinking also helps identify the stages where interventions—such as material substitution, process redesign, or recycling—can generate the highest sustainability gains (Bartolomé et al., 2016; Dewulf et al., 2023; Gargalo et al., 2023; Gheewala, 2020; Schmidt et al., 2016; Tatsi et al., 2016). For net-zero factories, this framework ensures that carbon reductions are genuine, comprehensive, and aligned with global environmental standards.

2.4 Industrial Energy Flow, Carbon Accounting, and Efficiency Metrics

Understanding how energy moves through a factory is essential for designing effective net-zero pathways (Geng et al., 2023). Industrial operations typically involve complex interactions between heating systems, electrical equipment, process machinery, and auxiliary services, each contributing to the overall carbon footprint (Gailani et al., 2024). Mapping these flows allows decision-makers to locate inefficiencies, quantify losses, and explore opportunities for renewable integration (Johannsen et al., 2023; Samadi et al., 2023). Carbon accounting methods convert these energy patterns into measurable emission profiles using standardized metrics, enabling organizations to track progress, benchmark performance, and comply with environmental regulations (Dhiman et al., 2025; Abbass & Akhai 2025). Efficiency indicators—such as energy intensity, process yield, and equipment utilization—further help industries evaluate operational performance and prioritize interventions that deliver the greatest carbon reductions.

2.5 Role of Digitalization in Achieving Climate Resilience

Climate or environment effects human health (Akhai et al., 2016; Akhai et al., 2020). Digitalization acts as a catalyst that strengthens the ability of factories to adapt to climate-related risks and sustainability pressures. Technologies such as the Industrial Internet of Things, cloud platforms, digital twins, and advanced analytics create real-time visibility into equipment behavior, energy consumption, and material movements (Kumar & Akhai 2022; Hasan et al., 2024). This continuous flow of information enables factories to anticipate disruptions, quickly adjust production schedules, and optimize resource use under varying environmental conditions (Pal & Joshi, 2023). Digital tools also support the integration of renewable energy by forecasting demand, balancing loads, and coordinating storage systems (Verpoort et al., 2024). As a result, digitalization enhances flexibility, reduces emissions, and supports long-term resilience against regulatory, economic, and environmental uncertainties.

2.6 Barriers to Traditional Sustainability Approaches

Conventional sustainability initiatives often struggle to produce meaningful improvements due to structural, technical, and organizational barriers (Gebler et al., 2024). Many industries rely on fragmented data systems that limit real-time monitoring and prevent accurate carbon assessment. Existing processes are frequently optimized for throughput rather than resource efficiency, leading to avoidable waste and energy losses (Gebler et al., 2024). Supply chains can be opaque, making it difficult to track emissions across multiple tiers of suppliers. Additionally, workforce skills, investment constraints, and resistance to technological change often slow the adoption of advanced sustainability solutions (Kanda et al., 2013). Without integrated digital tools, predictive capabilities, and adaptive automation, traditional approaches remain reactive and insufficient for achieving net-zero goals (Thareja et al., 2016; . Overcoming these barriers requires a shift toward intelligent, interconnected systems capable of supporting continuous improvement and long-term environmental performance.

Table 1 presents a multi-dimensional comparative assessment between traditional manufacturing systems and AI-enabled net-zero manufacturing frameworks, highlighting differences in energy management, process control, waste reduction, emissions tracking, supply-chain transparency, and decision-making capabilities.

Table 1. Traditional vs. AI-Enabled Net-Zero Manufacturing-Multi-Dimensional Comparison.

Parameter	Traditional Manufacturing	AI-Enabled Net-Zero Manufacturing	Impact on Sustainability	Example Technologies
Energy Management	Fixed operation schedules, high idle energy	Predictive energy forecasting, dynamic load balancing	15–30% energy reduction	ML-based forecasting, digital meters
Process Control	Manual, reactive adjustments	Autonomous, learning-based control loops	Lower process variability, stabilised energy use	RL controllers, smart sensors
Waste Generation	High due to inconsistent quality	Predictive quality + automated inspection	Minimises scrap, avoids rework	DL-based defect detection
Emissions Monitoring	Periodic audits	Real-time carbon-intensity monitoring	Accurate carbon accounting	Smart meters, IIoT
Maintenance Strategy	Breakdown or preventive	Predictive + prescriptive maintenance	Extends equipment life, reduces downtime	Vibration sensors, ML
Supply Chain Visibility	Fragmented and opaque	End-to-end digital traceability	Reduced logistics emissions	Blockchain, digital twins
Human Involvement	Manual-intensive	Human–AI collaboration, cobots	Ergonomic improvements	Cobots, AR
Decision-Making	Experience-driven	Data-driven, autonomous	Optimized sustainability outcomes	AI dashboards, analytics

3. ARTIFICIAL INTELLIGENCE IN SUSTAINABLE MANUFACTURING

Artificial intelligence has become a central driver of sustainable industrial transformation, enabling factories to shift from fixed and reactive operating modes to dynamic, predictive, and resource-efficient systems ("Artificial Intelligence in Manufacturing – Systematic Literature Review," 2025; Ikumapayi et al., 2022; Kumar & Shahin, 2025). By interpreting complex data patterns, automating decision processes, and continuously improving process performance, AI helps manufacturers reduce energy demand, minimize waste, and strengthen operational stability (Miele et al., 2024; Rane et al., 2024; Walter, 2023). This section discusses the major AI techniques used in modern manufacturing and explains how they support the transition to environmentally responsible and climate-aligned production systems.

3.1 Overview of AI Techniques Used in Manufacturing

AI in manufacturing encompasses a broad collection of computational approaches that mimic cognitive functions such as learning, reasoning, and prediction (Kumar & Shahin, 2025). Techniques such as machine learning, deep learning, and reinforcement learning allow production systems to interpret sensor data, detect patterns, and adjust operations autonomously (Cruz et al., 2024; Elahi et al., 2023; Ghahramani et al., 2020a, 2020b; Lee et al., 2023; Sinha & Dhanalakshmi, 2021; Verma, 2018; Wan et al., 2020). These tools support tasks ranging from fault detection and energy scheduling to material handling, process control, and long-term planning (Abadi et al., 2024; Waltersmann et al., 2021). Integrated with industrial automation, sensors, and digital platforms, AI serves as the analytical core of smart factories, guiding decisions that improve efficiency, reduce environmental impact, and enhance product reliability ("Artificial Intelligence in Manufacturing – Systematic Literature Review," 2025; Kumar & Shahin, 2025).

Figure 3. AI Across the Net-Zero Manufacturing Value Chain

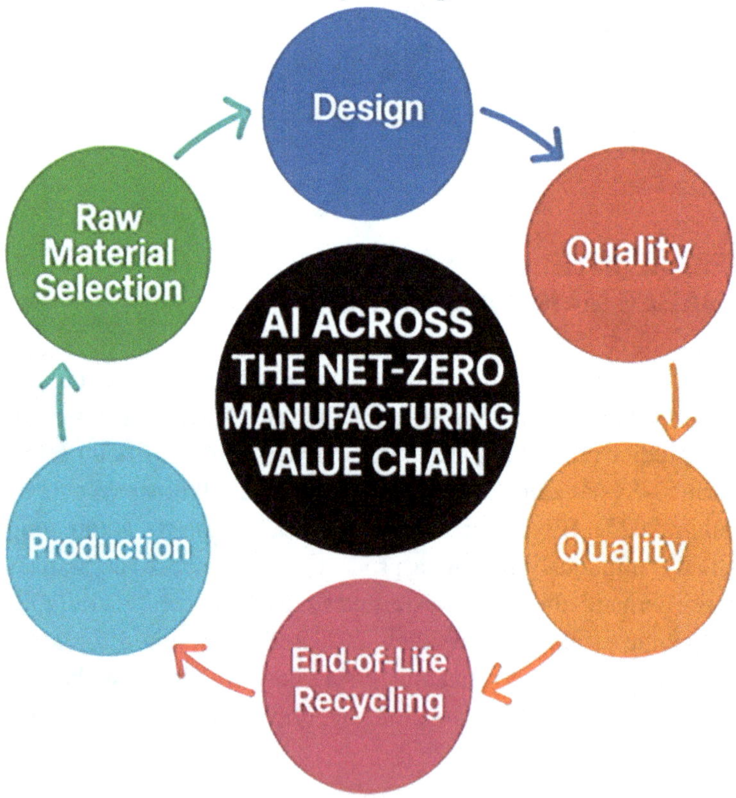

Figure 3 presents the integration points of artificial intelligence across the manufacturing value chain, showing how machine learning, robotics, and predictive analytics contribute to sustainability improvements at each stage.

3.2 Machine Learning for Energy Forecasting and Load Optimization

Machine learning plays an essential role in predicting energy consumption and managing loads in real time (Abadi et al., 2024; Wicaksono et al., 2024). By analyzing historical production data, environmental conditions, equipment behavior, and utility pricing trends, ML models can anticipate future energy requirements with high accuracy (Bertsimas et al., 2023; Mawson & Hughes, 2021; Trenz et al., 2023). These forecasts enable factories to schedule energy-intensive activities during low-carbon or low-cost periods, optimize the operation of renewable sources (Wicaksono et al., 2024), and reduce peak-load stress on equipment (Mawson & Hughes, 2021; Sarswatula et al., 2022). In addition, ML-driven load balancing helps maintain stable power usage, avoids inefficient start-stop cycles, and reduces the carbon intensity of manufacturing operations (Sarswatula et al., 2022). Through continuous learning, these models evolve to reflect changing production patterns and operational conditions (Alghieth, 2025).

3.3 Deep Learning for Process Automation and Anomaly Detection

Deep learning offers powerful capabilities for interpreting complex sensor signals, images, and machine states, making it invaluable for advanced automation (Li et al., 2025; Yan et al., 2024). In manufacturing, deep neural networks can recognize subtle process deviations, classify abnormal behaviors, and detect early signs of equipment deterioration long before failures occur (Akhavan et al., 2023; Alli et al., 2024; Ardito et al., 2024; Arellano-Espitia et al., 2021; Dimitriou et al., 2019; Domb et al., 2023; Lu et al., 2024; Maschler et al., 2021; Qiu et al., 2023; Wang et al., 2023; Zeiser et al., 2022). This level of insight enables more reliable process control, proactive maintenance, and reduced material waste. Deep learning also automates tasks such as visual inspection, robotic coordination, and real-time process adjustments, ensuring that operations remain efficient even under variable environmental or material conditions. Its capacity to handle unstructured data gives manufacturers a richer understanding of hidden inefficiencies and potential sustainability improvements.

3.4 Natural Language Processing for Manufacturing Knowledge Extraction

Natural Language Processing (NPL) allows factories to convert unstructured text such as manuals, inspection reports, operator notes, maintenance logs, and safety documents into structured knowledge that can guide decision-making (Akhbardeh et al., 2020; Brundage et al., 2020; Mandelli et al., 2023; Sala et al., 2022; Sundaram & Zeid, 2024). By extracting key insights from this information, NLP tools help manufacturers understand recurring problems, codify expert knowledge, and improve training and troubleshooting practices (Sala et al., 2022). NLP also supports intelligent assistants that provide operators with real-time guidance, reducing downtime and human error. In sustainability contexts, NLP can analyze regulatory documents, emissions reports, and environmental guidelines, ensuring that factories remain aligned with evolving climate compliance requirements (Hassani, 2024a, 2024b; Klessascheck et al., 2025; Ni et al., 2023; Ong et al., 2024; Thimm, 2023).

3.5 Reinforcement Learning for Self-Optimizing Production Lines

Reinforcement learning introduces the ability for production systems to learn optimal strategies through trial-and-error interactions with their environment (Gao &Schweidtmann, 2024). Instead of relying on fixed rules, RL algorithms continually refine their decisions to achieve objectives such as minimizing energy use, maximizing throughput, or reducing waste (Jabeur et al., 2023; Loffredo, 2023; Loffredo et al., 2023; Mazgualdi et al., 2022; Paraschos et al., 2023; Ponse et al., 2024; Zhang et al., 2024). This makes RL particularly useful for highly dynamic and multivariable processes where conventional optimization methods fall short. In sustainable manufacturing, RL can coordinate robotic cells, fine-tune process parameters, balance resource consumption, and adjust equipment behavior based on changing conditions. Over time, RL-enabled systems evolve into self-optimizing production lines capable of sustained low-carbon operations (Mazgualdi et al., 2022).

3.6 AI-Enabled Quality Inspection and Defect Prediction

Ensuring consistent product quality is essential for reducing waste and achieving sustainable production. AI-enabled inspection systems use sensors, cameras, and learning algorithms to detect defects with greater speed and accuracy than manual inspection (Kumbhar et al., 2023; Rožanec et al., 2021; Sundaram & Zeid, 2023). These systems identify micro-level deviations, surface anomalies, dimensional inconsistencies, and material flaws, preventing defective products from moving further down the production chain (Hafiz et al., 2025; Lario et al., 2024). In addition to real-time inspection, predictive quality models analyze process conditions to anticipate when defects are likely to occur, allowing manufacturers to intervene early (Msakni et al., 2023; Silva et al., 2024; Tercan & Meisen, 2022; Zoubaidi et al., 2025). This proactive approach reduces scrap generation, avoids rework, conserves raw materials, and improves overall efficiency (Catti et al., 2024; Hafiz et al., 2025; Schimanek et al., 2023).

3.7 Ethical, Transparent, and Explainable AI Practices in Industry

As AI becomes deeply embedded in manufacturing, ethical and transparent usage becomes critical (- & -, 2024; Brintrup et al., 2023). Explainable AI frameworks help ensure that decision-making processes can be interpreted by engineers, managers, and regulatory bodies (Abhilash et al., 2024; Agostinho et al., 2023; Galetić& Nottle, 2023; Moosavi et al., 2024; Rožanec et al., 2021; Sofianidis et al., 2021). Transparency is necessary not only for compliance but also for building trust among workers who interact with intelligent systems (Chen et al., 2023; Galetić& Nottle, 2023; Jayakumar & Park, n.d.; Narkar & Bandivadekar, n.d.). Ethical practices require careful handling of data privacy, avoidance of biased models, and clear guidelines for human-machine collaboration (Callari et al., 2024; Narkar & Bandivadekar, n.d.). Sustainable manufacturing also demands that AI systems themselves remain energy-efficient, secure, and aligned with broader social and environmental goals (Meena et al., 2025). The integration of responsible AI practices ensures that technological advancements contribute meaningfully to long-term industrial sustainability.

Table 2 presents an expanded classification of major AI and automation technologies used in sustainable manufacturing, detailing their core industrial functions, sustainability benefits, application domains, technology readiness levels, and estimated performance improvements.

Table 2. AI & Automation Technologies and Their Sustainability Contributions.

Technology	Core Function	Sustainability Benefit	Key Industrial Applications	Maturity Level (TRL)	Typical Reduction Achieved
Machine Learning	Forecasting, pattern discovery	Cuts peak loads, optimizes energy use	Energy scheduling, HVAC optimization	8–9	10–25% energy savings
Deep Learning	Image/signal interpretation	Reduces scrap, enhances quality	Visual inspection, anomaly detection	7–9	30–60% defect reduction
Reinforcement Learning	Self-learning optimization	Minimizes process variability	Real-time process tuning, robotics	6–7	10–20% material savings
NLP	Knowledge extraction	Faster maintenance, fewer breakdowns	Document mining, digital instructions	6–8	5–10% downtime reduction
Robotics &Cobots	Material handling, assembly	Precision → reduced waste	Assembly lines, welding, packaging	9	20–30% productivity boost
IIoT Sensors	Real-time data acquisition	Continuous monitoring → low emissions	Energy, vibration, temperature tracking	9	Accurate carbon profiling
Digital Twins	Virtual replication	Avoids physical trials & waste	Virtual commissioning, process simulation	7–8	15–40% efficiency gain
Autonomous Logistics	Smart routing, AGV/AMR systems	Lower fuel use, reduced transport emissions	Warehouses, shopfloor movement	8	10–25% emission reduction
Additive Manufacturing	Layer-wise fabrication	Minimal raw material wastage	Tooling, prototyping	8	Up to 70% material reduction

4. AUTOMATION TECHNOLOGIES FOR NET-ZERO OPERATIONS

Automation plays a decisive role in the shift toward net-zero factories by integrating precision, consistency, and resource-conscious operation (Badhoutiya et al., 2023; Psarommatis et al., 2023; Su & Xu, 2025). Traditional manufacturing systems often rely heavily on manual interventions, leading to variability, inefficiencies, and resource-intensive practices. In contrast, modern automation technologies create highly controlled environments where processes operate with minimal waste, optimized energy use, and predictable outputs (Badhoutiya et al., 2023; Setyadi et al.,

2025; Su & Xu, 2025; Todescato et al., 2023). By harmonizing robotics, control systems, digital platforms, and material-flow technologies, automation forms the operational spine of low-carbon production. This section elaborates on the major automation components that enable industries to meet sustainability commitments without compromising productivity or reliability.

4.1 Smart Industrial Robotics for Low-Emission Production

Smart industrial robots represent a new generation of automated systems designed to perform tasks with exceptional accuracy, speed, and environmental awareness (Adedokun et al., 2025; Heredia et al., 2023). Unlike earlier robotic systems that operated on rigid, pre-defined routines, modern robots incorporate adaptive algorithms, sensor feedback, and machine learning techniques that allow them to optimize their movements and energy consumption (Boscariol & Richiedei, 2019; Gadaleta et al., 2019; Pellicciari et al., 2015; Ružarovský et al., 2024). Their precise material handling contributes to reducing scrap production (Adedokun et al., 2025). Robotic systems also support sustainability by enabling advanced manufacturing strategies such as energy-conscious scheduling, optimized motion planning, and dynamic speed adjustments (Boscariol & Richiedei, 2019; Gadaleta et al., 2019; Heredia et al., 2023). When integrated with renewable-powered factory floors, smart robots can synchronize their operations with availability of clean energy, reducing reliance on carbon-intensive inputs (Koul, 2024). Ultimately, robotics contributes to net-zero goals by improving throughput, minimizing resource waste, and ensuring consistently low-emission operations (Adedokun et al., 2025).

4.2 Autonomous Material Handling and Warehouse Automation

Material movement is one of the most energy-consuming and error-prone activities in manufacturing environments (Minashkina, 2024; Soufi et al., 2021; Trstenjak et al., 2023). Autonomous material-handling systems—including automated guided vehicles, autonomous mobile robots, robotic forklifts, and conveyor intelligence—address this by orchestrating efficient, predictable, and low-energy material flow (Bechtsis et al., 2016; Miletić, 2023; Rubio et al., 2021). These systems navigate using onboard vision systems, LIDAR, proximity sensors, and mapped layouts, allowing them to adapt to dynamic shopfloor conditions without unnecessary detours or idling (Fottner et al., 2021). Warehouse automation systems further enhance sustainability by optimizing storage density, reducing temperature control loads, and minimizing redundant inventory that ties up resources (Cannava et al., 2024; Minashkina& Happonen, 2020; Tappia et al., 2015; Trstenjak et al., 2023). Automated sorting, retrieval, and replenishment ensure minimal physical waste and reduced risk of product damage (Minashkina& Happonen, 2020). By removing inefficiencies inherent in manual handling processes, autonomous logistics reduce fuel usage, operational delays, and environmental impact across the material flow chain (Cannava et al., 2024; Rubio et al., 2021).

4.3 Process Automation Using PLC, SCADA, and Distributed Control Systems

Programmable Logic Controllers, Supervisory Control and Data Acquisition systems, and Distributed Control Systems form the essential control architecture of modern factories (Banaulikar, 2020; Yao et al., 2024). These systems automate complex operations by continuously monitoring equipment performance, environmental factors, and process parameters. They regulate heating, pressure, flow, chemical dosing, and machining processes, ensuring that each step operates within its most efficient range (Banaulikar, 2020; Öner & Solak, 2020; Roseline et al., 2019). The sustainability benefits of these systems stem from their ability to capture real-time data, detect deviations instantly, and adjust operations before inefficiencies escalate (C. & Ocampo-Martínez, 2019; Diaz et al., 2019; Öner & Solak, 2020; Puig et al., 2019). For instance, automated control reduces energy-intensive overshooting or repeated calibration, minimizes losses associated with human error, and ensures consistent product quality that reduces rework (Banaulikar, 2020; Öner & Solak, 2020; Ramya, 2019). When combined with AI-driven analytics, these systems evolve from passive controllers into intelligent frameworks capable of predictive intervention and continuous optimization (Alemão et al., 2019; Bányai & Tóth, 2025; Görür et al., 2021; Paraschos et al., 2023; Plapper, 2022).

4.4 Human-Robot Collaboration for Ergonomic and Efficient Shopfloors

Human-robot collaboration brings together human adaptability and robotic precision in a shared workspace. Collaborative robots, or cobots, incorporate safety sensors, force control mechanisms, and responsive algorithms that allow them to work safely beside humans without the need for rigid fencing (Javernik et al., 2023; Miranda et al., 2025; Schöberl et al., 2023; Yitmen & Almusaed, 2024). This synergy not only improves productivity but also enhances sustainability by ensuring that tasks are carried out with minimal errors, reduced material use, and consistent quality (Adedokun et al., 2025; Calvo & Gil, 2022; Ojstersek et al., 2022; Pizoń et al., 2022). Cobots take over repetitive, strenuous, or hazardous tasks, reducing fatigue-related inefficiencies (Calvo & Gil, 2022; Schöberl et al., 2023). Workers, in turn, focus on decision-making, inspection, and problem-solving roles, creating an environment where human intelligence and machine consistency reinforce each other (Calvo & Gil, 2022; Yitmen & Almusaed, 2024). In the context of net-zero manufacturing, such collaboration reduces waste, optimizes cycle times, and enables smoother integration of sustainable practices across shopfloor operations (Ojstersek et al., 2022; Ojsteršek et al., 2023; Pizoń et al., 2022).

4.5 Additive Manufacturing and Automated Fabrication for Waste Reduction

Additive manufacturing transforms traditional material-intensive fabrication into a highly efficient layer-by-layer process that uses only the required feedstock (Frăţilă& Rotaru, 2017; Jayawardane et al., 2023; Maware et al., 2024; ÖZGÜNER & ÖZGÜNER, 2022). Automated AM systems enable the creation of lightweight structures, topology-optimized components, and complex geometries that minimize raw material use without compromising mechanical performance (Chyr& DeSimone, 2022; Dimitrijević et al., 2023; Frăţilă& Rotaru, 2017). This contributes directly to carbon reduction by lowering material extraction, transportation, and machining waste (Chyr& DeSimone, 2022; Dimitrijević et al., 2023; Frăţilă& Rotaru, 2017; Rezaie et al., 2021). Automated AM pipelines—integrating pre-processing, material feeding, environmental control, and post-processing—ensure repeatability and high-quality output (Lehmhus et al., 2018; Li & Pattinson, 2025; Rotari, 2023; Yang et al., 2020). Moreover, AM supports circular principles by enabling component repair, refurbishment, and customized low-volume production without the environmental burden of traditional tooling or batch processing (Chen, 2022; Dimitrijević et al., 2023; Jawahir & Bradley, 2016; Prasad et al., 2024; Zhong & Pearce, 2017). AM's alignment with digital design workflows also promotes dematerialization and rapid prototyping, reducing the environmental costs of iterative development (Chen, 2022; Chyr& DeSimone, 2022).

4.6 Industrial Internet of Things for Energy and Resource Monitoring

The Industrial Internet of Things establishes the digital nervous system of net-zero factories (Gontarz et al., 2015; Hanifi et al., 2025). Distributed sensors, actuators, gateways, and cloud platforms create a continuous feedback loop that captures data from equipment, utilities, environmental conditions, and material flows (Abdelmoneim et al., 2025; Hasan et al., 2024; Kaur et al., 2022). This real-time visibility allows factories to pinpoint inefficiencies, detect energy wastage, and coordinate equipment in response to fluctuating energy availability (Gontarz et al., 2015; Hanifi et al., 2025; Hasan et al., 2024). IIoT plays a vital role in integrating renewable energy sources by monitoring capacity, regulating loads, and coordinating power distribution (Abdelmoneim et al., 2025). Smart devices can autonomously adjust heating, ventilation, motors, and utilities to maintain efficient operation with minimal human intervention (Abdelmoneim et al., 2025; Sinha & Dhanalakshmi, 2021). When combined with advanced analytics, IIoT enables predictive maintenance, minimizes downtime, and supports adaptive scheduling, all of which contribute directly to sustainable and resource-conscious manufacturing (Alex & Johnson, 2025; Bányai & Tóth, 2025; Görür et al., 2021; Liu et al., 2021).

5. DIGITAL TWINS AND CYBER-PHYSICAL PRODUCTION SYSTEMS (CPPS)

Digital twins and cyber-physical production systems provide manufacturers with unprecedented insight, control, and predictability. They serve as the analytical and operational core of smart factories by creating deep synchrony between the physical world and its digital representation. When integrated effectively, these technologies allow industries to explore operational possibilities, anticipate disruptions, and optimize resource use with exceptional precision, forming a key pillar of the net-zero transition.

5.1 Concept and Architecture of Digital Twins

A digital twin is far more than a simulation model—it is a living, evolving replica of a physical entity, continuously updated through real-time data streams (Collan & Michelsen, 2020; Hammadi, 2021; Monek & Fischer, 2024). Its architecture typically comprises three layers:

1. **Physical Layer:** machines, systems, or products equipped with sensors (Rolofs et al., 2024);
2. **Virtual Layer:** models, simulations, and analytical frameworks (Rolofs et al., 2024);
3. **Integration Layer:** communication interfaces and data pipelines that synchronize physical and virtual behavior (Rolofs et al., 2024).

This dynamic link allows decision-makers to observe system performance, evaluate alternative strategies, and optimize operations without interrupting production (Collan & Michelsen, 2020). Digital twins serve as virtual laboratories where manufacturers test process changes, analyze root causes, and evaluate sustainability interventions (Ding & Li, 2024; Luo et al., 2025).

Figure 4. Digital Twin Framework for Sustainable Manufacturing

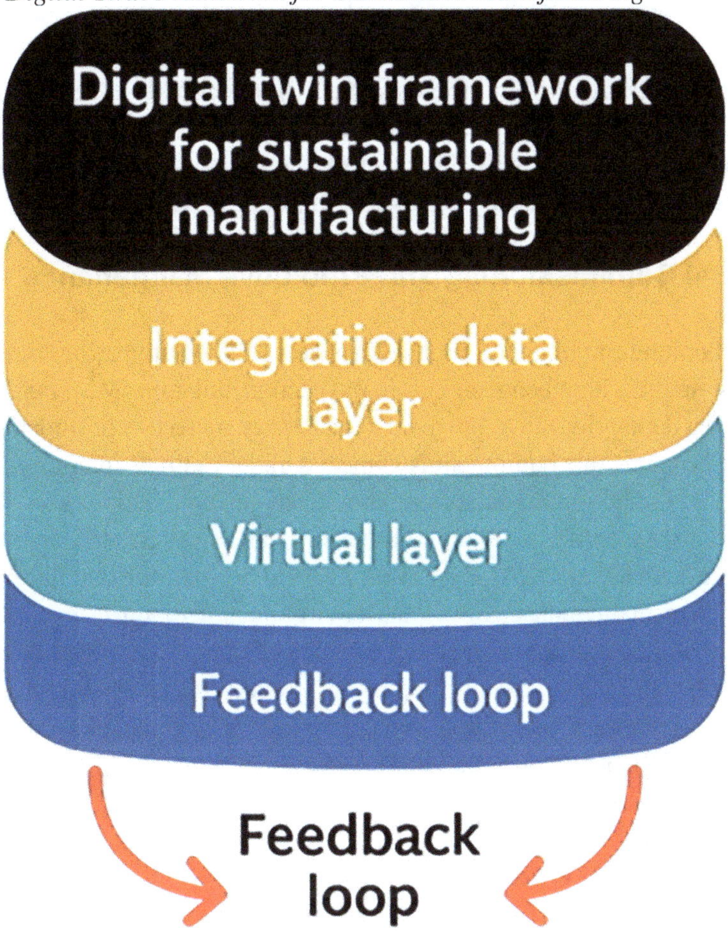

Figure 4 provides a conceptual framework of a sustainability-oriented digital twin, highlighting real-time data flows, simulation models, and the feedback mechanisms that optimize energy use and emissions.

5.2 Twin-Enabled Energy Optimization of Equipment and Factories

Digital twins enhance energy efficiency through their ability to emulate the energy behavior of equipment under diverse scenarios (Assad et al., 2023; Ding & Li, 2024; Rolofs et al., 2024; Yakovenko & Shaptala, 2024). They can simulate operating conditions, load variability, and control strategies with fine precision, helping engineers identify optimal energy configurations (Assad et al., 2023; Ding & Li,

2024). This capability supports renewable energy integration by enabling factories to simulate solar, wind, or storage inputs before applying them in real operations (Ding & Li, 2024; Rolofs et al., 2024). Twin-enabled energy management ensures that machines operate within efficient ranges, reduces idle losses, and prevents peak-load spikes that burden the grid (Assad et al., 2023; Yakovenko & Shaptala, 2024). Over time, these insights allow factories to redesign processes for inherently low-carbon operation (Ding & Li, 2024; Rolofs et al., 2024).

5.3 Virtual Commissioning and Predictive Maintenance

Virtual commissioning uses digital twins to validate automation logic, control sequences, and machine behavior before physical installation (Mayer et al., 2025). This reduces debugging time, prevents unnecessary material consumption during tuning, and ensures that equipment is launched under optimal conditions (Resman et al., 2025). Predictive maintenance, another major application, uses the twin's real-time data to identify patterns that indicate wear, misalignment, or failure risks (Yakovenko & Shaptala, 2024). By preventing breakdowns and extending equipment life, predictive maintenance lowers the environmental footprint associated with spare parts, unplanned downtime, and inefficient operations (Pater &Stadnicka, 2021; Resman et al., 2025). It also improves resource efficiency by replacing reactive, high-cost interventions with planned, sustainable maintenance routines (Pater &Stadnicka, 2021; Resman et al., 2025).

5.4 Real-Time Process Simulation and Carbon Footprint Analytics

Digital twins act as real-time simulation engines that allow factories to evaluate the environmental consequences of operational decisions (Ding & Li, 2024). By modeling energy flows, waste generation, cycle times, and material consumption, they help quantify the carbon footprint of various production scenarios (Ding & Li, 2024; Luo et al., 2025). This supports climate-conscious scheduling, load distribution, and process tuning (Ding & Li, 2024; Luo et al., 2025). Carbon analytics integrated with the twin reveal where emissions originate, how they evolve, and where reduction opportunities lie (Ding & Li, 2024; Luo et al., 2025). This granularity transforms environmental assessment from an annual reporting exercise into a continuous improvement mechanism (Ding & Li, 2024; Luo et al., 2025).

5.5 Integration of Digital Twins with CPPS and Smart Controllers

Cyber-physical production systems create a tightly integrated network of physical machinery, communication systems, and computational intelligence (Gallab et al., 2024; Vidyalakshmi et al., 2025). When paired with digital twins, CPPS creates a closed-loop environment where insights from the virtual model are directly applied to the physical process (Collan & Michelsen, 2020; Jia et al., 2022). Smart controllers adjust machine parameters, allocate resources, and respond to anomalies based on twin recommendations (Villalonga et al., 2021). This synergy supports autonomous operation, rapid adaptation to disturbances, and multi-objective optimization—balancing cost, productivity, and environmental constraints (Jia et al., 2022; Villalonga et al., 2021). Over time, twin-CPPS ecosystems evolve toward self-regulating production systems capable of sustaining net-zero operations (Borangiu et al., 2020).

5.6 Cybersecurity Considerations in Twin-Based Manufacturing

As digital twins and CPPS rely on continuous data exchange and networked architectures, cybersecurity becomes central to safe and sustainable operation (Chen et al., 2023). Security breaches can disrupt production, corrupt sustainability metrics, or compromise proprietary designs (Sharma et al., 2022). Comprehensive cybersecurity strategies must include encrypted communication, authenticated access, anomaly detection, and regular auditing (Chen et al., 2023). Protecting the integrity of digital-twin systems ensures that sustainability data remains reliable and that operational decisions are based on accurate information (Chen et al., 2023). Robust cyber governance preserves not only industrial security but also the credibility of environmental reporting. (Esiri et al., 2024)

6. SENSOR NETWORKS, DATA INFRASTRUCTURE, AND PREDICTIVE ANALYTICS

Sensor networks and data infrastructures constitute the informational backbone of net-zero factories. They gather real-time measurements, support analytics, and provide the transparency needed to make sustainability-focused decisions. Without robust sensing and data management, the insights required for carbon-conscious manufacturing remain inaccessible. This section examines how sensing technologies, analytics frameworks, and data governance collectively support the shift toward sustainable manufacturing.

6.1 Role of Advanced Sensor Technologies in Net-Zero Factories

Advanced sensors provide granular visibility into machine health, energy consumption, process parameters, environmental conditions, and emissions (Alghieth, 2025; Badhoutiya et al., 2023). By capturing high-resolution, time-sensitive data, these sensors enable factories to detect inefficiencies early and make data-driven adjustments (Badhoutiya et al., 2023). They support energy audits, waste reduction strategies, and predictive reliability programs (Badhoutiya et al., 2023).

Modern sensors often incorporate multi-modal capabilities—combining thermal readings, vibration analytics, acoustic sensing, and chemical detection—allowing a comprehensive understanding of factory operations (Alghieth, 2025). Their integration into automation systems ensures that sustainability goals are woven into daily decision-making rather than treated as external targets (Badhoutiya et al., 2023).

6.2 Smart Meters for Real-Time Carbon and Energy Tracking

Smart meters go beyond measuring energy usage; they quantify its carbon intensity based on real-time grid conditions or on-site renewable generation. By providing up-to-the-minute insights into how much carbon is associated with each unit of electricity consumed, these meters help factories plan production around periods of low-carbon supply (Román & Kabir, 2024).Real-time metering allows dynamic load shifting, optimized scheduling, and targeted efficiency improvements. When combined with AI models, smart meters can forecast demand peaks, identify abnormal patterns, and suggest corrective actions that minimize energy-related emissions (Román & Kabir, 2024).

6.3 Predictive Analytics for Downtime, Quality Control, and Waste Reduction

Predictive analytics interprets historical data, live sensor outputs, and behavioral trends to forecast future events (Bataille, 2019; Brundage et al., 2020). This supports sustainability in several ways:

1. **Downtime Reduction:** Predicting failures reduces unplanned stoppages that waste energy and resources (Badhoutiya et al., 2023; Brundage et al., 2020; Hansen, 2022; Ni et al., 2023).
2. **Quality Control:** Anticipating quality deviations minimizes scrap generation and reprocessing (Badhoutiya et al., 2023; Brundage et al., 2020; Hansen, 2022; Ni et al., 2023).

3. **Resource Optimization:** Forecasting material consumption avoids over-ordering and reduces storage burdens (Brundage et al., 2020; Ni et al., 2023).

Predictive analytics also enhances process stability, enabling manufacturers to maintain tight control over variables that influence emissions and waste (Brundage et al., 2020).

6.4 Edge, Fog, and Cloud Computing for Sustainable Data Processing

Net-zero factories require distributed computing architectures capable of processing vast sensor data efficiently (Alli et al., 2024; Anjum et al., 2025; Assad et al., 2023; Patouillard et al., 2017).

1. *Edge computing* processes data near the source, reducing latency and preventing excessive data transmission (Alli et al., 2024; Assad et al., 2023; Patouillard et al., 2017).
2. *Fog computing* acts as an intermediate layer, filtering and aggregating data to support regional decision-making (Assad et al., 2023; Patouillard et al., 2017).
3. *Cloud computing* handles large-scale analytics, storage, and historical data modeling (Anjum et al., 2025; Assad et al., 2023; Patouillard et al., 2017).

Together, these layers create a scalable, energy-conscious computing ecosystem. They support high-frequency monitoring, advanced simulations, and carbon accounting without overburdening local systems or increasing energy use unnecessarily (Alli et al., 2024; Assad et al., 2023; Patouillard et al., 2017).

6.5 Data Pipelines, Storage Architectures, and Interoperability Standards

Efficient data management is essential for sustainability-driven decision-making (Rolofs et al., 2024). Data pipelines collect, clean, standardize, and deliver information across platforms, ensuring that analytics teams and automated systems receive high-quality inputs (Rolofs et al., 2024). Storage architectures must balance speed, cost, energy use, and data longevity (Rolofs et al., 2024).Interoperability standards—such as OPC UA, MQTT, and ISO frameworks—ensure that heterogeneous machines and software tools communicate seamlessly (Li et al., 2023; Pal & Joshi, 2023; Walter, 2023). Eliminating data silos enhances transparency, reduces redundancies, and supports holistic carbon optimization across the entire manufacturing ecosystem (Rolofs et al., 2024).

6.6 Managing Data Privacy, Security, and Industrial IP Protection

As digital systems expand, protecting data authenticity and intellectual property becomes critical (Schimanek et al., 2023). Net-zero strategies rely on accurate environmental data, secure process models, and confidential performance indicators. Breaches or corrupted data can undermine sustainability initiatives and expose sensitive information (Schimanek et al., 2023; Verpoort et al., 2024).

Organizations must implement secure communication protocols, role-based access controls, encrypted storage, and continuous security monitoring (Boscariol & Richiedei, 2019; Schimanek et al., 2023; Verpoort et al., 2024). Strong governance frameworks ensure compliance with regulatory requirements while preserving trust in digital infrastructure (Schimanek et al., 2023; Verpoort et al., 2024). Ultimately, data security and privacy remain foundational pillars for sustainable, digitally integrated manufacturing (Schimanek et al., 2023).

7. GREEN SUPPLY CHAIN AUTOMATION

Achieving net-zero manufacturing requires a systemic transformation that extends far beyond the boundaries of individual factories. A substantial portion of industrial emissions originates from supply chains—raw material extraction, component manufacturing, logistics, warehousing, packaging, transportation, and product end-of-life handling (Xavier et al., 2024; Yang et al., 2025). Green supply chain automation addresses these challenges by creating intelligent, integrated, and resource-efficient networks that support environmentally responsible production from source to disposal. Automated systems enhance visibility, reduce inefficiencies, and enable carbon-conscious decision-making across every stage of the value chain (Lei, 2024; Ojadi et al., 2025). This section explores how AI-driven technologies, robotics, digital platforms, and circular-resource strategies collectively enable greener and more resilient supply networks.

Figure 5. Components of a Green Automated Supply Chain

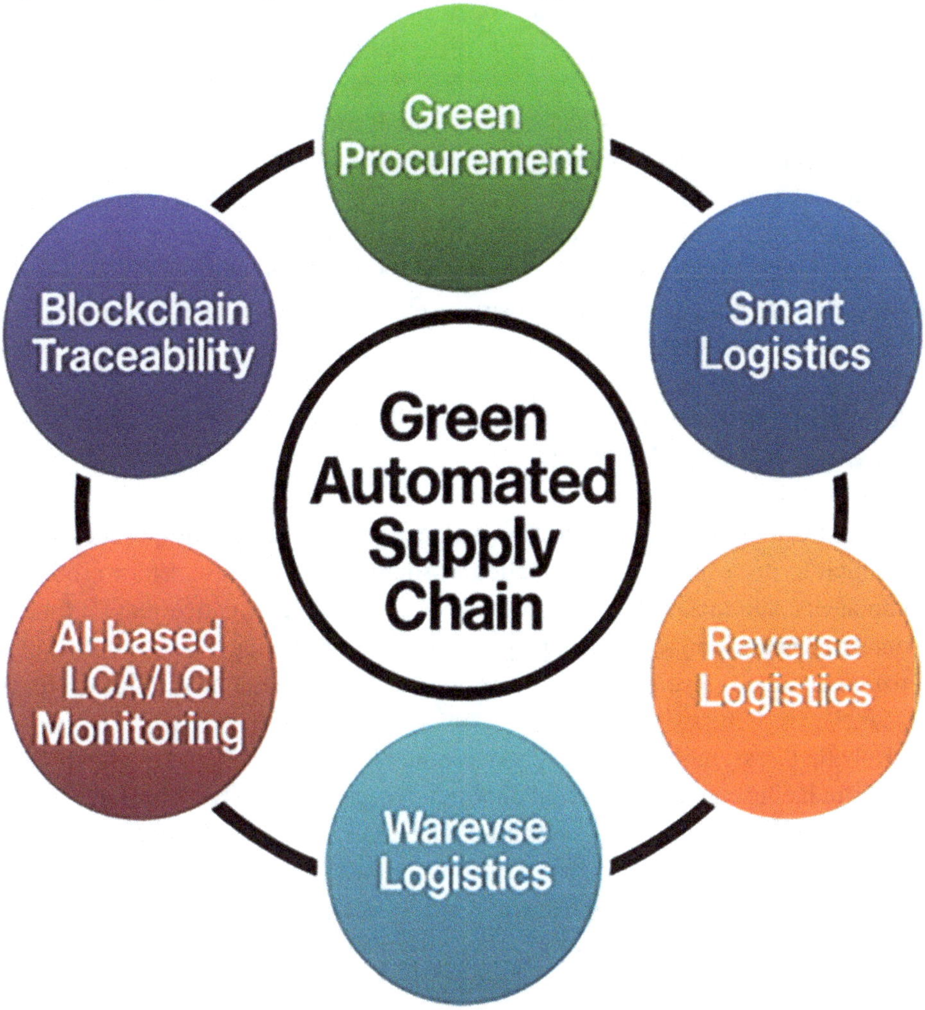

Figure 5 shows the major components of a green automated supply chain, emphasizing how circularity, traceability, and AI-enabled logistics contribute to reduced environmental impact.

7.1 AI-Driven Green Procurement and Sustainable Material Selection

Sustainable production begins with selecting raw materials that minimize environmental impact across their life cycles. AI-powered procurement platforms analyze supplier data, carbon footprints, ethical sourcing indicators, and life-cycle assessments to evaluate the environmental performance of potential inputs (Hianje& Sahbani, 2025; Khanna et al., 2022). These systems compare different material options, detect greenwashing attempts, and prioritize suppliers with verified sustainability credentials (Khanna et al., 2022).

Machine learning models can forecast supply risks, identify environmentally preferable substitutes, and optimize procurement schedules to avoid carbon-intensive logistics (Khanna et al., 2022; Okogwu et al., 2023). Automated negotiation agents streamline sourcing processes while ensuring compliance with environmental, social, and governance standards (Khanna et al., 2022). By embedding intelligence into procurement, factories can reduce upstream emissions and support global sustainability commitments (Hasan, 2024).

7.2 Automation for Circular Supply Chains and Closed-Loop Systems

Traditional supply chains operate on linear principles, moving products from extraction to disposal. In contrast, circular supply chains aim to return products, components, and materials back into industrial cycles. Automation plays a fundamental role in enabling closed-loop operations by coordinating repair, refurbishing, remanufacturing, recycling, and reverse logistics (Khanna et al., 2022; Meena et al., 2025).

Robotic disassembly systems separate valuable components and materials efficiently, reducing contamination and improving recovery rates (Khanna et al., 2022). Sensor-tagging technologies—such as RFID, NFC, and digital product passports—enable automated tracking of material flow, product health, and life-cycle history (Bosona & Gebresenbet, 2023; Khanna et al., 2022; Santos et al., 2021). Automated sorting systems in recycling facilities improve material purity and reduce landfill dependence (Khanna et al., 2022). These innovations collectively extend product life, minimize virgin material use, and reduce waste-related emissions (Khanna et al., 2022; Meena et al., 2025).

7.3 Route Optimization and Carbon-Aware Logistics

Transportation is one of the largest contributors to supply chain emissions (Khanna et al., 2022). Automated logistics platforms use AI, IoT data, and advanced optimization algorithms to design routes that minimize fuel consumption, reduce travel distances, and avoid congestion (Khanna et al., 2022). These systems balance constraints such as delivery time windows, energy availability, and vehicle load capacity to ensure optimal routing (Khanna et al., 2022).

Electric and hybrid vehicle fleets can be managed through automated charging schedules synchronized with renewable energy peaks (Khanna et al., 2022). Real-time load balancing ensures that vehicles carry materials efficiently, reducing partial-load trips (Khanna et al., 2022). Combined with automated warehouse coordination, route optimization reduces carbon intensity at every stage of material movement (Khanna et al., 2022).

7.4 Blockchain for Transparent and Sustainable Supply Networks

Blockchain brings transparency, traceability, and accountability to global supply chains (Hasan et al., 2024; Khanna et al., 2022; Okogwu et al., 2023). Distributed ledgers provide tamper-proof records of material origins, environmental impacts, and compliance with sustainability certifications (Khanna et al., 2022; Santos et al., 2021). Automated smart contracts initiate processes—such as payments, shipments, or customs documentation—only when predefined sustainability conditions are met (Hasan et al., 2024).

For manufacturers, blockchain ensures that carbon-neutral claims, ethical sourcing, and waste management practices are validated and trustworthy (Santos et al., 2021). It also enables precise tracking of embedded carbon in complex supply networks, supporting accurate carbon accounting and regulatory reporting (Khanna et al., 2022). When integrated with IoT devices, blockchain enhances the credibility and traceability of sustainability-related data (Hasan et al., 2024; Kaur et al., 2022).

7.5 Automated Life-Cycle Inventory and Environmental Impact Tracking

Life-Cycle Inventory is crucial for assessing the environmental burdens of products across their entire lifespan (Khanna et al., 2022). Automation accelerates LCI generation by collecting real-time data from suppliers, factories, transport providers, and end-of-life handlers (Khanna et al., 2022). Data integration pipelines automatically structure this information for detailed environmental analysis (Khanna et al., 2022).

Automated LCI systems track indicators such as energy use, greenhouse gas emissions, water consumption, particulate emissions, toxicity profiles, and waste volumes (Khanna et al., 2022). Such real-time environmental tracking allows manufacturers to compare design alternatives, optimize product specifications, and ensure that sustainability targets remain aligned with actual supply-chain behavior (Khanna et al., 2022).

7.6 Sustainable Packaging, Distribution, and Warehouse Operations

Packaging and distribution play critical roles in both product protection and environmental performance. automation supports sustainability in these areas by enabling:

- *Automated packaging design tools* that minimize material usage while maintaining structural integrity (Dixit et al., 2023; Khanna et al., 2022).
- *Robotic packaging systems* that ensure precise material handling, reducing excess filler and waste (Khanna et al., 2022).
- *Automated storage and retrieval systems (AS/RS)* that optimize warehouse space, reduce energy-intensive cooling/heating needs, and minimize handling errors (Khanna et al., 2022; Minashkina, 2024).
- *Digital twins of distribution centers*, enabling simulation of energy patterns and the optimization of layout and equipment operation (Khanna et al., 2022; Yakovenko & Shaptala, 2024).

Sustainable distribution also benefits from automated load consolidation, multimodal transport scheduling, and temperature-controlled logistics systems that adjust operation based on environmental conditions (Khanna et al., 2022; Sharma et al., 2022). Together, these systems contribute to lower resource use, reduced emissions, and improved operational resilience.

7.7 End-of-Life (EoL) Automation and Responsible Product Recovery

End-of-life handling determines whether a product becomes waste or a source of valuable resources. Automated EoL systems use advanced robotics, AI-driven material identification, and sensor-enriched inspection tools to sort, disassemble, and recover components efficiently. These systems are particularly valuable in industries such as electronics, automotive, and consumer goods, where products contain mixed materials and complex assemblies (Papalia et al., 2025; Zocco et al., 2024).

AI-driven models guide disassembly sequences, identify reusable parts, and determine optimal recovery pathways. Robotic systems reduce contamination in recycling streams, elevating the quality of recovered materials and improving their suitability for new manufacturing cycles. By integrating automated EoL handling, factories close the resource loop and achieve significant reductions in landfill waste and virgin material consumption (Capetillo et al., 2022; Friege & Kümmerer, 2022).

7.8 Integrated Supply Chain Dashboards and Carbon Intelligence Platforms

Green supply chain operations thrive on real-time visibility. Integrated dashboards consolidate data from suppliers, logistics partners, factories, warehouses, and customers to offer a holistic view of sustainability performance. Carbon intelligence platforms use AI to analyze transport emissions, embedded material carbon, waste indicators, and energy footprints across the entire value chain (Ameh, 2024; Parthasarathy, 2024) .

Automated alerts identify unsustainable practices, bottlenecks, or deviations from environmental targets. These platforms support scenario planning—evaluating the carbon impact of sourcing changes, alternative routes, new packaging choices, or different raw material suppliers. With these insights, manufacturers can manage their supply networks proactively and align operations with long-term net-zero objectives (Dasaklis & Pappis, 2013; Lei, 2024).

7.9 Strategic Role of Automation in Future Green Supply Chains

As global industries pursue net-zero commitments, supply chains will increasingly rely on automation to balance cost, efficiency, resilience, and sustainability. Future green supply chains will incorporate autonomous transport corridors, self-optimizing logistics hubs, AI-driven procurement strategies, and highly traceable circular-resource pathways. These systems will minimize entropy in material flow, reduce overall emissions, and enhance the adaptability of supply networks under variable market and environmental conditions (Abaku et al., 2024).

Automation will also empower manufacturers to internalize environmental costs, respond swiftly to climate regulations, and integrate sustainability as a default operational principle rather than an external requirement. Ultimately, green supply chain automation represents a critical enabler in shaping manufacturing ecosystems that are competitive, transparent, and aligned with global environmental priorities (Awain et al., 2023; Green Supply Chain Management: Strategies for Sustainability and Efficiency, 2025).

8. PRACTICAL IMPLEMENTATIONS

Real-world implementations of AI and automation provide compelling evidence of how advanced technologies enhance energy efficiency, reduce emissions, and promote resource-conscious manufacturing. Siemens' Amberg Electronics Plant in Germany exemplifies a fully digitalized smart factory with 75% of processes automated, AI-powered anomaly detection that minimizes waste, and digital twins optimizing energy consumption, resulting in near-zero defect rates and reduced electricity usage. Schneider Electric's Lexington smart factory in the USA integrates IoT-enabled EcoStruxure™, predictive maintenance via machine learning, and real-time energy dashboards, leading to a 26% reduction in energy consumption and lower machine downtime. BMW's Regensburg and Munich plants utilize AI for paint defect detection, robotics for precise material handling, and autonomous logistics, which lowers energy use in paint shops and reduces scrap rates. Toyota combines Lean manufacturing principles with modern automation and AI, deploying autonomous robots and energy sensors to cut electricity consumption and CO_2 emissions per vehicle. Unilever's factories in India and Europe apply digitalization and AI for utilities monitoring, waste heat recovery, and production optimization, achieving over 50% CO_2 emissions reduction along with significant water and plastic waste savings. Bosch's Industry 4.0 plants in Germany and China employ predictive maintenance, digital twins, and electric automated guided vehicles to reach climate-neutral operations by minimizing downtime and resource waste. Fujitsu's Shimane semiconductor factory leverages AI for cleanroom environment prediction, automated wafer handling, and process simulation to decrease chemical use and emissions. Common themes include automation reducing waste via precision, AI optimizing energy by forecasting and detecting inefficiencies, digital twins strengthening decisions, and IIoT providing transparency. Importantly, many legacy factories have achieved notable sustainability gains, with improved productivity and lower costs simultaneously.

Table 3 highlights how artificial intelligence (AI) and automation are applied in sustainable manufacturing by leading smart factories globally, with an emphasis on optimizing production, minimizing waste, and improving energy efficiency for environmentally responsible manufacturing.

Table 3. Practical Implementations of AI and Automation for Sustainable Manufacturing in Leading Global Smart Factories.

Company & Plant	Technologies & Strategies	Sustainability Features & Impacts	Highlights	Source URL
Siemens Amberg Electronics Plant (DE)	Industrial automation, AI-enabled quality control, high-density sensors, digital twins, real-time energy monitoring	75% automated processes; near-zero defect rates; reduced electricity & compressed-air consumption; optimized energy use	AI-driven anomaly detection, predictive maintenance, digital twin simulations	https://www.siemens.com/global/en/products/automation/topic-areas/digital-enterprise/dex/dex-locations/the-impulse-amberg.html
Schneider Electric Lexington (USA)	EcoStruxure™ IoT monitoring system, IIoT, real-time analytics, machine learning predictive maintenance	26% energy consumption reduction; reduced downtime and waste; real-time carbon tracking	Recognized as WEF Sustainability Lighthouse; legacy plant modernized with 4IR tech	https://www.se.com/ww/en/about-us/newsroom/news/press-releases/schneider-electric-launches-first-smart-factory-in-the-u-s-demonstrating-sustainability-leadership/
BMW Regensburg & Munich Plants (DE)	AI-based image recognition, robotics for material handling, autonomous logistics	Reduced energy in paint shops; lower scrap rates through defect detection	AI integration into quality control and supply chain; internal data lake for production data	https://www.automotivemanufacturingsolutions.com/digitalisation/bmw-reinvents-its-munich-plant-with-ai-automation-and-digital-logs/
Toyota "Smart Factory" Initiative	Lean automation, AI, autonomous material-handling robots, energy monitoring sensors, additive manufacturing	Reduced electricity use; lower CO2 emissions per vehicle; lightweight tooling	Merging Lean principles with digital tech; Jidoka and PDCA cycles enhanced by AI	https://www.automotivemanufacturingsolutions.com/editors-pick/inside-toyotas-2025-leaner-manufacturing-system/645325
Unilever Manufacturing (India & Europe)	Real-time utilities monitoring, IoT waste heat recovery, AI optimization of mixing/heating/packaging	Over 50% CO2 emission reduction; water and plastic waste reduction	Emphasis on digital twin tech and automated real-time monitoring for high-volume manufacturing	https://energydigital.com/articles/unilever-combines-lean-agile-scm
Bosch Industry 4.0 Plants (DE & China)	Predictive maintenance, digital twins for HVAC, electric automated guided vehicles	Achieved climate-neutral operations in many sites; reduced downtime and material waste	Industry 4.0 portfolio-driven energy platform, intelligent algorithms for load predictions	https://www.bosch.com/stories/10-years-industry-4-0-at-bosch/
Fujitsu Shimane Factory (Japan)	AI models for cleanroom condition prediction, automated wafer handling, digital twins	Lower specialty chemical consumption; emissions reduction from HVAC systems	Use of deep learning in defect detection; focus on semiconductor production sustainability	https://corporate-blog.global.fujitsu.com/fgb/2024-03-27/03/

9. KEY CHALLENGES AND LIMITATIONS

Although AI, automation, and digitalization offer powerful pathways toward net-zero factories, their adoption is far from straightforward. Manufacturers must navigate a complex landscape of technological, financial, organizational, and regulatory barriers that influence the pace and scale of sustainability-oriented transformation. Understanding these limitations provides essential context for evaluating what is realistically achievable within industrial environments and where further innovation and policy support are required (Anitha et al., 2023). The following subsections outline the most significant challenges that continue to shape the digital sustainability journey.

9.1 Technical and Integration Barriers

A core challenge lies in integrating advanced digital technologies with existing industrial infrastructure. Manufacturing environments are often built on legacy machinery and decades-old control systems that were never designed for data-driven operations or real-time digital connectivity. Retrofitting these assets with sensors, communication modules, or IIoT gateways can be both technically demanding and financially burdensome. Many machines lack standardized interfaces, forcing engineers to rely on customized adapters or middleware that increase system complexity. Integration across heterogeneous platforms—robotics, PLC networks, SCADA systems, ERP environments, and cloud platforms—often leads to compatibility issues that disrupt workflow continuity. These obstacles slow down automation deployment, limit data availability, and hinder the full use of AI-driven optimization tools (Abdelmoneim et al., 2025; Kantaros & Ganetsos, 2024; Torab-Miandoab et al., 2023).

9.2 High Implementation Costs and Infrastructure Gaps

Digital transformation requires substantial capital investment in robotics, sensing technologies, high-speed communication networks, cloud infrastructure, cybersecurity architecture, and advanced analytics software. For many small and medium-sized enterprises, these costs exceed available budgets or do not align with short-term financial priorities. Even when companies acknowledge the long-term environmental and productivity gains, they may hesitate due to unclear return-on-investment timelines. Infrastructure gaps—such as limited broadband connectivity, unreliable power supply, or inadequate industrial networking hardware—further restrict technology deployment, particularly in developing regions. As a result, the financial hurdle remains one of the most persistent barriers to achieving net-zero manufacturing at scale (Bandara et al., 2025; Beccarello & Foggia, 2023; Development, 2023; Gupta et al., 2021).

9.3 Skills Shortage and Workforce Transformation Challenges

The shift toward digitally enabled sustainability demands a workforce skilled in data interpretation, AI system management, robotics maintenance, cybersecurity, and digitally assisted decision-making. However, many industrial workforces are trained primarily in mechanical or manual operations, creating a significant mismatc (Bühler et al., 2022; Remer et al., 2023)h between required and available competencies. The rapid evolution of digital technologies means that workers must continuously learn new tools, yet many organizations lack structured upskilling programs or adequate digital training frameworks. Additionally, resistance to technological change—rooted in fears of job displacement or unfamiliarity with digital systems—can inhibit adoption. Effective transformation requires not only training but also cultural adaptation, where workers view digital technologies as enablers rather than threats (Adikwu et al., 2024; BOIKANYO, 2025; Mou et al., 2022).

9.4 Issues of Data Availability, Quality, and Governance

AI and automation systems rely on abundant, accurate, and well-structured data. Unfortunately, many factories struggle to generate the data volume and quality required for advanced analytics. Older machines may produce little or no digital data, while newer systems may generate fragmented information across isolated environments. Poor data quality—characterized by inconsistencies, missing values, noise, or non-standard formats—undermines the reliability of predictive algorithms and energy-optimization models. In many cases, organizations lack strong governance frameworks to regulate data access, define data ownership, enforce quality controls, and maintain cybersecurity. Without coherent data governance, sustainability strategies fail to produce reliable and actionable insights (OECD, 2023; Pesqueira et al., 2023; Ribeiro et al., 2022).

9.5 Cybersecurity and Digital Risk in Automated Factories

As manufacturing systems become increasingly connected, exposure to cyber threats expands dramatically. Automated factories rely on continuous communication between sensors, controllers, digital twins, cloud platforms, and enterprise systems. Any breach—whether data theft, ransomware, unauthorized control manipulation, or malicious interference—can compromise operational stability, worker safety, and environmental performance. Cyberattacks on industrial systems can lead to unplanned downtime, faulty machine behavior, and even catastrophic energy waste or material losses. Ensuring cybersecurity requires continuous monitoring, robust access controls, encryption, and a well-defined response strategy—all of which demand significant investment and expertise. These risks make cybersecurity an indispensable component of sustainable digital transformation (SERAC, 2023; Thomas & Sule, 2022).

9.6 Regulatory, Ethical, and Compliance Concerns

The regulatory environment surrounding AI, automation, and environmental performance is evolving rapidly. Manufacturers must comply with diverse regional policies related to carbon accounting, renewable-energy usage, emissions reporting, waste management, data protection, and AI transparency. The lack of harmonized global standards creates uncertainty, making long-term planning difficult. Ethical considerations also arise when AI systems influence decisions related to safety, resource allocation, labor practices, or environmental compliance. Ensuring fairness, transparency, and accountability in algorithmic decision-making is essential to maintain trust among workers, regulators, and the public. Compliance challenges intensify when data crosses borders or involves third-party systems, further complicating the regulatory landscape. These factors collectively underscore the need for clear governance frameworks that balance innovation with responsibility (Chukwurah et al., 2024; Gandhi et al., 2025; Meimandi et al., 2025).

10. EMERGING TRENDS AND FUTURE RESEARCH OPPORTUNITIES

As industries intensify their pursuit of net-zero targets, a new generation of technologies and research directions is beginning to reshape how factories will operate in the coming decades. Beyond current AI, automation, and digitalization efforts, the next phase of sustainable transformation will rely on deeper system intelligence, cross-disciplinary integration, nature-inspired algorithms, and computational models capable of handling unprecedented complexity. This section examines the emerging trends that are expected to redefine net-zero manufacturing, along with the research gaps that necessitate continued scientific inquiry (Fernández-Miguel et al., 2024; Jaouhari et al., 2023).

10.1 AI-Integrated Renewable Energy Microgrids

Manufacturing facilities are increasingly adopting localized renewable-energy systems—solar arrays, wind clusters, bioenergy plants, and energy storage modules—to reduce dependence on carbon-intensive grid power. Emerging microgrid models incorporate AI-based forecasting tools that predict renewable generation patterns, optimize battery charging cycles, and schedule high-energy tasks during periods of surplus clean power. Research is expanding toward fully automated microgrids that adjust load profiles autonomously, reroute power in real time, and coordinate multiple distributed energy resources with minimal human intervention. Integrating machine learning with microgrid controllers offers immense potential for stable, self-correcting low-carbon energy ecosystems (Alenazi, 2025; Singh et al., 2024).

10.2 Quantum Computing for Ultra-Complex Manufacturing Optimization

Quantum computing represents a transformational frontier in solving optimization problems that exceed the capacity of classical computers. Sustainable manufacturing involves thousands of interconnected variables—energy flows, material substitutions, logistics permutations, machine interactions, and carbon trade-offs. Traditional algorithms often struggle to navigate such multi-dimensional complexity in real time. Quantum algorithms, however, can evaluate enormous solution spaces simultaneously, supporting breakthroughs in energy scheduling, chemical-process simulation, supply-chain optimization, and multi-objective sustainability modeling. Although still in the early stages, quantum computing research promises future factories capable of computationally exploring highly complex operational scenarios with unprecedented speed (Hwang et al., 2021; Maaß et al., 2024).

10.3 Autonomous Net-Zero Plants with Self-Learning Systems

The concept of fully autonomous net-zero plants envisions factories where machines, energy systems, logistics networks, and digital platforms collaborate through self-learning mechanisms. Reinforcement learning, evolutionary algorithms, and advanced robotics could allow production systems to adapt continuously based on their environmental impact, operational constraints, and real-time performance. Such plants would automatically adjust production rates, resource consumption, and maintenance schedules to maintain carbon-neutral operations without manual oversight. Future research is moving toward integrating real-time carbon feedback loops into control algorithms so that sustainability becomes a fundamental parameter driving autonomous decision-making (Papalia et al., 2025; Zhao et al., 2025).

10.4 Hybrid Human-AI Decision-Making Frameworks

Although automation and AI bring extraordinary capabilities, human insight remains indispensable—especially in strategic planning, ethical considerations, and high-stakes environmental decisions. The next generation of manufacturing systems will rely on hybrid human-AI frameworks where AI performs high-speed analysis and humans provide contextual judgment, creativity, and oversight. These systems could enable dynamic collaboration where operators receive augmented-reality (AR) support, natural-language explanations of AI recommendations, and visualized sustainability implications. Research is expanding toward models that integrate human expertise directly into AI training cycles, ensuring transparent, interpretable, and trust-aligned decision systems (He et al., 2025; Zschech et al., 2025) (He et al., 2025; Zschech et al., 2025).

10.5 Bio-Inspired and Green AI Technologies

Nature-inspired computational methods—such as swarm intelligence, genetic algorithms, and neural morphogenesis—are beginning to influence energy optimization, waste minimization, and production scheduling. These biologically grounded algorithms are designed to solve complex problems efficiently, often with lower computational overhead. At the same time, the idea of Green AI is gaining prominence, promoting energy-efficient machine-learning models that reduce the carbon footprint of training and inference processes. As sustainability expectations rise, future research will focus on low-power AI chips, environmentally responsible algorithm design, and hybrid bio-digital optimization strategies (Pachot & Patissier, 2023; Tabbakh et al., 2024).

10.6 Next-Generation Digital Twins with Real-Time Carbon Prediction

Digital twins are evolving beyond process simulation toward real-time environmental intelligence. Next-generation models will be able to calculate carbon emissions for every operational state, material interaction, and equipment configuration. These twins will capture dynamic relationships between energy use, machine health, supply-chain changes, and emission profiles. Integrating carbon-prediction analytics directly into operational controls will allow factories to optimize production based on both economic and environmental objectives simultaneously. Research is advancing toward twins capable of multi-scale modeling, combining machine-level detail with plant-level and supply-chain-level environmental analytics (Cao et al., 2025; Johansen et al., 2023).

10.7 Research Gaps and Open Problems

Despite significant progress, several research gaps remain open. These include the need for universally accepted carbon-accounting standards that integrate real-time data; improved interoperability frameworks that unify different digital ecosystems; affordable and scalable digitalization pathways for SMEs; and advanced methods for ensuring cybersecurity in highly automated environments. Additionally, more research is needed in socially responsible AI design, ethical transparency, and long-term sustainability modeling that accounts for climate uncertainty. Addressing these gaps will be crucial for enabling practical, equitable, and globally scalable net-zero manufacturing systems (Brintrup et al., 2023; Cai & Hong, 2024).

11. STRATEGIC ROADMAP FOR INDUSTRY ADOPTION

Transitioning toward net-zero manufacturing requires a structured and multi-layered roadmap that integrates technology, workforce development, policy alignment, and organizational readiness. A carefully planned strategy ensures that sustainability ambitions translate into actionable, measurable, and financially viable industrial transformation. This section presents a comprehensive roadmap outlining how industries can adopt AI-enabled, environmentally responsible manufacturing systems in a phased and coherent manner.

11.1 Step-Wise Implementation Framework for Industry 4.0-Enabled Sustainability

A successful transformation begins with a systematic assessment of existing operations, identifying digital readiness, energy hotspots, equipment health, and major emission sources. Once baseline diagnostics are completed, industries can develop a phased implementation plan that prioritizes high-impact areas such as predictive maintenance, energy monitoring, and supply-chain digitization. Pilot projects serve as controlled environments to validate the feasibility of AI and automation solutions before scaling them across the factory. Over time, incremental integration of robotics, digital twins, IIoT platforms, and carbon-tracking systems leads toward a fully connected and sustainability-optimized manufacturing ecosystem (Cao et al., 2025; Rahmani et al., 2024).

11.2 Technology Selection, Cost-Benefit Analysis, and Scalability Factors

Choosing appropriate technologies requires a balanced evaluation of financial viability, operational impact, technological maturity, and compatibility with existing infrastructure. Factories must perform detailed cost-benefit analyses that consider not only immediate efficiency gains but also long-term energy savings, emission reductions, and resilience enhancements. Scalability is critical: technologies adopted today must remain adaptable to future environmental regulations, higher automation levels, and evolving digital ecosystems. Modular, interoperable solutions often provide the flexibility needed for long-term sustainability growth (Neger et al., 2025; Trump et al., 2025).

11.3 Cross-Sector Collaboration Models and Policy Support

Achieving net-zero manufacturing is a systemic challenge that extends beyond individual companies. Collaborative models involving government agencies, academic institutions, technology providers, supply-chain partners, and industry associations play a central role in accelerating progress. Public-private partnerships can support innovation, reduce financial risk, and create shared infrastructure for renewable energy and digital platforms. Policy frameworks—including emission standards, financial incentives, carbon taxes, and green-technology subsidies—provide an enabling environment that encourages industries to adopt sustainable digital technologies (Nauman et al., 2023; Pop et al., 2023).

11.4 Workforce Upskilling and Organizational Readiness

A digitally enabled sustainable factory relies heavily on a skilled and adaptable workforce. Training programs must cover AI literacy, data analytics, digital maintenance, process automation, energy management, and cybersecurity. Reskilling initiatives should integrate experiential learning—hands-on digital labs, simulation exercises, and collaborative problem-solving environments. Organizational readiness also involves cultural transformation, where sustainability is embedded into decision-making, and employees are encouraged to participate in digital innovation. Strong leadership commitment and transparent communication are crucial for building a workforce that embraces technological change (Dang-Pham et al., 2022; Valtonen & Holopainen, 2025; Wahdaniah et al., 2023).

11.5 Performance Monitoring and Sustainability KPIs

Measuring progress is essential for ensuring continuous improvement. Manufacturers must develop a set of sustainability KPIs that track energy use, emission intensity, waste levels, water consumption, equipment efficiency, renewable-energy penetration, and supply-chain environmental performance(Taneja et al., 2024; Taneja et al., 2025). Real-time dashboards and automated reporting tools allow managers to visualize changes, identify deviations, and intervene promptly. Establishing long-term sustainability targets—aligned with national and global climate commitments—helps maintain momentum and ensures that technology investments produce measurable environmental benefits (Abatan et al., 2024; Pazienza et al., 2024).

11.6 Building Resilient, Climate-Aligned Manufacturing Systems

Ultimately, the goal of AI-enabled sustainability is to create manufacturing systems that are resilient, adaptive, and capable of maintaining high performance under evolving environmental, regulatory, and economic conditions(Arif & Akhai 2024; Khang & Akhai 2025). Resilient systems integrate renewable energy, digital redundancy, predictive maintenance, supply-chain transparency, robust cybersecurity, and diversified material sourcing. Climate-aligned factories anticipate disruptions, adjust their energy and resource strategies proactively, and contribute to long-term environmental stability. Building such systems requires a holistic approach that combines technological innovation with strategic foresight and ecosystem-level collaboration (Rathore, 2016).

12. CONCLUSION

This chapter examined how global climate commitments and rising sustainability pressures are reshaping manufacturing, positioning AI, automation, and digital technologies as core drivers of net-zero industrial systems. It highlighted the importance of life-cycle thinking, circular production principles, and real-time carbon intelligence in achieving meaningful emission reductions. AI techniques—supported by robotics, digital twins, sensor networks, and automated supply chains—were shown to enhance energy efficiency, reduce waste, and improve operational visibility. Real-world case studies demonstrated that intelligent technologies not only strengthen environmental performance but also improve productivity and process reliability. The chapter also addressed the structural and organizational barriers that complicate digital sustainability efforts, setting the stage for future technological advancements and strategic pathways.

Achieving net-zero manufacturing requires coordinated action across policy, technology, and workforce development. Strong regulatory frameworks, clear carbon-accounting rules, and targeted incentives can accelerate digital adoption and renewable-energy integration. Technological progress—including AI-enabled microgrids, advanced analytics, quantum optimization, and next-generation digital twins—must be supported by robust data governance, cybersecurity, and interoperability standards. Equally, human capabilities remain central; workers must be equipped with digital skills, sustainability literacy, and opportunities to collaborate with intelligent systems. When these elements align, industries gain the stability, transparency, and resilience needed to meet long-term climate targets.

REFERENCES

Abadi, M., Liu, C., Zhang, M., Hu, Y., & Xu, Y. (2024). Leveraging AI for energy-efficient manufacturing systems: Review and future prospectives. *Journal of Manufacturing Systems*, *78*, 153–177. DOI: 10.1016/j.jmsy.2024.11.017

Abaku, E. A., Edunjobi, T. E., & Odimarha, A. C.Emmanuel Adeyemi AbakuTolulope Esther EdunjobiAgnes Clare Odimarha. (2024). Theoretical approaches to AI in supply chain optimization: Pathways to efficiency and resilience. *International Journal of Science and Technology Research Archive*, *6*(1), 92. DOI: 10.53771/ijstra.2024.6.1.0033

Abatan, A., Adeyinka, M. A., Sodiya, E. O., Jacks, B. S., Ugwuanyi, E. D., Daraojimba, O. H., & Lottu, O. A. (2024). The role of IT in sustainable environmental management: A global perspective review. *International Journal of Science and Research Archive*, *11*(1), 1874–1886. DOI: 10.30574/ijsra.2024.11.1.0288

Abbass, M., Abbas, U., Jafri, R., Arif, S. M., & Akhai, S. (2025). AI and machine learning applications in sustainable smart cities. In *Sustainable smart cities and the future of urban development* (pp. 1–32). IGI Global., DOI: 10.4018/979-8-3693-6740-7.ch001

Abbass, M., Abbas, U., Wani, S. A., & Akhai, S. (2025). Rethinking global food governance from historical perspectives for a sustainable future. In *Innovative trends shaping food marketing and consumption* (pp. 553–578). IGI Global., DOI: 10.4018/979-8-3693-8542-5.ch022

Abbass, M., & Akhai, S. (2025). Eco-friendly geopolymer bricks synthesized from alkali-activated sugarcane bagasse ash and rice husk ash. *Industrial Crops and Products*, *233*, 121410. DOI: 10.1016/j.indcrop.2025.121410

Abbass, M., & Akhai, S. (2025). Exploring the sustainability potential of geopolymer concrete with coal bottom ash and basalt fibres. *Multiscale and Multidisciplinary Modeling. Experiments and Design*, *8*(3), 199. DOI: 10.1007/s41939-025-00794-3

Abbass, M., Akhai, S., Abbas, U., Jafri, R., & Arif, S. M. (2025). AI-enabled sustainable urban planning and management. In *Real-world applications of AI innovation* (pp. 233–260). IGI Global., DOI: 10.4018/979-8-3693-4252-7.ch012

Abdelmoneim, A. A., Kimaita, H. N., Kalaany, C. M. A., Derardja, B., Dragonetti, G., & Khadra, R. (2025). IoT Sensing for Advanced Irrigation Management: A Systematic Review of Trends, Challenges, and Future Prospects [Review of *IoT Sensing for Advanced Irrigation Management: A Systematic Review of Trends, Challenges, and Future Prospects*]. *Sensors, 25*(7), 2291. Multidisciplinary Digital Publishing Institute. DOI: 10.3390/s25072291

Abhilash, P. M., Luo, X., Liu, Q., Madarkar, R., & Walker, C. (2024). Towards next-gen smart manufacturing systems: The explainability revolution. *Npj Advanced Manufacturing, 1*(1), 8. Advance online publication. DOI: 10.1038/s44334-024-00006-9

Adedokun, O. J., Keshinro, B. I., Uwaoma, I. A., Akindoyin, M. O., Onyia, P. C., & Ofremu, G. O. (2025). AI-DRIVEN SMART MANUFACTURING FOR SUSTAINABLE PRODUCTION. *LEVERAGING AUTOMATION AND ROBOTICS FOR ENERGY EFFICIENCY AND WASTE REDUCTION., 4*(1), 1. DOI: 10.34218/IJAIM_04_01_001

Adikwu, F. E., Esiri, A. E., Aderamo, A. T., Akano, O. A., & Erhueh, O. V.Friday Emmanuel AdikwuAndrew Emuobosa EsiriAdeoye Taofik AderamoOluwaseyi Ayotunde AkanoOvie Vincent Erhueh. (2024). Leveraging digital technologies for health, safety, and environmental (HSE) management in industrial operations. *International Journal of Applied Research in Social Sciences, 6*(11), 2712–2726. DOI: 10.51594/ijarss.v6i11.1722

Agostinho, C., Dikopoulou, Z., Lavasa, E., Perakis, K., Pitsios, S., Branco, R., Reji, S., Hetterich, J., Biliri, E., Lampathaki, F., Rey, S. R. D., & Gkolemis, V. (2023). Explainability as the key ingredient for AI adoption in Industry 5.0 settings. *Frontiers in Artificial Intelligence, 6*, 1264372. Advance online publication. DOI: 10.3389/frai.2023.1264372 PMID: 38146276

Akella, G. K., Wibowo, S., Grandhi, S., & Mubarak, S. (2023). *Systematic Review A Systematic Review of Blockchain Technology Adoption Barriers and Enablers for Smart and Sustainable Agriculture*.

Akhai, S. (2023). Biogas plants for sustainable municipal waste management: A brief review. In *Handbook of research on safe disposal methods of municipal solid wastes for a sustainable environment* (pp. 162–179). IGI Global., DOI: 10.4018/978-1-6684-8117-2.ch012

Akhai, S. (2023). Healthcare record management for healthcare 4.0 via blockchain: A review of current applications, opportunities, challenges, and future potential. In *Blockchain for healthcare 4.0* (Vol. 1, pp. 211–223). CRC Press., DOI: 10.1201/9781003408246-11

Akhai, S. (2024). A Review on Optimizations in μ-EDM Machining of the Biomedical Material Ti6Al4V Using the Taguchi Method: Recent Advances Since 2020. *Latest Trends in Engineering and Technology*, 395-402. DOI: 10.1201/9781032665443-56

Akhai, S. (2024). Innovating healthcare delivery: Harnessing drones for medical services. In *Modelling of virtual worlds using the Internet of Things* (pp. 232–250). CRC Press., DOI: 10.1201/9781003480181-10

Akhai, S. (2024). Smart home environment for society: An overview of IoT-enabled air conditioners. In *Modelling of virtual worlds using the Internet of Things* (Vol. 1, pp. 74–87). CRC Press., DOI: 10.1201/9781003480181-3

Akhai, S. (2024). *Towards trustworthy and reliable AI: The next frontier.* In *Explainable artificial intelligence (XAI) in healthcare* (Vol. 1, pp. 119–129). CRC Press. DOI: 10.1201/9781003426073-7

Akhai, S., & Balu, V. (2022). *Code injection assault & mitigation model to prevent attacks.* In *Proceedings of the 2022 11th International Conference on System Modeling& Advancement in Research Trends (SMART)* (pp. 1324–1328). IEEE. DOI: 10.1109/SMART55829.2022.10046835

Akhai, S., Bansal, S. A., & Singh, S. (2020). A critical review of thermal insulators from natural materials for energy saving in buildings. *Journal of Critical Reviews, 7*(19), 278–283.

Akhai, S., & Khang, A. (2024). Energy efficiency and human comfort: AI and IoT integration in hospital HVAC systems. In *Medical robotics and AI-assisted diagnostics for a high-tech healthcare industry* (pp. 93–108). IGI Global., DOI: 10.4018/979-8-3693-2105-8.ch007

Akhai, S., & Khang, A. (2024). Innovations in medical manufacturing: A review of 3D printing, robotics, and Internet of Things (IoT). In *The quantum evolution* (Vol. 1, pp. 226–241). CRC Press., DOI: 10.1201/9781032642079-10

Akhai, S., & Singh, H. (2013). Design optimization for modification of trough belt conveyor to reduce material spillage used in clinker transport in cement plant. *International Journal of Research, 1*, 27–37.

Akhai, S., Singh, V. P., & John, S. (2016). Human performance in industrial design centers with small unit air conditioning systems. *Journal of Advanced Research in Production Industrial Engineering, 3*(2), 5–11.

Akhai, S., & Wadhwa, A. (2024). Enhancing radiator cooling with CuO nanofluid microchannels. *International Journal of Automotive Science and Technology, 8*(2), 201–211. DOI: 10.30939/ijastech..1399702

Akhai, S., & Wadhwa, A. S. (2024). Wireless networking solutions for climate control and comfort in modern air conditioning systems. In *Intelligent networks* (pp. 165–181). CRC Press., DOI: 10.1201/9781003541363-11

Akhavan, J., Lyu, J., & Manoochehri, S. (2023). A Deep Learning Solution for Real-time Quality Assessment and Control in Additive Manufacturing Using Point Cloud Data. *Research Square (Research Square)*. DOI: 10.21203/rs.3.rs-2129511/v5

Akhbardeh, F., Desell, T., & Zampieri, M. (2020). *MaintNet: A Collaborative Open-Source Library for Predictive Maintenance Language Resources*. 7. DOI: 10.18653/v1/2020.coling-demos.2

Alemão, D., Rocha, A. D., & Barata, J. (2019). Production Scheduling Requirements to Smart Manufacturing. In *IFIP advances in information and communication technology* (p. 227). Springer Science+Business Media. DOI: 10.1007/978-3-030-17771-3_19

Alenazi, M. M. (2025). Machine Learning-Optimized Energy Management for Resilient Residential Microgrids with Dynamic Electric Vehicle Integration. *Journal on Artificial Intelligence*, *7*(1), 143–176. DOI: 10.32604/jai.2025.066067

Alevizos, V., Georgousis, I., & Kapodistria, A.-M. (2023). Towards Climate Neutrality: A Comprehensive Overview of Sustainable Operations Management, Optimization, and Wastewater Treatment Strategies. *arXiv (Cornell University)*. https://doi.org//arxiv.2308.00808DOI: 10.48550

Alex, B. J. A. R., & Johnson, M. (2025). A Framework for IoT-Enabled Smart Manufacturing for Energy and Resource Optimization. *arXiv (Cornell University)*. https://doi.org//arxiv.2502.03040DOI: 10.48550

Alghieth, M. (2025). Sustain AI: A Multi-Modal Deep Learning Framework for Carbon Footprint Reduction in Industrial Manufacturing. *Sustainability (Basel)*, *17*(9), 4134. DOI: 10.3390/su17094134

Alli, Y. A., Anuar, H., Manshor, M. R., Okafor, C. E., Kamarulzaman, A. F., Akçakale, N., Nazeri, F. N. M., Bodaghi, M., Suhr, J., & Nasir, N. A. M. (2024). Optimization of 4D/3D printing via machine learning: A systematic review [Review of *Optimization of 4D/3D printing via machine learning: A systematic review*]. *Hybrid Advances, 6*, 100242. Elsevier BV. DOI: 10.1016/j.hybadv.2024.100242

Almusharraf, A. (2025). Automation and Its Influence on Sustainable Development: Economic, Social, and Environmental Dimensions. *Sustainability (Basel)*, *17*(4), 1754. DOI: 10.3390/su17041754

Ameh, B. (2024). Digital tools and AI: Using technology to monitor carbon emissions and waste at each stage of the supply chain, enabling real-time adjustments for sustainability improvements. *International Journal of Science and Research Archive, 13*(1), 2741–2757. DOI: 10.30574/ijsra.2024.13.1.1995

Anitha, D., Awasthi, A., Dwivedi, S. P., Kumar, R., Pahwa, S., Rajalakshmi, B., & Alkhafaji, M. A. (2023). Beyond Limits: Envisioning Sustainable Design and Production Ecosystems. *E3S Web of Conferences, 453*, 1039. DOI: 10.1051/e3sconf/202345301039

Anjum, M., Kraïem, N., Min, H., Daradkeh, Y. I., Dutta, A. K., & Shahab, S. (2025). Optimizing sustainability: Information aggregation in smart factories for energy efficiency enhancement. *Scientific Reports, 15*(1), 8603. DOI: 10.1038/s41598-024-83304-8 PMID: 40075102

Ardito, S., Setiawan, W., & Wibisono, A. (2024). Enhancing Predictive Maintenance in Manufacturing Using Deep Learning-Based Anomaly Detection. *Enhancing Predictive Maintenance in Manufacturing Using Deep Learning-Based Anomaly Detection., 3*(1), 12–23. DOI: 10.63876/ijtm.v3i1.112

Arellano-Espitia, F., Delgado-Prieto, M., Martínez-Viol, V., Fernandez-Sobrino, A., & Osornio-Rios, R. A. (2021). *Anomaly Detection in Electromechanical Systems by means of Deep-Autoencoder*. 1. DOI: 10.1109/ETFA45728.2021.9613529

Arif, S. M., & Akhai, S. (2024). Fostering a sustainable and disaster-resilient transportation infrastructure. In *Challenges, strategies, and resiliency in disaster and risk management* (pp. 205–219). IGI Global., DOI: 10.4018/979-8-3693-2721-0.ch009

(2025). Artificial intelligence in manufacturing – systematic literature review. (2025). *Scientific Papers of Silesian University of Technology Organization and Management Series, 2025*(215). Advance online publication. DOI: 10.29119/1641-3466.2025.215.39

Assad, F., Rushforth, E., & Harrison, R. (2023). A component-based design approach for energy flexibility in cyber-physical manufacturing systems. *Journal of Intelligent Manufacturing*. Advance online publication. DOI: 10.1007/s10845-023-02280-4

Awain, A. M. S. B., Al-Ansi, A. M., & Jaboob, M. (2023). Green Supply Chain Management: A Comprehensive Review of Research, Applications and Future Directions [Review of *Green Supply Chain Management: A Comprehensive Review of Research, Applications and Future Directions*]. *Management and Production Engineering Review*. De Gruyter Open. https://doi.org/DOI: 10.24425/mper.2023.147194

Ba, L., Tangour, F., Abbassi, I. E., & Absi, R. (2025). Analysis of Digital Twin Applications in Energy Efficiency: A Systematic Review [Review of *Analysis of Digital Twin Applications in Energy Efficiency: A Systematic Review*]. *Sustainability, 17*(8), 3560. Multidisciplinary Digital Publishing Institute. DOI: 10.3390/su17083560

Badhoutiya, A., Darokar, H., Verma, R. P., Saraswat, M., Devaraj, S., Raj, V. H., &Abdulhussain, Z. N. (2023). Regenerative Manufacturing: Crafting a Sustainable Future through Design and Production. *E3S Web of Conferences, 453*, 1038. DOI: 10.1051/e3sconf/202345301038

Badhoutiya, A., Meshram, V., Gori, Y., Yadav, D. K., Viswanath, G., Revathi, V., &Brayyich, M. (2023). Blueprints for Green Horizons: Sustainable Strategies in Design and Production. *E3S Web of Conferences, 453*, 1030. DOI: 10.1051/e3sconf/202345301030

Banaulikar, A. (2020). Real Time Monitoring and Control for Industrial Automation using PLC. *International Journal for Research in Applied Science and Engineering Technology, 8*(7), 1193–1200. DOI: 10.22214/ijraset.2020.30453

Bandara, J., Ray, R., Lu, J., & Gallagher, K. P. (2025). Developing countries locked out of low-carbon technology trade. *Science, 388*(6744), 248–250. DOI: 10.1126/science.adu7731 PMID: 40245136

Bányai, T., & Tóth, Á. B. (2025). Real-Time Maintenance Optimization with Industrial Internet of Things. *Applied Sciences (Basel, Switzerland), 15*(10), 5640. DOI: 10.3390/app15105640

Barbhuiya, S., & Das, B. B. (2023). Life Cycle Assessment of construction materials: Methodologies, applications and future directions for sustainable decision-making. *Case Studies in Construction Materials, 19*, e02326. Advance online publication. DOI: 10.1016/j.cscm.2023.e02326

Barteková, E., Bibas, R., Börkey, P., Brown, A., Buzzi, E., Château, J., Cornago, E., Dellink, R., Dubois, M., Helf, M., Lanzi, E., Laubinger, F., Livingstone, L., Mavroeidi, E., McCarthy, A., Aigues, N., Yamaguchi, S., Ijeoma, I., Dalles, K., . . . Perrière, E. (2022). Synergies and trade-offs in the transition to a resource-efficient and circular economy. In *OECD environment policy papers*. DOI: 10.1787/23097841

Bartolomé, N., Hilber, I., Schulin, R., Mayer, P., Witt, G., & Bucheli, T. D. (2016). Equilibrium sampling in a semi in situ pot experiment to measure freely dissolved concentrations of hydrophobic organic compounds. *Research Portal Denmark*, 260. https://local.forskningsportal.dk/local/dki-cgi/ws/cris-link?src=dtu&id=dtu-fd499a44-cb03-4954-a110-212f7de6c3ef&ti=Equilibrium%20sampling%20in%20a%20semi%20in%20situ%20pot%20experiment%20to%20measure%20freely%20dissolved%20concentrations%20of%20hydrophobic%20organic%20compounds

Barzegaran, M. (2021). Configuration Optimization of Fog Computing Platforms for Control Applications. *Research Portal Denmark*, 194. https://local.forskningsportal.dk/local/dki-cgi/ws/cris-link?src=dtu&id=dtu-92dc705a-b8f0-41d2-928e-86a949a69162&ti=Configuration%20Optimization%20of%20Fog%20Computing%20Platforms%20for%20Control%20Applications

Bataille, C. (2019). Physical and policy pathways to net-zero emissions industry. *Wiley Interdisciplinary Reviews: Climate Change*, *11*(2), e633. Advance online publication. DOI: 10.1002/wcc.633

Bataille, C., Åhman, M., Neuhoff, K., Nilsson, L. J., Fischedick, M., Lechtenböhmer, S., Solano-Rodriquez, B., Denis-Ryan, A., Stiebert, S., Waisman, H., Sartor, O., & Rahbar, S. (2018). A review of technology and policy deep decarbonization pathway options for making energy-intensive industry production consistent with the Paris Agreement [Review of *A review of technology and policy deep decarbonization pathway options for making energy-intensive industry production consistent with the Paris Agreement*]. *Journal of Cleaner Production, 187*, 960. Elsevier BV. DOI: 10.1016/j.jclepro.2018.03.107

Bataille, C., Nilsson, L. J., & Jotzo, F. (2021). Industry in a net-zero emissions world: New mitigation pathways, new supply chains, modelling needs and policy implications. *Energy and Climate Change*, *2*, 100059. DOI: 10.1016/j.egycc.2021.100059

Beccarello, M., & Foggia, G. D. (2023). The Strategic Role of Energy Efficiency and Industry 4.0 Interventions in Manufacturing. *Journal of Management and Sustainability*, *13*(1), 31. DOI: 10.5539/jms.v13n1p31

Bechtsis, D., Tsolakis, N., Vlachos, D., & Iakovou, E. (2016). Sustainable supply chain management in the digitalisation era: The impact of Automated Guided Vehicles. *Journal of Cleaner Production*, *142*, 3970–3984. DOI: 10.1016/j.jclepro.2016.10.057

Bertsimas, D., Boussioux, L., & Zeng, C. (2023). Reducing Air Pollution through Machine Learning. *arXiv (Cornell University)*. https://doi.org//arxiv.2303.12285DOI: 10.48550

Bharadwaj, G., Chopde, S., Taluja, R., Gummidi, L., Chandrashekar, R., &Dhahi, H. A. (2023). Closing the Loop: Advances in Materials, Energy, and Waste Management. *E3S Web of Conferences, 453*, 1024. DOI: 10.1051/e3sconf/202345301024

Blasco, S. V., & Lambán, M. P. (2025). Enhancing Sustainability with LCA: A Comparative Analysis of Design and Manufacturing Processes. *Processes (Basel, Switzerland), 13*(1), 195. DOI: 10.3390/pr13010195

BOIKANYO, D. H. (2025). HARNESSING DIGITAL TECHNOLOGIES FOR VALUE CHAIN OPTIMIZATION AND SOCIETAL TRANSFORMATION. *BUSINESS EXCELLENCE AND MANAGEMENT, 15*, 14. DOI: 10.24818/beman/2025.S.I.5-02

Borangiu, T., Morariu, O., Răileanu, S., Trentesaux, D., Leitão, P., & Barata, J. (2020). Digital transformation of manufacturing. Industry of the Future with Cyber-Physical Production Systems. *HAL (Le Centre Pour La Communication Scientifique Directe)*. https://uphf.hal.science/hal-03424105

Boscariol, P., & Richiedei, D. (2019). Trajectory Design for Energy Savings in Redundant Robotic Cells. *Robotics (Basel, Switzerland), 8*(1), 15. DOI: 10.3390/robotics8010015

Bosona, T., &Gebresenbet, G. (2023). The Role of Blockchain Technology in Promoting Traceability Systems in Agri-Food Production and Supply Chains [Review of *The Role of Blockchain Technology in Promoting Traceability Systems in Agri-Food Production and Supply Chains*]. *Sensors, 23*(11), 5342. Multidisciplinary Digital Publishing Institute. DOI: 10.3390/s23115342

Brintrup, A., Baryannis, G., Tiwari, A., Ratchev, S., Martínez-Arellano, G., & Singh, J. (2023). Trustworthy, responsible, ethical AI in manufacturing and supply chains: synthesis and emerging research questions. *arXiv (Cornell University)*. https://doi .org//arxiv.2305.11581DOI: 10.48550

Brundage, M. P., Sexton, T., Hodkiewicz, M., Dima, A., & Lukens, S. (2020). Technical language processing: Unlocking maintenance knowledge. *Manufacturing Letters, 27*, 42–46. DOI: 10.1016/j.mfglet.2020.11.001

Brusso, B. C. (2018). 50 Years of Industrial Automation [History]. *IEEE Industry Applications Magazine, 24*(4), 8–11. DOI: 10.1109/MIAS.2018.2820440

Bühler, M. M., Jelinek, T., & Nübel, K. (2022). Training and Preparing Tomorrow's Workforce for the Fourth Industrial Revolution. *Education Sciences, 12*(11), 782. DOI: 10.3390/educsci12110782

C., J. L. D., & Ocampo-Martínez, C. (2019). Energy efficiency in discrete-manufacturing systems: Insights, trends, and control strategies. *Journal of Manufacturing Systems, 52*, 131. DOI: 10.1016/j.jmsy.2019.05.002

Cai, T., & Hong, Z. (2024). Exploring the structure of the digital economy through blockchain technology and mitigating adverse environmental effects with the aid of artificial neural networks. *Frontiers in Environmental Science, 12*, 1315812. Advance online publication. DOI: 10.3389/fenvs.2024.1315812

Callari, T. C., Segate, R. V., Hubbard, E., Daly, A., & Lohse, N. (2024). An ethical framework for human-robot collaboration for the future people-centric manufacturing: A collaborative endeavour with European subject-matter experts in ethics. *Technology in Society, 78*, 102680. DOI: 10.1016/j.techsoc.2024.102680

Calvo, R., & Gil, P. M. (2022). Evaluation of Collaborative Robot Sustainable Integration in Manufacturing Assembly by Using Process Time Savings. *Materials (Basel), 15*(2), 611. DOI: 10.3390/ma15020611 PMID: 35057338

Cannava, L., Javan, F. D., Najafi, B., & Perotti, S. (2024). Green warehousing practices: Assessing the impact of PV self-consumption enhancement strategies in a logistics warehouse. *Sustainable Energy Technologies and Assessments, 72*, 104054. DOI: 10.1016/j.seta.2024.104054

Cao, H., Söderlund, H., Fang, Q., Chen, S., Erdal, L., Gubartalla, A., Lopes, P. V., Shao, G., Lonnehed, P., Putto, H., Sekh, A. A., Ekered, S., & Johansson, B. (2025). Towards AI-based Sustainable and XR-based human-centric manufacturing: Implementation of ISO 23247 for digital twins of production systems. *arXiv (Cornell University)*. https://doi.org//arxiv.2508.14580DOI: 10.48550

Capetillo, A. A., Bauer, F., & Chaminâde, C. (2022). Emerging Technologies Supporting the Transition to a Circular Economy in the Plastic Materials Value Chain [Review of *Emerging Technologies Supporting the Transition to a Circular Economy in the Plastic Materials Value Chain*]. *Circular Economy and Sustainability, 3*(2), 953. Springer Nature. DOI: 10.1007/s43615-022-00209-2

Catti, P., Freitas, A., Pereira, E., Gonçalves, G., Lopes, R. P., Nikolakis, N., & Alexopoulos, K. (2024). Data Analytics and AI for Quality Assurance in Manufacturing: Challenges and Opportunities. In *Lecture notes in networks and systems* (p. 205). Springer International Publishing. DOI: 10.1007/978-3-031-65411-4_25

Chen, H., Jeremiah, S. R., Lee, C., & Park, J. H. (2023). A Digital Twin-Based Heuristic Multi-Cooperation Scheduling Framework for Smart Manufacturing in IIoT Environment. *Applied Sciences (Basel, Switzerland), 13*(3), 1440. DOI: 10.3390/app13031440

Chen, Y. (2022). Advantages of 3D Printing for Circular Economy and Its Influence on Designers. *Proceedings of the Design Society, 2*, 991–1000. DOI: 10.1017/pds.2022.101

Chen, Y.-H., Sharma, K., Sharma, C., & Sharma, S. (2023). Integrating explainable artificial intelligence and blockchain to smart agriculture: Research prospects for decision making and improved security. *Smart Agricultural Technology, 6*, 100350. DOI: 10.1016/j.atech.2023.100350

Chukwurah, N., Ige, A. B., Adebayo, V. I., & Eyieyien, O. G. (2024). Frameworks for effective data governance: Best practices, challenges, and implementation strategies across industries. *Computer Science & IT Research Journal, 5*(7), 1666–1679. DOI: 10.51594/csitrj.v5i7.1351

Chyr, G., & DeSimone, J. M. (2022). Review of high-performance sustainable polymers in additive manufacturing. *Green Chemistry, 25*(2), 453–466. DOI: 10.1039/D2GC03474C

Collan, M., & Michelsen, K.-E. (2020). *Technical, Economic and Societal Effects of Manufacturing 4.0.* DOI: 10.1007/978-3-030-46103-4

Colledani, M., Tolio, T., Fischer, A., Iung, B., Lanza, G., Schmitt, R., & Váncza, J. (2014). Design and management of manufacturing systems for production quality. *CIRP Annals, 63*(2), 773–796. DOI: 10.1016/j.cirp.2014.05.002

Cruz, Y. J., Castaño, F., Haber, R. E., Villalonga, A., Ejsmont, K., Gładysz, B., Flores, Á., & Alemany, P. (2024). *Self-Reconfiguration for Smart Manufacturing Based on Artificial Intelligence: A Review and Case Study* [Review of *Self-Reconfiguration for Smart Manufacturing Based on Artificial Intelligence: A Review and Case Study*]. 121. DOI: 10.1007/978-3-031-46452-2_8

D'Eusanio, M., & Petti, L. (2024). Blockchain technology and Social Life Cycle Assessment: Synergies and implications. *The International Journal of Life Cycle Assessment.* Advance online publication. DOI: 10.1007/s11367-024-02338-2

Dang-Pham, D., Hoang, A.-P., Vo, D.-T., & Kautz, K. (2022). Digital Kaizen: An Approach to Digital Transformation. *AJIS. Australasian Journal of Information Systems/AJIS. Australian Journal of Information Systems. AJIS. Australasian Journal of Information Systems, 26.* Advance online publication. DOI: 10.3127/ajis.v26i0.3851

Dasaklis, T. K., & Pappis, C. P. (2013). Supply chain management in view of climate change: An overview of possible impacts and the road ahead. *Journal of Industrial Engineering and Management, 6*(4). Advance online publication. DOI: 10.3926/jiem.883

Dennison, M. S., Kumar, M. B., & Jebabalan, S. K. (2024). Realization of circular economy principles in manufacturing: Obstacles, advancements, and routes to achieve a sustainable industry transformation. *Discover Sustainability*, *5*(1), 438. Advance online publication. DOI: 10.1007/s43621-024-00689-2

Development, U. N. C. on T. and. (2023). Harnessing green technology and innovation for inclusive and sustainable development. In *United Nations eBooks* (p. 33). United Nations. DOI: 10.18356/9789210028295c011

Dewulf, J., Beylot, A., Monfort, D., Lai, F., Santillán-Saldivar, J., Muller, S., & Mathieux, F. (2023). Contribution to inaccessibility as resource impact method: A base for sustainable resource management along the life cycle. *Resources, Conservation and Recycling*, *202*, 107363. DOI: 10.1016/j.resconrec.2023.107363

Dhiman, S., Seema, S., & Akhai, S. (2025). Performance evaluation of OPC 43 concrete incorporating micro silica and recycled ceramic tile waste as partial cement and coarse aggregate replacements. *Asian Journal of Civil Engineering*, 1-19. DOI: 10.1007/s42107-025-01504-7

Diaz, J. L., Bermeo, M. A., Díaz-Rozo, J., & Ocampo-Martínez, C. (2019). An optimization-based control strategy for energy efficiency of discrete manufacturing systems. *ISA Transactions*, *93*, 399–409. DOI: 10.1016/j.isatra.2019.03.015 PMID: 30955833

Dimitrijević, B., Stanković, M., & Anđelković, T. (2023). Synergy between the circular economy and 3D printing. *Ekonomski Signali*, *18*(2), 49–68. DOI: 10.5937/ekonsig2302049D

Dimitriou, N., Leontaris, L., Vafeiadis, T., Ioannidis, D., Wotherspoon, T., Tinker, G., & Tzovaras, D. (2019). A Deep Learning framework for simulation and defect prediction applied in microelectronics. *Simulation Modelling Practice and Theory*, *100*, 102063. DOI: 10.1016/j.simpat.2019.102063

Ding, Y., & Li, L. (2024). Optimization Method for Digital Twin Manufacturing System Based on NSGA-II. *International Journal of Advanced Computer Science and Applications*, *15*(4). Advance online publication. DOI: 10.14569/IJACSA.2024.01504100

Dip, H. S. (2024). Circular Economy Initiatives in Food Manufacturing Industry. *Research Square (Research Square)*. DOI: 10.21203/rs.3.rs-4996705/v1

Dixit, K. K., Dharme, A., Singh, D. P., Kalra, R., Bhavani, B., Asha, V., &Abdulsada, Zainab. R. (2023). Sustainability by Design: Innovative Ways of Revolutionizing Production Practices for a Better Tomorrow. *E3S Web of Conferences, 453*, 1026. DOI: 10.1051/e3sconf/202345301026

Domb, M., Joshi, S., & Khn, A. (2023). Anomaly Detection in IoT: Recent Advances, AI and ML Perspectives and Applications. In *Artificial intelligence*. IntechOpen. DOI: 10.5772/intechopen.111944

dos Santos, R. B., Torrisi, N. M., & Pantoni, R. P. (2021). Third Party Certification of Agri-Food Supply Chain Using Smart Contracts and Blockchain Tokens. *Sensors (Basel), 21*(16), 5307. DOI: 10.3390/s21165307 PMID: 34450749

Elahi, M., Afolaranmi, S. O., Lastra, J. L. M., & García, J. A. P. (2023). A comprehensive literature review of the applications of AI techniques through the lifecycle of industrial equipment. *Discover Artificial Intelligence, 3*(1), 43. Advance online publication. DOI: 10.1007/s44163-023-00089-x

Elmegaard, B., Kermani, N. A., Bühler, F., Nguyen, T.-V., Bergamini, R., Zühlsdorf, B., Holdt, F. D., Holm, F. M., Petersen, M. S., Helk, A., Mogensen, N. B., Jürgensen, P., Hansen, L. I., Munk, P., Godiksen, E. L., Hetting, V., & Jacobsen, E. (2021). Electrification of processes and technologies for Danish Industry: Elforsk project 350-038 : Final Report. In *Research Portal Denmark* (p. 348). Technical University of Denmark., https://local.forskningsportal.dk/local/dki-cgi/ws/cris-link?src=dtu&id=dtu-39c8ef2f-cd96-4e85-bc7f-9e5135da93ca&ti=Electrification%20of%20processes%20and%20technologies%20for%20Danish%20Industry%3A%20Elforsk%20project%20350-038%20%3A%20Final%20Report

Esiri, A. E., Sofoluwe, O. O., & Ukato, A. (2024). Digital twin technology in oil and gas infrastructure: Policy requirements and implementation strategies. *Engineering Science & Technology Journal, 5*(6), 2039–2049. DOI: 10.51594/estj.v5i6.1221

Fernández-Miguel, A., Muiña, F. E. G., Settembre-Blundo, D., Tarantino, S. C., & Riccardi, M. P. (2024). Exploring Systemic Sustainability in Manufacturing: Geoanthropology's Strategic Lens Shaping Industry 6.0. *Global Journal of Flexible Systems Managment, 25*(3), 579–600. DOI: 10.1007/s40171-024-00404-0

Filho, W. L., Aina, Y. A., Gatto, A., Rampasso, I. S., Sigahi, T. F. A. C., Anholon, R., Bella, R. L. F., Quelhas, O. L. G., Vasconcelos, C. R. P., & Luetz, J. M. (2025). Greening the factory floor and reducing the climate impact of the manufacturing sector. *Discover Sustainability, 6*(1), 1204. Advance online publication. DOI: 10.1007/s43621-025-02056-1

Fottner, J., Clauer, D., Hormes, F., Freitag, M., Beinke, T., Overmeyer, L., Gottwald, S. N., Elbert, R., Sarnow, T., Schmidt, T., Reith, K.-B., Zadek, H., & Thomas, F. (2021). Autonomous Systems in Intralogistics – State of the Art and Future Research Challenges. [Technical University of Darmstadt]. *TUbilio*, *14*, 2. DOI: 10.23773/2021_1

Frăţilă, D., & Rotaru, H. (2017). Additive manufacturing – a sustainable manufacturing route. *MATEC Web of Conferences*, *94*, 3004. DOI: 10.1051/matecconf/20179403004

Friege, H., & Kümmerer, K. (2022). Practising circular economy. In *Routledge eBooks* (p. 259). Informa., DOI: 10.4324/9781003244196-27

Gadaleta, M., Pellicciari, M., & Berselli, G. (2019). Optimization of the energy consumption of industrial robots for automatic code generation. *Robotics and Computer-integrated Manufacturing*, *57*, 452–464. DOI: 10.1016/j.rcim.2018.12.020

Gailani, A., Cooper, S., Allen, S., Pimm, A., Taylor, P., & Gross, R. (2024). Assessing the potential of decarbonization options for industrial sectors. *Joule*, *8*(3), 576–603. DOI: 10.1016/j.joule.2024.01.007

Galetić, V., & Nottle, A. (2023). Flexible and Inherently Comprehensible Knowledge Representation for Data-Efficient Learning and Trustworthy Human-Machine Teaming in Manufacturing Environments. *arXiv (Cornell University)*. https://doi.org//arxiv.2305.11597DOI: 10.48550

Gallab, M., Nardo, M. D., & Naciri, L. (2024). Navigating contemporary challenges and future prospects in digital industry evolution. *Discover Applied Sciences*, *6*(5), 259. Advance online publication. DOI: 10.1007/s42452-024-05913-2

Gandhi, D., Joshi, H., Hartman, L., & Hassani, S. (2025). Approaches to Responsible Governance of GenAI in Organizations. *arXiv (Cornell University)*. https://doi.org//arxiv.2504.17044DOI: 10.48550

Gao, Q., & Schweidtmann, A. M. (2024). Deep reinforcement learning for process design: Review and perspective. *Current Opinion in Chemical Engineering*, *44*, 101012. DOI: 10.1016/j.coche.2024.101012

Gargalo, C. L., Yu, H., Vollmer, N. I., Arabkoohsar, A., Gernaey, K. V., & Sin, G. (2023). A process systems engineering view of environmental impact assessment in renewable and sustainable energy production: Status and perspectives. *Computers & Chemical Engineering*, *180*, 108504. DOI: 10.1016/j.compchemeng.2023.108504

Gebler, M., Cerdas, F., Thiede, S., & Herrmann, C. (2020). Life cycle assessment of an automotive factory: Identifying challenges for the decarbonization of automotive production – A case study. *Journal of Cleaner Production, 270*, 122330. DOI: 10.1016/j.jclepro.2020.122330

Gebler, M., Warsen, J., Meininghaus, R., Baudis, M., Cerdas, F., & Herrmann, C. (2024). Implementing Zero Impact Factories in Volkswagen's Global Automotive Manufacturing System: A Discussion of Opportunities and Challenges from Integrating Current Science into Strategic Management. *Sustainability (Basel), 16*(7), 3011. DOI: 10.3390/su16073011

Geng, D., Evans, S., & Kishita, Y. (2023). The identification and classification of energy waste for efficient energy supervision in manufacturing factories. *Renewable & Sustainable Energy Reviews, 182*, 113409. DOI: 10.1016/j.rser.2023.113409

Ghahramani, M., Qiao, Y., Zhou, M., O'Hagan, A., & Sweeney, J. F. (2020b). AI-based Modeling and Data-driven Evaluation for Smart Manufacturing\n Processes. *arXiv (Cornell University)*. https://doi.org//arxiv.2008.12987DOI: 10.48550

Ghahramani, M., Qiao, Y., Zhou, M. C., O'Hagan, A., & Sweeney, J. (2020a). AI-based modeling and data-driven evaluation for smart manufacturing processes. *IEEE/CAA Journal of Automatica Sinica, 7*(4), 1026. DOI: 10.1109/JAS.2020.1003114

Ghayal, A. (2025). *Embedding ESG in Cost Accounting: A Simulation Framework for Financial Reporting and Transparency*. DOI: 10.31235/osf.io/4ge2z_v1

Gheewala, S. H. (2020). Life Cycle Thinking for Sustainable Consumption and Production towards a Circular Economy. *E3S Web of Conferences, 202*, 1003. DOI: 10.1051/e3sconf/202020201003

Glenk, G., Meier, R., & Reichelstein, S. (2024). Assessing the Costs of Industrial Decarbonization. SSRN *Electronic Journal*. DOI: 10.2139/ssrn.4998042

Gontarz, A., Hampl, D., Weiss, L., & Wegener, K. (2015). Resource Consumption Monitoring in Manufacturing Environments. *Procedia CIRP, 26*, 264–269. DOI: 10.1016/j.procir.2014.07.098

Görür, O. C., Yu, X., & Sivrikaya, F. (2021). Integrating Predictive Maintenance in Adaptive Process Scheduling for a Safe and Efficient Industrial Process. *Applied Sciences (Basel, Switzerland), 11*(11), 5042. DOI: 10.3390/app11115042

Green Supply Chain Management: Strategies for Sustainability and Efficiency. (2025). DOI: 10.52783/eel.v15i1.2433

Gupta, H., Lawal, J. N., Orji, I. J., & Kusi-Sarpong, S. (2021). Closing the Gap: The Role of Distributed Manufacturing Systems for Overcoming the Barriers to Manufacturing Sustainability. *IEEE Transactions on Engineering Management, 70*(5), 1754–1773. DOI: 10.1109/TEM.2021.3059231

Gutmann, T., Nyffenegger, F., Pellegrini, M., Cabrucci, A., & Guzzini, A. (2023). A Digital Twin-Based Approach for the Optimization of Floor-Ball Manufacturing. *Electronics (Basel), 12*(24), 4979. DOI: 10.3390/electronics12244979

Hafiz, M., Azmi, E. D. S. K., Azlie, M. I. N. N., Zaki, M. J., Mazilan, M. A. H., Mazuki, M. S. M., Nor, M. I. N. M., Ghani, J. A., Rahmat, M. S., & Khamis, N. K. (2025). Application of Artificial Intelligence in Electronics and Semiconductor Industries. *JurnalKejuruteraan, 37*(7), 3245. DOI: 10.17576/jkukm-2025-37(7)-11

Hammadi, M. (2021). Multidisciplinary collaborative design of complex systems. *HAL (Le Centre Pour La Communication Scientifique Directe)*. https://hal.science/tel-03614326 1. Hanenkamp, N., & Zipse, O. (2023). Sustainability in Manufacturing Transforming. In *Springer eBooks* (p. 187). Springer Nature. DOI: 10.1007/978-3-031-42224-9_7

Hanifi, S., Alkali, B., Lindsay, G., & McGlinchey, D. (2025). Optimizing Energy and Air Consumption in Smart Manufacturing: An Industrial Internet of Things-Based Monitoring and Efficiency Enhancement Solution. *Applied Sciences (Basel, Switzerland), 15*(6), 3222. DOI: 10.3390/app15063222

Hansen, E. B. (2022). AI and IoT for Production Data Analytics in SMEs : Ph.d.-afhandling. In *Research Portal Denmark*. Technical University of Denmark. https://local.forskningsportal.dk/local/dki-cgi/ws/cris-link?src=ucn&id=ucn-94c7b983-1dc0-46e6-a136-fc130c9ff48b&ti=AI%20and%20IoT%20for%20Production%20Data%20Analytics%20in%20SMEs%20%3A%20Ph.d.-afhandling

Hasan, H. R., Musamih, A., Salah, K., Jayaraman, R., Omar, M., Arshad, J., & Boscovic, D. (2024). Smart agriculture assurance: IoT and blockchain for trusted sustainable produce. *Computers and Electronics in Agriculture, 224*, 109184. DOI: 10.1016/j.compag.2024.109184

Hasan, M. R. (2024). AI-Enhanced Decision-Making for Sustainable Supply Chains: Reducing Carbon Footprints in the USA. *arXiv (Cornell University)*. https://doi.org//arxiv.2501.10364DOI: 10.48550

Hassani, S. (2024a). Enhancing Legal Compliance and Regulation Analysis with Large Language Models. *arXiv (Cornell University)*. /arxiv.2404.17522DOI: 10.1109/RE59067.2024.00065

Hassani, S. (2024b)... *Enhancing Legal Compliance and Regulation Analysis with Large Language Models, 507*, 507–511. Advance online publication. DOI: 10.1109/RE59067.2024.00065

He, G., Hemmer, P., Vössing, M., Schemmer, M., & Gadiraju, U. (2025). Fine-Grained Appropriate Reliance: Human-AI Collaboration with a Multi-Step Transparent Decision Workflow for Complex Task Decomposition. *arXiv (Cornell University)*. https://doi.org//arxiv.2501.10909DOI: 10.48550

Hellweg, S., Benetto, E., Huijbregts, M. A. J., Verones, F., & Wood, R. (2023). Life-cycle assessment to guide solutions for the triple planetary crisis [Review of *Life-cycle assessment to guide solutions for the triple planetary crisis*]. *Nature Reviews Earth & Environment, 4*(7), 471. Nature Portfolio. DOI: 10.1038/s43017-023-00449-2

Heredia, J., Kirschner, R. J., Schlette, C., Abdolshah, S., Haddadin, S., & Kjægaard, M. B. (2023). ECDP: Energy Consumption Disaggregation Pipeline for Energy Optimization in Lightweight Robots. *IEEE Robotics and Automation Letters, 8*(10), 6107–6114. DOI: 10.1109/LRA.2023.3301311

Hianje, A., & Sahbani, S. (2025). *Reviewing the Roles of Artificial Intelligence-Based Technologies in Sustainable Purchasing and Supply Management: Research propositions for future Directions*. DOI: 10.20944/preprints202507.1743.v1

Höse, K., Amaral, A., Götze, U., & Peças, P. (2023). Manufacturing Flexibility through Industry 4.0 Technological Concepts—Impact and Assessment. *Global Journal of Flexible Systems Managment, 24*(2), 271–289. DOI: 10.1007/s40171-023-00339-y

Hwang, M.-L. T., Collin, W.-L., & Wang-xu, L. S. (2021). Quantum computing and supply chain optimization: Addressing complexity and efficiency challenges. *International Journal of Enterprise Modelling, 15*(3), 148–162. DOI: 10.35335/emod.v15i3.49

Ikumapayi, O. M., Kayode, J. F., Afolalu, S. A., Nnochiri, E. S., Olowe, K., &Bodunde, O. P. (2022). *A study on AI and ICT for Sustainable Manufacturing*. DOI: 10.46254/AF03.20220259

Imdahl, C., Blume, S., Blume, S., Zellmer, S., Gensicke, M., & Herrmann, C. (2021). Potentials of Hydrogen Technologies for Sustainable Factory Systems. *Procedia CIRP, 98*, 583–588. DOI: 10.1016/j.procir.2021.01.158

Indicated, N. (2010). *International Reference Life Cycle Data System (ILCD) Handbook - General guide for Life Cycle Assessment - Detailed guidance. Joint Research Centre*. European Commission., DOI: 10.2788/38479

Jabeur, M. H., Mahjoub, S., Toublanc, C., & Cariou, V. (2023). A reinforcement learning approach for a lot sizing and production scheduling problem with energy consideration. *IFAC-PapersOnLine, 56*(2), 11141–11147. DOI: 10.1016/j.ifacol.2023.10.832

Jaouhari, A. E., Arif, J., Samadhiya, A., & Kumar, A. (2023). Net zero supply chain performance and industry 4.0 technologies: Past review and present introspective analysis for future research directions [Review of *Net zero supply chain performance and industry 4.0 technologies: Past review and present introspective analysis for future research directions*]. *Heliyon, 9*(11). Elsevier BV. DOI: 10.1016/j.heliyon.2023.e21525

Javernik, A., Buchmeister, B., & Ojsteršek, R. (2023). The NASA-TLX Approach to Understand Workers Workload in Human-Robot Collaboration. *International Journal of Simulation Modelling, 22*(4), 574–585. DOI: 10.2507/IJSIMM22-4-658

Jawahir, I. S., & Bradley, R. (2016). Technological Elements of Circular Economy and the Principles of 6R-Based Closed-loop Material Flow in Sustainable Manufacturing. *Procedia CIRP, 40*, 103–108. DOI: 10.1016/j.procir.2016.01.067

Jayakumar, K., & Park, J. P. (n.d.). *Stimuli-Responsive Polymer– Metal Organic Framework Materials Composites: Advances in Biomedical Applications and Future Design Strategies.*

Jayawardane, H., Davies, I. J., Gamage, J. R., John, M., & Biswas, W. K. (2023). Sustainability perspectives – a review of additive and subtractive manufacturing [Review of *Sustainability perspectives – a review of additive and subtractive manufacturing*]. *Sustainable Manufacturing and Service Economics, 2*, 100015. Elsevier BV. DOI: 10.1016/j.smse.2023.100015

Jia, Q., Panetto, H., Macchi, M., Siri, S., Weichhart, G., & Xu, Z. (2022). Control for smart systems: Challenges and trends in smart cities. *Annual Reviews in Control, 53*, 358–369. DOI: 10.1016/j.arcontrol.2022.04.010

Johannsen, R. M., Mathiesen, B. V., Kermeli, K., Crijns-Graus, W., & Østergaard, P. A. (2023). Exploring pathways to 100% renewable energy in European industry. *Energy, 268*, 126687. DOI: 10.1016/j.energy.2023.126687

Johansen, S. T., Ünal, P., Albayrak, Ö., Ikonen, E., Linnestad, K., Jawahery, S., Srivastava, A., & Løvfall, B. T. (2023). Hybrid and cognitive digital twins for the process industry. *Open Engineering, 13*(1), 20220418. Advance online publication. DOI: 10.1515/eng-2022-0418

Kanda, R., Akhai, S., & Bansal, R. (2013). Analysis of MOST technique for elimination of ideal time by synchronization of different lines. *International Journal of Research, 1*(4).

Kantaros, A., & Ganetsos, T. (2024). Integration of Cyber-Physical Systems, Digital Twins and 3D Printing in Advanced Manufacturing: A Synergistic Approach. *American Journal of Engineering and Applied Sciences, 17*(1), 1–22. DOI: 10.3844/ajeassp.2024.1.22

Kaur, A., Singh, G., Kukreja, V., Sharma, S., Singh, S., & Yoon, B. (2022). Adaptation of IoT with Blockchain in Food Supply Chain Management: An Analysis-Based Review in Development, Benefits and Potential Applications [Review of *Adaptation of IoT with Blockchain in Food Supply Chain Management: An Analysis-Based Review in Development, Benefits and Potential Applications*]. *Sensors, 22*(21), 8174. Multidisciplinary Digital Publishing Institute. DOI: 10.3390/s22218174

Kerdlap, P., Low, J. S. C., & Ramakrishna, S. (2019). Zero waste manufacturing: A framework and review of technology, research, and implementation barriers for enabling a circular economy transition in Singapore. *Resources, Conservation and Recycling, 151*, 104438. DOI: 10.1016/j.resconrec.2019.104438

Khang, A., & Akhai, S. (2025). E-Waste and Lithium-ion Battery Recycling Insights for Sustainable Transportation. In *Driving Green Transportation System Through Artificial Intelligence and Automation: Approaches, Technologies and Applications* (pp. 203-230). Cham: Springer Nature Switzerland. DOI: 10.1007/978-3-031-72617-0_11

Khanna, A., Jain, S., Burgio, A., Bolshev, V., & Панченко, В. (2022). Blockchain-Enabled Supply Chain platform for Indian Dairy Industry: Safety and Traceability. *Foods, 11*(17), 2716. DOI: 10.3390/foods11172716 PMID: 36076901

Kleinekorte, J., Fleitmann, L., Bachmann, M., Kätelhön, A., Barbosa-Póvoa, A. P., Aßen, N. von der, &Bardow, A. (2020). Life Cycle Assessment for the Design of Chemical Processes, Products, and Supply Chains [Review of *Life Cycle Assessment for the Design of Chemical Processes, Products, and Supply Chains*]. *Annual Review of Chemical and Biomolecular Engineering, 11*(1), 203. Annual Reviews. DOI: 10.1146/annurev-chembioeng-011520-075844

Klessascheck, F., Fahrenkrog-Petersen, S. A., Mendling, J., & Pufahl, L. (2025). Unlocking Sustainability Compliance: Characterizing the EU Taxonomy for Business Process Management. In *Lecture notes in computer science* (p. 339). Springer Science+Business Media. DOI: 10.1007/978-3-031-78338-8_18

Koul, P. (2024). Enhancing solar energy systems: The role of computational and automation techniques in achieving net zero by 2050. *World Journal of Advanced Engineering Technology and Sciences, 13*(1), 909–924. DOI: 10.30574/wjaets.2024.13.1.0500

Kronberger, G., Bachinger, F., & Affenzeller, M. (2020). Smart Manufacturing and Continuous Improvement and Adaptation of Predictive Models. *Procedia Manufacturing*, *42*, 528–531. DOI: 10.1016/j.promfg.2020.02.037

Kumar, H., Wadhwa, A. S., Akhai, S., & Kaushik, A. (2024). Parametric analysis, modeling and optimization of the process parameters in electric discharge machining of aluminium metal matrix composite. *Engineering Research Express*, *6*(2), 025542. DOI: 10.1088/2631-8695/ad4ba9

Kumar, P., & Akhai, S. (2022). Effective energy management in smart buildings using VRV/VRF systems. In *Additive manufacturing in Industry 4.0* (pp. 27–35). CRC Press., DOI: 10.1201/9781003360001-2

Kumar, V., & Shahin, K. (2025). Artificial Intelligence and Machine Learning for Sustainable Manufacturing: Current Trends and Future Prospects. *Intelligent and Sustainable Manufacturing*, *2*(1), 10002. DOI: 10.70322/ism.2025.10002

Kumbhar, A. D., Chougule, A., Lokhande, P., Navaghane, S., Burud, A., & Nimbalkar, S. (2023). DeepInspect: An AI-Powered Defect Detection for Manufacturing Industries. *arXiv (Cornell University)*. https://doi.org//arxiv.2311.03725DOI: 10.48550

Lario, J., Mateos, J., Psarommatis, F., & Ortíz, Á. (2024). *Towards Zero-Defect and Zero-Waste Manufacturing by Implementing Non-Destructive Inspection Technologies*. DOI: 10.2139/ssrn.4876982

Lee, J.-S., Kim, T., Jeon, S.-H., Park, S.-H., Kim, S.-H., Lee, E., & Lee, J.-H. (2023). Automation of trimming die design inspection by zigzag process between AI and CAD domains. *Engineering Applications of Artificial Intelligence*, *127*, 107283. DOI: 10.1016/j.engappai.2023.107283

Lehmhus, D., Busse, M., von Hehl, A., & Jägle, E. A. (2018). State of the Art and Emerging Trends in Additive Manufacturing: From Multi-Material processes to 3D printed Electronics. *MATEC Web of Conferences, 188*, 3013. DOI: 10.1051/matecconf/201818803013

Lei, J. (2024). Efficient Strategies on Supply Chain Network Optimization for Industrial Carbon Emission Reduction. *arXiv (Cornell University)*. https://doi.org//arxiv.2404.16863DOI: 10.48550

Leonzio, G. (2023). Life Cycle Assessment of Carbon Dioxide Supply Chains: State of the Art and Methodology Description. *Applied Sciences (Basel, Switzerland), 14*(1), 385. DOI: 10.3390/app14010385

Li, C., Ge, W., Huang, Z., Zhang, Q., Li, H., & Cao, H. (2024). Digital twin-driven modeling and application of carbon emission for machine tool. *International Journal of Advanced Manufacturing Technology, 133*(11-12), 5595–5609. DOI: 10.1007/s00170-024-13788-1

Li, H., Pangborn, H. C., & Kovalenko, I. (2023). *A System-Level Energy-Efficient Digital Twin Framework for Runtime Control of Batch Manufacturing Processes.* 1. DOI: 10.1109/CASE56687.2023.10260642

Li, X., & Pattinson, S. W. (2025). An efficient and uncertainty-aware reinforcement learning framework for quality assurance in extrusion additive manufacturing. *Additive Manufacturing, 110*, 104912. DOI: 10.1016/j.addma.2025.104912

Li, Z., Yan, Y., Xiangheng, W., Ge, Y., & Meng, L. (2025). A survey of deep learning for industrial visual anomaly detection. *Artificial Intelligence Review, 58*(9), 279. Advance online publication. DOI: 10.1007/s10462-025-11287-7

Liu, C., Su, Z., Xu, X., & Lu, Y. (2021). Service-oriented industrial internet of things gateway for cloud manufacturing. *Robotics and Computer-integrated Manufacturing, 73*, 102217. DOI: 10.1016/j.rcim.2021.102217

Loffredo, A. (2023). Reinforcement learning for energy-efficient control of multi-stage production lines with parallel machine workstations. *Materials Research Proceedings, 35*, 428–436. DOI: 10.21741/9781644902714-51

Loffredo, A., May, M. C., Matta, A., & Lanza, G. (2023). Reinforcement learning for sustainability enhancement of production lines. *Journal of Intelligent Manufacturing, 35*(8), 3775–3791. DOI: 10.1007/s10845-023-02258-2

Lu, X.-L., Hsu, H.-J., Chou, C.-W., Kung, H. T., Lee, C. K. H., & Cheng, S.-M. (2024). DeepMachining: Online Prediction of Machining Errors of Lathe Machines. *arXiv (Cornell University)*. https://doi.org//arxiv.2403.16451DOI: 10.48550

Luo, Y., Madarkar, R., Luo, X., & Ball, P. (2025). Leveraging Digital Twins and Dynamic Life Cycle Assessment for Sustainable Manufacturing: A Conceptual Framework. In *Lecture notes in mechanical engineering* (p. 285). Springer Nature. DOI: 10.1007/978-3-031-93891-7_32

Maaß, W., Agrawal, A., Ciani, A., Danz, S., Delgadillo, A., Ganser, P., Kienast, P., Kulig, M., König, V., Rodellas-Gràcia, N., Rughubar, R., Schröder, S., Stautner, M., Stein, H., Stollenwerk, T., Zeuch, D., & Wilhelm, F. K. (2024). QUASIM: Quantum Computing Enhanced Service Ecosystem for Simulation in Manufacturing. *KI - Künstliche Intelligenz, 38*(4), 361. DOI: 10.1007/s13218-024-00860-x

Mandelli, D., Wang, C., & Agarwal, V. (2023). Development of Analysis Methods that Integrate Numeric and Textual Equipment Reliability Data. In *OSTI OAI (U.S. Department of Energy Office of Scientific and Technical Information)*. Office of Scientific and Technical Information. https://www.osti.gov/biblio/2008354

Maschler, B., Knodel, T., & Weyrich, M. (2021)... *Towards Deep Industrial Transfer Learning for Anomaly Detection on Time Series Data, 1*, 01–08. Advance online publication. DOI: 10.1109/ETFA45728.2021.9613542

Maware, C., Muvunzi, R., Machingura, T., & Daniyan, I. (2024). Examining the Progress in Additive Manufacturing in Supporting Lean, Green and Sustainable Manufacturing: A Systematic Review [Review of *Examining the Progress in Additive Manufacturing in Supporting Lean, Green and Sustainable Manufacturing: A Systematic Review*]. *Applied Sciences, 14*(14), 6041. Multidisciplinary Digital Publishing Institute. DOI: 10.3390/app14146041

Mawson, V. J., & Hughes, B. R. (2021). Optimisation of HVAC control and manufacturing schedules for the reduction of peak energy demand in the manufacturing sector. *Energy, 227*, 120436. DOI: 10.1016/j.energy.2021.120436

May, M. C., Oberst, J., & Lanza, G. (2024). Managing product-inherent constraints with artificial intelligence: Production control for time constraints in semiconductor manufacturing. *Journal of Intelligent Manufacturing, 35*(8), 4259–4276. DOI: 10.1007/s10845-024-02472-6

Mayer, A., Greif, L., Häußermann, T., Otto, S., Kastner, K., Bobbou, S. E., Chardonnet, J.-R., Reichwald, J., Fleischer, J., & Ovtcharova, J. (2025). Digital Twins, Extended Reality, and Artificial Intelligence in Manufacturing Reconfiguration: A Systematic Literature Review. *Sustainability (Basel), 17*(5), 2318. DOI: 10.3390/su17052318

Mazgualdi, C. E., Masrour, T., Barka, N., & Hassani, I. E. (2022). A Learning-Based Decision Tool towards Smart Energy Optimization in the Manufacturing Process. *Systems, 10*(5), 180. DOI: 10.3390/systems10050180

Meena, R., Sahoo, S., Malik, A., Kumar, S., & Nguyen, M. (2025). Artificial intelligence and circular supply chains: Framework for applications and deployment from the triple bottom line model perspective. *Annals of Operations Research, 354*(1), 71–101. DOI: 10.1007/s10479-025-06510-1

Meimandi, K. J., Reuel, A., Aranguiz-Dias, G., Rahama, H., Ayadi, A.-E., Boullier, X., Verdo, J., Montanie, L., & Kochenderfer, M. J. (2025). An Adaptive Responsible AI Governance Framework for Decentralized Organizations. *arXiv (Cornell University)*. https://doi.org//arxiv.2510.03368DOI: 10.48550

Meinrenken, C. J., Chen, D., Esparza, R. A., Iyer, V., Paridis, S. P., Prasad, A., & Whillas, E. (2020). Carbon emissions embodied in product value chains and the role of Life Cycle Assessment in curbing them. *Scientific Reports, 10*(1), 6184. Advance online publication. DOI: 10.1038/s41598-020-62030-x PMID: 32277082

Mendonça, P., & Vieira, C. (2022). Embodied carbon and economic cost analysis of a contemporary house design using local and reused materials. *Sustainable Futures : An Applied Journal of Technology, Environment and Society, 4*, 100100. DOI: 10.1016/j.sftr.2022.100100

Miele, L., Fraile, F., Esteso, A., & Rocca, R. L. (2024). Integration of Artificial Intelligence in Manufacturing Companies for Achieving Zero Waste – A Systematic Literature Review. In *IFIP advances in information and communication technology* (p. 212). Springer Science+Business Media. DOI: 10.1007/978-3-031-71743-7_14

Mihai, F., & Minea, I. (2021). Sustainable Alternative Routes versus Linear Economy and Resources Degradation in Eastern Romania. *Sustainability (Basel), 13*(19), 10574. DOI: 10.3390/su131910574

Miletić, S. D. (2023). Ranking of Autonomous Alternatives for the Realization of Intralogistics Activities in Sustainable Warehouse Systems using the TOPSIS Method. *Spectrum of Engineering and Management Sciences, 1*(1), 48–57. DOI: 10.31181/sems1120234m

Minashkina, D. (2024). A review and research agenda for recent socially and environmentally sustainable practices for warehouse management systems [Review of *A review and research agenda for recent socially and environmentally sustainable practices for warehouse management systems*]. *The International Journal of Logistics Management, 35*(7), 60. Emerald Publishing Limited. DOI: 10.1108/IJLM-07-2023-0265

Minashkina, D., & Happonen, A. (2020). Decarbonizing warehousing activities through digitalization and automatization with WMS integration for sustainability supporting operations. *E3S Web of Conferences, 158*, 3002. DOI: 10.1051/e3sconf/202015803002

Miranda, S., Vázquez, C. R., & Navarro-Gutiérrez, M. (2025). Energy consumption analysis and optimization in collaborative robots. *Frontiers in Robotics and AI, 12*, 1671336. Advance online publication. DOI: 10.3389/frobt.2025.1671336 PMID: 41245194

Mo, F., Monetti, F. M., Torayev, A., Rehman, H. U., Alberola, J. A. M., Minango, N. R., Nguyen, H. N., Maffei, A., & Chaplin, J. C. (2023). A maturity model for the autonomy of manufacturing systems. *International Journal of Advanced Manufacturing Technology*, *126*(1-2), 405–428. DOI: 10.1007/s00170-023-10910-7

Monek, G. D., & Fischer, S. (2024). Expert Twin: A Digital Twin with an Integrated Fuzzy-Based Decision-Making Module. *Decision Making Applications in Management and Engineering*, *8*(1), 1–21. DOI: 10.31181/dmame8120251181

Moosavi, S., Farajzadeh-Zanjani, M., Razavi-Far, R., Palade, V., & Saif, M. (2024). Explainable AI in Manufacturing and Industrial Cyber–Physical Systems: A Survey. *Electronics (Basel)*, *13*(17), 3497. DOI: 10.3390/electronics13173497

Morel, G., Pereira, C. E., & Nof, S. Y. (2019). Historical survey and emerging challenges of manufacturing automation modeling and control: A systems architecting perspective. *Annual Reviews in Control*, *47*, 21–34. DOI: 10.1016/j.arcontrol.2019.01.002

Motia, K., Kumar, R., & Akhai, S. (2024). *AI and smart manufacturing: Building Industry 4.0*. In *Modern management science practices in the age of AI* (pp. 1–28). IGI Global. DOI: 10.4018/979-8-3693-6720-9.ch001

Motia, K., Kumar, R., & Akhai, S. (2025). *Navigating the Fourth Industrial Revolution: A comprehensive review of Industry 4.0 technologies and their global adoption*. In *Advances in materials and manufacturing: Select proceedings of ICDMT 2024* (pp. 317–333). Springer. DOI: 10.1007/978-981-96-7659-0_28

Mou, A. J., Hossain, M., & Siddiqui, N. A. (2022). DIGITAL TRANSFORMATION IN MARKETING: EVALUATING THE IMPACT OF WEB ANALYTICS AND SEO ON SME GROWTH. *American Journal of Interdisciplinary Studies*, *3*(4), 61–90. DOI: 10.63125/8t10v729

Msakni, M. K., Risan, A., & Schütz, P. (2023). Using machine learning prediction models for quality control: A case study from the automotive industry. *Computational Management Science*, *20*(1), 14. DOI: 10.1007/s10287-023-00448-0 PMID: 36942085

Murugesan, R. (2006). Evolution of industrial automation. *International Journal of Computer Applications in Technology*, *25*(4), 169. DOI: 10.1504/IJCAT.2006.009389

Narkar, S., &Bandivadekar, T. (n.d.). *TRANSFORMING EPILEPSY CARE: APPLICATION OF AI FROM DETECTION TO PERSONALIZED TREATMENT.*

Nauman, M., Naheed, R., & Khan, J. (2023). Sustainable Development and Financial Resilience: Linking Greenhouse Gas Emissions, Risk, and Green Growth with Innovation, Inclusion, Renewables, and Infrastructure. *Research Square (Research Square)*. DOI: 10.21203/rs.3.rs-3246034/v1

Neger, M., Rahid, A. O., & Alnour, M. (2025). Moderating effects of energy poverty for sustainable tourism, policy, innovation, and environmental resilience: Evidence from SEM-ANN approaches. *Discover Sustainability*, *6*(1), 103. Advance online publication. DOI: 10.1007/s43621-025-00904-8

Ni, J., Bingler, J., Colesanti-Senni, C., Kraus, M., Gostlow, G., Schimanski, T., Stammbach, D., Vaghefi, S. A., Wang, Q., Webersinke, N., Wekhof, T., Yu, T., &Leippold, M. (2023). *CHATREPORT: Democratizing Sustainability Disclosure Analysis through LLM-based Tools*. DOI: 10.18653/v1/2023.emnlp-demo.3

Obydenkova, S. V., Kouris, P. D., Smeulders, D., Boot, M. D., & van der Meer, Y. (2021). Modeling life-cycle inventory for multi-product biorefinery: Tracking environmental burdens and evaluation of uncertainty caused by allocation procedure. *Biofuels, Bioproducts & Biorefining*, *15*(5), 1281–1300. DOI: 10.1002/bbb.2214

OECD. (2023). Smart City Data Governance. In *OECD urban studies*. Organization for Economic Cooperation and Development., DOI: 10.1787/e57ce301-

Ojadi, J. O., Owulade, O. A., Odionu, C. S., & Onukwulu, E. C.Jessica Obianuju OjadiOlumide Akindele OwuladeChinekwu Somtochukwu OdionuEkene Cynthia Onukwulu. (2025). Blockchain and IoT for Transparent Carbon Tracking and Emission Reduction in Global Supply Chains. *International Journal of Scientific Research in Science, Engineering and Technology*, *12*(2), 78–118. DOI: 10.32628/IJSRSET25122110

Ojsteršek, R., Buchmeister, B., & Javernik, A. (2023). Human-Robot Collaboration, Sustainable Manufacturing Perspective. In *Lecture notes in mechanical engineering* (p. 71). Springer Nature., DOI: 10.1007/978-3-031-38241-3_9

Ojstersek, R., Javernik, A., & Buchmeister, B. (2022). Importance of Sustainable Collaborative Workplaces – Simulation Modelling Approach. *International Journal of Simulation Modelling, 21*(4), 627–638. DOI: 10.2507/IJSIMM21-4-623

Okogwu, C., Agho, M. O., Adeyinka, M. A., Ayodeji, S. A., & Daraojimba, C. (2023). THE ROLE OF DECARBONISATION IN ENHANCING SUSTAINABLE PROCUREMENT PRACTICES IN THE U.S. *Environmental Contaminants Reviews, 6*(2), 116–125. DOI: 10.26480/ecr.02.2023.116.125

Öner, M. A., & Solak, S. (2020). PLC Tabanlı Uygulamalar: Mill Makinelerinde Oluşacak İş Kazalarının PLC Kullanılarak Önlenmesi. *European Journal of Science and Technology*. DOI: 10.31590/ejosat.754688

Ong, K., Mao, R., Satapathy, R., Filho, R. S., Cambria, E., Sulaeman, J., & Mengaldo, G. (2024). Explainable natural language processing for corporate sustainability analysis. *Information Fusion, 115*, 102726. DOI: 10.1016/j.inffus.2024.102726

ÖZGÜNER, M., & ÖZGÜNER, Z. (2022). Evaluation of the importance of additive manufacturing technology in terms of sustainable production with the DEMATEL method. *International Journal of ADVANCED AND APPLIED SCIENCES, 9*(10), 116. DOI: 10.21833/ijaas.2022.10.015

Pachot, A., & Patissier, C. (2023). *Towards Sustainable Artificial Intelligence: An Overview of Environmental Protection Uses and Issues*. Green and Low-Carbon Economy., DOI: 10.47852/bonviewGLCE3202608

Pal, D., & Joshi, S. (2023). AI, IoT and Robotics in Smart Farming: Current Applications and Future Potentials. *2022 International Conference on Sustainable Computing and Data Communication Systems (ICSCDS)*, 1096. DOI: 10.1109/ICSCDS56580.2023.10105101

Panagiotopoulou, V. C., Stavropoulos, P., & Chryssolouris, G. (2021). A critical review on the environmental impact of manufacturing: a holistic perspective [Review of *A critical review on the environmental impact of manufacturing: a holistic perspective*]. *The International Journal of Advanced Manufacturing Technology, 118*, 603. Springer Science+Business Media. DOI: 10.1007/s00170-021-07980-w

Papalia, A., Dawson, C., Antón, L. J. M., Bayomi, N. M., Champenois, B., Cho, J.-H., Cai, L., DelPreto, J., Edwards, K., Githinji, B.-C., Hickert, C., Jayawardana, V., Krämer, M., Raghavan, S., Russell, D. L., Salimi, S., Shi, J., Sudhakar, S., Wang, Y., . . . Singh, H. (2025). A Roadmap for Climate-Relevant Robotics Research. *arXiv (Cornell University)*. https://doi.org//arxiv.2507.11623DOI: 10.48550

Paraschos, P. D., Koulinas, G. K., & Koulouriotis, D. E. (2023). A reinforcement learning/ad-hoc planning and scheduling mechanism for flexible and sustainable manufacturing systems. *Flexible Services and Manufacturing Journal, 36*(3), 714–736. DOI: 10.1007/s10696-023-09496-9

Parthasarathy, V. (2024). AI-Driven Carbon Footprint Tracking and Emission Reduction in Logistics Networks. *International Journal of Artificial Intelligence Data Science and Machine Learning, 5*, 47–56. DOI: 10.63282/3050-9262.IJAIDSML-V5I2P106

Pater, J., & Stadnicka, D. (2021). Towards Digital Twins Development and Implementation to Support Sustainability – Systematic Literature Review. *Management and Production Engineering Review*. https://doi.org/DOI: 10.24425/mper.2021.138531

Patouillard, L., Bulle, C., Querleu, C., Maxime, D., Osset, P., & Margni, M. (2017). Critical review and practical recommendations to integrate the spatial dimension into life cycle assessment. *Journal of Cleaner Production, 177*, 398–412. DOI: 10.1016/j.jclepro.2017.12.192

Pazienza, A., Baselli, G., Vinci, D. C., & Trussoni, M. V. (2024). A holistic approach to environmentally sustainable computing. *Innovations in Systems and Software Engineering, 20*(3), 347–371. DOI: 10.1007/s11334-023-00548-9

Pellicciari, M., Avotiņš, A., Bengtsson, K., Berselli, G., Bey, N., Lennartson, B., & Meike, D. (2015). AREUS — Innovative hardware and software for sustainable industrial robotics. *IEEE International Conference on Automation Science and Engineering (CASE)*, 1325. DOI: 10.1109/CoASE.2015.7294282

Pesqueira, A., Sousa, M. J., & Costa, J. M. (2023). Exploring the Role of Big Data Analytics and Dynamic Capabilities in ESG Programs within Pharmaceuticals. *Research Square (Research Square)*. DOI: 10.21203/rs.3.rs-3315014/v1

Pizoń, J., Cioch, M., Kański, Ł., & Sánchez-García, E. (2022). Cobots Implementation in the Era of Industry 5.0 Using Modern Business and Management Solutions. *Advances in Science and Technology – Research Journal, 16*(6), 166. DOI: 10.12913/22998624/156222

Plapper, P. (2022). *Digitization of the work environment for sustainable production*. DOI: 10.30844/wgab_2022

Ponse, K., Kleuker, F., Fejer, M. E., Serra-Gómez, Á., Plaat, A., & Moerland, T. M. (2024). Reinforcement Learning for Sustainable Energy: A Survey. *arXiv (Cornell University)*. https://doi.org//arxiv.2407.18597DOI: 10.48550

Pop, P., Törngren, M., Yeh, S., & Thompson, H. A. (2023). Digital Technologies for a Sustainable Nordic Future: Challenges, Opportunities, and Policy Implications. *Research Portal Denmark*, 5. https://local.forskningsportal.dk/local/dki-cgi/ws/cris-link?src=dtu&id=dtu-3f4cae2e-ed19-4403-b0c8-376873525b43&ti=Digital%20Technologies%20for%20a%20Sustainable%20Nordic%20Future%3A%20Challenges%2C%20Opportunities%2C%20and%20Policy%20Implications

Prasad, G., Arunav, H., Dwight, S. S., Ghosh, M., Jayadev, A., & Nair, D. I. (2024). Advancing Sustainable Practices in Additive Manufacturing: A Comprehensive Review on Material Waste Recyclability [Review of *Advancing Sustainable Practices in Additive Manufacturing: A Comprehensive Review on Material Waste Recyclability*]. *Sustainability, 16*(23), 10246. Multidisciplinary Digital Publishing Institute. DOI: 10.3390/su162310246

Prinz, F., Min, S., & An, J. (2023). Preface for NetZero, Achievable by Manufacturing? *International Journal of Precision Engineering and Manufacturing-Green Technology*, *10*(3), 635–636. DOI: 10.1007/s40684-023-00510-x

Psarommatis, F., Fraile, F., Mendonça, J. P., Meyer, O., Lázaro, Ó., & Kiritsis, D. (2023). Editorial: Zero defect manufacturing in the era of industry 4.0 for achieving sustainable and resilient manufacturing. *Frontiers in Manufacturing Technology*, *3*, 1124624. Advance online publication. DOI: 10.3389/fmtec.2023.1124624

Puig, M. P., Paniego, J. M., Medina, S., Eguren, S. R., Libutti, L., Lanciotti, J., de Antueno, J., Estrebou, C. A., Chichizola, F., & Giusti, L. D. (2019). Intelligent Distributed System for Energy Efficient Control. In *Communications in computer and information science* (p. 51). Springer Science+Business Media. DOI: 10.1007/978-3-030-27713-0_5

Qiu, S., Cui, X., Ping, Z., Shan, N., Li, Z., Bao, X., & Xu, X. (2023). Deep Learning Techniques in Intelligent Fault Diagnosis and Prognosis for Industrial Systems: A Review [Review of *Deep Learning Techniques in Intelligent Fault Diagnosis and Prognosis for Industrial Systems: A Review*]. *Sensors, 23*(3), 1305. Multidisciplinary Digital Publishing Institute. DOI: 10.3390/s23031305

Rahmani, R., Jesus, C., & Lopes, S. I. (2024). Implementations of Digital Transformation and Digital Twins: Exploring the Factory of the Future. *Processes (Basel, Switzerland)*, *12*(4), 787. DOI: 10.3390/pr12040787

Ramya, M. (2019). Automation of Material Hanging Unit of a Cement Plant using ECS and PLC. *International Journal for Research in Applied Science and Engineering Technology*, *7*(3), 2322–2327. DOI: 10.22214/ijraset.2019.3426

Rane, N. L., Kaya, Ö., & Rane, J. (2024). *Advancing the Sustainable Development Goals (SDGs) through artificial intelligence, machine learning, and deep learning*. DOI: 10.70593/978-81-981271-8-1_4

Rathore, B. (2016). The Next Frontier: How the Integration of AI Transforms Manufacturing for a Sustainable Future. *International Journal of New Media Studies*, *3*(2), 1. DOI: 10.58972/eiprmj.v3i2y16.74

Remer, S. J., Hall, A., Spielman, Z., Vollmer, J. L., & Hansen, J. (2023). FY23 ION Based Approaches to Address Labor and Knowledge Retention. In *OSTI OAI (U.S. Department of Energy Office of Scientific and Technical Information)*. Office of Scientific and Technical Information. https://www.osti.gov/biblio/2204962

Ren, Y., Liu, Y., Ji, T., & Xu, X. (2025). *AI Agents and Agentic AI-Navigating a Plethora of Concepts for Future Manufacturing*. https://doi.org/DOI: 10.48550/ARXIV.2507.01376

Resman, M., Herakovič, N., & Debevec, M. (2025). Integrating Digital Twin Technology to Achieve Higher Operational Efficiency and Sustainability in Manufacturing Systems. *Systems*, *13*(3), 180. DOI: 10.3390/systems13030180

Rezaie, B., Anderson, A., Gallegos, S., & Azarmi, F. (2021). Present and Future Sustainability Development of 3D Metal Printing. *European Journal of Sustainable Development Research*, *5*(3), em0168. Advance online publication. DOI: 10.21601/ejosdr/11132

Ribeiro, V., Barata, J., & da Cunha, P. R. (2022). Sustainable Data Governance: A Systematic Review and a Conceptual Framework [Review of *Sustainable Data Governance: A Systematic Review and a Conceptual Framework*]. *Proceedings of the International Conference on Information Systems Development*. DOI: 10.62036/ISD.2022.44

Rissman, J., Bataille, C., Masanet, E., Aden, N., Morrow, W. R., Zhou, N., Elliott, N., Dell, R., Heeren, N., Huckestein, B., Cresko, J., Miller, S. A., Roy, J., Fennell, P. S., Cremmins, B., Blank, T. K., Hone, D., Williams, E. D., & Can, S. (2020). Technologies and policies to decarbonize global industry: Review and assessment of mitigation drivers through 2070. *Applied Energy*, *266*, 114848. DOI: 10.1016/j.apenergy.2020.114848

Rolofs, G., Wilking, F., Goetz, S., & Wartzack, S. (2024). Integrating Digital Twins and Cyber-Physical Systems for Flexible Energy Management in Manufacturing Facilities: A Conceptual Framework. *Electronics (Basel)*, *13*(24), 4964. DOI: 10.3390/electronics13244964

Román, Á. F. G., & Kabir, G. (2024). Assessing Carbon Dioxide Emissions in Manufacturing Industries: A Systematic Review [Review of *Assessing Carbon Dioxide Emissions in Manufacturing Industries: A Systematic Review*]. *Energies, 17*(20), 5119. Multidisciplinary Digital Publishing Institute. DOI: 10.3390/en17205119

Roseline, J. F., Leela, G. V. S., & Pravallika, K. R. S. (2019). Automation Control of Metro Traction Using Plc and Scada. *Journal of Physics: Conference Series, 1362*(1), 12079. DOI: 10.1088/1742-6596/1362/1/012079

Rotari, M. (2023). Production Analytics for a novel additive manufacturing system. *Research Portal Denmark*, 130. https://local.forskningsportal.dk/local/dki-cgi/ws/cris-link?src=dtu&id=dtu-5b4fb73a-9c12-44f7-8f20-bf9aa37006be&ti=Production%20Analytics%20for%20a%20novel%20additive%20manufacturing%20system

Rožanec, J. M., Trajkova, E., Dam, P., Fortuna, B., & Mladenić, D. (2021). Streaming Machine Learning and Online Active Learning for Automated\n Visual Inspection. *arXiv (Cornell University)*. https://doi.org//arxiv.2110.09396DOI: 10.48550

Rožanec, J. M., Zajec, P., Kenda, K., Novalija, I., Fortuna, B., Mladenić, D., Veliou, E., Papamartzivanos, D., Giannetsos, T., Menesidou, S. A., Alonso, R., Cauli, N., Recupero, D. R., Kyriazis, D., Sofianidis, G., Theodoropoulos, S., & Soldatos, J. (2021). STARdom: an architecture for trusted and secure human-centered\n manufacturing systems. *arXiv (Cornell University)*. /arxiv.2104.00983DOI: 10.1007/978-3-030-85910-7_21

Rubio, F., Llopis-Albert, C., & Valero, F. (2021). Multi-objective optimization of costs and energy efficiency associated with autonomous industrial processes for sustainable growth. *Technological Forecasting and Social Change, 173*, 121115. DOI: 10.1016/j.techfore.2021.121115

Ružarovský, R., Horák, T., & Bočák, R. (2024). Evaluating Energy Efficiency and Optimal Positioning of Industrial Robots in Sustainable Manufacturing. *Journal of Manufacturing and Materials Processing, 8*(6), 276. DOI: 10.3390/jmmp8060276

Sala, R., Pirola, F., Pezzotta, G., & Cavalieri, S. (2022). NLP-based insights discovery for industrial asset and service improvement: An analysis of maintenance reports. *IFAC-PapersOnLine, 55*(2), 522–527. DOI: 10.1016/j.ifacol.2022.04.247

Samadi, S., Fischer, A., & Lechtenböhmer, S. (2023). The renewables pull effect: How regional differences in renewable energy costs could influence where industrial production is located in the future. *Energy Research & Social Science, 104*, 103257. DOI: 10.1016/j.erss.2023.103257

Samuel, S.Solarin Adebayo Samuel. (2025). Carbon pricing mechanisms for reducing greenhouse gas emissions and encouraging sustainable industrial practices. *World Journal of Advanced Research and Reviews, 25*(2), 1. DOI: 10.30574/wjarr.2025.25.2.0350

Sarswatula, S. A., Pugh, T., & Prabhu, V. V. (2022). Modeling Energy Consumption Using Machine Learning. *Frontiers in Manufacturing Technology, 2*, 855208. Advance online publication. DOI: 10.3389/fmtec.2022.855208

Sartal, A., Bellas, R., Mejías, A. M., & Garcia, A. (2014). The sustainable manufacturing concept, evolution and opportunities within Industry 4.0: A literature review [Review of *The sustainable manufacturing concept, evolution and opportunities within Industry 4.0: A literature review*]. *Advances in Mechanical Engineering, 12*(5). SAGE Publishing. DOI: 10.1177/1687814020925232

Schimanek, R., Bilge, P., & Dietrich, F. (2023). Data fusion for improved circularity through higher quality of prediction and increased reliability of inspection. *International Journal of Sustainable Manufacturing, 1*(1), 10058051. Advance online publication. DOI: 10.1504/IJSM.2023.10058051

Schmidt, S. N., Bridges, T. S., Gidley, P. T., Lotufo, G. R., Ruiz, C. E., Kennedy, A. R., Rogers, N. S., Ghosh, U., Burgess, R. M., Fernandez, L. A., & Mayer, P. (2016). Remediation of PCB contaminated sediments using activated carbon: Assessing efficiency and robustness using equilibrium sampling. *Research Portal Denmark*, 258. https://local.forskningsportal.dk/local/dki-cgi/ws/cris-link?src=dtu&id=dtu-cdda31c3-d615-4b57-a7d4-6834b91c0937&ti=Remediation%20of%20PCB%20contaminated%20sediments%20using%20activated%20carbon%3A%20Assessing%20efficiency%20and%20robustness%20using%20equilibrium%20sampling

Schöberl, M., Huber, A., Kreppold, S., Dirnaichner, J., Keßler, S., & Fottner, J. (2023). Cobot uptake in construction: Embedding collaborative robots in digital construction processes. *Construction Robotics, 7*(1), 89–103. DOI: 10.1007/s41693-023-00098-3

Serac, C. A.SERAC. (2023). DIGITAL TRANSFORMATION VULNERABILITIES: ASSESSING THE RISKS AND STRENGTHENING CYBER SECURITY. *The Annals of the University of Oradea Economic Sciences, 3*(1), 771–781. DOI: 10.47535/1991AUOES32(1)059

Setyadi, A., Soekotjo, S., Lestari, S. D., Pawirosumarto, S., & Damaris, A. (2025). Trends and Opportunities in Sustainable Manufacturing: A Systematic Review of Key Dimensions from 2019 to 2024 [Review of *Trends and Opportunities in Sustainable Manufacturing: A Systematic Review of Key Dimensions from 2019 to 2024*]. *Sustainability, 17*(2), 789. Multidisciplinary Digital Publishing Institute. DOI: 10.3390/su17020789

Sharma, A., Priya, G. S. K., & Bandyopadhyay, S. (2023). Industrial decarbonization: A revolution ahead. *Clean Technologies and Environmental Policy, 25*(8), 2467–2468. DOI: 10.1007/s10098-023-02602-3

Sharma, V., Tripathi, A. K., & Mittal, H. (2022). *Technological revolutions in smart farming: Current trends, challenges & future directions.*

Siddique, I. (2024). Sustainable Manufacturing Practices: Reducing Waste and Energy Consumption in Production Systems. SSRN *Electronic Journal.* DOI: 10.2139/ssrn.4919617

Silva, A. S. S., Machado, J., & Sampaio, P. (2024). Predictive quality model for customer defects. *The TQM Journal, 36*(9), 155–174. DOI: 10.1108/TQM-09-2023-0302

Singh, A. R., Kumar, R. S., Bajaj, M., Khadse, C. B., & Зайцев, Є. (2024). Machine learning-based energy management and power forecasting in grid-connected microgrids with multiple distributed energy sources. *Scientific Reports, 14*(1). Advance online publication. DOI: 10.1038/s41598-024-70336-3 PMID: 39160194

Singh, B., Kumar, N., Kaur, G., Akhai, S., Abbass, M., & Kaur, P. (2025). Emerging trends in electric automotive instrumentation and control systems: A review of current trends and future technologies. *AIP Conference Proceedings, 3343*, 020012. DOI: 10.1063/5.0292567

Sinha, B. B., & Dhanalakshmi, R. (2021). Recent advancements and challenges of Internet of Things in smart agriculture: A survey. *Future Generation Computer Systems, 126*, 169–184. DOI: 10.1016/j.future.2021.08.006

Sofianidis, G., Rožanec, J. M., Mladenić, D., & Kyriazis, D. (2021). A Review of Explainable Artificial Intelligence in Manufacturing [Review of *A Review of Explainable Artificial Intelligence in Manufacturing*]. *arXiv (Cornell University).* Cornell University. https://doi.org//arxiv.2107.02295DOI: 10.48550

Soufi, Z., David, P., & Yahouni, Z. (2021). A methodology for the selection of Material Handling Equipment in manufacturing systems. *IFAC-PapersOnLine, 54*(1), 122–127. DOI: 10.1016/j.ifacol.2021.08.193

Su, Y., & Xu, G. (2025). The Influencing Mechanism of Intelligent Production on Carbon Emission Efficiency of Equipment Manufacturing Industry in China. *Processes (Basel, Switzerland)*, *13*(4), 1102. DOI: 10.3390/pr13041102

Sundaram, S., & Zeid, A. (2023). Artificial Intelligence-Based Smart Quality Inspection for Manufacturing. *Micromachines*, *14*(3), 570. DOI: 10.3390/mi14030570 PMID: 36984977

Sundaram, S., & Zeid, A. (2024). Technical language processing for Prognostics and Health Management: Applying text similarity and topic modeling to maintenance work orders. *Journal of Intelligent Manufacturing*. Advance online publication. DOI: 10.1007/s10845-024-02323-4

Tabbakh, A., Amin, A., Islam, M., Mahmud, G., Chowdhury, I. K., & Mukta, M. S. H. (2024). Towards sustainable AI: A comprehensive framework for Green AI. *Discover Sustainability*, *5*(1), 408. Advance online publication. DOI: 10.1007/s43621-024-00641-4

Taneja, T., Bhatia, B. S., & Akhai, S. (2024). *Assessing groundwater quality in Rampura-Bathinda for physicochemical parameters & heavy metal concentrations using electronic equipment*. In *2024 1st International Conference on Sustainability and Technological Advancements in Engineering Domain (SUSTAINED)* (pp. 791–796). IEEE. DOI: 10.1109/SUSTAINED63638.2024.11073880

Taneja, T., Bhatia, B. S., & Akhai, S. (2025). Agricultural resilience: Water quality and human well-being. In *Emerging smart agricultural practices using artificial intelligence* (pp. 1–23). Wiley., DOI: 10.1002/9781394274277.ch1

Tappia, E., Marchet, G., Melacini, M., & Perotti, S. (2015). Incorporating the environmental dimension in the assessment of automated warehouses. *Production Planning and Control*, *26*(10), 824–838. DOI: 10.1080/09537287.2014.990945

Tatsi, K., Hutchinson, T. H., Ehrlich, N., Correia, M., & Handy, R. D. (2016). Toxicity of Pristine and Aged Coated Copper Oxide Engineered Nanomaterials (CuO ENMs) to the Earthworm E. fetida. *Research Portal Denmark*, 120. https://local.forskningsportal.dk/local/dki-cgi/ws/cris-link?src=dtu&id=dtu-fba0fc31-7b81-4f4d-96f0-421cc3b994bc&ti=Toxicity%20of%20Pristine%20and%20Aged%20Coated%20Copper%20Oxide%20Engineered%20Nanomaterials%20(CuO%20ENMs)%20to%20the%20Earthworm%20E.%20fetida

Tercan, H., & Meisen, T. (2022). Machine learning and deep learning based predictive quality in manufacturing: a systematic review [Review of *Machine learning and deep learning based predictive quality in manufacturing: a systematic review*]. *Journal of Intelligent Manufacturing, 33*(7), 1879. Springer Science+Business Media. DOI: 10.1007/s10845-022-01963-8

Thareja, P., & Akhai, S. (2016). Processing Aluminum Fly Ash Composites via Parametric Analysis of Stir Casting. *Journal of Advanced Research in Manufacturing. Material Science & Metallurgical Engineering*, *3*(3&4), 21–28.

Thimm, H. (2023). Data modeling and NLP-based scoring method to assess the relevance of environmental regulatory announcements. *Environment Systems & Decisions*, *43*(3), 416–432. DOI: 10.1007/s10669-023-09900-7

Thomas, G., & Sule, M. (2022). A service lens on cybersecurity continuity and management for organizations' subsistence and growth. *Organizational Cybersecurity Journal Practice Process and People*, *3*(1), 18–40. DOI: 10.1108/OCJ-09-2021-0025

Todescato, M., Braholli, O., Chaltsev, D., Blasio, I. D., Don, D., Egger, G., Emig, J., Monizza, G. P., Sacco, P., Siegele, D., Steiner, D., Terzer, M., Riedl, M., Giusti, A., & Matt, D. T. (2023). Sustainable manufacturing through application of reconfigurable and intelligent systems in production processes: A system perspective. *Scientific Reports*, *13*(1), 22374. Advance online publication. DOI: 10.1038/s41598-023-49727-5 PMID: 38104211

Torab-Miandoab, A., Samad-Soltani, T., Jodati, A., & Rezaei-Hachesu, P. (2023). Interoperability of heterogeneous health information systems: a systematic literature review [Review of *Interoperability of heterogeneous health information systems: a systematic literature review*]. *BMC Medical Informatics and Decision Making, 23*(1). BioMed Central. DOI: 10.1186/s12911-023-02115-5

Trenz, A., Hoffmann, C. T., Lange, C., & Öchsner, R. (2023). Increasing Energy Efficiency and Flexibility by Forecasting Production Energy Demand Based on Machine Learning. In *Lecture notes in mechanical engineering* (p. 449). Springer Nature., DOI: 10.1007/978-3-031-28839-5_50

Triebe, M. J., Deng, S., Pérez-Cardona, J. R., Joung, B. G., Wu, H., Shakelly, N., Pieper, J. P., Zhou, X. K., Maani, T., Zhao, F., & Sutherland, J. W. (2023). Perspectives on future research directions in green manufacturing for discrete products. *Green Manufacturing Open*, *1*(1), 10. DOI: 10.20517/gmo.2022.11

Trstenjak, M., Mustapić, M., Gregurić, P., & Opetuk, T. (2023). Use of Green Industry 5.0 Technologies in Logistics Activities. *TehničkiGlasnik*, *17*(3), 471–477. DOI: 10.31803/tg-20230518185836

Trump, B. D., Mitoulis, S. A., Argyroudis, S., Kiker, G. A., Palma-Oliveira, J. M., Horton, R. A., Pescaroli, G., Pinigina, E., Trump, J., & Linkov, I. (2025). Threat-Agnostic Resilience: Framing and Application for Critical Infrastructure. *arXiv (Cornell University)*. https://doi.org//arxiv.2501.01318DOI: 10.48550

Tsai, W.-H., & Lin, W.-H. (2024). Production Decision Model for the Cement Industry in Pursuit of Carbon Neutrality: Analysis of the Impact of Carbon Tax and Carbon Credit Costs. *Sustainability (Basel), 16*(6), 2251. DOI: 10.3390/su16062251

Turner, C., Oyekan, J., Garn, W., Duggan, C., & Abdou, K. (2022). Industry 5.0 and the Circular Economy: Utilizing LCA with Intelligent Products. *Sustainability (Basel), 14*(22), 14847. DOI: 10.3390/su142214847

Valdivia, S., Backes, J. G., Traverso, M., Sonnemann, G., Cucurachi, S., Guinée, J. B., Schaubroeck, T., Finkbeiner, M., Leroy-Parmentier, N., Ugaya, C. M. L., Peña, C., Zamagni, A., Inaba, A., Amaral, M., Berger, M., Dvarionienė, J., Vakhitova, T., Benoît-Norris, C., Prox, M., & Goedkoop, M. (2021). Principles for the application of life cycle sustainability assessment. *The International Journal of Life Cycle Assessment, 26*(9), 1900–1905. DOI: 10.1007/s11367-021-01958-2

Valtonen, A., & Holopainen, M. (2025). Mitigating employee resistance and achieving well-being in digital transformation. *Information Technology & People, 38*(8), 42–72. DOI: 10.1108/ITP-05-2024-0701

Vance, D., Jin, M., Wenning, T., Nimbalkar, S., & Price, C. (2025). Next-Level Energy Management in Manufacturing: Facility-Level Energy Digital Twin Framework Based on Machine Learning and Automated Data Collection. *Energies, 18*(13), 3242. DOI: 10.3390/en18133242

Vater, J., Harscheidt, L., & Knoll, A. (2019).. . *Smart Manufacturing with Prescriptive Analytics, 224*, 224–228. Advance online publication. DOI: 10.1109/ICITM.2019.8710673

Verma, D. K. (2018). Analysis of Smart Manufacturing Technologies for Industry Using AI Methods. *Türk BilgisayarveMatematikEğitimi Dergisi, 9*(2), 529–540. DOI: 10.17762/turcomat.v9i2.13857

Verpoort, P. C., Gast, L., Hofmann, A., & Ueckerdt, F. (2024). Impact of global heterogeneity of renewable energy supply on heavy industrial production and green value chains. *Nature Energy, 9*(4), 491–503. DOI: 10.1038/s41560-024-01492-z

Vidyalakshmi, G. V., Gopikrishnan, S., Boulila, W., Koubâa, A., & Srivastava, G. (2025). Digital Twins and Cyber-Physical Systems: A New Frontier in Computer Modeling. *Computer Modeling in Engineering & Sciences, 143*(1), 51–113. DOI: 10.32604/cmes.2025.057788

Villalonga, A., Negri, E., Biscardo, G., Castaño, F., Haber, R. E., Fumagalli, L., & Macchi, M. (2021). A decision-making framework for dynamic scheduling of cyber-physical production systems based on digital twins. *Annual Reviews in Control, 51*, 357–373. DOI: 10.1016/j.arcontrol.2021.04.008

Wadhwa, A. S., Abbass, M., Akhai, S., Kumar, D., & Kumar, P. (2025). Integrating Taguchi optimization for multi-criteria decision making in engineering applications. In *Recent theories and applications for multi-criteria decision-making* (pp. 125–150). IGI Global., DOI: 10.4018/979-8-3693-6502-1.ch005

Wadhwa, A. S., Akhai, S., Abbass, M., Chouksey, A., Tiwari, S., & Taneja, T. (2025). Computational intelligence-driven design and optimization of polyurethane belt-type oil skimmer for sustainable manufacturing using SolidWorks 3D CAD. In *Computational intelligence for sustainable manufacturing of advanced materials* (pp. 445–464). IGI Global., DOI: 10.4018/979-8-3693-7974-5.ch019

Wahdaniah, S., Sucianti, R., Ambalele, E., & Tellu, A. H. (2023). Human Resource Management Transformation in the Digital Age: Recent Trends and Implications. *International Journal of Applied Research and Sustainable Sciences, 1*(3), 239–258. DOI: 10.59890/ijarss.v1i3.902

Walter, S. (2023). AI impacts on supply chain performance : A manufacturing use case study. *Discover Artificial Intelligence*, *3*(1), 18. Advance online publication. DOI: 10.1007/s44163-023-00061-9 PMID: 40476886

Waltersmann, L., Kiemel, S., Stuhlsatz, J., Sauer, A., & Miehe, R. (2021). Artificial Intelligence Applications for Increasing Resource Efficiency in Manufacturing Companies—A Comprehensive Review. *Sustainability (Basel), 13*(12), 6689. DOI: 10.3390/su13126689

Wan, J., Li, X., Dai, H.-N., Kusiak, A., Martinez-Garcia, M., Li, D., Wan, J., Li, X., Dai, H.-N., Kusiak, A., Martinez-Garcia, M., & Li, D. (2020). Artificial-Intelligence-Driven Customized Manufacturing Factory: Key Technologies, Applications, and Challenges. *Proceedings of the IEEE, 109*(4), 377–398. DOI: 10.1109/JPROC.2020.3034808

Wang, W., Wang, P., Zhang, H., Chen, X., Wang, G., Lu, Y., Chen, M., Liu, H., & Li, J. (2023). A Real-Time Defect Detection Strategy for Additive Manufacturing Processes Based on Deep Learning and Machine Vision Technologies. *Micromachines, 15*(1), 28. DOI: 10.3390/mi15010028 PMID: 38258148

Wicaksono, H., Trat, M., Bashyal, A., Boroukhian, T., Felder, M., Ahrens, M., Bender, J., Groß, S., Steiner, D., July, C., Dorus, C., & Zoerner, T. (2024). Artificial-intelligence-enabled dynamic demand response system for maximizing the use of renewable electricity in production processes. *International Journal of Advanced Manufacturing Technology, 138*(1), 247–271. DOI: 10.1007/s00170-024-13372-7

Xavier, B. M., Thollander, P., Hilletofth, P., & Johansson, M. (2024). Exploring energy management integration into upstream supply chains: A systematic literature review. *Frontiers in Energy Research, 12*, 1425795. Advance online publication. DOI: 10.3389/fenrg.2024.1425795

Xia, T., Sun, H., Ding, Y., Han, D., Qin, W., Seidelmann, J., & Xi, L. (2023). Digital twin-based real-time energy optimization method for production line considering fault disturbances. *Journal of Intelligent Manufacturing, 36*(1), 569–593. DOI: 10.1007/s10845-023-02219-9

Xia, Y., & Ahmad, M. (2025). Industrial Robotics, Resource Efficiency, Energy Transition, and Environmental Quality: Designing a Sustainable Development Goals Framework for G7 Countries in the Presence of Geopolitical Risk. *Sustainability (Basel), 17*(5), 1960. DOI: 10.3390/su17051960

Yakovenko, Y., & Shaptala, R. (2024). Study of digital twins as the driving force of digital transformation and achieving the goals of sustainable development. *Technology Audit and Production Reserves, 2*(4(76)), 11–20. DOI: 10.15587/2706-5448.2024.301423

Yan, P., Abdulkadir, A., Luley, P.-P., Rosenthal, M., Schatte, G. A., Grewe, B. F., & Stadelmann, T. (2024). A Comprehensive Survey of Deep Transfer Learning for Anomaly Detection in Industrial Time Series: Methods, Applications, and Directions. *IEEE Access : Practical Innovations, Open Solutions, 12*, 3768–3789. DOI: 10.1109/ACCESS.2023.3349132

Yang, H., Rao, P., Simpson, T. W., Lu, Y., Witherell, P., Nassar, A. R., Reutzel, E. W., & Kumara, S. (2020). Six-Sigma Quality Management of Additive Manufacturing. *Proceedings of the IEEE, 109*(4), 347–376. DOI: 10.1109/JPROC.2020.3034519 PMID: 34248180

Yang, X., Smith, T. M., Prado, A., & Yang, Y. (2025). Net-zero greenhouse gas mitigation potential across multi-tier supply chains. *Communications Earth & Environment, 6*(1), 230. Advance online publication. DOI: 10.1038/s43247-025-02173-9

Yao, K., Lin, C.-L., & Pan, C.-H. (2024). Industrial Sustainable Development: The Development Trend of Programmable Logic Controller Technology. *Sustainability (Basel), 16*(14), 6230. DOI: 10.3390/su16146230

Yitmen, İ., & Almusaed, A. (2024). Synopsis of Industry 5.0 Paradigm for Human-Robot Collaboration. In *Artificial intelligence*. IntechOpen., DOI: 10.5772/intechopen.1005583

Zeiser, A., van Stein, B., & Bäck, T. (2022). Deep Learning based pipeline for anomaly detection and quality enhancement in industrial binder jetting processes. *arXiv (Cornell University)*. https://doi.org//arxiv.2209.10178DOI: 10.48550

Zhang, C., Juraschek, M., & Herrmann, C. (2024). Deep reinforcement learning-based dynamic scheduling for resilient and sustainable manufacturing: A systematic review [Review of *Deep reinforcement learning-based dynamic scheduling for resilient and sustainable manufacturing: A systematic review*]. *Journal of Manufacturing Systems, 77*, 962. Elsevier BV. DOI: 10.1016/j.jmsy.2024.10.026

Zhao, G., Ruan, D., Wang, Q., Zhang, X., & Wang, Y. (2018). Systemic boundaries in industrial systems: A new concept defined to improve LCA for metallurgical and manufacturing systems. *Journal of Cleaner Production, 187*, 717–729. DOI: 10.1016/j.jclepro.2018.03.263

Zhao, Y., Zhang, W., Xiang, C., Du, H., Niyato, D., & Gao, S. (2025). Diffusion-Modeled Reinforcement Learning for Carbon and Risk-Aware Microgrid Optimization. *arXiv (Cornell University)*. https://doi.org//arxiv.2507.16867DOI: 10.48550

Zhong, S., & Pearce, J. M. (2017). Tightening the loop on the circular economy: Coupled distributed recycling and manufacturing with recyclebot and RepRap 3-D printing. *Resources, Conservation and Recycling, 128*, 48–58. DOI: 10.1016/j.resconrec.2017.09.023

Zhu, S., Hu, H., Yang, H., Qu, Y., & Li, Y. (2023). Mini-Review of Best Practices for Greenhouse Gas Reduction in Singapore's Semiconductor Industry. *Processes (Basel, Switzerland), 11*(7), 2120. DOI: 10.3390/pr11072120

Zocco, F., Haddad, W. M., Corti, A., & Malvezzi, M. (2024). A Unification Between Deep-Learning Vision, Compartmental Dynamical Thermodynamics, and Robotic Manipulation for a Circular Economy. *arXiv (Cornell University)*. https://doi.org//arxiv.2405.14406DOI: 10.48550

Zou, J., Chang, Q., Arinez, J., & Xiao, G. (2017). Data-driven modeling and real-time distributed control for energy efficient manufacturing systems. *Energy, 127*, 247–257. DOI: 10.1016/j.energy.2017.03.123

Zoubaidi, Z., Herrou, B., & Sekkat, S. (2025). Towards Intelligent Manufacturing: AI Based Defect Prediction and MES Integration. *Research Square (Research Square)*. DOI: 10.21203/rs.3.rs-7521361/v1

Zschech, P., Weinzierl, S., & Kraus, M. (2025). Inherently Interpretable Machine Learning: A Contrasting Paradigm to Post-hoc Explainable AI. *Business & Information Systems Engineering*. Advance online publication. DOI: 10.1007/s12599-025-00964-0

KEY TERMS AND DEFINITIONS

Additive Manufacturing (AM): A layer-by-layer fabrication method that minimizes material waste and allows lightweight, resource-efficient product designs aligned with sustainable manufacturing principles.

Artificial Intelligence (AI): A set of computational techniques that enable machines to learn patterns, make predictions, and support autonomous decision-making for energy optimization, quality improvement, and resource-efficient operations.

Automation: The use of control systems, robotics, and intelligent machines to perform industrial tasks with precision and consistency, reducing waste, variability, and manual dependence.

Circular Production: A manufacturing approach that maximizes resource reuse, recycling, repair, and refurbishment, reducing dependence on virgin materials and lowering overall environmental burden.

Cyber-Physical Production System (CPPS): An integrated environment where physical machinery and digital intelligence interact continuously, enabling adaptive control, autonomous operations, and sustainability-oriented optimization.

Deep Learning (DL): A subset of ML using multi-layer neural networks capable of interpreting complex signals, images, and sensor data for precision inspection, anomaly detection, and process automation.

Digital Twin (DT): A real-time virtual replica of physical equipment, processes, or entire factories that supports simulation, optimization, predictive maintenance, and carbon-footprint evaluation.

Energy Management System (EMS): A coordinated set of tools, controls, and analytics that measure, monitor, and optimize industrial energy use to support low-carbon factory operations.

Green Supply Chain: A supply-chain model emphasizing sustainable procurement, circular logistics, transparent traceability, optimized transport routes, and environmentally responsible end-of-life handling.

Industrial Internet of Things (IIoT): A network of interconnected sensors, devices, and machines that collect and exchange real-time data for monitoring energy consumption, emissions, equipment health, and process performance.

Machine Learning (ML): A branch of AI that uses data-driven algorithms to recognize patterns and generate predictions for tasks such as energy forecasting, defect detection, and process optimization.

Net-Zero Manufacturing: A manufacturing paradigm in which total greenhouse gas emissions are reduced and balanced through energy efficiency, circular production, renewable integration, and carbon-aware operational planning.

Predictive Maintenance: A data-driven strategy that forecasts equipment failures before they occur, reducing downtime, energy loss, material waste, and unnecessary part replacements.

Reinforcement Learning (RL): A learning method where an AI agent improves decisions through trial-and-error interactions, enabling self-optimizing production lines and low-carbon process adjustments.

Smart Sensors: Advanced sensing devices that monitor temperature, vibration, energy use, emissions, and machine behavior, providing real-time inputs for sustainable decision-making.

Chapter 7
Integrating Drones into Manufacturing:
Real Time Inspection, Environmental Monitoring, and Quality Assurance

Shalom Akhai
http://orcid.org/0000-0002-7533-457X
MMEC, Maharishi Markandeshwar University, Mullana, India

Mahapara Abbass
http://orcid.org/0000-0002-0838-2598
MMEC, Maharishi Markandeshwar University, Mullana, India

ABSTRACT

This chapter examines the expanding role of drones in modern manufacturing by outlining their contributions to real-time inspection, environmental monitoring, and quality assurance. The chapter explains how drones equipped with optical, thermal, and gas-sensing technologies enhance operational visibility, detect anomalies, and strengthen regulatory compliance. It further discusses how drone-generated data supports digital twins, predictive maintenance, and smart-factory integration. Key considerations related to safety, workforce transformation, risk management, and implementation strategies are highlighted. Overall, the chapter demonstrates how drone-enabled systems are transforming manufacturing into a safer, more efficient, and data-driven industrial environment.

DOI: 10.4018/979-8-3373-5756-0.ch007

CHAPTER HIGHLIGHTS

1. *Drones enhance real-time industrial inspection* by providing high-resolution visual, thermal, and sensor-based monitoring, reducing downtime and enabling predictive maintenance.
2. *Environmental monitoring becomes more accurate and responsive*, as drones map emissions, air quality, noise levels, and water contamination across large factory zones in real time.
3. *Quality assurance processes are strengthened* through automated defect detection using AI-enabled imaging, ensuring higher product consistency and minimizing waste.
4. *Drone-collected data fuels digital twins and analytics,* enabling smarter decision-making, improved layout optimization, and more responsive manufacturing operations.
5. *Workplace safety and risk management improve significantly*, as drones replace humans in hazardous inspections, support emergency response, and promote a safer work culture.
6. *Implementation requires structured strategies*, including regulatory compliance, operator training, robust data governance, and pilot testing to overcome technical and organizational challenges.

ABBREVIATIONS

AI: Artificial Intelligence
AR: Augmented Reality
CAD: Computer-Aided Design
CCTV: Closed-Circuit Television
CV: Computer Vision
DT: Digital Twin
EMS: Environmental Monitoring System
ERP: Enterprise Resource Planning
GPS: Global Positioning System
HD: High Definition
IIoT: Industrial Internet of Things
IoT: Internet of Things
IR: Infrared
LED: Light Emitting Diode
LiDAR: Light Detection and Ranging
ML: Machine Learning

PM: Particulate Matter
PPE: Personal Protective Equipment
QA: Quality Assurance
QC: Quality Control
RGB: Red-Green-Blue (Imaging Mode)
RMP: Revolutions Per Minute *(Correct abbreviation: RPM)*
ROI: Return on Investment
RPM: Revolutions Per Minute
RTI: Real-Time Inspection
SOP: Standard Operating Procedure
UAS: Unmanned Aerial System
UAV: Unmanned Aerial Vehicle
VOCs: Volatile Organic Compounds
VR: Virtual Reality

1. INTRODUCTION

The manufacturing sector is undergoing rapid transformation as industries adopt automation, robotics, and advanced digital technologies (Askerbekov et al., 2024; Sharma et al., 2022). These developments are changing how factories operate, how products are made, and how efficiency is evaluated (Akhai & Khang, 2024; Akhai et al., 2025). Among the many technologies contributing to this shift, drones have moved from being experimental devices to essential industrial tools (Askerbekov et al., 2024; Nooralishahi et al., 2021). Although originally used for photography and remote surveys, drones have evolved into versatile systems that can observe processes, collect data, and strengthen both safety and quality in modern production environments (Almadani et al., 2020; Askerbekov et al., 2024; Kas & Johnson, 2019; Nooralishahi et al., 2021).

The evolution from Industry 4.0 to Industry 5.0 reflects a shift toward human-centric, resilient, and sustainable manufacturing ecosystems. While Industry 4.0 focused on automation, cyber-physical systems, and data connectivity, Industry 5.0 emphasizes collaboration between advanced technologies and human expertise. In this framework, drones function as cooperative cyber-physical agents that enhance worker safety, improve decision support, and enable adaptive, intelligent operations aligned with Industry 5.0's human-machine partnership principles (Motia et al., 2024; Motia et al., 2025).

Manufacturing relies heavily on accurate and timely information. Machines must operate within precise settings, products must meet defined tolerances, and managers depend on real-time insights to make decisions that influence cost and productivity

(Gerencheal & Lee, 2025; Lee et al., 2022). Drones support this information-driven environment by moving freely across industrial spaces and gathering continuous visual and sensor-based data (Akhai 2024; Baumann et al., 2020; Nooralishahi et al., 2021). They can enter confined or hazardous areas that are difficult for workers to access and provide uninterrupted monitoring of production activities (Almadani et al., 2020; Gerencheal & Lee, 2025; Gómez-Rosal et al., 2023; Kas & Johnson, 2019). This helps companies detect performance declines or potential risks early, reducing the likelihood of costly disruptions (Askerbekov et al., 2024; Lee et al., 2022).

One of the major strengths of drones is their flexibility (Baumann et al., 2020). Depending on the task, they can carry high-definition cameras, thermal imagers, gas-detection sensors, or lightweight manipulators (Davuluri, 2024; Gerencheal & Lee, 2025; Nooralishahi et al., 2021). A single flight may capture heat patterns inside a steel plant, while another may track dust levels or emissions in a chemical facility (Gómez-Rosal et al., 2023; Stroud & Weinel, 2020). At a time when industries aim to balance productivity with sustainability, drones serve as objective observers that can document and verify conditions beyond the reach of traditional monitoring systems (Kurbanzade et al., 2023; Sun et al., 2022).

As factories become more digitally connected, drones also act as physical extensions of digital networks, linking real-world processes with systems based on the Internet of Things, cloud analytics, and automated decision-making tools (Askerbekov et al., 2024; Awasthi et al., 2022). For example, video captured by a drone can be streamed directly into a machine-vision system that identifies surface defects or alignment issues within seconds (Lee et al., 2022). Corrections can then be made immediately, reducing delays and supporting smoother production flow.

This integration of mobile robotics, advanced sensing, and digital intelligence marks an important step in industrial modernisation (Akhai and Balu 2022; Lee et al., 2022; Sun et al., 2022). Drones help create continuous feedback loops within the factory, allowing equipment, sensors, and decision-making systems to interact more effectively. This leads to more efficient operations, lower defect rates, and faster responses to emerging problems (Lee et al., 2022). However, integrating drones into manufacturing requires careful planning, including an understanding of plant layout, safety considerations, and data requirements.

The following sections examine how drones contribute to three important areas of manufacturing: real-time inspection, environmental monitoring, and quality assurance. Together, these areas show that drones are not peripheral tools but essential components of future-ready manufacturing systems.

1.1 The Role of Drones in Modern Manufacturing

Manufacturing facilities are often complex, densely arranged, and filled with continuously operating machinery (Jaroš et al., 2022; Layer et al., 2023; Pellicer et al., 2023; Wadhwa et al., 2025). Traditional monitoring approaches—such as fixed cameras, manual inspections, and stationary sensors—offer only limited coverage (Li et al., 2022; Múñiz et al., 2022; Purohit et al., 2019). They cannot adjust when production layouts change or when new processes are introduced, facing challenges with scalability and adaptability in evolving contexts (De Gasperis & Facchini, 2025; Malanchini et al., 2023). Drones address these limitations by adding autonomous mobility to industrial monitoring networks (Lim et al., 2023; Liu et al., 2025). Their ability to move freely throughout the facility provides managers and engineers with a more complete and dynamic understanding of operational conditions (Baumann et al., 2020; González-Castaño et al., 2020).

Scalability is another strength of drone systems. Small factories may begin with one or two drones performing visual surveys, while larger facilities can operate coordinated drone fleets using centralized management platforms (Jacobsen et al., 2023; Leet et al., 2022; Schilling et al., 2022). These systems can handle automated scheduling, route planning, and battery supervision (Kim et al., 2018; Yi & Sutrisna, 2020). As production grows or shifts, the drone network adapts without major structural changes, making it a sustainable long-term solution (Todescato et al., 2023; Zhang et al., 2022).

Drones also contribute significantly to workplace safety (Nooralishahi et al., 2021; Zhong et al., 2025). Many industrial zones expose workers to risks such as chemical vapours, high temperatures, heavy machinery, or electrical hazards (Bos et al., 2015; Kas & Johnson, 2019). Conducting manual inspections in these areas increases the possibility of accidents due to human error, fatigue, or exposure to hazardous conditions (Bao et al., 2023; Brogaard & Boukas, 2023; Chen et al., 2024; Luo et al., 2025). Drones remove this risk by collecting thermal images, spectral readings, or gas-concentration data while operators remain safely at control stations (Al-Battbootti et al., 2021; Benyeogor et al., 2020; Lim et al., 2023). This reduces exposure, strengthens safety compliance, and enhances overall risk management (Alsayed et al., 2021; Kas & Johnson, 2019).

Beyond their monitoring role, drones can also support internal logistics (Baumann et al., 2020; Olivares et al., 2015). They can transport small tools, documents, or lightweight components between departments, reducing delays during urgent maintenance or production tasks (Awasthi et al., 2022; Zhong et al., 2020). Although this application is still emerging, it suggests that drones may eventually take on more interactive responsibilities within manufacturing operations (Mészáros et al., 2018; Rejeb et al., 2021).

1.2 Real-Time Inspection: A Turning Point

Inspection plays a vital role in manufacturing because it ensures that processes remain within expected limits and that deviations are identified promptly (Barkman, 1989; Papananias et al., 2020; Sundaram & Zeid, 2023; Westkämper & Warnecke, 1994). Traditionally, inspection work has been carried out manually by quality inspectors or maintenance teams (Hütten et al., 2024; Lin et al., 2024; Ma et al., 2024; Magalhães & Ferreira, 2022). While effective, manual inspection requires time, may involve subjective judgement, and often detects issues only after they begin affecting production (Alzarooni et al., 2025; Hütten et al., 2024; Lindblad, 2006; Stephani et al., 2023). Drone-supported real-time inspection changes this by enabling continuous, automated observation (Davuluri, 2024; Hoang et al., 2019; Kim et al., 2020; Salhaoui et al., 2019).

A facility equipped with drones can schedule routine flights along production lines (González-Castaño et al., 2020; Yi & Sutrisna, 2020). During each mission, drones collect high-resolution images and sensor data from machines, conveyor systems, and storage areas (Lai et al., 2024; Zhong et al., 2020). Advanced vision systems can interpret this information instantly to identify overheating components, leaks, mechanical misalignment, or unusual noise patterns when equipped with audio sensors (Bretschneider et al., 2024; Ghazal et al., 2020; Kilickaya et al., 2025; Lim et al., 2023; Midkiff et al., 2024; Purohit et al., 2019; Salvati et al., 2019; Sledz et al., 2020; Yi et al., 2023). Some drones can even operate inside large industrial halls without GPS by using computer-vision algorithms to navigate based on surrounding surfaces and structural landmarks (Border et al., 2023; Filho et al., 2017; Gago et al., 2021; Karnes et al., 2024; Wang et al., 2023; Waqas et al., 2023).

For example, in a metal fabrication facility where welding stations operate continuously, a drone equipped with thermal sensors can periodically monitor whether welding temperatures follow the expected distribution (Singh and Akhai 2024; Lim et al., 2023). If one area shows excessive heat or uneven thermal spread, it may indicate tool wear or improper alignment. Instead of stopping the production line for manual checks, supervisors receive real-time alerts and can schedule targeted maintenance during planned breaks (Ayvaz & Alpay, 2021; Maghazei et al., 2021; Shahin et al., 2023). This approach prevents unnecessary downtime and improves overall productivity (Alsayed et al., 2021; Askerbekov et al., 2024; Kas & Johnson, 2019; Mészáros et al., 2018; Yuan et al., 2019).

Real-time inspection also supports better utilisation of machinery. Flight data, when combined with predictive-maintenance algorithms, helps identify early signs of wear (Ayvaz & Alpay, 2021; Kammerer et al., 2019; Shahin et al., 2023; Yuan et al., 2019). Abnormal vibration patterns, thermal irregularities, or unusual electrical loads can signal problems long before a breakdown occurs (Hermansa et al., 2021;

Kammerer et al., 2019; Yan et al., 2023). This reduces unexpected shutdowns, which are among the most costly interruptions in heavy manufacturing (Ayvaz & Alpay, 2021; Shahin et al., 2023; Yuan et al., 2019).

Another important benefit of drone-based inspection is the creation of a detailed visual archive (Maboudi et al., 2021). Each flight produces time-stamped records that can be saved and used to trace long-term equipment behaviour (Davuluri, 2024; Lin et al., 2020). These records are valuable during audits, troubleshooting, or investigations of recurring issues. Instead of relying on written descriptions or memory, managers can refer to precise visual evidence (Lin et al., 2020; Maboudi et al., 2021).

Drones also enhance communication across departments. Maintenance teams may focus on machine health, safety officers may review environmental conditions, and production teams may use the same data to evaluate workflow efficiency. A shared data platform ensures that different teams have access to a common set of observations, improving coordination and decision-making (Ahmed & Choudhury, 2025; Awasthi et al., 2022; Jacobsen et al., 2023; Lai et al., 2024).

Table 1 presents the key applications of drones in manufacturing, while *Figure 1* illustrates the drone-enabled smart manufacturing flow, showing how visual, thermal, and environmental data collected by drones support defect detection, predictive maintenance, digital twin updates, and automated decision-making processes.

Table 1. Applications of drones in manufacturing.

Manufacturing Function	Drone Capability	Key Outputs	Benefits
Real-Time Inspection	HD imaging, thermal scans, audio sensors	Overheating detection, misalignment, vibration anomalies	Reduced downtime, faster maintenance
Environmental Monitoring	Gas detectors, thermal/ multispectral sensors	Air quality, noise mapping, emission hotspots	Stronger compliance, early risk detection
Quality Assurance & Defect Detection	High-resolution imaging, AI-based pattern recognition	Surface defects, cracks, coating inconsistencies	Higher product consistency, lower waste
Safety Surveillance	Live video, hazard-zone scans	Heat zones, leaks, structural risks	Lower worker exposure, improved safety
Internal Logistics Support	Autonomous navigation	Tool/document delivery	Reduced delays, faster communication
Layout and Asset Mapping	LiDAR mapping, 3D scanning	Structural profiles, clearance checks	Better space utilisation, digital twin updates

Figure 1. Drone-Enabled smart manufacturing flow.

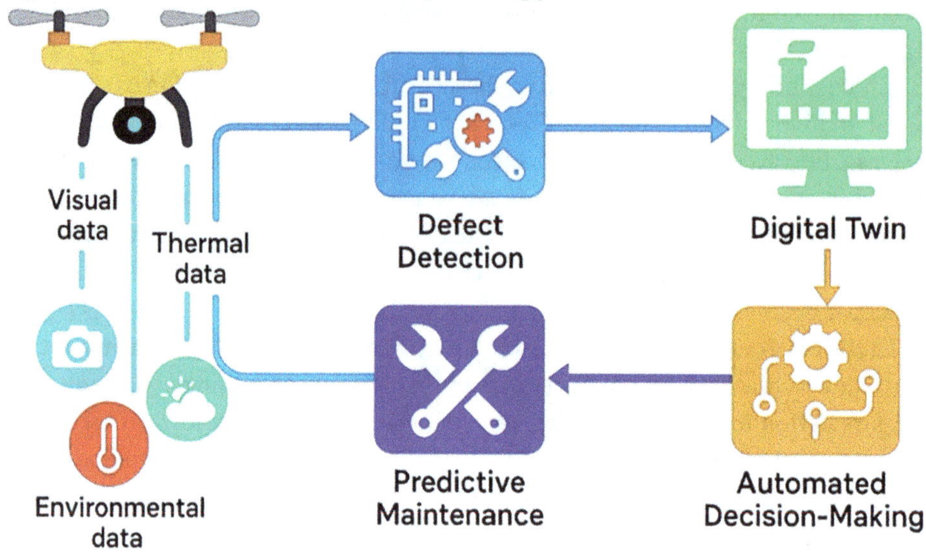

2. DRONE-BASED ENVIRONMENTAL MONITORING IN MANUFACTURING SYSTEMS

Manufacturing facilities have long been associated with significant environmental pressures (Kumar et al., 2024). Their activities influence air quality, water resources, energy use, and surrounding ecosystems (Akhai et al., 2016; Akhai and Wadhwa 2024). Traditionally, environmental monitoring in factories relied on fixed sensors, periodic manual sampling, and laboratory analysis (Kumar & Akhai 2022; Akhai et al., 2022). While these methods provide useful information, they offer limited spatial coverage, are slow to update, and often cannot respond immediately to changing conditions (Nazarov et al., 2024; Nycz & Pietrucha-Urbanik, 2024). Drone technology is helping overcome these limitations by introducing new, flexible, and more comprehensive ways of observing environmental conditions around industrial sites (Fascista, 2022).

One of the most important advantages of drones is their ability to combine an elevated viewpoint with precise sensor measurements (Guebsi et al., 2024; Urama et al., 2020). Equipped with suitable payloads, drones can cover large factory areas rapidly, enter confined or hazardous zones, and capture data that stationary monitors cannot reach (Gerencheal & Lee, 2025; Kas & Johnson, 2019). For example, air-quality measurement is normally restricted to selected points inside and outside a plant. With drones, manufacturers can create detailed aerial maps showing temperature distribution, humidity levels, particulate matter, and chemical traces in real time (Allers et al., 2023; Liang & Shen, 2024; Nycz & Pietrucha-Urbanik, 2024). This provides a clearer understanding of how environmental factors vary throughout the day and across different parts of the site.

Noise and vibration are additional areas where drones offer significant value. Industrial plants often contain heavy motors, grinding units, stamping machines, and conveyor systems that generate constant noise. Excessive sound levels pose both regulatory and occupational health concerns (Akhai 2024). Drones fitted with audio sensors can measure noise intensity across the entire facility within minutes (Salom et al., 2023; Salvati et al., 2019). This information helps identify high-noise zones and supports better planning of work schedules, equipment placement, and maintenance routines (Carroll et al., 2021; Watson et al., 2021).

Water-related monitoring also benefits greatly from drone use. Many industrial sites operate near natural water bodies or use water in their processes. Drones can assist by inspecting water surfaces using imaging sensors or small sample collectors that detect oil films, chemical traces, or biological growth (Duan et al., 2019; Букин et al., 2021; Соколов et al., 2022). Regular drone flights over treatment ponds, drainage channels, and discharge outlets allow early detection of irregularities, reducing the risk of environmental contamination (Asensio et al., 2025; Isgró et al., 2021; Nasiri et al., 2025).

Emissions monitoring—such as greenhouse gases, dust, or volatile organic compounds—is one of the most closely regulated areas (Taneja et al., 2024; Taneja et al., 2025). Drones have changed the way emissions are assessed by enabling sensor-equipped flights near chimneys, vents, and process outlets without exposing workers to hazards (Burgués & Marco, 2020; Cheng & Yuan, 2020; Акматова et al., 2024). The collected readings can be linked to exact coordinates, elevation, and, in some cases, thermal data (Bonne et al., 2024; Dooley et al., 2024; Gullett et al., 2021). This makes it easier to identify emission hotspots and develop targeted mitigation strategies (Burgués & Marco, 2020; Kistner et al., 2024; Midkiff et al., 2024).

Drones are also improving how industries demonstrate regulatory compliance. Instead of relying solely on annual inspections, factories can conduct automated environmental scans at regular intervals (Burgués & Marco, 2020; Reid & Castka, 2023). This creates a continuous feedback system that highlights potential issues early and shows regulators that the company is actively monitoring its impact (Cabianca et al., 2022; Mahariya et al., 2023; Mahroof et al., 2023). In highly regulated sectors such as chemical production and electronics recycling, this proactive approach strengthens both operational control and public trust.

There are practical benefits as well. Fixed monitoring systems require careful installation, extensive wiring, and continual calibration. Drones, in contrast, can be deployed on demand and reach any area of the facility quickly (Urama et al., 2020). Their mobility is especially important during emergencies such as chemical leaks, fires, or equipment failures, where rapid situational awareness is essential for effective containment (Gămulescu et al., 2020; Kordoni et al., 2023; Lim et al., 2022).

Facilities located near residential areas or sensitive ecological zones also benefit from drone-based monitoring. The visual reports produced by drones—such as aerial images or time-lapse sequences—can be shared with community groups to communicate environmental efforts in a transparent manner. This helps build trust and demonstrates the company's commitment to protecting its surroundings (Joyce et al., 2022; Kleinschroth et al., 2022; Sauls et al., 2023).

Advances in battery performance, sensor quality, and autonomous navigation continue to expand the potential of drone-based monitoring (Khang and Akhai 2025; Caballero-Martin et al., 2024; Davuluri, 2024; Xiao et al., 2025). Newer systems can conduct longer flights, carry a wider range of instruments, and even collaborate with ground robots or fixed monitoring stations (Burgués & Marco, 2020). Some drones use onboard AI to process data instantly, allowing them to identify unusual patterns or risks before returning to base (Guebsi et al., 2024; Xiao et al., 2025).

Another important development is the integration of drone-collected data into centralized digital platforms. Environmental readings, flight logs, and sensor outputs can be combined with production data, maintenance records, and safety reports. This unified view helps managers understand how environmental conditions relate to overall factory performance and can uncover trends that traditional methods might miss (Hollenbeck & Chen, 2022; Salem et al., 2024; To et al., 2021).

Despite these benefits, some challenges remain. Weather conditions such as strong winds, heavy rain, or extreme temperatures can limit drone flights (Joyce et al., 2022; Qin et al., 2022; Rahman & Setiawan, 2020; Urama et al., 2020). Manufacturers must plan suitable schedules and backup strategies to ensure consistent data collection. Data privacy and system security are additional concerns, especially as drones interact with cloud-based platforms and broader networks (Nazarov et al., 2024; Qin et al., 2022).

Even with these challenges, the advantages of drones in environmental monitoring are substantial. They provide faster insights, richer information, and greater flexibility than conventional methods (Giordano et al., 2025; Nazarov et al., 2024; Nycz & Pietrucha-Urbanik, 2024; Tran & Nguyen, 2022). By offering a clearer view of environmental impacts and enabling quicker response, drone systems strengthen both operational readiness and regulatory compliance (Ayamga et al., 2021; Dronova et al., 2021; Tmušić et al., 2020).

Drones also encourage better collaboration within organizations. Environmental managers, safety teams, and operations engineers can all work from the same set of drone-generated data. This shared perspective helps set realistic improvement targets, track progress more accurately, and respond effectively to unexpected events (Ahmed & Choudhury, 2025; Lai et al., 2024; Mammarella et al., 2021).

As manufacturing becomes more intelligent and adaptive, drone-enabled environmental monitoring is becoming a benchmark for responsible industrial practice. It transforms factories from closed, difficult-to-observe systems into transparent, well-monitored environments (Sun et al., 2022). In doing so, it reinforces the idea that technological progress and environmental stewardship can advance together.

Table 2 presents the key sensors and digital technologies integrated into industrial drones—including optical, thermal, gas-detection, multispectral, and LiDAR systems—along with their manufacturing applications and operational advantages. *Figure 2* illustrates the major applications of drones in smart manufacturing, highlighting their roles in preventive maintenance, warehouse management, safety inspections, and security surveillance across industrial environments.

Table 2. Key Sensors and Technologies Used in Industrial Drones.

Sensor / Technology	Purpose	Typical Manufacturing Use	Advantages
High-Resolution Optical Cameras	Visual inspection	Surface defects, alignment, assembly verification	High clarity, repeatable inspection
Thermal Infrared Sensors	Heat distribution mapping	Furnace checks, welds, motor overheating	Early fault detection
Gas/Emission Detectors	Air-quality monitoring	VOCs, dust, toxic gases	Regulatory compliance
Multispectral / Hyperspectral Sensors	Emission pattern analysis	Detecting chemical signatures	High sensitivity, precise mapping
LiDAR	3D mapping & navigation	Layout modeling, warehouse scanning	Supports digital twins
AI / Machine Vision	Defect detection, anomaly recognition	Automated QA workflows	Fast, objective decision-making
Integrated Digital Connectivity (IoT, Cloud)	Data transmission & integration	Dashboard updates, maintenance logs	Real-time analysis & coordination

Figure 2. Applications of drones in smart manufacturing.

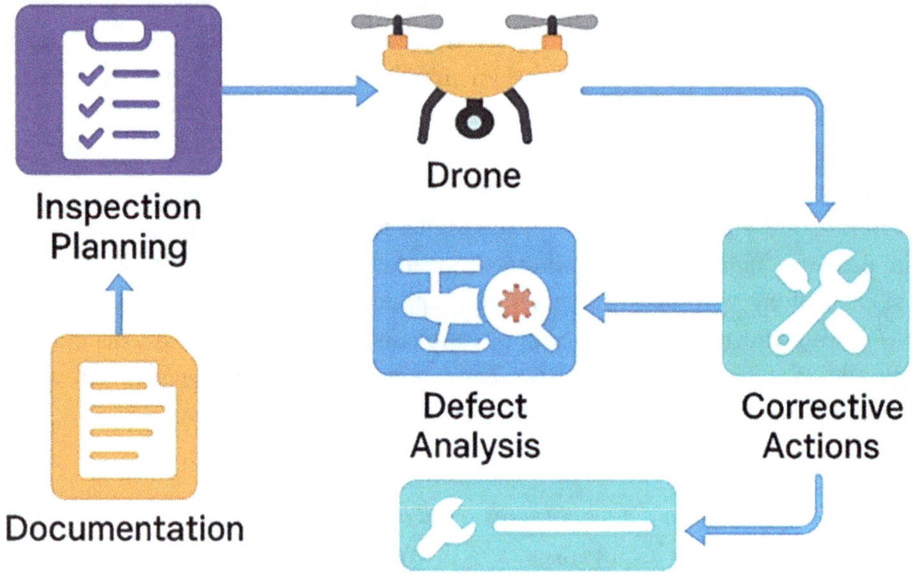

3. QUALITY ASSURANCE AND DEFECT DETECTION WITH DRONES

Quality assurance is a central pillar of effective manufacturing. Without strong quality controls, factories risk releasing defective products, disappointing customers, and weakening their competitive position (Abbass et al., 2025; Kanda et al., 2013; Akhai & Khang, 2024; Akhai & Singh, 2013). Traditionally, quality checks rely on manual inspection, fixed cameras, sampling, and periodic testing. While these methods are useful, they may fail to detect subtle defects or may identify issues only after production is complete, leading to expensive rework or product recalls (Ma et al., 2024). Drone technology is helping address these limitations by introducing faster, more flexible, and more comprehensive inspection capabilities. Drones can collect detailed visual data, thermal signatures, and even structural integrity assessments, providing a holistic view of product quality and manufacturing processes (Heo & Na, 2025). For example, drones equipped with high-resolution cameras and advanced imaging software can identify microscopic flaws, surface irregularities, or structural weaknesses that might be missed by human inspectors. (Askerbekov et al., 2024)

Modern production lines operate at high speed, often producing thousands of parts each hour. Human inspectors can examine only a limited portion of this output, and fatigue or oversight can affect accuracy (Lin et al., 2024; Magalhães & Ferreira, 2022). Drones provide a practical alternative. They can move autonomously along production lines or within storage areas, capturing detailed images of components from multiple angles (Lai et al., 2024; Yi & Sutrisna, 2020; Zhong et al., 2020). High-resolution cameras allow close inspection of surfaces and edges, while software tools analyse the captured images to detect very small defects—such as cracks, scratches, colour inconsistencies, or shape deviations—that may not be noticeable to the human eye (Lim et al., 2023; Shafi et al., 2023; Szira et al., 2023)

Drones are valuable not only for visual assessment. They can carry advanced sensors to detect issues that cannot be seen directly (Gerencheal & Lee, 2025; Suresh et al., 2025; Meydan et al., 2024). Infrared cameras can reveal temperature variations in electronic assemblies or metal structures, helping identify potential short circuits, poor welds, or insulation failures (Ghazal et al., 2020; Sledz et al., 2020). In assembly plants, drones can highlight areas where adhesives have not cured properly or where coatings have been applied unevenly. This early detection prevents minor irregularities from turning into major problems later in the production process (Hermansa et al., 2021).

Efficiency is another significant advantage. Drones gather inspection data without interrupting production, avoiding the downtime associated with manual checks that require equipment to be paused (Li et al., 2022; Watson et al., 2021). Data captured by drones can be transmitted instantly to quality teams and supervisors, allowing rapid decision-making (Salhaoui et al., 2019). If a recurring fault appears, adjustments can be made immediately, reducing waste and protecting raw materials (Ayvaz & Alpay, 2021; Shahin et al., 2023). Over time, this rapid responsiveness contributes to substantial improvements in cost efficiency.

Modern supply chains increasingly expect strong traceability. Customers and regulators want clear evidence that products meet all required standards. Drones support this by creating a comprehensive digital archive of inspection data (Maboudi et al., 2021). Each flight produces time-stamped records that document the condition of products and processes (Lin et al., 2020). This repository acts as a visual reference for each batch and can be used to investigate customer complaints or warranty claims (Maboudi et al., 2021). Instead of relying solely on written notes, companies have precise visual documentation that strengthens quality assurance and regulatory compliance (Reid & Castka, 2023).

Drones also excel in inspecting areas that are difficult or unsafe for workers to reach (Bos et al., 2015; Kas & Johnson, 2019). These may include the tops of large tanks, interior surfaces of storage silos, elevated platforms, or the underside of heavy machinery. Traditionally, accessing these areas requires scaffolding, ladders, or safety harnesses, exposing workers to risks (Alsayed et al., 2021). Drones can enter such environments easily and safely, using optical zoom and lighting systems to capture clear images (Brogaard & Boukas, 2023). This improves both safety and the thoroughness of inspections.

A practical example can be seen in factories producing injection-moulded components. Defects such as voids, surface irregularities, or incomplete filling can be difficult to detect using conventional methods, especially when production volumes are high. By conducting scheduled drone inspections with high-resolution imaging and colour analysis tools, these defects become more visible. Drones can compare new products with an approved reference sample, immediately flagging deviations. This information feeds into manufacturing execution systems, creating a continuous loop between production and quality control (González-Castaño et al., 2020).

Drones also help standardize inspection practices across multiple facilities. Companies with factories in different regions often face challenges in maintaining consistent quality due to variations in equipment, workforce skills, or environmental conditions (Malanchini et al., 2023). Using identical drone systems with unified inspection protocols ensures consistent quality standards across all sites. This harmonisation simplifies performance comparisons and supports the global rollout of improvements (Jacobsen et al., 2023).

Another valuable contribution of drone-based inspection is predictive capability. Over time, the large volume of data collected by drones helps identify patterns (Ayvaz & Alpay, 2021; Kammerer et al., 2019; Shahin et al., 2023). For example, certain machines may begin producing defective parts after a specific operating duration, or certain environmental conditions may cause defects in certain materials. With enough data, manufacturers can shift from reactive quality management to predictive quality planning (Hermansa et al., 2021; Yan et al., 2023). Maintenance teams can intervene before machinery performance declines, improving overall reliability (Yuan et al., 2019).

Small and medium-scale manufacturers also benefit from these technologies. Drones offer extended inspection reach without the need to hire additional staff (Awasthi et al., 2022). This makes high-quality oversight more accessible across different scales of production.

Team collaboration also improves with drone-supported inspections. When a defect is detected, engineers, operators, and quality managers can review the same footage or data, even from remote locations (Salem et al., 2024). Experts in other factories or at company headquarters can join troubleshooting discussions through live drone feeds, reducing travel requirements and accelerating problem resolution (Ahmed & Choudhury, 2025).

However, successful use of drones requires careful planning. Factories must map flight paths, ensure safe drone operation near workers and machinery, and protect sensitive design information (Urama et al., 2020). Teams must define inspection priorities, determine inspection frequency, and establish actions to take when issues are identified (Kim et al., 2018; Yi & Sutrisna, 2020). Some facilities may also require drones built to withstand conditions such as dust, high humidity, or extreme temperatures (Qin et al., 2022; Rahman & Setiawan, 2020). Specialised software may be needed to detect very specific defects, such as micro-cracks in ceramic components (Noğay et al., 2021) or subtle colour variation in textiles (Carvalho et al., 2025; Özek et al., 2025).

Integrating drones into quality assurance is therefore not simply about adding a new device; it represents a broader commitment to continuous improvement. By identifying issues early, improving documentation, and strengthening collaboration, drones play a key role in helping manufacturers enhance quality, reduce risks, and increase confidence in their products. In the pursuit of reliable and fault-free production, drones are emerging as essential partners in modern industrial operations.

Figure 3 shows the drone-based inspection and monitoring cycle, illustrating the sequential operations from mission planning and data capture to AI-based analysis, reporting, and continuous feedback.

Figure 3. Inspection and monitoring cycle.

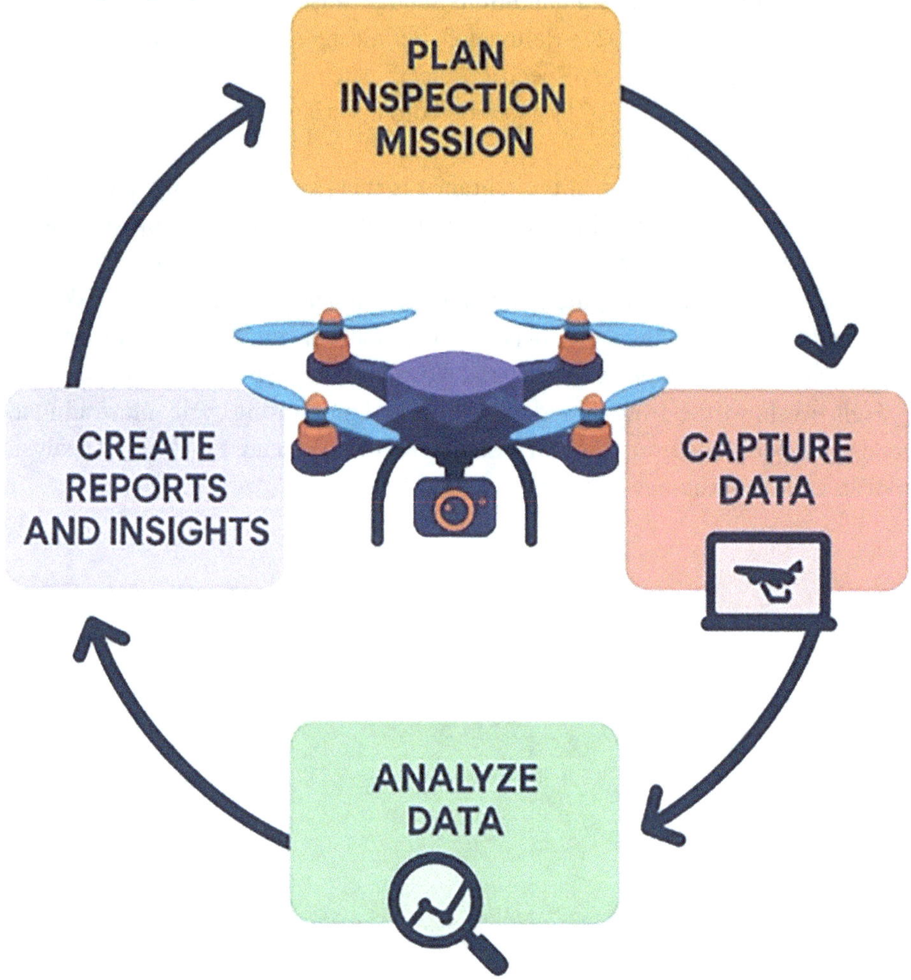

4. DATA INTEGRATION, DIGITAL TWINS, AND SMART MANUFACTURING WITH DRONES

Drones are not just standalone tools flying through the factory skies. Their greatest impact emerges Drones are becoming valuable components of modern smart factories not because they operate independently, but because they integrate smoothly with existing digital systems (Askerbekov et al., 2024). Their most meaningful contribution emerges when the information they collect is linked with advanced tools such as data-integration platforms, predictive analytics, and digital twins (Sun et al., 2022; To et al., 2021). These technologies together help factories operate with more accuracy, adaptability, and transparency (Salem et al., 2024).

Every drone flight generates a wide range of data—images, sensor readings, flight paths, and environmental measurements (Guebsi et al., 2024; Lai et al., 2024). On their own, these datasets are useful, but when connected to the broader information systems within a manufacturing plant, they become far more powerful (Lai et al., 2024; Salhaoui et al., 2019). Modern factories typically rely on layered digital infrastructures that include machine-control systems, production-monitoring software, inventory databases, and predictive-maintenance applications (Jaroš et al., 2022; Li et al., 2022). When drone data flows into these platforms, it enriches them with real-time information and creates a more complete picture of factory operations (Lai et al., 2024; Salhaoui et al., 2019).

Drone–PLC/SCADA Data Flow and Interface Integration - Drones equipped with industrial sensors generate visual, thermal, positional, and environmental data during each inspection flight. This data is transmitted either wirelessly through Wi-Fi/5G or via edge gateways that convert drone outputs into standardized protocols such as OPC-UA, MQTT, or Modbus—ensuring compatibility with existing factory automation systems. Once received, PLCs and SCADA platforms interpret the processed data to update dashboards, trigger alarms, or initiate automated responses such as adjusting machine parameters or scheduling maintenance tasks. This seamless interface allows drone-generated insights to merge directly with plant control networks, enabling real-time monitoring, rapid anomaly detection, and synchronized decision-making across the production environment (Meydan et al., 2024; Suresh et al., 2025).

A clear example of this integration is the use of digital twins. A digital twin is a virtual model of a manufacturing system that mirrors the real environment as closely as possible (Layer et al., 2023; To et al., 2021). It includes representations of equipment, workflows, workers, and even changes in temperature or humidity (Hollenbeck & Chen, 2022). The accuracy of a digital twin depends on continuous, up-to-date information (Hollenbeck & Chen, 2022). Drones play a key role by sending fresh imagery and sensor data directly into these virtual models (To et al., 2021). By keeping the digital twin synchronized with actual factory conditions, operators can test new scenarios, predict failures, optimise layouts, and explore process changes safely and efficiently (Salem et al., 2024).

As drones move throughout the plant, they act as mobile data collectors, creating a living record of factory activities (Baumann et al., 2020). When new machinery is introduced, drones can quickly examine how it interacts with existing equipment, checking clearances, safety distances, and potential bottlenecks. This information helps engineers update the digital twin without interrupting production. Before making any physical changes, managers can test different layouts and operating conditions virtually, reducing time, cost, and risk (Salem et al., 2024).

Integrating drones into existing IT systems requires thoughtful coordination (Lee et al., 2022). Many older industrial platforms were not designed for mobile robots (Malanchini et al., 2023). To address this, factories often use integration software that converts drone-generated data into standardized formats understandable by manufacturing systems (Lee et al., 2022). As a result, information such as vibration readings, thermal patterns, or image analysis can be automatically incorporated into maintenance logs or production dashboards (Kammerer et al., 2019; Lim et al., 2023). For instance, if a drone detects an overheated motor, the system can immediately generate a maintenance request or alert technical staff (Ghazal et al., 2020).

The flow of information unlocked by drones also supports advanced data analytics (Salhaoui et al., 2019). Machine-learning models improve when they receive large volumes of varied, high-quality data (Lim et al., 2023). Over time, analytical systems use this data to recognize patterns, identify inefficiencies, and recommend adjustments that would be difficult to detect manually (Hermansa et al., 2021; Yuan et al., 2019).

This integrated ecosystem contributes to faster decision-making. For example, if a drone detects unusual heat distribution on a furnace surface, the system can compare these readings against historical data and propose immediate corrective measures (Ghazal et al., 2020; Sledz et al., 2020). Inventory management also benefits. Drones can scan warehouse shelves, verify stock levels, and feed the information directly into ordering systems (Almadani et al., 2020; Gago et al., 2021). This helps reduce shortages, lower inventory costs, and maintain a steady flow of materials.

Drone integration also supports improvements in security. Facilities that handle sensitive designs or critical components often require strict safety protocols. Drones can perform perimeter checks, inspect restricted zones, and send live video to security teams (Almadani et al., 2020). Automated patrol routes reduce blind spots and create a more reliable record of security events.

Collaboration across factory teams becomes easier when drone-collected data is shared. Previously, maintenance, production, and management teams often relied on different sets of reports. With drones providing consistent, real-time information, all departments can work from the same data (Ahmed & Choudhury, 2025; Salem et al., 2024). Shared dashboards help synchronise actions, align goals, and solve problems more effectively. Even remote experts can join discussions through live drone feeds, reducing the need for travel and speeding up responses (Ahmed & Choudhury, 2025).

Digital transformation driven by drone integration is not a one-time shift but a continuous journey (Akhai, 2023; Akhai, 2024; Singh et al., 2025; Arif & Akhai, 2024). Manufacturers must update workflows, train staff, and refine protocols to support combined human-drone-digital systems. Drone operators need to understand the technical meaning of the data they collect, while engineers must ensure that systems can communicate with growing fleets and more advanced sensors (Xiao et al., 2025).

Looking ahead, drones are expected to play even larger roles in manufacturing optimisation. Emerging technologies such as coordinated drone swarms can inspect large or complex areas more quickly (Leet et al., 2022; Schilling et al., 2022). Some plants already use mixed fleets, combining aerial drones for elevated spaces with ground robots for heavier or repetitive tasks. As these systems communicate with one another through smart networks, the factory becomes a highly connected and responsive environment.

The advantages extend beyond productivity. Drones also support sustainability goals by collecting data on energy consumption, air quality, emissions, and waste flow (Kurbanzade et al., 2023; Pal & Joshi, 2023). This helps identify opportunities for reducing environmental impact and ensures that compliance with environmental standards is maintained consistently—not just during scheduled audits (Mahroof et al., 2023; Reid & Castka, 2023). Such improvements build trust among regulators, customers, and local communities (Kleinschroth et al., 2022; Sauls et al., 2023).

Manufacturing is inherently competitive and constantly evolving for better human comfort and safety (Akhai, 2023; Abbass et al., 2025). Integrating drones into digital workflows helps factories respond to new requirements with speed and confidence. Whether adjusting layouts, meeting stricter regulations, or adapting to shifts in demand, drones strengthen the resilience and adaptability of industrial operations.

In essence, drone integration represents both a practical improvement and a cultural shift. It encourages employees to view their environment more holistically and to rely on richer, real-time information. As factories continue their transition toward smart manufacturing, drones contribute significantly to creating systems that are more connected, more informed, and more capable of continuous improvement (Sun et al., 2022).

Figure 4 shows the digital twin integration framework, illustrating how real-time drone data from visual, thermal, and emission sensors feeds into simulation and analysis modules to update virtual factory models and generate actionable insights.

Figure 4. Digital twin integration.

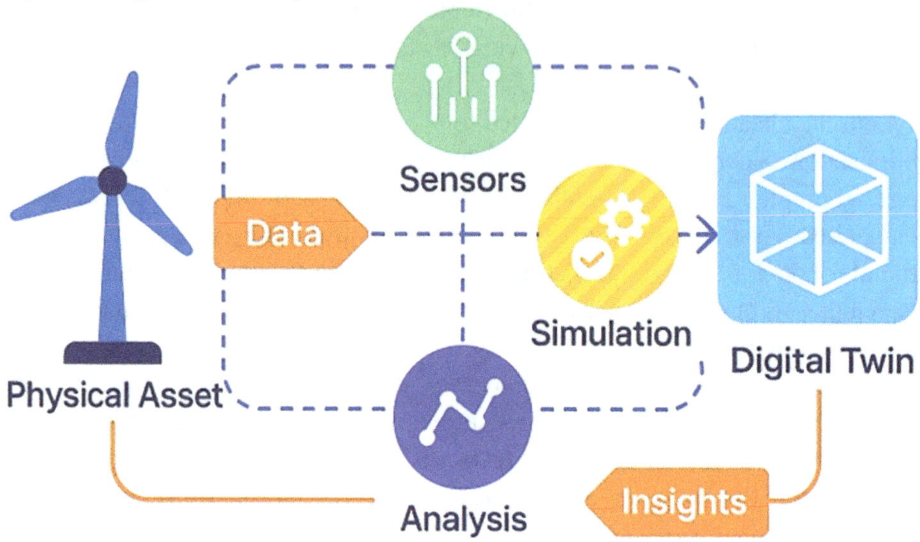

5. SAFETY, RISK MANAGEMENT, AND WORKFORCE IMPACTS OF DRONES IN MANUFACTURING

The introduction of drones into manufacturing facilities brings important changes to safety practices, risk management strategies, and workforce responsibilities. Industrial environments traditionally contain many hazards, including heavy machinery, confined spaces, toxic materials, and high-temperature operations (Bos et al., 2015; Chen et al., 2024). Protecting workers has always been a core priority. Drones support this goal by reducing direct exposure to dangerous conditions, while also introducing new factors that must be managed carefully (Alsayed et al., 2021; Brogaard & Boukas, 2023; Kas & Johnson, 2019; Szira et al., 2023; Zhong et al., 2025).

One of the most immediate safety benefits of drones is the ability to limit worker entry into hazardous areas. Tasks that previously required climbing scaffolding, working at heights, or entering restricted spaces can now be performed remotely. For example, inspections on the tops of storage tanks, along elevated pipelines, or inside chemical processing units traditionally required workers to use ladders, platforms, or protective equipment (Kas & Johnson, 2019). Drones can complete these inspections while operators remain safely on the ground, significantly reducing the risk of falls, chemical exposure, and heat-related injuries (Alsayed et al., 2021; Bao et al., 2023; Bos et al., 2015).

Similar advantages appear in environments where toxic gases, hot surfaces, or explosive atmospheres are present. Drones equipped with thermal sensors, gas detectors, or high-resolution cameras can investigate leaks, measure temperature extremes, or capture detailed images without placing workers at risk (Al-Battbootti et al., 2021; Allers et al., 2023; Benyeogor et al., 2020; Bretschneider et al., 2024; Dooley et al., 2024; Lim et al., 2023; Акматова et al., 2024). Over time, reducing human exposure leads to fewer accidents and illnesses and strengthens a company's safety record.

However, introducing drones into manufacturing requires thoughtful planning to ensure that they operate safely around people and equipment. Designated flight zones, operating schedules, and speed limits help prevent drones from interfering with machinery or personnel (Kim et al., 2018; Yi & Sutrisna, 2020). Many plants schedule drone operations during shift changes or in low-activity periods to minimise the chance of accidental contact. These measures ensure drones enhance, rather than compromise, overall safety (Urama et al., 2020).

Worker training becomes an essential part of this transition. Employees must understand when drones will operate, where flight paths are located, and what actions to take in case of unexpected behaviour or system faults. Safety programs increasingly include drone-awareness sessions, visual signage, and emergency response guidelines. As drone use becomes more common, organisations develop standard operating procedures and update safety manuals to reflect new responsibilities (Cheng et al., 2022; Mendes et al., 2022; Nwaogu et al., 2024; Onososen et al., 2023).

Risk management also evolves with drone integration. While drones remove certain risks, they create new ones. Mechanical failures, such as battery issues or motor malfunctions, could cause drones to fall or collide with equipment (Qin et al., 2022). Software errors or loss of connectivity may pose similar concerns (Caballero-Martin et al., 2024; Khan, 2021). To address these risks, manufacturers use high-quality hardware, perform routine maintenance checks, and impose strict limits on flight paths (Namian et al., 2021; Tiusanen et al., 2023). Redundancy is also common: backup drones are kept on hand, critical inspections are monitored closely, and sensitive tasks are conducted under controlled conditions.

Cybersecurity becomes increasingly important as drones rely more on autonomous navigation, wireless communication, and cloud-based data storage (Akhai, 2024; Akhai & Wadhwa, 2024; Felix, 2023; Omolara et al., 2023). Protecting drone systems from hacking or unauthorized access requires strong passwords, encryption protocols, and regular software updates. Many companies adopt strict digital-security policies to safeguard operational data and prevent tampering (Adil et al., 2023; Ashraf et al., 2023; Durfey & Sajal, 2022; Sihag et al., 2023).

Drones also support emergency-response and incident-management activities. During fires, equipment failures, or chemical releases, drones can be deployed immediately to capture real-time images and sensor readings (Gămulescu et al., 2020; Kordoni et al., 2023; Lim et al., 2022). These aerial perspectives help emergency teams assess the scale of the situation, identify hazards, and plan safer and more effective responses. Drones can pinpoint leak locations, observe structural damage, or reveal blocked escape routes, giving responders vital information when time is critical.

Beyond safety and risk management, drones influence the roles and expectations of the manufacturing workforce. Drone integration creates new job opportunities—such as drone pilots, data analysts, sensor technicians, and maintenance specialists (Maghazei, 2022; Stroud & Weinel, 2020). Existing workers often participate in upskilling programs, learning about flight protocols, sensor interpretation, and regulatory requirements (Nwaogu et al., 2024). This shift does not eliminate jobs; instead, it changes the nature of work. Routine, physically strenuous, or hazardous tasks are reduced, and employees take on roles that require analysis, decision-making, and creative problem-solving (Stroud & Weinel, 2020).

Workforce morale may improve as a result. Employees who are no longer required to perform dangerous or repetitive tasks often report greater job satisfaction (Barbu et al., 2025; "Proceedings of the 19th Conference on Computer Science and Intelligence Systems (FedCSIS)," 2024). With drones handling high-risk activities, workers can focus on innovation, troubleshooting, and continuous improvement. Companies that invest in training and communication around drone adoption tend to build stronger employee engagement and loyalty (Callari et al., 2024).

Drone use also contributes to a more transparent and collaborative manufacturing culture. The data collected—such as images, measurements, or safety observations—can be shared across departments, allowing maintenance teams, production managers, and executives to review the same information (Ahmed & Choudhury, 2025; Awasthi et al., 2022; Jacobsen et al., 2023; Lai et al., 2024). This shared visibility builds trust, reduces misunderstandings, and encourages collective problem-solving. Transparent documentation also supports regulatory compliance and internal audits (Salem et al., 2024).

Challenges do remain. Some workers may initially worry about privacy, job displacement, or changes in daily routines (Nazarov et al., 2024; Qin et al., 2022). Addressing these concerns through open discussion, clear communication, and participatory training helps build acceptance (Callari et al., 2024; Stroud & Weinel, 2020). Involving employees early in planning and decision-making turns potential resistance into constructive engagement (Shankar et al., 2024).

As drones become more widespread, regulatory bodies and insurance providers continue to update guidelines on safe operation near people and equipment (Joyce et al., 2022; Urama et al., 2020). Manufacturers work closely with these agencies to ensure compliance, define liability boundaries, and support continuous improvement in drone technology and operational practices (Tiusanen et al., 2023).

Overall, drones are reshaping the landscape of safety and risk management in manufacturing. They minimize exposure to hazardous environments, support fast and informed decision-making during emergencies, and enable workers to focus on higher-value tasks. Their influence extends beyond physical safety, promoting a more informed, confident, and future-ready workforce.

6. IMPLEMENTATION STRATEGIES, CHALLENGES, AND BEST PRACTICES FOR DRONES IN MANUFACTURING

Introducing drones into a manufacturing facility is a significant organizational step that requires careful planning, thoughtful execution, and continuous improvement (Shankar et al., 2024). While the benefits—faster inspections, safer working conditions, richer data, and stronger quality control—are substantial, a successful drone program must overcome a range of technical, regulatory, and cultural challenges (Askerbekov et al., 2024; Callari et al., 2024; Nooralishahi et al., 2021). This section outlines key strategies, common obstacles, and best practices to support manufacturers in this transition.

A foundational step is identifying the specific needs and objectives of the manufacturing operation. Each facility differs in layout, processes, and safety requirements. A drone integration plan should begin with a detailed assessment of existing inspection routines, environmental monitoring practices, and quality-control processes. For example, drones may provide clear advantages in facilities where workers frequently access high platforms, confined spaces, or hazardous environments (Alsayed et al., 2021; Bos et al., 2015; Brogaard & Boukas, 2023; Kas & Johnson, 2019). Similarly, if manual visual inspections are slow or inconsistent, drones can deliver faster and more reliable results (Alzarooni et al., 2025; Lin et al., 2024; Ma et al., 2024; Magalhães & Ferreira, 2022). Clear goals help determine the appropriate drone models, sensor payloads, and operational procedures (Heo & Na, 2025).

Compliance with legal and regulatory requirements is another essential consideration. Drone operations in industrial zones must follow aviation and safety regulations designed to protect workers, infrastructure, and surrounding airspace (Szira et al., 2023; Tiusanen et al., 2023; Urama et al., 2020). Manufacturers should consult local aviation authorities early, secure necessary permits, and understand operational restrictions or no-fly areas (Bhat et al., 2024; Habibi et al., 2023; Joyce et al., 2022; Krook et al., 2024; Öz et al., 2022). Establishing strong communication with regulators helps streamline approvals and minimizes future compliance issues.

Internally, companies should create policies governing flight schedules, approved zones, and data-handling practices to ensure alignment with organizational standards and employee rights (Callari et al., 2024; Hovsha & Brien, 2022; Pillai et al., 2022, 2024; Qin et al., 2022; Tianyuan et al., 2020; Ventura et al., 2023).

Selecting suitable hardware and supporting technologies is a critical part of the planning process. Consumer-grade drones are often unsuitable for industrial use due to dust, heat, humidity, or electromagnetic interference (Gyrichidi et al., 2024; Rahman & Setiawan, 2020). Industrial-grade drones offer sturdier construction, longer battery life, and compatibility with specialized sensors such as thermal imagers, gas detectors, or high-resolution optical cameras (Burgués & Marco, 2020; Cheng & Yuan, 2020; Davuluri, 2024). Some manufacturers choose modular drone platforms that allow easy sensor changes depending on the mission, reducing long-term costs (Cameron et al., 2022; Heřt et al., 2023; McQueen & Drennan, 2024; Panarotto et al., 2022; Walsh et al., 2023; Wang et al., 2023).

Operational integration requires structured flight schedules, mapped routes, and trained personnel (Kim et al., 2018; Yi & Sutrisna, 2020). Autonomous flight software can plan optimal paths, avoid obstacles, and maximise coverage while conserving battery power (Border et al., 2023; Caballero-Martin et al., 2024; Liu et al., 2025). Facilities also need trained operators or drone pilots who understand not only how to fly safely but also how to collect, interpret, and apply drone data effectively (Cheng et al., 2022; Mendes et al., 2022; Nwaogu et al., 2024; Onososen et al., 2023; Plaza et al., 2020). A centralized drone-operations center strengthens coordination by managing missions, resolving technical issues, and monitoring real-time performance.

Data management is another important factor. Drone operations generate large volumes of images, temperature readings, flight logs, and other sensor data (Lin et al., 2020; Maboudi et al., 2021). Manufacturers should establish reliable data pipelines that store information securely on cloud platforms or local servers. Automated analysis tools—such as AI-driven image recognition or anomaly-detection algorithms—reduce manual workloads and allow quicker identification of defects or environmental changes (Alzarooni et al., 2025; Lim et al., 2023; Yan et al., 2023). Clear guidelines must ensure secure storage, controlled data access, and responsible long-term retention (Callari et al., 2024; Hovsha & Brien, 2022; Pillai et al., 2022, 2024; Qin et al., 2022; Tianyuan et al., 2020; Ventura et al., 2023).

Technical and cultural challenges can arise during implementation (Shankar et al., 2024). Manufacturing environments may cause interference with drones, such as electromagnetic fields, GPS signal loss, or physical obstructions (Gerencheal & Lee, 2025; Urama et al., 2020). Continuous testing and calibration help overcome these issues. If needed, drones that use visual or LiDAR navigation can operate effectively in GPS-restricted zones (Filho et al., 2017; Karnes et al., 2024; Wang et al., 2023). Cultural resistance may also appear among workers who worry about job security or privacy (Callari et al., 2024; Nazarov et al., 2024; Qin et al., 2022; Stroud & Weinel, 2020). Open communication, transparent objectives, and hands-on training help build trust and encourage acceptance (Callari et al., 2024; Shankar et al., 2024; Stroud & Weinel, 2020).

Cost is another consideration. While drones can reduce inspection time and minimize downtime, expenses such as equipment purchases, operator training, and software subscriptions must be planned carefully (Bakacsy & Zakar, 2025; Grote et al., 2024; Mishra & Mishra, 2024; Nielsen et al., 2023; Nooralishahi et al., 2021; Triantafyllou et al., 2020). Many manufacturers begin with small pilot projects on selected production lines, measuring returns before scaling up (Maghazei et al., 2021). This staged approach improves confidence, reduces risk, and ensures that investments deliver meaningful benefits.

Table 3. Challenges & Best Practices.

Area	Key Challenges / Risks	Best Practices / Solutions	Expected Impact
Safety & Workforce	Worker concern, privacy issues, hazardous zones	Awareness training, defined flight paths, signage	Higher acceptance, safer operations
Technical Constraints	GPS loss, battery limits, dust/heat	Industrial-grade drones, LiDAR navigation, redundancy	Reliable performance
Regulatory Compliance	Aviation rules, safety codes	Early coordination with authorities	Smooth approval process
Data Management	Large data volumes, security risks	Cloud servers, encryption, access control	Secure, fast analytics
Operational Planning	Poor scheduling, untrained pilots	Standard operating procedures, pilot certification	Smooth daily drone operations
Cost Considerations	Hardware + software cost	Pilot testing before scaling, modular sensors	Higher ROI, controlled investment
Emergency Use	Fire, chemical leaks	Rapid-deployment protocols	Faster incident response

- Experience from successful drone programs highlights several best practices:
  Start with Pilot Projects: Begin on a small scale to learn operational needs and improve procedures before expanding (Maghazei et al., 2021).
- **Prioritize Data Quality:** High-quality images and sensor readings must be paired with strong analytical tools and accessible dashboards (Lim et al., 2023; Lin et al., 2020; Maboudi et al., 2021).

- **Maintain Strong Safety Protocols:** Define safe flight zones, provide clear signage, and implement emergency shut-off procedures (Tiusanen et al., 2023).
- **Invest in Workforce Training:** Include operators, maintenance staff, and quality teams in training sessions to build competence and support (Cheng et al., 2022; Mendes et al., 2022; Nwaogu et al., 2024; Onososen et al., 2023; Plaza et al., 2020).
- **Integrate Drones into Existing Processes:** Align flight schedules with shift patterns, maintenance cycles, and existing inspection workflows (Kim et al., 2018; Yi & Sutrisna, 2020).
- **Commit to Continuous Improvement:** Regularly review drone operations, update software, and reassess regulatory compliance (Krook et al., 2024; Stanton et al., 2021).
- **Prepare for Emergency Use:** Establish rapid-deployment protocols for fires, spills, or equipment failures (Gămulescu et al., 2020; Kordoni et al., 2023; Lim et al., 2022).
- **Ensure Data Governance and Privacy:** Policies should address data storage, sharing, deletion, and protection of worker privacy (Callari et al., 2024; Hovsha & Brien, 2022; Pillai et al., 2022, 2024; Qin et al., 2022; Tianyuan et al., 2020; Ventura et al., 2023).
- **Collaborate with Vendors:** Strong partnerships with drone manufacturers and software providers support troubleshooting and upgrades.

When implemented strategically, drone programs help factories shift from reactive to proactive operations. Early detection of defects, real-time environmental monitoring, and consistent quality control contribute to lower operational costs and better customer satisfaction (Ayvaz & Alpay, 2021; Shahin et al., 2023; Yuan et al., 2019). The insights gained from drone data also enhance flexibility, allowing companies to adapt more quickly to regulatory updates, market changes, or supply chain disruptions (Todescato et al., 2023).

In the long term, effective drone integration becomes a competitive advantage. Manufacturers that master safe and efficient drone use demonstrate innovation, agility, and strong technological capacity (Askerbekov et al., 2024). As the technology continues to advance, drones will move from being optional tools to essential resources that shape the future of intelligent, adaptive, and resilient manufacturing systems (Sun et al., 2022).

The ongoing development of advanced technologies, such as the Internet of Drones and sophisticated AI-driven analytics, promises to further revolutionize manufacturing by enabling interconnected drone systems for real-time data collection and analysis (Askerbekov et al., 2024). This transformative potential extends beyond operational efficiency to include significant improvements in occupational safety by minimizing human exposure to hazardous environments and tasks (Maghazei & Netland, 2019).

7. CONCLUSIONS AND THE FUTURE SCOPE OF DRONES IN MANUFACTURING

The integration of drones into manufacturing is still evolving, yet it is already reshaping how factories conduct inspection, monitoring, logistics, and quality management. What once appeared to be a novel concept has developed into a practical and influential technology, supported by rapid advancements in artificial intelligence, sensor engineering, connectivity, and battery performance. As these technologies mature, drones are expected to become more capable, more autonomous, and more deeply embedded in manufacturing operations.

One promising direction is the emergence of highly autonomous drone fleets that operate with limited human involvement. Current drones can already navigate complex indoor environments, avoid obstacles, and complete predefined inspection tasks using integrated sensors and onboard intelligence. In the future, with improved fleet-management platforms, drones will communicate with one another and with production machinery. They will share data instantly, coordinate missions, and adjust their routes based on operational needs or emerging risks.

Such capabilities will support the development of self-optimizing manufacturing environments. When a defect or anomaly arises, multiple drones may automatically assess the affected area while others monitor surrounding conditions. These insights can activate maintenance robots or alert technicians before a failure escalates. This coordinated system will significantly reduce unplanned downtime and foster factory ecosystems that continuously learn, adapt, and improve with minimal manual intervention.

Drones are also likely to expand beyond observation into more active roles. Some facilities already use drones to transport small tools and components across large areas faster than traditional systems. With progress in payload capacity, stability, and precision, drones may assist in assembly tasks, material handling, equipment cleaning, and micro-level adjustments in hard-to-reach areas. These roles will further integrate aerial robotics into the production workflow.

The combination of drones with the industrial Internet of Things (IoT), digital twins, and augmented reality will strengthen human-machine collaboration. Workers equipped with augmented reality glasses could view live drone feeds, receive guided repair instructions, or visualize equipment diagnostics directly in their field of view. Supervisors may rely on dashboards enriched with real-time drone data to oversee large facilities efficiently. These applications support safer operations and enhance workforce effectiveness.

Sustainability will benefit significantly from advanced drone use. Continuous drone-enabled monitoring of emissions, energy use, heat distribution, and waste streams will allow manufacturers to optimize resource consumption. Instead of periodic environmental audits, factories will use constant data flows to maintain compliance, reduce environmental impact, and pursue long-term sustainability targets.

However, realizing these possibilities requires overcoming challenges related to regulation, cybersecurity, data governance, and workforce adaptation. As drones become more autonomous and more connected, ensuring secure communication, protecting intellectual property, and maintaining responsible data management will be essential. Ethical considerations, including workforce displacement or concerns about surveillance, must be addressed through reskilling programs, transparent communication, and supportive policies.

For manufacturers, future success depends on viewing drones as strategic assets rather than isolated tools. This involves investment in digital infrastructure, staff training, and organizational culture. Effective leadership will balance innovation with risk management, ensuring that drone operations remain safe, compliant, and aligned with organizational goals.

In summary, drones represent one of the most transformative developments in modern manufacturing. They enhance real-time inspection, environmental monitoring, quality assurance, and logistics, helping factories achieve higher visibility, fewer errors, and improved safety. As technology progresses, drones will evolve from supplementary devices into essential collaborators, enabling more intelligent, efficient, and sustainable production systems.

Manufacturers who adopt drones thoughtfully will be well-positioned to lead in a competitive and rapidly changing industrial landscape. Their facilities will showcase advanced human–machine collaboration, agile decision-making, and continuous innovation. The role of drones in manufacturing is only beginning to unfold, and their future impact promises to redefine how factories operate for decades to come.

Table 4 presents summary of the key themes, functions, and impacts of drone adoption in modern manufacturing systems, including real-time inspection, environmental monitoring, quality assurance, digital integration, safety enhancements, and implementation strategies.

Table 4. Summary of drone integration in manufacturing.

Aspect	Key Highlights
Why Drones Matter in Manufacturing	• Provide mobile, flexible monitoring across complex factory layouts • Access confined and hazardous areas safely • Deliver real-time images, thermal and sensor data • Improve operational visibility, safety, and efficiency
Real-Time Inspection	**Monitors:** overheating, leaks, misalignment, vibrations, weld stations, conveyors **Benefits:** early fault detection, reduced downtime, predictive maintenance, faster decision-making
Environmental Monitoring	**Sensors:** gas detectors, thermal, multispectral, optical **Measures:** emissions, VOCs, dust levels, air quality, noise, temperature, water contamination **Impact:** better compliance, hotspot detection, transparent environmental reporting
Quality Assurance & Defect Detection	• High-resolution imaging for cracks, scratches, dimensional errors • Infrared for curing issues, insulation faults, overheating • AI-based automated defect recognition **Outcome:** higher consistency, less rework, strong traceability
Digital Twins & Smart Integration	**Drone Data Feeds Into:** digital twins, IoT dashboards, maintenance logs **Enables:** layout optimization, real-time condition updates, predictive analytics, faster operational decisions
Safety & Workforce Impact	• Reduces worker exposure to heights, toxins, heat zones • Supports emergency response (fire, leaks, breakdowns) • Creates new job roles: drone pilot, data analyst, sensor specialist • Improves workforce confidence and safety culture
Implementation Requirements	• Industrial-grade drones, modular sensors • SOPs, flight zones, regulatory compliance • Cybersecurity, cloud data governance • Operator training and phased pilot deployment
Overall Insight	**Drones are reshaping manufacturing**, enabling safer operations, smarter inspections, higher product quality, and data-driven decision-making, moving factories from reactive to predictive.

9. CHAPTER SUMMARY

Drones are becoming essential in modern manufacturing by enhancing inspection, monitoring, and quality-management processes. Their mobility and advanced sensors—capturing high-resolution images, thermal data, gas levels, and multispectral readings—provide real-time insights that fixed sensors and manual inspections cannot. They enable early detection of faults such as overheating, misalignment, leaks, and vibration issues, reducing downtime and supporting predictive maintenance.

The chapter also highlights drones' role in environmental monitoring, mapping emissions, air quality, noise, water contamination, and temperature variations to improve compliance and sustainability. In quality assurance, AI-enabled drone

REFERENCES

Abbass, M., Abbas, U., Jafri, R., Arif, S. M., & Akhai, S. (2025). AI and Machine Learning Applications in Sustainable Smart Cities. In *Sustainable Smart Cities and the Future of Urban Development* (pp. 1–32). IGI Global., DOI: 10.4018/979-8-3693-6740-7.ch001

Abbass, M., Abbas, U., Wani, S. A., & Akhai, S. (2025). Rethinking global food governance from historical perspectives for a sustainable future. In *Innovative trends shaping food marketing and consumption* (pp. 553–578). IGI Global., DOI: 10.4018/979-8-3693-8542-5.ch022

Abbass, M., Akhai, S., Abbas, U., Jafri, R., & Arif, S. M. (2025). AI-enabled sustainable urban planning and management. In *Real-World Applications of AI Innovation* (pp. 233–260). IGI Global., DOI: 10.4018/979-8-3693-4252-7.ch012

Adil, M., Song, H., Mastorakis, S., Abulkasim, H., Farouk, A., & Jin, Z. (2023). UAV-Assisted IoT Applications, Cybersecurity Threats, AI-Enabled Solutions, Open Challenges With Future Research Directions. *IEEE Transactions on Intelligent Vehicles*, *9*(4), 4583–4605. DOI: 10.1109/TIV.2023.3309548

Ahmed, T., & Choudhury, S. (2025). *AI and Semantic Communication for Infrastructure Monitoring in 6G-Driven Drone Swarms*. DOI: 10.48550/ARXIV.2503.00053

Akhai, S. (2023). Biogas Plants for Sustainable Municipal Waste Management: A Brief Review. *Handbook of Research on Safe Disposal Methods of Municipal Solid Wastes for a Sustainable Environment*, 162-179. DOI: 10.4018/978-1-6684-8117-2.ch012

Akhai, S. (2023). Healthcare record management for healthcare 4.0 via blockchain: a review of current applications, opportunities, challenges, and future potential. In *Blockchain for Healthcare 4.0* (Vol. 1, pp. 211–223). CRC Press., DOI: 10.1201/9781003408246-11

Akhai, S. (2024). Innovating Healthcare Delivery: Harnessing Drones for Medical Services. In *Modelling of Virtual Worlds Using the Internet of Things* (pp. 232-250). CRC Press. DOI: 10.1201/9781003480181-10

Akhai, S. (2024). Smart Home Environment for Society: An Overview of IoT-Enabled Air Conditioners. *Modelling of Virtual Worlds Using the Internet of Things*, 74-87. DOI: 10.1201/9781003480181-3

Akhai, S. (2024). Smart Home Environment for Society: An Overview of IoT-Enabled Air Conditioners. In *Modelling of Virtual Worlds Using the Internet of Things* (Vol. 1, pp. 74-87). CRC Press. DOI: 10.1201/9781003480181-3

Akhai, S. (2024). Towards Trustworthy and Reliable AI The Next Frontier. In *Explainable Artificial Intelligence (XAI) in Healthcare* (Vol. 1, pp. 119-129). CRC Press, Taylor & Francis Group. DOI: 10.1201/9781003426073-7

Akhai, S., Abbass, M., Kaur, P., & Kaur, T. (2025). Digital Transformation Across Generations: Robotics and AI in Action. In *Impacts of Digital Technologies Across Generations* (pp. 23-40). IGI Global. DOI: 10.4018/979-8-3693-6366-9.ch002

Akhai, S., & Balu, V. (2022). *Code injection assault & mitigation model to prevent attacks*. In Proceedings of the 2022 11th International Conference on System Modeling & Advancement in Research Trends (SMART) (pp. 1324–1328). IEEE. DOI: 10.1109/SMART55829.2022.10046835

Akhai, S., & Khang, A. (2024). Energy efficiency and human comfort: AI and IoT integration in hospital HVAC systems. Medical Robotics and AI-Assisted Diagnostics for a High-Tech Healthcare Industry, 93-108. DOI: 10.4018/979-8-3693-2105-8.ch007

Akhai, S., & Khang, A. (2024). Energy efficiency and human comfort: AI and IoT integration in hospital HVAC systems. Medical Robotics and AI-Assisted Diagnostics for a High-Tech Healthcare Industry, 93-108. DOI: 10.4018/979-8-3693-2105-8.ch007

Akhai, S., & Khang, A. (2024). Innovations in Medical Manufacturing: A Review of 3D Printing, Robotics, and Internet of Things (IoT). In *The Quantum Evolution* (Vol. 1, pp. 226-241). CRC Press. DOI: 10.1201/9781032642079-10

Akhai, S., & Singh, H. (2013). Design optimization for modification of trough belt conveyor to reduce material spillage used in clinker transport in cement plant. *Int. J. Res*, *1*, 27–37.

Akhai, S., Singh, V. P., & John, S. (2016). Human performance in industrial design centers with small unit air conditioning systems. *Journal of Advanced Research in Production Industrial Engineering*, *3*(2), 5–11.

Akhai, S., & Wadhwa, A. (2024). Enhancing Radiator Cooling with CuO Nanofluid Microchannels. *International Journal of Automotive Science And Technology*, *8*(2), 201–211. DOI: 10.30939/ijastech..1399702

Akhai, S., & Wadhwa, A. S. (2024). Wireless networking solutions for climate control and comfort in modern air conditioning systems. In *Intelligent Networks* (pp. 165–181). CRC Press., DOI: 10.1201/9781003541363-11

Al-Battbootti, M. J. H., Marin, I., Goga, N., & Popa, R. (2021). OIL AND GAS PIPELINE MONITORING DURING COVID-19 PANDEMIC VIA UNMANNED AERIAL VEHICLE. *ICERI Proceedings, 1*, 5756. DOI: 10.21125/iceri.2021.1299

Allers, M., Ahrens, A., Hitzemann, M., Bock, H., Wolf, T., Radunz, J., Meyer, F., Wilsenack, F., Zimmermann, S., & Ficks, A. (2023). Real-Time Remote Detection of Airborne Chemical Hazards—An Unmanned Aerial Vehicle (UAV) Carrying an Ion Mobility Spectrometer. *IEEE Sensors Journal, 23*(15), 16562–16570. DOI: 10.1109/JSEN.2023.3287448

Almadani, Y., Plets, D., Bastiaens, S., Joseph, W., Ijaz, M., Ghassemlooy, Z., & Rajbhandari, S. (2020). Visible Light Communications for Industrial Applications—Challenges and Potentials. *Electronics (Basel), 9*(12), 2157. DOI: 10.3390/electronics9122157

Alsayed, A., Yunusa-Kaltungo, A., Quinn, M. K., Arvin, F., & Nabawy, M. R. A. (2021). Drone-Assisted Confined Space Inspection and Stockpile Volume Estimation. *Remote Sensing (Basel), 13*(17), 3356. DOI: 10.3390/rs13173356

Alzarooni, A., Iqbal, E., Khan, S. U., Javed, S., Moyo, B., & Abdulrahman, Y. (2025). Anomaly Detection for Industrial Applications, Its Challenges, Solutions, and Future Directions: A Review [Review of *Anomaly Detection for Industrial Applications, Its Challenges, Solutions, and Future Directions: A Review*]. *arXiv (Cornell University)*. Cornell University. https://doi.org//arxiv.2501.11310DOI: 10.48550

Arif, S. M., & Akhai, S. (2024). Fostering a sustainable and disaster-resilient transportation infrastructure. In *Challenges, strategies, and resiliency in disaster and risk management* (pp. 205–219). IGI Global., DOI: 10.4018/979-8-3693-2721-0.ch009

Asensio, O. I., Moore, C. E., Ulibarrí, N., Simsekler, M. C. E., Lan, T., & Rivero, G. (2025). Data technologies and analytics for policy and governance: a landscape review [Review of *Data technologies and analytics for policy and governance: a landscape review*]. *Data & Policy, 7*. Cambridge University Press. DOI: 10.1017/dap.2024.49

Ashraf, S. N., Manickam, S., Zia, S. S., Abro, A. A., Obaidat, M., Uddin, M., Abdelhaq, M., & Alsaqour, R. (2023). IoT empowered smart cybersecurity framework for intrusion detection in internet of drones. *Scientific Reports, 13*(1), 18422. Advance online publication. DOI: 10.1038/s41598-023-45065-8 PMID: 37891186

Askerbekov, D., Garza-Reyes, J. A., Ghatak, R. R., Joshi, R., Kandasamy, J., & Nascimento, D. L. de M. (2024). Embracing drones and the Internet of drones systems in manufacturing – An exploration of obstacles. *Technology in Society, 78*, 102648. DOI: 10.1016/j.techsoc.2024.102648

Awasthi, S., Gramse, N., Reining, C., & Roidl, M. (2022). UAVs for Industries and Supply Chain Management. *arXiv (Cornell University)*. https://doi.org//arxiv.2212.03346DOI: 10.48550

Ayamga, M., Tekinerdoğan, B., & Kassahun, A. (2021). Exploring the Challenges Posed by Regulations for the Use of Drones in Agriculture in the African Context. *Land (Basel)*, *10*(2), 164. DOI: 10.3390/land10020164

Ayvaz, S., & Alpay, K. (2021). Predictive maintenance system for production lines in manufacturing: A machine learning approach using IoT data in real-time. *Expert Systems with Applications*, *173*, 114598. DOI: 10.1016/j.eswa.2021.114598

Bakacsy, L., & Zakar, T. (2025). An investigation into the spatial patterns of invasive common milkweed (Asclepias syriaca L.) stands through the utilization of drone images. *Scientific Reports*, *15*(1), 28889. Advance online publication. DOI: 10.1038/s41598-025-14034-8 PMID: 40775501

Bao, N., Fan, Y., Li, C., & Simeone, A. (2023). A computer vision approach to improve maintenance automation for thermal power plants lubrication systems. *Journal of Quality in Maintenance Engineering*, *29*(5), 120–137. DOI: 10.1108/JQME-01-2023-0007

Barbu, A., Ichimov, M. A. M., Costea-Marcu, I. C., Militaru, G., Deselnicu, D. C., & Moiceanu, G. (2025). Exploring Employee Perspectives on Workplace Technology: Usage, Roles, and Implications for Satisfaction and Performance. *Behavioral Sciences (Basel, Switzerland)*, *15*(1), 45. DOI: 10.3390/bs15010045 PMID: 39851850

Barkman, W. E. (1989). In-Process Quality Control for Manufacturing. In *CRC Press eBooks*. Informa., DOI: 10.1201/b15229

Baumann, D., Mager, F., Wetzker, U., Thiele, L., Zimmerling, M., & Trimpe, S. (2020). Wireless Control for Smart Manufacturing: Recent Approaches and Open Challenges. *Proceedings of the IEEE*, *109*(4), 441–467. DOI: 10.1109/JPROC.2020.3032633

Benyeogor, M. S., Olatunbosun, A., & Kumar, S. A. (2020). Airborne System for Pipeline Surveillance Using an Unmanned Aerial Vehicle. *European Journal of Engineering and Technology Research*, *5*(2), 178–182. DOI: 10.24018/ejers.2020.5.2.1761

Bhat, G. R., Dudhedia, M. A., Panchal, R., Shirke, Y. S., Angane, N. R., Khonde, S. R., Khedkar, S., Pansare, J. R., Bere, S. S., Wahul, R. M., & Gawande, S. H. (2024). Autonomous drones and their influence on standardization of rules and regulations for operating–A brief overview. *Results in Control and Optimization*, *14*, 100401. DOI: 10.1016/j.rico.2024.100401

Bonne, J., Donnat, L., Albora, G., Burgalat, J., Chauvin, N., Combaz, D., Cousin, J., Decarpenterie, T., Duclaux, O., Dumelié, N., Galas, N., Juéry, C., Parent, F., Pineau, F., Maunoury, A., Ventre, O., Bénassy, M.-F., & Joly, L. (2024). A measurement system for CO2 and CH4 emissions quantification of industrial sites using a new in situ concentration sensor operated on board uncrewed aircraft vehicles. *Atmospheric Measurement Techniques*, *17*(14), 4471–4491. DOI: 10.5194/amt-17-4471-2024

Border, R., Chebrolu, N., Tao, Y., Gammell, J. D., & Fallon, M. (2023). Osprey: Multi-Session Autonomous Aerial Mapping with LiDAR-based SLAM and Next Best View Planning. *arXiv (Cornell University)*. https://doi.org//arxiv.2311.03484DOI: 10.48550

Bretschneider, L., Bollmann, S., Houssin-Agbomson, D., Shaw, J. T., Howes, N., Nguyen, L. T. H., Robinson, R., Helmore, J., Lichtenstern, M., Nwaboh, J. A., Pogány, A., Ebert, V., & Lampert, A. (2024). Concepts for drone based pipeline leak detection. *Frontiers in Robotics and AI*, *11*, 1426206. DOI: 10.3389/frobt.2024.1426206 PMID: 39211418

Brogaard, R. Y., & Boukas, E. (2023). Autonomous GPU-based UAS for inspection of confined spaces: Application to marine vessel classification. *Robotics and Autonomous Systems*, *172*, 104590. DOI: 10.1016/j.robot.2023.104590

Burgués, J., & Marco, S. (2020). Environmental chemical sensing using small drones: A review [Review of *Environmental chemical sensing using small drones: A review*]. *The Science of The Total Environment, 748*, 141172. Elsevier BV. DOI: 10.1016/j.scitotenv.2020.141172

Caballero-Martin, D., López-Guede, J. M., Estévez, J., & Graña, M. (2024). Artificial Intelligence Applied to Drone Control: A State of the Art. *Drones (Basel)*, *8*(7), 296. DOI: 10.3390/drones8070296

Cabianca, M., Clemente, M. L., Gatto, G., Impagliazzo, C., Leoni, L., Masia, M., & Piras, R. (2022). An Application of IoT in a Drone Inspection Service for Environmental Control. *IoT*, *3*(3), 366–380. DOI: 10.3390/iot3030020

Callari, T. C., Segate, R. V., Hubbard, E., Daly, A., & Lohse, N. (2024). An Ethical Framework for Human-Robot Collaboration for the Future People-Centric Manufacturing: A Collaborative Endeavour with European subject-matter experts in Ethics. *Technology in Society*, *78*, 102680. DOI: 10.1016/j.techsoc.2024.102680

Cameron, C., Fredin, Z., & Gershenfeld, N. (2022). Discrete Assembly of Unmanned Aerial Systems. *2022 International Conference on Unmanned Aircraft Systems (ICUAS)*, 339. DOI: 10.1109/ICUAS54217.2022.9836082

Carroll, S., Satme, J. N., Alkharusi, S., Vitzilaios, N., Downey, A., & Rizos, D. (2021). Drone-Based Vibration Monitoring and Assessment of Structures. *Applied Sciences (Basel, Switzerland), 11*(18), 8560. DOI: 10.3390/app11188560

Chen, Z., Chen, H., Imani, M., Chen, R., & Imani, F. (2024). Vision language model for interpretable and fine-grained detection of safety compliance in diverse workplaces. *Expert Systems with Applications, 125769*. Advance online publication. DOI: 10.1016/j.eswa.2024.125769

Cheng, J.-Y., Mendes, E. L. S., Gheisari, M., & Jeelani, I. (2022). Construction Worker-Drone Safety Training in a 360 Virtual Reality Environment: A Pilot Study. *EPiC Series in Built Environment, 3*, 19–8. DOI: 10.29007/8cjv

Cheng, W., & Yuan, C. (2020). Using a Micro Sampler on a Drone to Extract Organic Vapors—A Case Study of Monitoring Industrial Pollution. *Aerosol and Air Quality Research, 21*(2), 200359. DOI: 10.4209/aaqr.2020.06.0359

Davuluri, K. (2024). Facilitating Real-time Road Event Detection on Collaborative UAV and Ground Vehicle. *Deep Blue (University of Michigan)*. DOI: 10.7302/24661

de Carvalho, P. T., Lopes, J. D., & Raimundo, R. (2025). Innovation Impact in the Textile Industry: From the Toyota Production System to Artificial Intelligence. *Sustainability (Basel), 17*(3), 1170. DOI: 10.3390/su17031170

De Gasperis, G., & Facchini, S. D. (2025). *A Comparative Study of Rule-Based and Data-Driven Approaches in Industrial Monitoring*. DOI: 10.48550/ARXIV.2509.15848

Dooley, J. F., Minschwaner, K., Dubey, M. K., Abbadi, S. H. E., Sherwin, E. D., Meyer, A. G., Follansbee, E., & Lee, J. E. (2024). A new aerial approach for quantifying and attributing methane emissions: Implementation and validation. *Atmospheric Measurement Techniques, 17*(17), 5091–5111. DOI: 10.5194/amt-17-5091-2024

Dronova, I., Kislik, C., Dinh, Z., & Kelly, M. (2021). A Review of Unoccupied Aerial Vehicle Use in Wetland Applications: Emerging Opportunities in Approach, Technology, and Data [Review of *A Review of Unoccupied Aerial Vehicle Use in Wetland Applications: Emerging Opportunities in Approach, Technology, and Data*]. *Drones, 5*(2), 45. Multidisciplinary Digital Publishing Institute. DOI: 10.3390/drones5020045

Duan, Z., Li, Y., Wang, J., Zhao, G., & Svanberg, S. (2019). Aquatic environment monitoring using a drone-based fluorosensor. *Applied Physics. B, Lasers and Optics, 125*(6), 108. Advance online publication. DOI: 10.1007/s00340-019-7215-y

Durfey, N., & Sajal, S. (2022, May 1). A Comprehensive Survey: Cybersecurity Challenges and Futures of Autonomous Drones. *2022 Intermountain Engineering, Technology and Computing (IETC)*. DOI: 10.1109/IETC54973.2022.9796881

Fascista, A. (2022). Toward Integrated Large-Scale Environmental Monitoring Using WSN/UAV/Crowdsensing: A Review of Applications, Signal Processing, and Future Perspectives [Review of *Toward Integrated Large-Scale Environmental Monitoring Using WSN/UAV/Crowdsensing: A Review of Applications, Signal Processing, and Future Perspectives*]. *Sensors, 22*(5), 1824. Multidisciplinary Digital Publishing Institute. DOI: 10.3390/s22051824

Felix, O. O.Oroo Oyondi Felix. (2023). Securing the skies: A comprehensive survey on internet of drones security challenges and solutions. *World Journal of Advanced Research and Reviews, 20*(3), 780–800. DOI: 10.30574/wjarr.2023.20.3.2491

Filho, P. S., Shiguemori, E. H., & Saotome, O. (2017). UAV VISUAL AUTOLOCALIZATON BASED ON AUTOMATIC LANDMARK RECOGNITION. *ISPRS Annals of the Photogrammetry, Remote Sensing and Spatial Information Sciences, 89*(W3), 89–94. Advance online publication. DOI: 10.5194/isprs-annals-IV-2-W3-89-2017

Gago, R. M., Pereira, M. Y. A., & Pereira, G. A. S. (2021). An aerial robotic system for inventory of stockpile warehouses. *Engineering Reports, 3*(9), e12396. Advance online publication. DOI: 10.1002/eng2.12396

Gămulescu, O.-M., Rosca, S., Panaite, F. A., Costandoiu, A., & Riurean, S. (2020). Accident sites management using drones. *MATEC Web of Conferences, 305*, 4. DOI: 10.1051/matecconf/202030500004

Gerencheal, T. A., & Lee, J. H. (2025). Development of a Collision Avoidance System Using Multiple Distance Sensors for Indoor Inspection Drone. *Applied Sciences (Basel, Switzerland), 15*(16), 8791. DOI: 10.3390/app15168791

Ghazal, M., Basmaji, T., Yaghi, M., Alkhedher, M., Mahmoud, M., & El-Baz, A. (2020). Cloud-Based Monitoring of Thermal Anomalies in Industrial Environments Using AI and the Internet of Robotic Things. *Sensors (Basel), 20*(21), 6348. DOI: 10.3390/s20216348 PMID: 33171714

Giordano, C., Girelli, V. A., Lambertini, A., Tini, M. A., & Zanutta, A. (2025). UAV Data Collection Co-Registration: LiDAR and Photogrammetric Surveys for Coastal Monitoring. *Drones (Basel), 9*(1), 49. DOI: 10.3390/drones9010049

Gómez-Rosal, D. A., Bergau, M., Fischer, G., Wachaja, A., Gräter, J., Odenweller, M., Piechottka, U., Hoeflinger, F., Gosala, N., Wetzel, N., Btischer, D., Valada, A., & Burgard, W. (2023). *A Smart Robotic System for Industrial Plant Supervision*. 1. DOI: 10.1109/SENSORS56945.2023.10325162

González-Castaño, F. J., Gil-Castiñeira, F., Rodriguez-Pereira, D., Regueiro-Janeiro, J. A., García-Méndez, S., & Candal-Ventureira, D. (2020). Self-Corrective Sensor Fusion for Drone Positioning in Indoor Facilities. *IEEE Access : Practical Innovations, Open Solutions*, *9*, 2415–2427. DOI: 10.1109/ACCESS.2020.3048194

Grote, M., Oakey, A., Pilko, A., Krol, J., Blakesley, A., Cherrett, T., Scanlan, J. P., Anvari, B., & Martínez-Sykora, A. (2024). The effects of costs on drone uptake in multi-modal logistics systems within a healthcare setting. *Transport Economics and Management*, *2*, 58–75. DOI: 10.1016/j.team.2024.03.001

Guebsi, R., Mami, S., & Chokmani, K. (2024). Drones in Precision Agriculture: A Comprehensive Review of Applications, Technologies, and Challenges [Review of *Drones in Precision Agriculture: A Comprehensive Review of Applications, Technologies, and Challenges*]. *Drones, 8*(11), 686. Multidisciplinary Digital Publishing Institute. DOI: 10.3390/drones8110686

Gullett, B. K., Aurell, J., Mitchell, W. J., & Richardson, J. L. (2021). Use of an unmanned aircraft system to quantify NO_{<i>x</i>} emissions from a natural gas boiler. *Atmospheric Measurement Techniques*, *14*(2), 975. DOI: 10.5194/amt-14-975-2021 PMID: 34122665

Gyrichidi, N., Khalyasmaa, A. I., Eroshenko, S. A., & Romanov, A. M. (2024). Are Modern Market-Available Multi-Rotor Drones Ready to Automatically Inspect Industrial Facilities? *Drones (Basel)*, *8*(10), 549. DOI: 10.3390/drones8100549

Habibi, H., Rao, D. M. K. K. V., Sánchez-López, J. L., & Voos, H. (2023). On SORA for High-Risk UAV Operations under New EU Regulations. *Perspectives for Automated Approach*, *213*, 213–220. Advance online publication. DOI: 10.1109/ICUAS57906.2023.10156517

Heo, S.-J., & Na, W. S. (2025). Review of Drone-Based Technologies for Wind Turbine Blade Inspection. *Electronics (Basel)*, *14*(2), 227. DOI: 10.3390/electronics14020227

Hermansa, M., Kozielski, M., Michalak, M., Szczyrba, K., Wróbel, Ł., & Sikora, M. (2021). Sensor-Based Predictive Maintenance with Reduction of False Alarms—A Case Study in Heavy Industry. *Sensors (Basel)*, *22*(1), 226. DOI: 10.3390/s22010226 PMID: 35009777

Heřt, D., Báča, T., Petráček, P., Krátký, V., Pěnička, R., Spurný, V., Petrlík, M., Vrba, M., Žaitlík, D., Stoudek, P., Walter, V., Štěpán, P., Horyna, J., Pritzl, V., Šrámek, M., Ahmad, A., Silano, G., Licea, D. B., Štibinger, P., & Saska, M. (2023). MRS Drone: A Modular Platform for Real-World Deployment of Aerial Multi-Robot Systems. *Journal of Intelligent & Robotic Systems, 108*(4), 64. Advance online publication. DOI: 10.1007/s10846-023-01879-2

Hoang, V. T., Phung, M. D., Dinh, T. H., & Ha, Q. P. (2019). System Architecture for Real-Time Surface Inspection Using Multiple UAVs. *IEEE Systems Journal, 14*(2), 2925–2936. DOI: 10.1109/JSYST.2019.2922290

Hollenbeck, D., & Chen, Y. (2022). A Digital Twin Framework for Environmental Sensing with sUAS. *Journal of Intelligent & Robotic Systems, 105*(1), 1. Advance online publication. DOI: 10.1007/s10846-021-01542-8

Hovsha, J., & Brien, D. O. (2022). Drone Use Cases and their Privacy Impacts: A Taxonomy. *Tilburg Law Review, 27*(1), 60–80. DOI: 10.5334/tilr.281

Hütten, N., Gomes, M. A., Hölken, F., Andricevic, K., Meyes, R., & Meisen, T. (2024). Deep Learning for Automated Visual Inspection in Manufacturing and Maintenance: A Survey of Open- Access Papers. *Applied System Innovation, 7*(1), 11. DOI: 10.3390/asi7010011

Isgró, M. A., Basallote, M. D., & Barbero, L. (2021). Unmanned Aerial System-Based Multispectral Water Quality Monitoring in the Iberian Pyrite Belt (SW Spain). *Mine Water and the Environment, 41*(1), 30–41. DOI: 10.1007/s10230-021-00837-4

Jacobsen, R. H., Matleković, L., Shi, L., Malle, N. H., Ayoub, N., Hageman, K., Hansen, S., Nyboe, F. F., & Ebeid, E. (2023). Design of an Autonomous Cooperative Drone Swarm for Inspections of Safety Critical Infrastructure. *Applied Sciences (Basel, Switzerland), 13*(3), 1256. DOI: 10.3390/app13031256

Jaroš, R., Byrtus, R., Dohnal, J., Danys, L., Baroš, J., Koziorek, J., Zmij, P., & Martínek, R. (2022). Advanced Signal Processing Methods for Condition Monitoring. *Archives of Computational Methods in Engineering, 30*(3), 1553–1577. DOI: 10.1007/s11831-022-09834-4

Joyce, K. E., Fickas, K. C., & Kalamandeen, M. (2022). The unique value proposition for using drones to map coastal ecosystems. *Cambridge Prisms: Coastal Futures, 1*, e6. Advance online publication. DOI: 10.1017/cft.2022.7

Kammerer, K., Hoppenstedt, B., Pryss, R., Stökler, S., Allgaier, J., & Reichert, M. (2019). Anomaly Detections for Manufacturing Systems Based on Sensor Data—Insights into Two Challenging Real-World Production Settings. *Sensors (Basel), 19*(24), 5370. DOI: 10.3390/s19245370 PMID: 31817471

Kanda, R., Akhai, S., & Bansal, R. (2013). Analysis of MOST technique for elimination of ideal time by synchronization of different lines. *International Journal of Research, 1*(4).

Karnes, M., Riffel, J., & Yılmaz, A. (2024). Key-Region-Based UAV Visual Navigation. *"The œinternational Archives of the Photogrammetry, Remote Sensing and Spatial Information Sciences. The International Archives of the Photogrammetry, Remote Sensing and Spatial Information Sciences, 173*, 173–179. Advance online publication. DOI: 10.5194/isprs-archives-XLVIII-2-2024-173-2024

Kas, K., & Johnson, G. K. (2019). Using unmanned aerial vehicles and robotics in hazardous locations safely. *Process Safety Progress, 39*(1), e12066. Advance online publication. DOI: 10.1002/prs.12066

Khan, A. (2021). Risk Assessment, Prediction, and Avoidance of Collision in Autonomous Drones. *arXiv (Cornell University)*. https://doi.org//arxiv.2108.12770DOI: 10.48550

Khang, A., & Akhai, S. (2025). E-Waste and Lithium-ion Battery Recycling Insights for Sustainable Transportation. In Khang, A. (Ed.), *Driving Green Transportation System Through Artificial Intelligence and Automation. Lecture Notes in Intelligent Transportation and Infrastructure*. Springer., DOI: 10.1007/978-3-031-72617-0_11

Kilickaya, S., Ahishali, M., Celebioglu, C., Sohrab, F., Eren, L., İnce, T., Aşkar, M., & Gabbouj, M. (2025).. . *Audio-based Anomaly Detection in Industrial Machines Using Deep One-Class Support Vector Data Description, 1*, 1–5. Advance online publication. DOI: 10.1109/CIESCompanion65073.2025.11010815

Kim, H., Ham, S., & Lee, I. (2020). REAL-TIME DRONE MAPPING BASED ON REFERENCE IMAGES FOR VEHICLE FACILITY MONITORING. *"The œinternational Archives of the Photogrammetry, Remote Sensing and Spatial Information Sciences. The International Archives of the Photogrammetry, Remote Sensing and Spatial Information Sciences, 43*. Advance online publication. DOI: 10.5194/isprs-archives-XLIII-B2-2020-43-2020

Kim, S., Lim, G. J., & Cho, J. (2018). Drone flight scheduling under uncertainty on battery duration and air temperature. *Computers & Industrial Engineering, 117*, 291–302. DOI: 10.1016/j.cie.2018.02.005

Kistner, L., Schmoll, R., & Kroll, A. (2024). An airborne measurement system to detect, locate, and quantify methane emissions with attitude-based wind speed estimation. *Journal of Sensors and Sensor Systems*, *13*(2), 211–218. DOI: 10.5194/jsss-13-211-2024

Kleinschroth, F., Banda, K., Zimba, H., Dondeyne, S., Nyambe, I., Spratley, S., & Winton, R. S. (2022). Drone imagery to create a common understanding of landscapes. *Landscape and Urban Planning*, *228*, 104571. DOI: 10.1016/j.landurbplan.2022.104571

Kordoni, A., Gavidia-Calderon, C., Levine, M., Bennaceur, A., & Nuseibeh, B. (2023). "Are we in this together?": Embedding social identity detection in drones improves emergency coordination. *Frontiers in Psychology*, *14*, 1146056. Advance online publication. DOI: 10.3389/fpsyg.2023.1146056 PMID: 37744604

Krook, J., Bossens, D. M., Winter, P., Araujo-Estrada, S. A., Downer, J., & Windsor, S. P. (2024). Mapping the complexity of legal challenges for trustworthy drones on construction sites in the United Kingdom. *ACM Journal on Responsible Computing*, *1*(3), 1–26. DOI: 10.1145/3664617

Kumar, H., Wadhwa, A. S., Akhai, S., & Kaushik, A. (2024). Parametric analysis, modeling and optimization of the process parameters in electric discharge machining of aluminium metal matrix composite. *Engineering Research Express*, *6*(2), 025542. DOI: 10.1088/2631-8695/ad4ba9

Kumar, P., & Akhai, S. (2022). Effective energy management in smart buildings using VRV/VRF systems. In *Additive manufacturing in Industry 4.0* (pp. 27–35). CRC Press., DOI: 10.1201/9781003360001-2

Kurbanzade, A. K., Baig, A. M., & Mehrotra, S. (2023). Elevating Industries with Unmanned Aerial Vehicles: Integrating Sustainability and Operational Innovation. *arXiv (Cornell University)*. https://doi.org//arxiv.2312.13551DOI: 10.48550

Lai, Q., Xie, R., Yang, Z., Wu, G., Hong, Z.-Y., & Yang, C. (2024). Efficient multiple unmanned aerial vehicle-assisted data collection strategy in power infrastructure construction. *Frontiers in Communications and Networks*, *5*, 1390909. Advance online publication. DOI: 10.3389/frcmn.2024.1390909

Layer, M., Leidich, J., Schwoch, S., Saske, B., Neubert, S., Robl, P., & Paetzold-Byhain, K. (2023). Data management of process plants as complex systems: Systematic literature review and identification of challenges and opportunities. *Reviews in Chemical Engineering*, *40*(3), 329–349. DOI: 10.1515/revce-2022-0077

Lee, G., Saad, W., Bennis, M., Kim, C., & Jung, M. (2022). An Online Framework for Ephemeral Edge Computing in the Internet of Things. *IEEE Transactions on Wireless Communications*, *22*(3), 1992–2007. DOI: 10.1109/TWC.2022.3208096

Leet, C., Li, J., & Koenig, S. (2022). Shard Systems: Scalable, Robust and Persistent Multi-Agent Path Finding with Performance Guarantees. *Proceedings of the AAAI Conference on Artificial Intelligence*, *36*(9), 9386–9395. DOI: 10.1609/aaai.v36i9.21170

Li, Z., Fei, F., & Zhang, G. (2022). Edge-to-Cloud IIoT for Condition Monitoring in Manufacturing Systems with Ubiquitous Smart Sensors. *Sensors (Basel)*, *22*(15), 5901. DOI: 10.3390/s22155901 PMID: 35957460

Liang, C.-W., & Shen, C.-H. (2024). An integrated uncrewed aerial vehicle platform with sensing and sampling systems for the measurement of air pollutant concentrations. *Atmospheric Measurement Techniques*, *17*(9), 2671–2686. DOI: 10.5194/amt-17-2671-2024

Lim, D. Y., Jin, I. J., & Bang, I. C. (2023). Heat-vision based drone surveillance augmented by deep learning for critical industrial monitoring. *Scientific Reports*, *13*(1), 22291. Advance online publication. DOI: 10.1038/s41598-023-49589-x PMID: 38097731

Lim, E., Jung, Y. H., Shin, D. Y., & Kim, S. S. (2022). SMALL UAVS-BASED DISASTER DAMAGE INVESTIGATION: FOCUSED ON FLOODING DISASTER AND CHEMICAL ACCIDENT. *˜The œinternational Archives of the Photogrammetry, Remote Sensing and Spatial Information Sciences. The International Archives of the Photogrammetry, Remote Sensing and Spatial Information Sciences*, *1121*. Advance online publication. DOI: 10.5194/isprs-archives-XLIII-B3-2022-1121-2022

Lin, J. J., Ibrahim, A., Sarwade, S., & Golparvar-Fard, M. (2020). Bridge Inspection with Aerial Robots: Automating the Entire Pipeline of Visual Data Capture, 3D Mapping, Defect Detection, Analysis, and Reporting. *Journal of Computing in Civil Engineering*, *35*(2), 04020064. Advance online publication. DOI: 10.1061/(ASCE)CP.1943-5487.0000954

Lin, Y., Chang, Y., Tong, X., Yu, J., Liotta, A., Huang, G., Song, W., Zeng, D., Wu, Z., Wang, Y., & Zhang, W. (2024). *A Survey on Rgb, 3d, and Multimodal Approaches for Unsupervised Industrial Anomaly Detection*. DOI: 10.2139/ssrn.5005467

Lindblad, M. (2006). *Human Inspection Work: A case study of why faults are missed?* https://lup.lub.lu.se/student-papers/record/1327412/file/1327413.pdf

Liu, W., Wu, H., Shi, L., Zhu, F., Zou, Y., Kong, F., & Zhang, F. (2025). LiDAR-Based Quadrotor Autonomous Inspection System in Cluttered Environments. *IEEE Transactions on Field Robotics.*, *2*, 753–767. DOI: 10.1109/TFR.2025.3604716

Luo, S., Liu, Q., Guo, X., Yin, M., Li, Z., & Lang, X. (2025). Chemical Safety Inspection Path Optimization Problems Using Improved Multi-Objective Discrete Growth Optimization Algorithm. *Processes (Basel, Switzerland)*, *13*(5), 1445. DOI: 10.3390/pr13051445

Ma, Y., Yin, J., Huang, F., & Li, Q. (2024). Surface defect inspection of industrial products with object detection deep networks: a systematic review [Review of *Surface defect inspection of industrial products with object detection deep networks: a systematic review*]. *Artificial Intelligence Review*, *57*(12). Springer Science+Business Media. DOI: 10.1007/s10462-024-10956-3

Maboudi, M., Alamouri, A., López, V. D. A., Bajauri, M. S., Berger, C., & Gerke, M. (2021). DRONE-BASED CONTAINER CRANE INSPECTION: CONCEPT, CHALLENGES AND PRELIMINARY RESULTS. *ISPRS Annals of the Photogrammetry, Remote Sensing and Spatial Information Sciences*, *121*, 121–128. Advance online publication. DOI: 10.5194/isprs-annals-V-1-2021-121-2021

Magalhães, P. S. G., & Ferreira, N. M. F. (2022). Inspection Application in an Industrial Environment with Collaborative Robots. *Automation*, *3*(2), 258–268. DOI: 10.3390/automation3020013

Maghazei, O. (2022). Drones in Manufacturing. *Opportunities and Challenges*, *290*, 290–298. Advance online publication. DOI: 10.46254/EU05.20220073

Maghazei, O., & Netland, T. H. (2019). Drones in manufacturing: Exploring opportunities for research and practice. *Journal of Manufacturing Technology Management*, *31*(6), 1237–1259. DOI: 10.1108/JMTM-03-2019-0099

Maghazei, O., Netland, T. H., Frauenberger, D., & Thalmann, T. (2021). Automatic Drones for Factory Inspection: The Role of Virtual Simulation. In *IFIP advances in information and communication technology* (p. 457). Springer Science+Business Media. DOI: 10.1007/978-3-030-85910-7_48

Mahariya, S. K., Kumar, A., Singh, R., Gehlot, A., Akram, S. V., Twala, B., Iqbal, M. I., & Priyadarshi, N. (2023). Smart Campus 4.0: Digitalization of University Campus with Assimilation of Industry 4.0 for Innovation and Sustainability. *Journal of Advanced Research in Applied Sciences and Engineering Technology*, *32*(1), 120–138. DOI: 10.37934/araset.32.1.120138

Mahroof, K., Omar, A., Yaroson, E. V., Tenebe, S. A., Rana, N. P., Sivarajah, U., & Weerakkody, V. (2023). Evaluating the intention to use Industry 5.0 (I5.0) drones for cleaner production in Sustainable Food Supply Chains: An emerging economy context. *Supply Chain Management, 29*(3), 468–496. DOI: 10.1108/SCM-01-2023-0045

Malanchini, I., Michailow, N., Agostini, P., Ali-Tolppa, J., Hock, D., Kasparick, M., Lieto, A., Marchenko, N., Alba, A. M., Pries, R., & Zhou, Q. (2023). Convergence of Manufacturing and Networking in Future Factories. *arXiv (Cornell University)*. https://doi.org//arxiv.2312.08708DOI: 10.48550

Mammarella, M., Comba, L., Biglia, A., Dabbene, F., & Gay, P. (2021). Cooperation of unmanned systems for agricultural applications: A theoretical framework. *Biosystems Engineering, 223*, 61–80. DOI: 10.1016/j.biosystemseng.2021.11.008

McQueen, N., & Drennan, D. (2024). The use of warehouse automation technology for scalable and low-cost direct air capture. *Frontiers in Climate, 6*, 1415642. Advance online publication. DOI: 10.3389/fclim.2024.1415642

Mendes, E., Albeaino, G., Brophy, P., Gheisari, M., & Jeelani, I. (2022). Working Safely with Drones: A Virtual Training Strategy for Workers on Heights. *Construction Research Congress 2022, 3*, 622. DOI: 10.1061/9780784483985.063

Mészáros, B., Bóna, K., & Bertalan, M. (2018). MONITORING OF PRODUCTION LINE SUPPLY SYSTEMS WITH DRONES. *IOP Conference Series. Materials Science and Engineering, 448*, 12037. DOI: 10.1088/1757-899X/448/1/012037

Meydan, C. H., & Akkaş, H. (2024). The role of triangulation in qualitative research: Converging perspectives. In Elhami, A., Roshan, A., & Chandan, H. C. (Eds.), *Principles of conducting qualitative research in multicultural settings* (pp. 101–132). IGI Global., DOI: 10.4018/979-8-3693-3306-8.ch006

Midkiff, J., Hlaing, M., Fan, K.-C., Jain, S., Akella, M., & Chen, R. T. (2024). *Mid-IR UAV-based sensing platform with deep learning to Identify and Quantify Gaseous Emission in Gas Flares*. DOI: 10.2172/2405949

Mishra, H., & Mishra, D. (2024). *Economic Evaluation of UAV-Based Soil Sampling Approaches*., DOI: 10.1007/978-981-99-8684-2_15

Motia, K., Kumar, R., & Akhai, S. (2024). AI and smart manufacturing: Building Industry 4.0. In Wongmahesak, K., Phongkraphan, N., & Lekhavichit, N. (Eds.), *Modern management science practices in the age of AI* (pp. 1–28). IGI Global., DOI: 10.4018/979-8-3693-6720-9.ch001

Motia, K., Kumar, R., & Akhai, S. (2025). *Navigating the Fourth Industrial Revolution: A comprehensive review of Industry 4.0 technologies and their global adoption.* In S. Singh & I. Singh (Eds.), *Advances in Materials and Manufacturing: Select Proceedings of ICDMT 2024* (pp. 317–333). Springer. DOI: 10.1007/978-981-96-7659-0_28

Múñiz, R., Nuño, F., Díaz, J., González, M., Prieto, M. J., & Menéndez, Ó. (2022). Real-time monitoring solution with vibration analysis for industry 4.0 ventilation systems. *The Journal of Supercomputing*, *79*(6), 6203. DOI: 10.1007/s11227-022-04897-3

Namian, M., Khalid, M., Wang, G., & Türkan, Y. (2021). Revealing Safety Risks of Unmanned Aerial Vehicles in Construction. *Transportation Research Record: Journal of the Transportation Research Board, 2675*(11), 334–347. DOI: 10.1177/03611981211017134

Nasiri, N., Maashri, A. A., Bourdoucen, H., Al-Abri, S., & Aouf, N. (2025). Environmental Monitoring in the Oil and Gas Industry Using Machine Vision Associated with Robotics and AI. In *IntechOpen eBooks*. IntechOpen., DOI: 10.5772/intechopen.1010026

Nazarov, D., Sulimin, V., & Shvedov, V. (2024). Advancing environmental stewardship: The role of automation in enhanced environmental monitoring. *E3S Web of Conferences, 542*, 5005. DOI: 10.1051/e3sconf/202454205005

Nielsen, M. M., Thomasberger, A., Svane, N., Thompson, F., & Hansen, F. T. (2023). Development of new tools for eelgrass monitoring in Natura 2000 areas. In *Research Portal Denmark* (p. 96). Technical University of Denmark., https://local.forskningsportal.dk/local/dki-cgi/ws/cris-link?src=dtu&id=dtu-cc657e36-eac2-4373-ae75-c73dc9d91e30&ti=Development%20of%20new%20tools%20for%20eelgrass%20monitoring%20in%20Natura%202000%20areas

Noğay, H. S., Akıncı, T. Ç., & Yılmaz, M. (2021). Detection of invisible cracks in ceramic materials using by pre-trained deep convolutional neural network. *Neural Computing & Applications, 34*(2), 1423–1432. DOI: 10.1007/s00521-021-06652-w

Nooralishahi, P., Ibarra-Castanedo, C., Deane, S., López, F., Pant, S., Genest, M., Avdelidis, N. P., & Maldague, X. (2021). Drone-Based Non-Destructive Inspection of Industrial Sites: A Review and Case Studies [Review of *Drone-Based Non-Destructive Inspection of Industrial Sites: A Review and Case Studies*]. *Drones, 5*(4), 106. Multidisciplinary Digital Publishing Institute. DOI: 10.3390/drones5040106

Nwaogu, J. M., Yang, Y., Chan, A. P. C., & Wang, X. (2024). Enhancing Drone Operator Competency within the Construction Industry: Assessing Training Needs and Roadmap for Skill Development. *Buildings (Basel, Switzerland)*, *14*(4), 1153. DOI: 10.3390/buildings14041153

Nycz, B., & Pietrucha-Urbanik, K. (2024). Advancements in Air Quality Monitoring. *The Role of Drone Technology*, *19*, 19. Advance online publication. DOI: 10.3390/proceedings2024105019

Olivares, V., Córdova, F. M., Sepúlveda, J. M., & Dérpich, I. (2015). Modeling Internal Logistics by Using Drones on the Stage of Assembly of Products. *Procedia Computer Science*, *55*, 1240–1249. DOI: 10.1016/j.procs.2015.07.132

Omolara, A. E., Alawida, M., & Omolara, A. E. (2023). Drone cybersecurity issues, solutions, trend insights and future perspectives: A survey. *Neural Computing & Applications*, *35*(31), 23063–23101. DOI: 10.1007/s00521-023-08857-7

Onososen, A., Musonda, I., Ramabodu, M., & Dzuwa, C. (2023). Safety and Training Implications of Human-Drone Interaction in Industrialised Construction Sites. In *Lecture notes in civil engineering* (p. 281). Springer Nature., DOI: 10.1007/978-3-031-32515-1_20

Öz, E., Heikkilä, E., & Tiusanen, R. (2022). Development of an Organisational Certification Process for Specific Category Drone Operations. *Drones (Basel)*, *6*(10), 278. DOI: 10.3390/drones6100278

Özek, A., Seçkin, M., Demircioğlu, P., & Böğrekçi, İ. (2025). Artificial Intelligence Driving Innovation in Textile Defect Detection. *Textiles*, *5*(2), 12. DOI: 10.3390/textiles5020012

Pal, D., & Joshi, S. (2023). AI, IoT and Robotics in Smart Farming: Current Applications and Future Potentials. *2022 International Conference on Sustainable Computing and Data Communication Systems (ICSCDS)*, 1096. DOI: 10.1109/ICSCDS56580.2023.10105101

Panarotto, M., Isaksson, O., & Vial, V. (2022). Cost-efficient digital twins for design space exploration: A modular platform approach. *Computers in Industry*, *145*, 103813. DOI: 10.1016/j.compind.2022.103813

Papananias, M., McLeay, T. E., Obajemu, O., Mahfouf, M., & Kadirkamanathan, V. (2020). Inspection by exception: A new machine learning-based approach for multistage manufacturing. *Applied Soft Computing*, *97*, 106787. DOI: 10.1016/j.asoc.2020.106787

Pellicer, M. R., Tungekar, M. Y., & Carpitella, S. (2023). Where to Place Monitoring Sensors for Improving Complex Manufacturing Systems? Discussing a Real Case in the Food Industry. *Sensors (Basel), 23*(7), 3768. DOI: 10.3390/s23073768 PMID: 37050828

Pillai, G. M., Gupta, E., Suresh, A., Ganapathy, V., & Patra, A. (2022). Privadome: Protecting Citizen Privacy from Delivery Drones. *arXiv (Cornell University)*. https://doi.org//arxiv.2205.04961DOI: 10.48550

Pillai, G. M., Suresh, A., Gupta, E., Ganapathy, V., & Patra, A. (2024). Privadome: Delivery Drones and Citizen Privacy. *Proceedings on Privacy Enhancing Technologies. Privacy Enhancing Technologies Symposium, 2024*(2), 29–48. DOI: 10.56553/popets-2024-0039

Plaza, J. M., Martín, D., Arias-Perez, P., Vega, J., Roldán-Álvarez, D., García-Pérez, L., & Fernández-Conde, J. (2020). Open-Source Drone Programming Course for Distance Engineering Education. *Electronics (Basel), 9*(12), 2163. DOI: 10.3390/electronics9122163

Proceedings of the 19th Conference on Computer Science and Intelligence Systems (FedCSIS). (2024). *Annals of Computer Science and Information Systems, 39*. https://doi.org/DOI: 10.15439/978-83-969601-6-0

Purohit, H., Tanabe, R., Ichige, T., Endo, T., Nikaido, Y., Suefusa, K., & Kawaguchi, Y. (2019). *MIMII Dataset: Sound Dataset for Malfunctioning Industrial Machine Investigation and Inspection*. DOI: 10.33682/m76f-d618

Qin, Y., Kishk, M. A., & Alouini, M. (2022). Drone Charging Stations Deployment in Rural Areas for Better Wireless Coverage: Challenges and Solutions. *IEEE Internet of Things Magazine, 5*(1), 148–153. DOI: 10.1109/IOTM.001.2100083

Rahman, D. A., & Setiawan, Y. (2020). Possibility of applying unmanned aerial vehicle and thermal imaging in several canopy cover class for wildlife monitoring – preliminary results. *E3S Web of Conferences, 211*, 4007. DOI: 10.1051/e3sconf/202021104007

Reid, J. D., & Castka, P. (2023). The impact of remote sensing on monitoring and reporting - The case of conformance systems. *Journal of Cleaner Production, 393*, 136331. DOI: 10.1016/j.jclepro.2023.136331

Rejeb, A., Rejeb, K., Simske, S. J., & Treiblmaier, H. (2021). Drones for supply chain management and logistics: a review and research agenda [Review of *Drones for supply chain management and logistics: a review and research agenda*]. *International Journal of Logistics Research and Applications, 26*(6), 708. Taylor & Francis. DOI: 10.1080/13675567.2021.1981273

Salem, T., Dragomir, M., & Chatelet, É. (2024). Strategic Integration of Drone Technology and Digital Twins for Optimal Construction Project Management. *Applied Sciences (Basel, Switzerland), 14*(11), 4787. DOI: 10.3390/app14114787

Salhaoui, M., González, A. G., Arioua, M., Ortiz, F., Oualkadi, A. E., & Bonet, C. T. (2019). Smart Industrial IoT Monitoring and Control System Based on UAV and Cloud Computing Applied to a Concrete Plant. *Sensors (Basel), 19*(15), 3316. DOI: 10.3390/s19153316 PMID: 31357720

Salom, I., Dimić, G., Čelebić, V., Spasenović, M., Raičković, M., Mihajlović, M., & Todorović, D. (2023). An Acoustic Camera for Use on UAVs. *Sensors (Basel), 23*(2), 880. DOI: 10.3390/s23020880 PMID: 36679677

Salvati, D., Drioli, C., Ferrin, G., & Foresti, G. L. (2019). Acoustic Source Localization From Multirotor UAVs. *IEEE Transactions on Industrial Electronics, 67*(10), 8618–8628. DOI: 10.1109/TIE.2019.2949529

Sauls, L. A., Paneque-Gálvez, J., Amador-Jiménez, M., Vargas-Ramírez, N., & Laumonier, Y. (2023). Drones, communities and nature: Pitfalls and possibilities for conservation and territorial rights. *Global Social Challenges Journal, 2*, 24–46. DOI: 10.1332/AJHA9183

Schilling, F., Soria, E., & Floreano, D. (2022). On the Scalability of Vision-Based Drone Swarms in the Presence of Occlusions. *IEEE Access : Practical Innovations, Open Solutions, 10*, 28133–28146. DOI: 10.1109/ACCESS.2022.3158758

Shafi, I., Mazhar, M., Fatima, A., Álvarez, R. M., Miró, Y., Espinosa, J. C. M., & Ashraf, I. (2023). Deep Learning-Based Real Time Defect Detection for Optimization of Aircraft Manufacturing and Control Performance. *Drones (Basel), 7*(1), 31. DOI: 10.3390/drones7010031

Shahin, M., Chen, F. F., Hosseinzadeh, A., & Zand, N. (2023). Using machine learning and deep learning algorithms for downtime minimization in manufacturing systems: An early failure detection diagnostic service. *International Journal of Advanced Manufacturing Technology, 128*(9-10), 3857–3883. DOI: 10.1007/s00170-023-12020-w

Shankar, A., Behl, A., Pereira, V., Chavan, M., & Chirico, F. (2024). Exploring enablers and inhibitors of AI-enabled drones for manufacturing process audits: A mixed-method approach. *Business Strategy and the Environment, 33*(5), 3749–3768. DOI: 10.1002/bse.3679

Sharma, V., Tripathi, A. K., & Mittal, H. (2022). *Technological revolutions in smart farming: Current trends, challenges & future directions.*

Sihag, V., Choudhary, G., Choudhary, P. K., & Dragoni, N. (2023). Cyber4Drone: A Systematic Review of Cyber Security and Forensics in Next-Generation Drones [Review of *Cyber4Drone: A Systematic Review of Cyber Security and Forensics in Next-Generation Drones*]. *Drones, 7*(7), 430. Multidisciplinary Digital Publishing Institute. DOI: 10.3390/drones7070430

Singh, B., Kumar, N., Kaur, G., Akhai, S., Abbass, M., & Kaur, P. (2025). Emerging trends in electric automotive instrumentation and control systems: A review of current trends and future technologies. *AIP Conference Proceedings, 3343*, 020012. DOI: 10.1063/5.0292567

Singh, S., & Akhai, S. (2024). Electronics in precision control for mechanical engineering: A brief overview. *AIP Conference Proceedings, 3121*, 020007. DOI: 10.1063/5.0221464

Sledz, A., Unger, J., & Heipke, C. (2020). UAV-BASED THERMAL ANOMALY DETECTION FOR DISTRIBUTED HEATING NETWORKS. *˜The œinternational Archives of the Photogrammetry, Remote Sensing and Spatial Information Sciences. The International Archives of the Photogrammetry, Remote Sensing and Spatial Information Sciences, 499.* Advance online publication. DOI: 10.5194/isprs-archives-XLIII-B1-2020-499-2020

Stanton, M. C., Kalonde, P. K., Zembere, K., Spaans, R. H., & Jones, C. M. (2021). The application of drones for mosquito larval habitat identification in rural environments: A practical approach for malaria control? *Malaria Journal, 20*(1), 244. Advance online publication. DOI: 10.1186/s12936-021-03759-2 PMID: 34059053

Stephani, H., Weibel, T., Rösch, R., & Moghiseh, A. (2023). Challenges and approaches when realizing online surface inspection systems with deep learning algorithms. *Discover Data, 1*(1), 3. Advance online publication. DOI: 10.1007/s44248-023-00004-w

Stroud, D., & Weinel, M. (2020). A safer, faster, leaner workplace? Technical-maintenance worker perspectives on digital drone technology 'effects' in the European steel industry. *New Technology, Work and Employment, 35*(3), 297–313. DOI: 10.1111/ntwe.12174

Sun, Y., Fesenko, H., Kharchenko, V., Luo, Z., Kliushnikov, I., Illiashenko, O., Morozova, O., & Sachenko, A. (2022). UAV and IoT-Based Systems for the Monitoring of Industrial Facilities Using Digital Twins: Methodology, Reliability Models, and Application. *Sensors (Basel)*, *22*(17), 6444. DOI: 10.3390/s22176444 PMID: 36080903

Sundaram, S., & Zeid, A. (2023). Artificial Intelligence-Based Smart Quality Inspection for Manufacturing. *Micromachines*, *14*(3), 570. DOI: 10.3390/mi14030570 PMID: 36984977

Suresh, S., Velmurugan, D., Vasudevan, N., Surekha Rathi Samundi, D., & Elayaraja, R. (2025). Autonomous UAV swarms: Innovations and applications. In *Innovations and developments in unmanned aerial vehicles* (pp. 95–128). IGI Global., DOI: 10.4018/979-8-3693-8462-6.ch005

Szira, Z., Varga, E., Csegődi, T. L., & Milics, G. (2023). The Development of Drone Techology and its Regulation in the European Union. *EU Agrarian Law*, *12*(1), 35–41. DOI: 10.2478/eual-2023-0005

Taneja, T., Bhatia, B. S., & Akhai, S. (2024, December). Assessing Groundwater Quality in Rampura-Bathinda for Physicochemical Parameters & Heavy Metal Concentrations using Electronic Equipment. In *2024 1st International Conference on Sustainability and Technological Advancements in Engineering Domain (SUSTAINED)* (pp. 791-796). IEEE. DOI: 10.1109/SUSTAINED63638.2024.11073880

Taneja, T., Bhatia, B. S., & Akhai, S. (2025). Agricultural Resilience: Water Quality and Human Well-Being. *Emerging Smart Agricultural Practices Using Artificial Intelligence*, 1-23. DOI: 10.1002/9781394274277.ch1

Tianyuan, L., Guo, H., Danilov, C., & Nahrstedt, K. (2020). *A Privacy-Preserving Data Collection and Processing Framework for Third-Party UAV Services*. 683. DOI: 10.1109/TrustCom50675.2020.00095

Tiusanen, R., Heikkilä, E., Välisalo, T., & Öz, E. (2023). Identification of Emerging Safety and Security Risks in Drone Operations at Work Sites. In *Lecture notes in mechanical engineering* (p. 499). Springer Nature., DOI: 10.1007/978-3-031-25448-2_47

Tmušić, G., Manfreda, S., Aasen, H., James, M. R., Gonçalves, G., Dor, E. B., Brook, A., Polinova, M., Justel, J. J. A., Mészáros, J., Zhuang, R., Johansen, K., Malbéteau, Y., de Lima, M. I. P., Davids, C., Herban, S., & McCabe, M. F. (2020). Current Practices in UAS-based Environmental Monitoring. *Remote Sensing (Basel)*, *12*(6), 1001. DOI: 10.3390/rs12061001

To, A., Liu, M., Hairul, M. H. B. M., Davis, J. G., Lee, J. S. A., Hesse, H., & Nguyen, H. D. (2021). Drone-Based AI and 3D Reconstruction for Digital Twin Augmentation. In *Lecture notes in computer science* (p. 511). Springer Science+Business Media. DOI: 10.1007/978-3-030-77626-8_35

Todescato, M., Braholli, O., Chaltsev, D., Blasio, I. D., Don, D., Egger, G., Emig, J., Monizza, G. P., Sacco, P., Siegele, D., Steiner, D., Terzer, M., Riedl, M., Giusti, A., & Matt, D. T. (2023). Sustainable manufacturing through application of reconfigurable and intelligent systems in production processes: A system perspective. *Scientific Reports, 13*(1), 22374. DOI: 10.1038/s41598-023-49727-5 PMID: 38104211

Tran, T.-H., & Nguyen, D. D. (2022). Management and Regulation of Drone Operation in Urban Environment: A Case Study. *Social Sciences (Basel, Switzerland), 11*(10), 474. DOI: 10.3390/socsci11100474

Triantafyllou, A., Sarigiannidis, P., Bibi, S., Vakouftsi, F., & Pantzios, V. (2020). *Modelling deployment costs of Precision Agriculture Monitoring Systems. 5*, 252. DOI: 10.1109/DCOSS49796.2020.00048

Urama, J., Wiren, R., Galinina, O., Kauppi, J., Hiltunen, K., Erkkilä, J., Chernogorov, F., Eteläaho, P., Heikkilä, M., Torsner, J., Andreev, S., & Valkama, M. (2020). UAV-Aided Interference Assessment for Private 5G NR Deployments: Challenges and Solutions. *IEEE Communications Magazine, 58*(8), 89–95. DOI: 10.1109/MCOM.001.2000042

van Bos, B., Mallion, A., Wilson, C., Zwicker, E., Schler, A., & Black, T. (2015, November 3). Robotic Inspection Solutions for Petrochemical Pressure Vessels. *Abu Dhabi International Petroleum Exhibition and Conference*. DOI: 10.2118/177913-MS

Ventura, S. M., Bellagente, P., Rinaldi, S., Flammini, A., & Ciribini, A. L. C. (2023). Enhancing Safety on Construction Sites: A UWB-Based Proximity Warning System Ensuring GDPR Compliance to Prevent Collision Hazards. *Sensors (Basel), 23*(24), 9770. DOI: 10.3390/s23249770 PMID: 38139616

Wadhwa, A. S., Abbass, M., Akhai, S., Kumar, D., & Kumar, P. (2025). Integrating Taguchi optimization for multi-criteria decision making in engineering applications. In *Recent theories and applications for multi-criteria decision-making* (pp. 125–150). IGI Global., DOI: 10.4018/979-8-3693-6502-1.ch005

Wadhwa, A. S., Akhai, S., Abbass, M., Chouksey, A., Tiwari, S., & Taneja, T. (2025). Computational intelligence-driven design and optimization of polyurethane belt-type oil skimmer for sustainable manufacturing using SolidWorks 3D CAD. In *computational intelligence for sustainable manufacturing of advanced materials* (pp. 445–464). IGI Global. DOI: 10.4018/979-8-3693-7974-5.ch019

Walsh, G., Andersen, N. S., Stoianov, N., & Jänicke, S. (2023). Towards a New Symbology and Visual Interaction Design for Sub-Sea Military Operations. *Research Portal Denmark*, 24. https://local.forskningsportal.dk/local/dki-cgi/ws/cris-link?src=sdu&id=sdu-145ca22b-f48d-48cd-8ffd-bb523392d677&ti=Towards%20a%20New%20Symbology%20and%20Visual%20Interaction%20Design%20for%20Sub-Sea%20Military%20Operations

Wang, F., Zou, Y., Zhang, C., Buzzatto, J., Liarokapis, M., Castillo, E. del R., & Lim, J. B. P. (2023). UAV navigation in large-scale GPS-denied bridge environments using fiducial marker-corrected stereo visual-inertial localisation. *Automation in Construction, 156*, 105139. DOI: 10.1016/j.autcon.2023.105139

Wang, S., Smith, M., & Gershenfeld, N. (2023). Voxelcopter. *Modular Autonomous Aerial Systems, 1*, 1–2. Advance online publication. DOI: 10.1145/3623263.3629155

Waqas, A., Kang, D., & Cha, Y. (2023). Deep learning-based obstacle-avoiding autonomous UAVs with fiducial marker-based localization for structural health monitoring. *Structural Health Monitoring, 23*(2), 971–990. DOI: 10.1177/14759217231177314 PMID: 38405115

Watson, R., Kamel, M., Zhang, D., Dobie, G., MacLeod, C., Pierce, G., & Nieto, J. (2021). Dry Coupled Ultrasonic Non-Destructive Evaluation Using an Over-Actuated Unmanned Aerial Vehicle. *IEEE Transactions on Automation Science and Engineering, 19*(4), 2874–2889. DOI: 10.1109/TASE.2021.3094966

Westkämper, E., & Warnecke, H. J. (1994). Zero-Defect Manufacturing by Means of a Learning Supervision of Process Chains. *CIRP Annals, 43*(1), 405–408. DOI: 10.1016/S0007-8506(07)62240-9

Xiao, J., Zhang, R., Zhang, Y., & Feroskhan, M. (2025). Vision-Based Learning for Drones: A Survey. *IEEE Transactions on Neural Networks and Learning Systems, 36*(9), 15601–15621. DOI: 10.1109/TNNLS.2025.3564184 PMID: 40327484

Yan, W., Wang, J., Lu, S., Zhou, M., & Peng, X. (2023). A Review of Real-Time Fault Diagnosis Methods for Industrial Smart Manufacturing [Review of *A Review of Real-Time Fault Diagnosis Methods for Industrial Smart Manufacturing*]. *Processes, 11*(2), 369. Multidisciplinary Digital Publishing Institute. DOI: 10.3390/pr11020369

Yi, W., Choi, J., & Lee, J. W. (2023). Sound-based drone fault classification using multitask learning. *arXiv (Cornell University)*. https://doi.org//arxiv.2304.11708DOI: 10.48550

Yi, W., & Sutrisna, M. (2020). Drone scheduling for construction site surveillance. *Computer-Aided Civil and Infrastructure Engineering, 36*(1), 3–13. DOI: 10.1111/mice.12593

Yuan, Y., Ma, G., Cheng, C., Zhou, B., Zhao, H., Zhang, H., & Ding, H. (2019). *Artificial Intelligent Diagnosis and Monitoring in Manufacturing. arXiv*. Cornell University., DOI: 10.1093/nsr/nwz190

Zhang, K., Chermprayong, P., Xiao, F., Tzoumanikas, D., Dams, B., Kay, S., Koçer, B. B., Burns, A., Orr, L., Alhinai, T., Choi, C. Y., Darekar, D. D., Li, W., Hirschmann, S., Soana, V., Ngah, S. A., Grillot, C., Sareh, S., Choubey, A., & Kovač, M. (2022). Aerial additive manufacturing with multiple autonomous robots. *Nature*, *609*(7928), 709–717. DOI: 10.1038/s41586-022-04988-4 PMID: 36131037

Zhong, W., Rasouli, S., Singh, A. K., Mohandes, S. R., Antwi-Afari, M. F., Cheung, C., Manu, P., & Agrawal, U. S. (2025). The Adoption of UAVs for Enhancing Safety in Construction Industry: A Systematic Literature Review. *Intelligent Infrastructure and Construction*, *1*(1), 3. DOI: 10.3390/iic1010003

Zhong, Y., Wang, Z., Yalamanchili, A., Yadav, A., Srivatsa, B. N. R., Saripalli, S., & Bukkapatnam, S. (2020). Image-based flight control of unmanned aerial vehicles (UAVs) for material handling in custom manufacturing. *Journal of Manufacturing Systems*, *56*, 615–621. DOI: 10.1016/j.jmsy.2020.04.004

KEY TERMS AND DEFINITIONS

Autonomous Navigation: The ability of drones to fly without human control using onboard sensors, computer vision, LiDAR, obstacle detection, and pre-programmed flight paths.

Data Integration Platform: Software that merges drone data (images, sensor readings, flight logs) with existing factory systems such as maintenance dashboards, ERP systems, and analytics engines.

Defect Detection: The identification of cracks, scratches, coating inconsistencies, colour variations, or structural deviations using drone-captured images and automated vision algorithms.

Digital Twin: A virtual, continuously updated 3D or simulation model of a real manufacturing system that integrates drone data to reflect real-time conditions for planning, analysis, and optimisation.

Drone (Unmanned Aerial Vehicle – UAV): An aerial robotic system capable of remote or autonomous flight used for inspection, monitoring, data collection, and surveillance within industrial environments.

Drone Fleet Management: Centralised control and scheduling of multiple drones for inspection, monitoring, environmental scanning, and quality-assurance tasks across large industrial facilities.

Environmental Monitoring: The use of drone-mounted sensors to measure air quality, emissions, dust levels, noise intensity, temperature distribution, and water contamination across industrial zones.

Gas and Emission Detectors: Drone-mounted instruments used to identify and measure volatile organic compounds (VOCs), dust, toxic gases, and other industrial emissions for compliance and risk management.

Hazardous Area Inspection: Drone-enabled monitoring of locations unsafe for workers—such as chemical storage units, high-temperature zones, confined spaces, or elevated structures.

Industrial-Grade Drone: A rugged, high-performance drone designed to operate safely in dusty, hot, humid, or electromagnetically noisy industrial environments.

Internet of Things (IoT): A connected network of sensors, machines, and digital systems where drone-collected data is shared for real-time monitoring, analysis, and decision-making in smart factories.

Multispectral / Hyperspectral Sensors: Advanced sensors that capture data across multiple wavelengths, enabling detection of chemical signatures, emission patterns, and environmental anomalies with high precision.

Predictive Maintenance: A maintenance strategy that uses drone-generated data—such as vibration patterns, thermal profiles, and historical records—to predict equipment failures before they occur.

Quality Assurance (QA): A drone-supported process that ensures product consistency and defect-free production through high-resolution imaging, AI-based defect detection, and thermal analysis.

Real-Time Inspection: Continuous, sensor-based monitoring of machines, equipment, and production lines using drones to detect overheating, misalignment, leaks, vibrations, and operational faults instantly.

Regulatory Compliance: Adherence to aviation, safety, data privacy, and environmental regulations governing drone use in manufacturing spaces.

Risk Management: The systematic identification and mitigation of hazards in manufacturing using drone-supported inspections, emergency monitoring, and data-driven safety decisions.

Safety Surveillance: Monitoring of hazardous zones, elevated platforms, restricted spaces, or emergency situations using drones to enhance worker safety and situational awareness.

Thermal Imaging: A technology that captures temperature variations across surfaces or machinery using infrared sensors mounted on drones, enabling early detection of hotspots, inefficiencies, or faults.

Workforce Upskilling: Training workers in drone operation, data interpretation, sensor handling, and digital tools to support a technology-enabled manufacturing environment.

Chapter 8
Strategic Navigation in the VUCA Manufacturing Landscape:
Geopolitical, Technological, and Regulatory Change

Göran Roos
http://orcid.org/0000-0003-0943-3585
Adelaide University, Australia

ABSTRACT

Contemporary manufacturing organisations operate within volatile, uncertain, complex, and ambiguous (VUCA) environments characterised by rapid technological advancement, geopolitical instability, and regulatory transformation. This chapter examines strategic approaches enabling manufacturers to navigate these challenges through AI, automation, and robotics integration. Drawing upon dynamic capabilities theory, institutional theory, and strategic management literature, this analysis presents evidence-based frameworks for transforming operational volatility into competitive advantage. The chapter synthesises theoretical foundations with empirical evidence, demonstrating how smart manufacturing strategies enhance organisational resilience. Key contributions include a VUCA navigation framework, technology integration models, and guidance for managing institutional complexity. Analysis reveals successful VUCA navigation requires orchestration of technological capabilities, institutional adaptation strategies, and dynamic learning processes.

DOI: 10.4018/979-8-3373-5756-0.ch008

INTRODUCTION

In March 2021, the *Ever Given* container ship blocked the Suez Canal for six days, stranding 369 vessels and disrupting global supply chains. BMW's Leipzig plant halted production within 72 hours, not from missing ship components, but because alternative routing decisions created cascading uncertainties across 47 supplier tiers (Verschuur et al., 2021). This episode reveals how contemporary manufacturing operates within what Bennett and Lemoine (2014) characterise as VUCA environments—contexts where "volatility, uncertainty, complexity, and ambiguity" intersect to create challenges that transcend traditional operational problems. The BMW disruption illustrates all four dimensions simultaneously: volatile shipping routes, uncertain transit times, complex supplier interdependencies, and ambiguous decision criteria when choosing between air freight costs and production delays. Manufacturing enterprises worldwide now confront such environments routinely, as established operational frameworks prove inadequate for managing the compounding effects of geopolitical tensions, technological transformation, and regulatory evolution (Schoemaker et al., 2018; Ivanov & Dolgui, 2021). The question facing manufacturing leaders is not whether VUCA conditions exist, but rather how organisations develop capabilities to navigate them effectively whilst maintaining competitive advantage.

The military genesis of VUCA warrants critical examination before accepting its applicability to manufacturing contexts. Developed at the U.S. Army War College in 1987 under Commandant MG Thompson's curricular reforms, the framework addressed post-Cold War strategic challenges in military leadership (Mackey, 1992). However, military and manufacturing environments differ fundamentally in three respects. First, military VUCA addresses existential survival under adversarial conditions, whilst manufacturing VUCA concerns competitive positioning within market systems (Whiteman, 1998). Second, military operations accept mission failure rates that would constitute catastrophic quality defects in manufacturing—a 5% operation failure rate represents military success but manufacturing disaster (Weick & Sutcliffe, 2007). Third, military frameworks prioritise rapid decision-making under incomplete information, whilst manufacturing increasingly demands data-driven precision and continuous improvement (Deming, 2000). These distinctions suggest that whilst military VUCA provides useful conceptual vocabulary, manufacturing organisations require substantial adaptation rather than direct framework transfer. The critical question becomes not whether VUCA describes manufacturing environments—it clearly does—but whether military-derived response strategies suit manufacturing's emphasis on systematic process improvement, statistical control, and incremental innovation (Liker, 2004).

The VUCA manufacturing environment reflects wider economic shifts characterised by accelerated technological innovation cycles (Hofer et al., 2020; Mubarak & Petraite, 2020), evolving geopolitical tensions (Bednarski et al., 2025), and progressively stringent regulatory frameworks that fundamentally reshape competitive dynamics across manufacturing sectors (Dicken, 1992; Dechezleprêtre & Sato, 2017; O'Brien, 2002; OECD, 2025).

Contemporary manufacturing organisations encounter complex challenges transcending conventional operational boundaries. Industry 4.0 technologies—encompassing artificial intelligence, automation, and robotics—simultaneously present considerable opportunities for operational enhancement and substantial risks relating to technological complexity and cybersecurity vulnerabilities (Xu et al., 2018; Ghobakhloo, 2020). Concurrently, geopolitical tensions manifesting through trade disputes, supply chain disruptions, and technological export controls (Bown, 2020; Celestin & Sujatha, 2024) establish additional uncertainty layers requiring strategic navigation by manufacturing leaders (Doherty et al., 2014).

Environmental, social, and governance (ESG) requirements constitute a distinct VUCA dimension that manufacturing organisations can no longer treat as peripheral compliance obligations. The European Union's Corporate Sustainability Reporting Directive (CSRD), effective January 2024, mandates sustainability reporting for 50,000 companies operating in EU markets, creating cascading requirements through global supply chains (European Parliament & Council, 2022). This regulatory expansion generates compound VUCA effects. Volatility emerges through fluctuating carbon prices—the EU Emissions Trading System price ranged from €25 to €96 per tonne between 2020 and 2023—directly affecting manufacturing cost structures (European Commission, 2024). Uncertainty manifests in evolving taxonomies: what qualifies as "sustainable manufacturing" changes as scientific understanding advances and political coalitions shift, creating moving compliance targets (IFRS Foundation, 2023). Complexity arises from competing sustainability frameworks—GRI, SASB, TCFD, and now CSRD—each demanding different metrics and reporting structures, whilst manufacturing operations must reconcile these with existing quality, safety, and financial systems (Adams & Abhayawansa, 2022). Ambiguity pervades strategic decisions when sustainability and traditional performance metrics conflict, as when investing in circular economy initiatives that improve environmental outcomes but extend payback periods beyond conventional capital budgeting thresholds (Geissdoerfer et al., 2017).

Manufacturing leaders confront sustainability not as isolated initiative but as systemic transformation that intersects with technology adoption, supply chain configuration, and competitive positioning. Consider Unilever's commitment to achieve net-zero emissions across its value chain by 2039, requiring 1,500 manufacturing suppliers to implement verified carbon reduction programmes (Unilever, 2024). This single corporate commitment creates ripples throughout manufacturing networks, particularly for small and medium enterprises lacking resources for sophisticated carbon accounting. Research examining sustainability reporting in SMEs found that sustainability disclosure requirements create significant operational complexity, with many small and medium-sized manufacturing enterprises experiencing sustainability reporting demands as more burdensome than technology implementation initiatives (Galli et al., 2023). The critical insight for manufacturing strategy emerges: sustainability constitutes a primary VUCA driver rather than secondary consideration, fundamentally reshaping competitive dynamics, investment priorities, and capability requirements.

MISSION AND RESEARCH OBJECTIVES

This chapter's fundamental aim involves developing strategic frameworks enabling manufacturing organisations to navigate VUCA environments effectively through advanced technology integration whilst mitigating risks arising from geopolitical turbulence and regulatory transformation. The endeavour encompasses theoretical contributions to strategic management literature alongside practical guidance for manufacturing leaders seeking to transform operational challenges into competitive advantages.

Seven interconnected research objectives guide this investigation. Firstly, the chapter delivers systematic analysis of contemporary VUCA conditions affecting global manufacturing operations, examining how volatility, uncertainty, complexity, and ambiguity manifest within specific manufacturing contexts and generate cascading effects throughout organisational systems (Cernega et al., 2024; Mohanta et al., 2019). Secondly, it evaluates strategic implications of artificial intelligence, automation, and robotics integration within manufacturing operations, exploring how these technologies simultaneously mitigate and amplify VUCA conditions according to implementation approaches and organisational capabilities (Chiarini, 2020; Kang et al., 2020).

The third objective assesses geopolitical risk factors and their operational impacts upon manufacturing organisations, examining how trade tensions, regulatory changes, and technological export controls generate strategic challenges requiring sophisticated risk management approaches (U.S. Government Accountability Office, 2024; Doherty et al., 2014). Fourth, the analysis examines regulatory transformation requirements across major economic regions, investigating how evolving compliance frameworks create constraints and opportunities for manufacturing innovation (Khan et al., 2017; Kauppi & Hannibal, 2017).

Fifth, the chapter addresses supply chain resilience through technological integration, exploring how advanced technologies enhance supply chain visibility, flexibility, and robustness whilst creating new vulnerabilities requiring careful management (Tukamuhabwa et al., 2015; Negri et al., 2021; Ivanov, 2021). The sixth objective examines cybersecurity strategies for increasingly digital manufacturing environments, investigating how digitalisation creates novel risk profiles demanding integrated security approaches (WEF, 2019; Ghobakhloo, 2020; Wu et al., 2018; Ramya & Srinivasagan, 2024).

Finally, the chapter proposes integrated strategic frameworks for sustainable competitive positioning within turbulent environments, synthesising theoretical insights with practical guidance for manufacturing leaders seeking to enhance organisational resilience and adaptability (Beliaeva et al., 2020; Liu et al., 2013; Pratono, 2022).

THEORETICAL FOUNDATIONS AND SIGNIFICANCE

This chapter advances existing scholarly understanding by synthesising multiple theoretical perspectives within a comprehensive framework for VUCA navigation in manufacturing contexts. The investigation builds upon dynamic capabilities theory (Teece et al., 1997; Helfat & Peteraf, 2003) to elucidate how manufacturing organisations cultivate capacities for sensing, seizing, and transforming in response to environmental turbulence (Lin et al., 2020; Ahmed et al., 2024). This theoretical foundation receives enrichment through institutional theory perspectives (DiMaggio & Powell, 1983; Scott, 2008) that illuminate how manufacturing firms navigate multiple, frequently conflicting institutional pressures whilst maintaining legitimacy and operational effectiveness (Acquah et al., 2021; Kauppi, 2013).

The significance of this investigation transcends theoretical contributions to address urgent practical challenges confronting manufacturing leaders worldwide. Contemporary empirical evidence reveals increasing severity of VUCA conditions within manufacturing contexts, with capacity utilisation volatility, supply chain disruptions, and regulatory complexity attaining unprecedented levels (Board of Governors of the Federal Reserve System, 2025; Bednarski et al., 2025). Manufacturing capacity utilisation patterns demonstrate heightened instability, with recent data indicating substantial variations from historical baselines across major industrial economies.

Such volatility manifests through specific causal pathways creating compounding effects across manufacturing systems. Input volatility exhibits elevated coefficients of variation in raw material costs, reflecting broader commodity price instabilities and supply chain uncertainties (Moheb-Alizadeh & Handfield, 2018; Ashena et al., 2024). Demand volatility demonstrates significant quarterly fluctuations across key manufacturing sectors, whilst operational volatility emerges through capacity utilisation variance creating substantial challenges for production planning and resource allocation that conventional management approaches struggle to address effectively (Nitsche & Straube, 2020).

METHODOLOGICAL APPROACH

This chapter adopts a mixed-methods research design integrating systematic literature review, empirical analysis, and case study synthesis to generate comprehensive insights into VUCA navigation strategies (Fernando et al., 2023; Azorín & Cameron, 2010). The systematic literature review encompasses peer-reviewed publications spanning 2020-2025, concentrating on strategic management, operations management, and technology management literature pertinent to manufacturing contexts (Psomas, 2021). Search strategies employed multiple academic databases, including Web of Science, Scopus, and Business Source Premier, utilising keyword combinations encompassing VUCA, manufacturing strategy, artificial intelligence, automation, robotics, geopolitical risk, and regulatory transformation (Kumpulainen & Seppänen, 2022; Pranckutė, 2021).

The empirical analysis incorporates secondary data drawn from industry reports, governmental publications, and international organisation studies to furnish quantitative context for VUCA conditions within manufacturing environments (Halkos et al., 2021). Principal data sources encompass the Organisation for Economic Co-operation and Development (OECD) industrial statistics, the United Nations Industrial Development Organization (UNIDO) manufacturing reports, and regional manufacturing association publications that deliver sectoral and geographical insights into VUCA manifestations (Horvát & Webb, 2020; Pilat et al., 2008).

Case study synthesis utilises publicly available information from manufacturing organisations that have successfully navigated VUCA conditions through strategic technology integration and institutional adaptation (Hoon, 2013; Calabrese et al., 2021). This methodological approach furnishes practical illustrations of theoretical concepts whilst respecting commercial confidentiality constraints that restrict access to proprietary strategic information (Cruzes et al., 2015).

LITERATURE REVIEW AND THEORETICAL FOUNDATIONS

VUCA Conceptualisation in Manufacturing Contexts

The VUCA framework emerged from military strategic thinking at the U.S. Army War College in 1987, subsequently developing into a sophisticated theoretical construct that fundamentally challenges conventional management paradigms and organisational theories (Bennett & Lemoine, 2014; Bennis & Nanus, 1985). Recent theoretical advances highlight the complex interdependencies between VUCA dimensions rather than treating them as discrete constructs. These interdependencies generate cascading effects that magnify organisational challenges through dimensional interactions (Cernega et al., 2024).

Institutional records substantiate VUCA's formal adoption at the U.S. Army War College during autumn 1987, though its development involved multiple senior military leaders rather than a single originator. Oral histories credit General L. Glenn Orr (Army War College Class of 1987) with popularising the acronym amongst his cohort. Concurrently, Army Chief of Staff General Gordon R. Sullivan and General Maxwell Thurman (former TRADOC commander) introduced similar terminology around 1990-1991 (Sullivan & Harper, 1996; Mackey, 1992). This concurrent adoption suggests shared recognition of the emerging post-Cold War strategic environment rather than individual conceptual ownership.

The framework explicitly drew upon civilian leadership scholarship, particularly Bennis and Nanus's (1985) examination of “accelerating change”, “uncertainty”, and organisational complexity. This demonstrates early military-civilian knowledge transfer that would subsequently reverse as VUCA transitioned from military to manufacturing applications.

Bennett and Lemoine (2014) provide crucial theoretical clarification by deconstructing VUCA components, demonstrating that each dimension necessitates distinct organisational responses rather than generic adaptation strategies. Their framework reveals that volatility encompasses rapid, unpredictable environmental fluctuations requiring vision and values-based responses that anchor organisations during turbulent periods.

Uncertainty represents the unpredictability of events, demanding enhanced information-gathering and analytical capabilities that enable informed decision-making despite incomplete data. Complexity involves interconnected forces and multiple variables that obscure cause-effect relationships, requiring clarity through simplification and structural approaches for effective decision-making within complicated systems. Ambiguity reflects unclear realities and multiple event interpretations, necessitating agility and adaptive capacity to respond effectively when situations lack clear meaning or direction.

Manufacturing environments exhibit distinctive VUCA interaction patterns that generate sector-specific challenges requiring specialised strategic responses. The conceptual migration from military strategic leadership to manufacturing operations reveals both theoretical continuities and significant adaptations.

Military VUCA was introduced by the U.S. Army War College in 1987, adapting the leadership theories of Bennis and Nanus (1985) to address strategic leadership challenges in the volatile, uncertain, complex, and ambiguous post-Cold War environment where traditional doctrine proved inadequate (Barber, 1992). Manufacturing applications preserve this strategic orientation whilst incorporating operational dimensions. Military volatility (rapid threat changes) parallels manufacturing demand volatility; military uncertainty (unpredictable adversary behaviour) corresponds to supply chain uncertainty; military complexity (multi-domain operations) aligns with manufacturing network complexity; military ambiguity (unclear intelligence) translates to market signal ambiguity (Giones & Brem., 2022).

However, manufacturing VUCA differs fundamentally from military applications through its emphasis on measurable performance metrics, systematic process improvement, and economic optimisation rather than strategic survival. This necessitates adapting military frameworks to manufacturing operational excellence requirements (Troise et al., 2022).

Supply chain volatility triggers demand uncertainty through disrupted production schedules and inventory management challenges. This subsequently amplifies operational complexity as manufacturing systems struggle to maintain efficiency whilst accommodating variability. Operational complexity then creates strategic ambiguity in resource allocation decisions as manufacturing leaders confront multiple potential scenarios with unclear probability distributions (Shishodia et al., 2023; Chen, 2024).

Contemporary empirical research demonstrates VUCA conditions' practical relevance in manufacturing environments through quantifiable indicators of environmental turbulence. Systematic analysis reveals measurable performance benefits from VUCA-oriented manufacturing practices, validating military-derived strategic frameworks in operational contexts.

Performance data across supply chain agility metrics shows substantial advantages for VUCA-prepared organisations. Top-quartile firms in supply chain agility deliver orders 30% faster than bottom-quartile firms in high-volatility markets, experiencing 50% fewer stockouts during demand spikes (Braunscheidel & Suresh, 2009).

Resilience metrics demonstrate significant differences in disruption recovery. Highly resilient manufacturing firms lost less than 5% of annual sales due to disruptions versus over 10% for low-resilience firms across 300 companies over ten years, despite carrying 3-5% higher cost-to-serve due to holding costs and redundancy (McKinsey, 2024).

Quality performance, measured through parts-per-million defect rates, remained stable in firms maintaining high-reliability practices during volatile periods. Conversely, firms without such practices experienced deterioration in defect rates under volatility stress, with empirical evidence showing that firms facing quality disruptions experience significant declines in operating performance and do not quickly recover (Hendricks & Singhal, 2005).

These empirical findings demonstrate that military VUCA navigation principles, when appropriately adapted to manufacturing contexts, produce measurable operational improvements across quality, delivery, cost, and resilience dimensions.

Systematic diffusion followed a well-documented 25-year timeline with specific institutional bridges. War College alumni and military-to-business thought leaders served as initial conduits in the late 1990s, introducing VUCA terminology into corporate leadership training programmes led by retired officers (Cuban, 2010). Academic legitimacy followed through leadership and organisational behaviour literature, with management scholar Peter Drucker alluding to "periods of turbulence" by 1999, followed by Joel Kurtzman's MBA in a Box (2004) including sections on thriving in VUCA conditions.

Bob Johansen's influential works Get There Early (2007) and Leaders Make the Future (2009) established "VUCA Prime" as a business response framework, countering volatility with vision, uncertainty with understanding, complexity with clarity, and ambiguity with agility. The 2008 global financial crisis accelerated adoption as business writers sought frameworks for extreme market turbulence, with global consulting firms incorporating VUCA into white papers on risk management and strategy. Manufacturing adoption emerged through supply chain management literature in the mid-2010s, with operations journals accepting VUCA terminology when authors linked it to contemporary challenges like digital transformation and global disruption.

Manufacturing organisations report increasing frequency of supply chain disruptions, with 73% of global manufacturers experiencing significant supply chain interruptions in 2023 compared to 41% in 2019 (OECD, 2025). These disruptions exhibit VUCA characteristics through their unpredictable timing (volatility), unclear duration and severity (uncertainty), multiple interconnected causes (complexity), and ambiguous implications for strategic planning (ambiguity).

Dynamic Capabilities Theory in Manufacturing

Dynamic capabilities theory provides frameworks for understanding how manufacturing organisations develop capacities for navigating VUCA environments. Teece et al. (1997) conceptualise dynamic capabilities as firms' abilities to integrate, build, and reconfigure competences addressing rapidly changing environments through sensing, seizing, and transforming processes. Manufacturing organisations possess distinctive characteristics fundamentally shaping dynamic capability requirements. Capital intensity establishes path dependencies constraining rapid reconfiguration, whilst process complexity necessitates sophisticated coordination mechanisms reconciling efficiency with flexibility imperatives (Macher & Mowery, 2009). This creates what Adler et al. (1999) identify as the flexibility-efficiency paradox requiring ambidextrous organisational designs optimising current performance whilst cultivating future adaptation capabilities.

Helfat and Peteraf's (2003) extension introduces dynamic capability lifecycles demonstrating how capabilities emerge through systematic organisational learning. Manufacturing organisations require meta-capabilities—higher-order competences enabling dynamic capability development and modification themselves. The sensing dimension encompasses systematic environmental scanning identifying opportunities and threats. Manufacturing firms deploy diverse mechanisms including customer collaboration, supplier partnerships, technology scouting, and regulatory monitoring (Teece, 2007). Effective sensing requires cross-functional integration combining technical, commercial, and strategic expertise, addressing complexity where technological, market, and regulatory changes interact unpredictably (Vance et al., 2023). Capital-intensive manufacturing operations demand high-quality environmental intelligence informing significant investment decisions, consequently requiring substantial sensing capability investment.

Seizing capabilities enable value capture through strategic resource deployment and organisational reconfiguration involving complex investment decisions regarding technologies, market entries, partnerships, and capability development (Teece, 2007). Manufacturing contexts require careful timing and sequencing attention, as capital intensity and operational complexity create constraints on rapid strategic pivoting. Transforming capabilities encompass organisational reconfiguration maintaining alignment between internal capabilities and external demands. For manufacturing, transformation involves fundamental changes to production processes, supply chain configurations, organisational structures, and cultural practices enabling sustained advantage in altered conditions (Helfat & Peteraf, 2003). Contemporary evidence demonstrates successful transformation requires careful orchestration of technological, organisational, and human resource changes maintaining operational continuity whilst enabling strategic adaptation (Wagire et al., 2021). Unlike service organisations where transformation focuses primarily on human capital and structures, manufacturing transformation involves physical assets, technological systems, supply networks, and human capabilities in integrated change programmes. Long asset lifecycles and process interdependencies mean transformation initiatives span multiple years, requiring sustained commitment throughout extended implementation periods.

Building Sensing Capabilities Systematically

Effective environmental sensing requires systematic organisation, not just intelligence gathering. Manufacturing leaders should implement four mechanisms: (1) Cross-functional sensing teams—establish quarterly horizon scanning sessions combining R&D, operations, procurement, and strategy functions to synthesise diverse environmental signals (Teece, 2007); (2) Customer co-development programmes—engage lead customers in joint product development, gaining early visibility into emerging requirements before market-wide demand materialises; (3) Technology scouting networks—maintain relationships with university research groups and startup accelerators in relevant technology domains, accessing emerging capabilities before competitive diffusion; (4) Regulatory intelligence systems—monitor not just published regulations but regulatory agency consultation documents and industry working group discussions, anticipating requirement changes 12–18 months before implementation. Siemens' Technology department exemplifies systematic sensing—maintaining extensive university and startup collaborations provides continuous technology intelligence feeding business unit innovation pipelines (Siemens, 2022). The investment appears modest compared to value gained from early market moves enabled by superior environmental awareness

Strategic Perspectives: Mintzberg's Five Ps Framework

Mintzberg's (1987) framework conceptualising strategy through five perspectives—plan, ploy, pattern, position, and perspective—illuminates how manufacturing organisations navigate VUCA environments through multiple simultaneous strategic approaches rather than singular planned responses. Strategy as plan provides deliberate structural guidance through scenario planning and contingency frameworks that enable purposeful action despite uncertainty (Roos et al., 2025). Strategy as ploy encompasses competitive manoeuvres that leverage environmental turbulence to mask strategic intentions whilst generating temporal windows for repositioning (Thorén & Vendel, 2019). Strategy as pattern manifests through consistent organisational responses that emerge independently of formal planning, creating recognisable orientations through accumulated experiential learning (Dhillon & Nguyen, 2021).

Strategy as position involves organisational placement within competitive environments, requiring frequent repositioning as VUCA conditions alter competitive dynamics and value creation opportunities (Schuh et al., 2019). Manufacturing firms must balance strategic agility with core competitive advantage preservation, maintaining capabilities that distinguish them whilst adapting to turbulent conditions. Strategy as perspective encompasses shared mental models and cognitive frameworks guiding environmental interpretation, which must evolve continuously through learning processes that integrate emerging knowledge and adapt to changing realities (Baran & Woznyj, 2021; Peschl, 2023).

The framework's value for VUCA navigation lies in recognising that effective strategic responses require simultaneous attention to planning, positioning, pattern recognition, competitive manoeuvring, and perspective evolution. Manufacturing organisations that excel in uncertain environments develop capabilities across all five dimensions rather than privileging formal planning processes. This multidimensional approach proves particularly valuable when traditional strategic planning assumptions no longer hold, enabling organisations to recognise emergent patterns, adjust positioning dynamically, and maintain strategic coherence despite environmental turbulence.

Institutional Theory and Manufacturing Adaptation

Institutional theory illuminates how manufacturing organisations achieve legitimacy beyond technical efficiency within complex institutional environments characterised by competing logics—market efficiency imperatives, regulatory compliance requirements, environmental sustainability mandates, and social responsibility expectations (Thornton et al., 2012; Besharov & Smith, 2014). Manufacturing firms increasingly adopt hybrid approaches blending multiple institutional logics, particularly when addressing sustainability challenges within circular economy frameworks (Ebrahimi & Koh, 2021). These organisations confront three primary institutional pressures that fundamentally reshape operational practices.

Coercive pressures from regulatory bodies establish mandatory requirements such as environmental standards, workplace safety regulations, and trade compliance obligations (Scott, 2008). The European Union's environmental directives—RoHS, WEEE, and ELV—exemplify powerful coercive mechanisms creating cascading effects throughout global manufacturing networks (Khan et al., 2017; Van Rossem, 2008). Normative pressures manifest through professionalisation processes and industry associations that establish legitimate practices, often operating in conjunction with coercive mechanisms to influence supply chain management practices (Kauppi & Hannibal, 2017; Bhakoo & Choi, 2013). Mimetic pressures emerge when organisations confronting uncertainty model themselves upon successful peers, with manufacturing firms demonstrating strongest responses under elevated environmental uncertainty (Liu et al., 2013). This imitative behaviour creates strategic tensions between legitimacy-seeking conformity and differentiation-seeking competitive advantage, requiring sophisticated management approaches that selectively adopt industry practices whilst maintaining distinctive capabilities (Hoejmose et al., 2014; Zhu, 2016).

The European Union's environmental directives exemplify powerful coercive mechanisms that have fundamentally transformed global manufacturing practices. The RoHS (Restriction of Hazardous Substances), WEEE (Waste Electrical and Electronic Equipment), and ELV (End-of-Life Vehicles) directives demonstrate how regulatory frameworks extend beyond national boundaries to influence international supply chains and manufacturing processes (Khan et al., 2017; Yu et al., 2006). These directives have created cascading effects throughout global manufacturing networks, compelling organisations to redesign products, modify supplier relationships, and implement comprehensive compliance systems that extend far beyond immediate regulatory jurisdictions (Van Rossem, 2008).

Normative pressures manifest through professionalisation processes, industry associations, and educational institutions that establish legitimate organisational practices and behavioural expectations within manufacturing sectors. These pressures operate through the dissemination of professional standards, best practices, and industry norms that define appropriate organisational conduct (Gunningham & Rees, 1997). Research demonstrates that normative pressures from professional networks significantly influence supply chain management practices within manufacturing contexts, frequently operating in conjunction with coercive mechanisms rather than independently (Kauppi & Hannibal, 2017).

Professional associations and industry standards bodies create powerful normative forces that drive standardisation of operational practices across manufacturing supply chains (Bhakoo & Choi, 2013; Casper & Hancké, 1999). These organisations establish technical standards, ethical guidelines, and performance benchmarks that shape organisational behaviour through peer pressure and professional legitimacy considerations. The influence of such normative mechanisms often proves particularly pronounced in knowledge-intensive manufacturing sectors where professional expertise and technical competence serve as primary sources of organisational legitimacy.

Mimetic pressures emerge when organisations confronting uncertainty model themselves upon other organisations perceived as successful or legitimate within their institutional field. This imitative behaviour represents a risk-reduction strategy that enables organisations to acquire legitimacy by adopting practices associated with successful peers (Wang et al., 2018). Empirical evidence indicates that manufacturing firms demonstrate strongest responses to mimetic pressures when facing elevated levels of environmental uncertainty, leading to widespread adoption of similar technological solutions and organisational practices across industry sectors (Liu et al., 2013).

Such mimetic behaviour generates complex strategic implications for manufacturing organisations. Whilst imitation may enhance legitimacy and reduce uncertainty-related risks, it potentially diminishes competitive differentiation among industry participants (Hoejmose et al., 2014). This tension between legitimacy-seeking and differentiation-seeking behaviours creates strategic dilemmas for manufacturing managers who must balance conformity pressures with competitive positioning requirements. Research suggests that successful manufacturing organisations develop sophisticated approaches that selectively adopt industry practices whilst maintaining distinctive capabilities that preserve competitive advantage (Zhu, 2016).

Technology Integration and Digital Transformation

The convergence of artificial intelligence, automation, and robotics within manufacturing operations constitutes fundamental transformation reshaping competitive dynamics whilst generating novel VUCA sources. Digital transformation encompasses systematic integration of advanced technologies across organisational operations, fundamentally altering value creation processes and competitive positioning strategies (Vial, 2019; Verhoef et al., 2021). This technological revolution extends beyond operational efficiency to encompass business model innovation, particularly through digital servitization where manufacturers augment physical products with data-driven services creating recurring revenue streams and deepening customer relationships (Paschou et al., 2020). Such transformation demands careful navigation of complex implementation challenges whilst managing amplified uncertainty accompanying systemic changes.

Artificial intelligence applications within manufacturing environments span diverse operational domains, including predictive maintenance, quality assurance, supply chain optimisation, and production planning. These applications enable enhanced decision-making capabilities whilst creating demanding requirements for data management infrastructure and algorithmic governance frameworks (Kang et al., 2020; Plathottam et al., 2023). Contemporary AI implementations demonstrate particular effectiveness in predictive maintenance applications, where machine learning algorithms support real-time monitoring and failure prediction, significantly reducing unplanned downtime and maintenance costs (Keleko et al., 2022; Uçar et al., 2024).

Siemens' Amberg Electronics Plant exemplifies both the opportunities and challenges of AI integration in manufacturing. The facility, which produces SIMATIC programmable logic controllers, reports built-in quality of about 99.9989% and manufactures roughly 1,000 product variants, with machine-learning-enabled in-line monitoring supporting quality assurance (Siemens, 2017). The plant's experience reveals three critical lessons for manufacturing leaders. First, legacy equipment generated data in 47 incompatible formats, requiring custom middleware development before AI algorithms could function—a technical debt rarely acknowledged in vendor demonstrations (Kagermann et al., 2013). Second, operator resistance emerged not from technophobia but from legitimate concerns about accountability when AI recommendations conflicted with experiential knowledge, requiring extensive change management and revised decision protocols (Jarrahi, 2018). Third, cybersecurity vulnerabilities increased exponentially as connected systems grew from 400 to 3,200 endpoints, demanding parallel security infrastructure investment that matched AI implementation costs (Tuptuk & Hailes, 2018). The Siemens case demonstrates that AI delivers measurable manufacturing benefits, but only when organisations address integration complexity, workforce adaptation, and security implications systematically rather than treating them as secondary implementation details

The Hidden Costs Of AI Implementation

The Siemens Amberg experience reveals that AI implementation costs extend far beyond vendor quotations. Manufacturing leaders should budget implementation at 2-3 times initial technology costs to account for: (1) Data integration middleware—legacy equipment generates incompatible data formats requiring custom software development (Kagermann et al., 2013); (2) Change management processes—operator resistance stems from legitimate accountability concerns when AI recommendations conflict with experience, demanding revised decision protocols and training (Jarrahi, 2018); (3) Cybersecurity infrastructure—connected systems exponentially increase attack surfaces, requiring parallel security investment matching AI costs (Tuptuk & Hailes, 2018). Vendors demonstrating impressive AI capabilities in controlled environments rarely disclose these integration realities. Successful manufacturers treat AI as organisational transformation rather than technology purchase, budgeting accordingly and planning 3-5 year implementation horizons for complex systems.

Machine learning algorithms increasingly underpin manufacturing decision-making processes through sophisticated pattern recognition, anomaly detection, and predictive analytics capabilities that enhance operational efficiency (Andrianandrianina Johanesa et al., 2024). These intelligent systems require comprehensive data infrastructure and analytical capabilities that extend beyond traditional manufacturing competencies, demanding significant organisational learning and capability development. Research demonstrates that successful AI implementation in manufacturing contexts depends critically upon integrating predictive analytics with continuous quality control mechanisms, creating synergistic effects that enhance overall system performance (Lindström et al., 2017; Lee et al., 2020).

Automation technologies fundamentally transform manufacturing operations through enhanced precision, consistency, and processing speed whilst reducing labour intensity and minimising human error rates. However, automation implementation paradoxically generates new sources of complexity through system integration challenges, maintenance requirements, and skill development demands that may amplify VUCA conditions during transitional periods (Ghobakhloo, 2020). Successful automation deployment necessitates careful orchestration of technological, organisational, and human resource transformations that preserve operational continuity whilst enabling performance enhancement.

The complexity inherent in automated manufacturing systems stems from increased interdependencies between technological components, creating cascading effects when system failures occur (Cagliano et al., 2019; Abdullah et al., 2023). These interdependencies complicate troubleshooting processes and require sophisticated system integration approaches that account for complex cause-and-effect relationships within manufacturing environments (Efthymiou et al., 2016). Such complexity management becomes particularly challenging when organisations attempt to integrate multiple automation technologies simultaneously, creating compounding effects that may overwhelm existing organisational capabilities.

Managing the Automation Complexity Paradox

Manufacturing automation creates a performance paradox: systems optimised for normal operations become brittle during disruptions. Leaders should implement deliberate complexity management: (1) Preserve manual override capabilities—ensure automated systems include accessible manual controls rather than complete automation, maintaining operator agency during failures; (2) Implement modular architecture—design systems with clear boundaries enabling isolation of failures rather than cascading shutdowns across integrated processes; (3) Conduct regular degradation testing—intentionally disable system components during low-stakes periods to verify graceful degradation and train operators in partial-automation scenarios. Organisations with very high uptime during normal operations can still experience catastrophic failures during disruptions if partial-failure modes are never exercised and trained; this risk is highlighted in high-reliability and resilience literature (Weick & Sutcliffe, 2007; Perrow, 1999). The goal is not avoiding automation but designing resilient automation acknowledging that perfect reliability remains unattainable in complex systems.

Robotics applications within manufacturing enable flexible automation that adapts to varying production requirements whilst maintaining operational precision and consistency. Collaborative robotics (cobots) represent a particularly significant development, facilitating human-robot collaboration that combines human flexibility and problem-solving capabilities with robotic precision and reliability (Xu et al., 2018; Matheson et al., 2019). These systems create novel possibilities for manufacturing system design by enabling shared workspaces where humans and robots collaborate as partners rather than operating in isolation.

Collaborative robots demonstrate particular effectiveness in applications requiring adaptive responses to varying production requirements, enabling manufacturers to achieve flexibility whilst maintaining quality standards (Hentout et al., 2019; Liu et al., 2024). However, cobot implementation demands significant capital investment, technical expertise development, and comprehensive safety considerations that create substantial barriers for many manufacturing organisations (Vysocky & Novak, 2016; Faccio & Cohen, 2024). The successful integration of collaborative robotics requires careful attention to human factors, including operator training, workplace redesign, and cultural adaptation to human-machine collaboration paradigms.

Digital transformation journeys within manufacturing typically follow established maturity models that describe progressive stages of digital capability development. Santos and Martinho (2020) propose an Industry 4.0 maturity framework encompassing initial, repeatable, defined, managed, and optimised stages of digital transformation. Each developmental stage requires specific capabilities, investment priorities, and organisational modifications that enable progression toward fully integrated digital manufacturing ecosystems.

These maturity frameworks demonstrate that successful digital transformation represents a systematic process rather than discrete technological implementations. Organisations must develop capabilities incrementally, building upon foundational competencies whilst managing the complexity that accompanies technological integration (Momeni et al., 2024). The progression through maturity stages requires careful balance between technological advancement and organisational capacity development, ensuring that human resources can effectively utilise advancing technological capabilities.

Cybersecurity threats constitute perhaps the most acute VUCA dimension in contemporary manufacturing, transforming from a technical concern to an existential business risk. The Colonial Pipeline ransomware attack in May 2021 demonstrates this transformation: a single compromised password enabled attackers to disrupt fuel supplies across the eastern United States, and the company paid a US$4.4 million ransom, part of which was later recovered by U.S. authorities (U.S. Department of Justice, 2021; Axios, 2021). Manufacturing faces disproportionate cyber risk—ransomware incidents against industrial organisations increased by around 50% in 2023 compared with 2022, and the mean recovery cost (excluding any ransom) reached approximately US$2.73 million per incident (Axios, 2024; Sophos, 2024). The volatility dimension manifests through attack frequency and method evolution: manufacturing organisations must defend against threats that change tactics monthly, from ransomware to supply chain compromise to operational technology sabotage (Dragos, 2024). Uncertainty pervades attribution and motive—was the January 2024 attack on a German automotive supplier criminal extortion, state-sponsored espionage, or competitor intelligence gathering? Investigation required 11 weeks without conclusive attribution (Bundesamt für Sicherheit in der Informationstechnik, 2024). Complexity emerges from manufacturing's convergence of information technology and operational technology: the same network integration that enables predictive maintenance creates pathways for attackers to compromise physical production processes, not merely data systems (Miller & Rowe, 2012). Ambiguity characterises security investment decisions when calculating return on preventing events that may never occur, particularly when security measures may reduce operational efficiency or slow innovation cycles (Gordon & Loeb, 2002).

Manufacturing leaders must recognise that cybersecurity strategy determines operational resilience as much as supply chain diversification or technology selection. Norsk Hydro's response to the March 2019 LockerGoga ransomware attack illustrates effective navigation of cyber-induced VUCA conditions. Rather than pay ransom, the aluminium manufacturer managed an attack that affected around 22,000 computers by switching many sites to manual operations; a week after the incident most operations were running at normal capacity, with the hardest-hit Extruded Solutions business operating at about 70–80%, and by mid-April most operations were back to normal or near-normal even though some administrative systems took longer to restore (Norsk Hydro, 2019). This decision resulted in an estimated impact of NOK 550–650 million (about US$60–70 million) in the first half of 2019 but preserved operational knowledge and demonstrated resilience (Tøndel et al., 2014; Patterson et al., 2024). The critical lesson transcends technical controls: Norsk Hydro's ability to operate semi-manually revealed that workforce cross-training and documented manual procedures—seemingly obsolete in automated environments—provided resilience that no cybersecurity technology alone could deliver. Manufacturing organisations navigating VUCA environments require integrated approaches where cybersecurity strategy, operational resilience, and workforce capability development reinforce rather than compete with one another (Leng et al., 2021).

Cybersecurity as Operational Resilience Strategy

The Norsk Hydro response demonstrates that cybersecurity resilience depends on operational capabilities, not just technical controls. Manufacturing leaders should implement three parallel strategies: (1) Maintain manual operation capability—document manual procedures for critical processes and conduct quarterly manual operation drills ensuring workforce can sustain production without digital systems; (2) Invest in workforce cross-training—develop T-shaped skills where operators possess deep expertise in primary roles plus breadth across adjacent processes, enabling flexible response during system failures; (3) Calculate ransomware response economics—pre-determine decision criteria balancing ransom payment versus manual operation costs, considering reputation impacts and operational knowledge preservation. Norsk Hydro's NOK 550–650 million impact (approximately US$63.5–75.0 million) proved less costly than compromising operational knowledge or demonstrating willingness to pay ransomware (Meland et al., 2021). Manufacturing leaders should pre-position capabilities enabling similar choices rather than making desperate decisions during active attacks

System complexity escalates as multiple digital technologies integrate across manufacturing operations, generating interdependencies that may amplify failure modes and complicate diagnostic processes (Hermann et al., 2020). This technological complexity creates situations where localised failures can cascade throughout integrated systems, potentially causing widespread operational disruptions that exceed the impact of traditional manufacturing problems (Wang et al., 2024). Managing such complexity requires sophisticated system design approaches that incorporate redundancy, modularity, and robust failure recovery mechanisms whilst maintaining operational efficiency and cost effectiveness.

Supply Chain Resilience and Risk Management

Manufacturing organisations operating within VUCA environments increasingly recognise supply chain resilience as fundamental to sustained competitive advantage. This capability encompasses preparation, response, and recovery from disruptions whilst preserving operational continuity (Tukamuhabwa et al., 2015). The COVID-19 pandemic exposed fragility in interconnected global networks, with firms losing substantial supplier relationships and experiencing significant market value erosion (Aral et al., 2020). Post-pandemic analysis reveals that resilience strategies must address both acute disruptions and chronic uncertainty, requiring fundamental reconsideration of supply chain design principles including redundancy, flexibility, and visibility (Ivanov, 2021).

Traditional risk management paradigms focusing primarily on prevention and mitigation prove insufficient. Resilience transcends these approaches by emphasising adaptive capacity and organisational learning. Wieland and Wallenburg (2013) establish that supply chain resilience emerges through relational competencies enabling collaborative problem-solving and adaptive responses to unforeseen disruptions. Manufacturing enterprises cultivate resilience through strategic diversification, redundancy implementation, flexibility enhancement, and collaborative partnerships facilitating rapid response mechanisms. Digital technologies transform supply chain resilience by providing enhanced visibility, predictive analytics, and automated response capabilities (Ivanov, 2021). Real-time monitoring through Internet of Things sensors enables continuous surveillance, whilst artificial intelligence algorithms identify patterns forecasting potential disruptions before materialisation (Negri et al., 2021). Blockchain technologies facilitate transparency and traceability, enhancing trust and enabling swift disruption source identification (Al-Talib et al., 2020). However, technological solutions introduce novel vulnerabilities requiring careful consideration. Supply chain information systems face increasing cybersecurity threats compromising operational integrity (Creazza et al., 2022), whilst system failures propagate across interconnected networks more rapidly than traditional disruptions (Tang et al., 2016). Manufacturing organisations therefore require integrated resilience strategies combining technological capabilities with organisational practices and collaborative relationships, enabling effective responses to diverse disruption scenarios.

The Resilience-Efficiency Trade-off

Manufacturing leaders face a fundamental tension: resilience requires redundancy whilst competitiveness demands efficiency. Three approaches balance this trade-off: (1) Strategic buffer placement—concentrate inventory buffers at high-variability bottlenecks rather than uniform safety stock, optimising protection per dollar invested (Simchi-Levi et al., 2014); (2) Flexible capacity over dedicated capacity—invest in manufacturing equipment capable of producing multiple product families, accepting 10-15% efficiency penalty for 40-50% capacity flexibility during disruptions; (3) Supplier relationship portfolios—maintain dual sourcing for critical inputs whilst accepting single-source relationships for standardised commodities, calibrating relationship investment to disruption consequence. The optimal resilience investment depends on disruption tolerance: industries with high switching costs (aerospace, automotive) justify greater resilience investment than industries with commodity products and flexible customers. Manufacturing leaders should explicitly calculate their disruption cost per day and size resilience investments accordingly rather than pursuing resilience as absolute goal without cost-benefit discipline.

Geopolitical Risk and Regulatory Complexity

Manufacturing enterprises increasingly confront multifaceted geopolitical environments characterised by escalating trade tensions, regulatory fragmentation, and technological competition, which collectively generate unprecedented strategic uncertainties (Roscoe et al., 2022). These geopolitical risks manifest through government interventions, international conflicts, and policy transformations that substantially affect manufacturing operations via trade restrictions, technology export controls, and market access constraints (Doherty et al., 2014).

Technological competition amongst major economies has intensified, resulting in comprehensive export controls targeting critical manufacturing technologies. Such restrictions fundamentally disrupt global manufacturing supply chains dependent upon international technology transfers (Ernst, 2021). The U.S. Government Accountability Office (2024) demonstrates that semiconductor export controls have profoundly impacted global electronics manufacturing, necessitating extensive supply chain reconfiguration and technology substitution strategies that substantially increase operational costs and complexity for affected manufacturers.

Simultaneously, regulatory complexity proliferates as manufacturing organisations navigate multiple jurisdictions with divergent regulatory frameworks, creating intricate compliance challenges requiring sophisticated management approaches (Chima et al., 2022). Environmental regulations, safety standards, data protection requirements, and trade regulations generate overlapping compliance obligations. These frequently conflicting requirements constrain operational flexibility and create substantial administrative burdens for manufacturing organisations (Khan et al., 2017).

In response, manufacturing organisations develop comprehensive geopolitical risk management strategies encompassing scenario planning, supply chain diversification, technology independence initiatives, and robust regulatory compliance systems. These approaches enable operational continuity despite geopolitical turbulence through strategic redundancy and adaptive capacity building. However, empirical evidence indicates such strategies require substantial capital investments whilst potentially reducing short-term operational efficiency, creating significant trade-offs between risk mitigation and performance optimisation (Moradlou et al., 2025). Although these strategies provide essential protection against long-term geopolitical risks, organisations must carefully balance resilience investments against immediate competitive pressures.

Key Insights for Manufacturing Leaders from this section

- **VUCA dimensions interact dynamically**: Volatility triggers uncertainty, which amplifies complexity and creates strategic ambiguity. Managing these dimensions requires integrated rather than sequential responses (Bennett & Lemoine, 2014).
- **Dynamic capabilities are essential**: Sensing, seizing, and transforming capabilities enable organisations to identify environmental changes, capture emerging opportunities, and adapt organisational systems (Teece, 2007).
- **Institutional pressures shape strategy**: Manufacturing firms must navigate coercive regulations, normative industry standards, and mimetic competitive pressures whilst maintaining operational effectiveness (DiMaggio & Powell, 1983).

- **Technology creates paradoxes**: Digital transformation simultaneously enhances capabilities and amplifies vulnerabilities. Success depends more on implementation approach than technology selection (Ghobakhloo, 2020).
- **Military frameworks require adaptation**: Whilst VUCA originated in military contexts, manufacturing applications demand substantial modification to address systematic process improvement, quality control, and long-term capital requirements (Liker, 2004).

STRATEGIC FRAMEWORKS FOR VUCA NAVIGATION

Integrated VUCA Navigation Framework for Manufacturing

Building upon established theoretical foundations, this analysis presents an integrated framework enabling manufacturing organisations to navigate VUCA environments effectively through strategic technology integration and institutional adaptation. The framework synthesises dynamic capabilities theory, institutional theory, and strategic management perspectives, providing practical guidance for manufacturing leaders confronting environmental turbulence (Carayannis et al., 2025).

The integrated VUCA navigation framework shares conceptual similarities with Boyd's OODA loop—Observe, Orient, Decide, Act (Boyd, 1976, 1987, 1995)—originally developed for military tactical decision-making and subsequently adapted for business strategy (Richards, 2004). Whilst both frameworks emphasise iterative cycles of environmental scanning and adaptive response, the manufacturing VUCA framework extends beyond tactical decision-making to encompass strategic capability development, institutional adaptation, and technology integration dimensions (Osinga, 2007). This manufacturing-specific framework acknowledges military antecedents whilst addressing the distinctive requirements of manufacturing environments, including systematic process improvement, statistical control, and long-term capital investment horizons that differentiate industrial operations from military contexts (Womack et al., 1990).

This integrated approach encompasses four interconnected dimensions: environmental sensing and intelligence gathering, strategic option development and selection, implementation orchestration and coordination, and adaptive learning and capability evolution. Each dimension addresses distinct aspects of VUCA navigation whilst requiring cross-dimensional integration to achieve sustained competitive advantage in turbulent environments (Quotb, 2024).

Environmental sensing and intelligence gathering constitute the foundational element of effective VUCA navigation, enabling manufacturing organisations to develop comprehensive awareness of emerging opportunities and threats across technological, market, regulatory, and geopolitical domains (YahiaMarzouk & Jin, 2022).

Military strategic concepts provide valuable frameworks for manufacturing VUCA navigation. Military environmental sensing emphasises continuous scanning, threat assessment, and strategic early warning—capabilities directly applicable to manufacturing through market intelligence, competitive analysis, and supply chain monitoring (Yarger, 2006; Whiteman, 1998). The military "OODA loop" concept (Observe, Orient, Decide, Act) offers a particularly relevant framework for manufacturing VUCA responses, enabling rapid decision-making cycles that maintain competitive advantage under uncertainty (Richards, 2004).

Manufacturing organisations can adapt military sensing disciplines through systematic competitor monitoring, supplier risk assessment, and technology scouting that mirror military intelligence gathering whilst focusing on commercial outcomes (Wright et al., 2002; Gilad, 2004; Prescott, 1999). This competitive intelligence approach, with its historical roots in military command and control systems, enables manufacturers to anticipate market shifts, control risk, and create powerful strategic responses through structured information collection and analysis. This approach emphasises proactive environmental engagement rather than reactive crisis management, aligning with manufacturing requirements for anticipatory planning and risk mitigation.

Contemporary manufacturing firms establish sensing capabilities through systematic environmental scanning, stakeholder engagement, expert networks, and analytical systems providing early warning of environmental changes (Teece, 2007; Schoemaker et al., 2018). Effective sensing requires integration across multiple organisational functions and external partnerships to capture diverse environmental perspectives. Technical sensing focuses on emerging technologies and production innovations. Market sensing encompasses customer evolution, competitor strategies, and industry dynamics. Regulatory sensing monitors compliance requirements and policy developments. Geopolitical sensing tracks international relations and trade policies affecting global manufacturing supply chains.

Strategic option development enables manufacturing organisations to create and evaluate alternative responses to identified environmental changes through systematic analysis of strategic possibilities and their implications (Hussain et al., 2017). Manufacturing firms develop strategic options through scenario planning, technology roadmapping, capability assessment, and stakeholder analysis that illuminate potential pathways for addressing VUCA conditions (McGrath & MacMillan, 2000; Roos et al., 2025).

Option development demands both creativity and analytical rigour to generate diverse alternatives addressing different environmental scenarios whilst remaining feasible within organisational constraints (Saritas & Aylen, 2010). Manufacturing organisations create options for technology integration, market positioning, partnership formation, and capability development providing flexibility for environmental uncertainty responses. Option evaluation employs decision analysis techniques, financial modelling, and risk assessment to compare alternatives systematically whilst considering implementation requirements and potential outcomes.

Implementation orchestration encompasses systematic deployment of selected strategic options through project management, change management, and performance monitoring ensuring effective execution whilst maintaining operational continuity (Brodeur et al., 2023). Manufacturing organisations require sophisticated implementation capabilities to manage complex technology integration projects whilst maintaining production schedules and quality standards (Wagire et al., 2021; Mick et al., 2024).

Implementation success depends upon careful orchestration of technological, organisational, and human resource changes that complement each other whilst avoiding critical operational disruption (Lorenzo Ochoa et al., 2017). Manufacturing firms develop implementation capabilities through project management excellence, change management competencies, and stakeholder engagement enabling smooth transitions from current to future states. Implementation monitoring employs performance metrics, milestone tracking, and feedback systems enabling course correction whilst maintaining momentum toward strategic objectives.

Adaptive learning ensures manufacturing organisations continuously enhance VUCA navigation capabilities through systematic reflection, knowledge capture, and capability development improving future performance (Pradhan et al., 2017). Manufacturing firms establish learning mechanisms through post-implementation reviews, best practice identification, and knowledge management systems enabling organisational improvement (Helfat & Peteraf, 2003; Beliaeva et al., 2020).

Learning processes must balance exploitation of existing knowledge with exploration of new possibilities to maintain adaptive capacity whilst building upon accumulated experience (Appelbaum & Goransson, 1997). Manufacturing organisations develop learning capabilities through experimentation, collaboration, and reflection enhancing understanding of effective VUCA navigation approaches (Doskenov et al., 2025). Capability evolution requires systematic investment in skill development, technology advancement, and organisational improvement maintaining competitive advantage in changing environments.

Strategic Technology Integration Framework for Manufacturing

The strategic integration of artificial intelligence, automation, and robotics within manufacturing operations demands systematic approaches that maximise value creation whilst managing implementation risks and organisational disruption.

Manufacturing organisations face complex challenges when navigating digital transformation, requiring structured methodologies that ensure successful technology adoption whilst maintaining operational excellence and competitive advantage.

This framework presents a comprehensive five-phase approach to technology integration: strategic assessment and readiness evaluation, technology selection and architecture design, pilot implementation and validation, scaled deployment and integration, and continuous optimisation and evolution (Abdallah, Shehab & Al-Ashaab, 2021). Each phase demands specific capabilities, resources, and management commitment to ensure effective progression through the digital transformation journey.

Strategic assessment and readiness evaluation establish the foundation for successful technology integration initiatives. Manufacturing organisations must comprehensively understand their current capabilities, identify improvement opportunities, and evaluate their preparedness for digital transformation endeavours. This preliminary phase involves conducting thorough assessments of existing technological infrastructure, organisational capabilities, financial resources, and strategic objectives to establish baseline conditions and define transformation goals (Santos & Martinho, 2020; Vance et al., 2023).

Readiness evaluation encompasses multiple dimensions critical to implementation success. Technical readiness requires systematic examination of infrastructure capabilities, data quality standards, and systems integration requirements that determine the feasibility of proposed technology initiatives (Saad et al., 2021). Organisational readiness involves evaluating leadership commitment, change management capabilities, and cultural factors that influence technology adoption outcomes (Machado, Winroth & Almström, 2021). Financial readiness encompasses investment capacity assessment, return on investment projections, and funding mechanism identification to enable sustained technology integration efforts.

The selection of appropriate technologies and design of integrated architectures represent critical determinants of implementation success. Manufacturing firms must systematically evaluate technology alternatives through comprehensive functionality assessment, vendor analysis, total cost of ownership calculations, and strategic fit evaluation to guide selection decisions (Çınar et al., 2021; Spaltini et al., 2022).

Architecture design demands systematic planning across manufacturing systems, data flows, and organisational processes to ensure seamless operation and future scalability. Organisations develop robust technology architectures through detailed systems analysis, interface design, and integration planning that facilitate effective deployment whilst minimising disruption to existing operations. Critical considerations include cybersecurity requirements, data governance frameworks, and compliance obligations that fundamentally shape technology implementation approaches (Toussaint et al., 2024).

Pilot programmes provide essential opportunities for manufacturing organisations to test technology solutions within controlled environments before committing to full-scale deployment. Firms establish targeted pilot initiatives through limited scope implementation, rigorous performance measurement, and systematic learning capture that demonstrate technology value whilst identifying potential implementation challenges (Butt, 2020; Kırmızı & Kocaoglu, 2022).

Effective validation requires systematic performance measurement, comprehensive stakeholder feedback collection, and structured lessons learned documentation to inform scaled deployment decisions. Manufacturing organisations develop robust validation criteria through key performance indicator definition, success metrics establishment, and comprehensive evaluation frameworks that enable objective assessment of pilot outcomes. Validation processes must consider technical performance, organisational impact, and strategic value creation to ensure thorough evaluation of technology integration results.

Systematic rollout of validated technology solutions across manufacturing operations requires careful coordination to maintain operational continuity and performance standards. Manufacturing firms develop comprehensive deployment strategies incorporating phased implementation approaches, structured change management processes, and extensive training programmes that enable smooth transition to technology-enabled operations (Wagire et al., 2021; Mick et al., 2024).

Successful deployment demands careful coordination of technical implementation, organisational change initiatives, and continuous performance monitoring to ensure technology integration achieves intended benefits whilst minimising operational disruption. Manufacturing organisations must establish deployment capabilities through project management excellence, technical expertise development, and comprehensive stakeholder engagement that enable effective technology integration across complex manufacturing systems.

The Pilot-to-Scale Transition Gap

Most manufacturing digital transformations fail not in pilots but in scaling—pilot success rates exceed 70% whilst full deployment success rates remain below 30% (Joglekar et al., 2024). This "pilot-to-scale gap" stems from three factors: (1) Pilot enthusiasm versus organisational resistance—volunteer pilot participants possess motivation that broader workforce lacks, requiring different change management approaches for scaling; (2) Controlled pilot environments versus operational complexity—pilots in clean-sheet facilities avoid legacy system integration challenges that plague full deployment; (3) Pilot budgets versus scaling economics—pilots receive disproportionate resources and attention that cannot sustain across enterprise deployment. Manufacturing leaders should structure pilots as realistic scaling tests rather than proof-of-concept demonstrations: use representative facilities facing typical operational constraints, involve reluctant participants alongside enthusiasts, and validate economics at scale rather than subsidised pilot conditions. The goal is proving scalability rather than technical feasibility—technical feasibility is rarely the binding constraint in digital transformation failure.

Sustainable value creation from technology investments requires ongoing improvement, capability development, and technology advancement that maintain competitive advantage in dynamic environments. Manufacturing firms establish systematic optimisation processes through continuous performance monitoring, improvement opportunity identification, and enhancement implementation that enable sustained value creation from technology integration initiatives.

Effective optimisation demands systematic analysis of technology performance, comprehensive user feedback collection, and structured improvement opportunity assessment to guide enhancement investments and capability development. Manufacturing organisations develop robust optimisation capabilities through advanced data analytics, sustained user engagement, and continuous improvement methodologies that ensure technology investments continue delivering value whilst adapting to evolving operational requirements and competitive conditions.

This framework provides manufacturing organisations with a structured approach to technology integration that addresses the complexity and challenges inherent in digital transformation initiatives. Success requires systematic progression through each phase, with careful attention to technical, organisational, and strategic considerations that influence implementation outcomes. The framework emphasises the importance of continuous learning, adaptation, and improvement throughout the technology integration journey.

Institutional Adaptation Framework for Manufacturing Organisations

Manufacturing organisations confronting volatile, uncertain, complex, and ambiguous (VUCA) environments must cultivate sophisticated approaches to managing multiple, often conflicting institutional pressures whilst preserving operational effectiveness and competitive positioning. The complexity of contemporary institutional environments demands strategic frameworks that enable firms to navigate regulatory demands whilst transforming institutional challenges into sources of competitive advantage.

This framework presents a comprehensive four-dimensional approach to institutional adaptation: institutional mapping and analysis, compliance strategy development, stakeholder engagement and relationship management, and institutional entrepreneurship and change leadership. Each dimension addresses distinct aspects of institutional navigation whilst requiring integration to achieve sustained legitimacy and competitive advantage (DiMaggio & Powell, 1983; Oliver, 1991).

Institutional mapping and analysis provide manufacturing organisations with essential understanding of the complex web of institutional pressures, requirements, and opportunities that fundamentally shape their operating environments. Manufacturing firms must conduct systematic analysis of regulatory frameworks, industry standards, professional requirements, and stakeholder expectations that constitute their institutional landscape (Scott, 2008; Greenwood et al., 2011).

Effective institutional mapping requires identification of coercive pressures emanating from regulatory bodies, normative pressures from professional associations, and mimetic pressures from industry peers that influence organisational behaviour and strategic choices (Ebrahimi & Koh, 2021). Manufacturing organisations develop analytical capabilities through stakeholder analysis, regulatory monitoring, and industry benchmarking that illuminate institutional forces shaping competitive dynamics and strategic options.

The analysis must acknowledge the dynamic nature of institutional environments, particularly within VUCA contexts where regulatory frameworks, industry standards, and stakeholder expectations evolve rapidly in response to technological, social, and political transformations. Manufacturing firms develop robust analytical capabilities through environmental scanning, expert consultation, and scenario planning that enable anticipation of institutional changes and proactive adaptation strategies (Balzano et al., 2024).

Compliance strategy development enables manufacturing organisations to meet institutional requirements efficiently whilst minimising operational disruption and competitive disadvantage. Manufacturing firms construct compliance strategies through requirement analysis, cost-benefit evaluation, and implementation planning that balance compliance obligations with operational efficiency and strategic objectives (Khan et al., 2017; Kauppi & Hannibal, 2017).

Superior compliance strategies transcend minimal compliance requirements to identify opportunities for competitive advantage through exceptional compliance performance or innovative approaches that exceed requirements whilst creating operational benefits. Manufacturing organisations develop compliance capabilities through legal expertise acquisition, regulatory intelligence systems, and implementation excellence that enable efficient compliance whilst maintaining operational effectiveness (Étienne, 2010; Fiene, 2023).

Compliance strategy must consider the cumulative burden of multiple regulatory frameworks and their interactions to avoid conflicts and inefficiencies arising from overlapping or contradictory requirements. Manufacturing firms develop integrated compliance approaches through systems thinking, process optimisation, and technology deployment that streamline compliance activities whilst maintaining comprehensive coverage of institutional requirements.

Stakeholder engagement and relationship management enable manufacturing organisations to build productive relationships with institutional actors whilst influencing institutional development in directions that support organisational objectives. Manufacturing firms engage with regulatory bodies, industry associations, professional networks, and advocacy groups through participation, consultation, and collaboration that enhance understanding whilst building influence (Bhakoo & Choi, 2013; Liu et al., 2013).

Effective stakeholder engagement requires comprehensive understanding of stakeholder interests, influence patterns, and decision-making processes that enable strategic relationship building and influence development (Leonidou et al., 2020). Manufacturing organisations develop engagement capabilities through stakeholder analysis, communication excellence, and relationship management that enable productive interactions with diverse institutional actors.

Relationship management must carefully balance cooperation and competition considerations, particularly when engaging with industry peers through trade associations and standard-setting bodies. Manufacturing firms develop relationship strategies through strategic analysis, collaboration frameworks, and boundary management that enable beneficial engagement whilst protecting competitive advantages and proprietary information (Cohen et al., 2024).

Institutional entrepreneurship and change leadership enable manufacturing organisations to influence institutional development whilst positioning themselves advantageously within evolving institutional environments. Manufacturing firms engage in institutional entrepreneurship through advocacy, innovation, and demonstration that shape institutional evolution whilst creating competitive advantages (Battilana et al., 2009; Hardy & Maguire, 2008).

Institutional entrepreneurship demands vision, resources, and influence to pursue institutional change initiatives that benefit both individual organisations and broader industry development. Manufacturing organisations develop entrepreneurship capabilities through strategic vision articulation, coalition building, and change leadership that enable effective influence within institutional environments whilst maintaining legitimacy and stakeholder support (Battilana et al., 2009).

Change leadership must consider the long-term implications of institutional entrepreneurship initiatives, particularly their effects on competitive dynamics and industry structure. Manufacturing firms develop change strategies through systems thinking, stakeholder analysis, and impact assessment that enable beneficial institutional influence whilst avoiding unintended consequences that may harm organisational interests or industry development.

The successful implementation of this institutional adaptation framework requires sophisticated integration across all four dimensions. Manufacturing organisations must develop capabilities that enable simultaneous execution of mapping activities, compliance management, stakeholder engagement, and entrepreneurship initiatives. This integration demands organisational structures that support cross-functional coordination and strategic alignment.

Implementation success depends upon developing institutional intelligence systems that provide real-time monitoring of regulatory changes, stakeholder dynamics, and competitive responses. Manufacturing firms must establish governance structures that enable rapid response to institutional changes whilst maintaining strategic coherence across adaptation initiatives.

This institutional adaptation framework provides manufacturing organisations with a comprehensive approach to managing complex institutional environments whilst creating competitive advantage. Success requires systematic development of capabilities across all four dimensions, with particular attention to integration and strategic alignment. The framework emphasises the importance of viewing institutional pressures not merely as constraints but as opportunities for strategic differentiation and competitive positioning.

Manufacturing organisations that successfully implement this framework demonstrate superior ability to navigate regulatory complexity, build influential stakeholder relationships, and shape institutional development in ways that support their strategic objectives whilst contributing to industry advancement.

Regulatory Excellence as Market Entry Barrier

Sophisticated manufacturers transform regulatory compliance into competitive moats. Three strategies prove particularly effective: (1) Proactive standard-setting participation—engage with ISO technical committees and industry consortia during standard development rather than reactive compliance after publication, shaping requirements toward organisational capabilities (Casper & Hancké, 1999); (2) Exceed compliance thresholds—achieve certifications beyond customer requirements (e.g., ISO 14001 when only ISO 9001 required), differentiating in procurement processes where buyers increasingly evaluate supplier risk profiles; (3) Develop compliance expertise as service—large manufacturers with sophisticated compliance capabilities offer compliance consulting to smaller suppliers, simultaneously strengthening supply chain compliance whilst generating revenue streams. BMW's supplier development programme exemplifies this approach—extending quality and environmental management expertise to suppliers creates more reliable supply base whilst establishing BMW as preferred customer for high-performing suppliers (Wagner, 2006; Krause et al., 2007; Modi & Mabert, 2007). Regulatory excellence creates sustainable competitive advantage precisely because it requires systematic long-term capability investment that competitors cannot rapidly imitate.

Risk Management and Resilience Framework

Manufacturing enterprises today confront an unprecedented confluence of risks emanating from technological complexity, supply chain vulnerabilities, geopolitical turbulence, and regulatory uncertainty. These challenges necessitate sophisticated risk management approaches that transcend conventional assessment and mitigation methodologies (Bennett & Lemoine, 2014). The framework presented herein enables manufacturing firms to anticipate, respond to, and recover from diverse disruption scenarios whilst preserving operational continuity and competitive advantage.

The comprehensive risk management and resilience framework comprises four synergistic elements: systematic risk identification and assessment, adaptive mitigation and contingency planning, resilience capability development and maintenance, and continuous learning from risk events. These interconnected components address distinct dimensions of risk management whilst requiring integration to achieve robust organisational resilience.

Effective risk identification and assessment furnish manufacturing organisations with comprehensive understanding of their risk landscape whilst enabling resource prioritisation according to significance and probability. Manufacturing enterprises conduct systematic risk evaluation through hazard identification, vulnerability analysis, and impact assessment that illuminate potential threats to operational continuity and competitive positioning (Tukamuhabwa et al., 2015; Negri et al., 2021).

Contemporary risk identification must encompass multiple dimensions. Operational risks arise from equipment failures, supply chain disruptions, and quality deficiencies. Strategic risks emerge from competitive pressures, market shifts, and technological obsolescence. Financial exposures include currency volatility, credit risks, and liquidity constraints. Regulatory risks stem from compliance failures, policy modifications, and enforcement actions. Systemic risks originate from natural disasters, geopolitical developments, and economic disruption.

Assessment requires systematic evaluation of risk probability, potential impact, and interdependency effects that may amplify individual risks through cascading failures. Manufacturing organisations develop assessment capabilities through risk modelling, scenario analysis, and expert consultation that facilitate understanding of risk profiles and their implications for organisational performance (Fridgeirsson et al., 2021).

Risk assessment must acknowledge the dynamic nature of contemporary risk environments, particularly in volatile, uncertain, complex, and ambiguous (VUCA) contexts where risk profiles evolve rapidly due to technological change, geopolitical developments, and regulatory shifts (Bennett & Lemoine, 2014). Manufacturing firms establish dynamic assessment processes through continuous monitoring, regular review, and adaptive analysis that maintain current understanding of evolving risk conditions.

Adaptive risk mitigation and contingency planning enable manufacturing organisations to reduce risk exposure whilst preparing alternative responses when mitigation proves insufficient. Manufacturing firms develop mitigation strategies through risk reduction, risk transfer, risk acceptance, and risk avoidance that balance cost-effectiveness with risk reduction objectives (Wieland & Wallenburg, 2013; Balugani et al., 2020; NIST, 2024).

Mitigation strategies must consider the interconnected nature of manufacturing systems, where mitigation measures for one risk may create new vulnerabilities or amplify other risks. Manufacturing organisations develop integrated approaches through systems analysis, trade-off evaluation, and holistic planning that optimise overall risk profiles rather than addressing individual risks in isolation.

Contingency planning requires development of alternative response scenarios that enable rapid and effective action when risk events occur despite mitigation efforts. Manufacturing firms create contingency plans through scenario development, resource allocation, and response protocols that enable coordinated action during crisis situations whilst maintaining stakeholder communication and operational coordination.

Effective contingency planning must consider resource requirements, decision-making authority, and communication protocols that enable response implementation under stress conditions. Manufacturing organisations develop contingency capabilities through training, simulation, and exercise programmes that test response plans whilst building organisational confidence and competence in crisis management.

Resilience capability development encompasses the systematic building of organisational capacity to absorb disruption, adapt to changed conditions, and transform in response to fundamental environmental shifts. Manufacturing firms develop resilience capabilities through redundancy creation, flexibility enhancement, and adaptive capacity building that enable continued operation despite significant disruption (Molenda et al., 2023; Pertheban et al., 2023).

Resilience capabilities encompass several dimensions. Operational redundancy includes backup systems, alternative suppliers, and flexible production capacity. Financial resilience involves cash reserves, credit facilities, and insurance coverage. Technological resilience comprises system robustness, cybersecurity measures, and technology diversification. Organisational resilience encompasses leadership development, cultural strength, and learning capability (Gu et al., 2015).

Resilience development requires careful balance between efficiency and resilience considerations, as excessive redundancy may reduce operational efficiency whilst insufficient resilience creates vulnerability to disruption (Kristianto et al., 2014). Manufacturing organisations develop balanced approaches through trade-off analysis, optimisation modelling, and stakeholder consultation that achieve appropriate resilience levels whilst maintaining competitive efficiency.

Maintenance of resilience capabilities requires ongoing investment, testing, and renewal to ensure continued effectiveness as risk environments and organisational conditions evolve. Manufacturing firms establish maintenance processes through regular assessment, capability testing, and improvement implementation that sustain resilience capabilities whilst adapting to changing requirements.

Continuous learning and improvement from risk events and near-misses enable manufacturing organisations to enhance risk management and resilience capabilities through systematic reflection, knowledge capture, and capability development. Manufacturing firms establish learning processes through post-incident analysis, best practice identification, and knowledge sharing that improve future risk management performance whilst building organisational wisdom about effective risk navigation (Iborra et al., 2020; Zavala-Alcívar et al., 2020).

Learning requires systematic documentation of risk events, response effectiveness, and improvement opportunities that enable evidence-based enhancement of risk management and resilience capabilities. Manufacturing organisations develop learning capabilities through incident analysis, knowledge management, and improvement implementation that translate experience into enhanced organisational capability.

Learning processes must balance learning from negative events with proactive capability development that anticipates future challenges rather than simply responding to past experiences. Manufacturing firms develop anticipatory learning through scenario planning, simulation exercises, and external learning that enhance risk management and resilience capabilities beyond the lessons available from direct organisational experience.

Cross-Sector Analysis of Institutional Adaptation Strategies

The examination of institutional adaptation strategies across diverse manufacturing sectors reveals distinctive patterns in organisational navigation of complex regulatory environments whilst leveraging institutional changes as sources of competitive advantage. This cross-sectoral analysis investigates adaptation strategies within automotive, electronics, pharmaceuticals, chemicals, and food processing sectors to identify sector-specific approaches and universal principles for effective institutional navigation.

The automotive sector demonstrates sophisticated approaches to managing environmental regulations whilst maintaining operational efficiency and competitive positioning (Guo et al., 2025). Automotive manufacturers develop integrated strategies addressing emissions standards, safety requirements, and recycling obligations through coordinated technology development and supply chain management. Leading automotive firms establish dedicated regulatory intelligence capabilities that monitor policy developments across major markets whilst developing technology roadmaps anticipating future regulatory requirements.

Automotive sector institutional adaptation encompasses proactive engagement with regulatory bodies through industry associations, collaborative research programmes, and policy consultation processes that influence regulatory development whilst building industry expertise (Funk & Hirschman, 2017). Manufacturers develop compliance strategies integrating environmental performance with operational efficiency through technology selection, supplier development, and production process optimisation that achieve superior environmental performance whilst maintaining cost competitiveness.

Successful automotive adaptation requires balance between global standardisation and local compliance adaptation, as manufacturers operate across jurisdictions with different regulatory requirements and enforcement approaches (Del Río-Casasola et al., 2024). Leading firms develop modular compliance strategies enabling local adaptation whilst maintaining global platform efficiency through standardised components and processes that accommodate regulatory variation without compromising operational effectiveness.

The electronics sector confronts particularly complex institutional environments characterised by rapid technology evolution, environmental regulations, and trade policy uncertainty that require agile adaptation strategies (Zainol et al., 2019). Electronics manufacturers develop dynamic compliance capabilities enabling rapid response to regulatory changes whilst maintaining technology development momentum and market responsiveness.

Electronics sector adaptation encompasses supply chain compliance management through vendor certification, material tracking, and environmental impact assessment ensuring regulatory compliance whilst maintaining supply chain efficiency and cost competitiveness. Manufacturers develop particular expertise in managing conflict minerals regulations, electronic waste requirements, and chemical restriction directives affecting material selection and supply chain configuration (Coglianese & Nash, 2021).

Technology export controls create additional complexity for electronics manufacturers, requiring sophisticated trade compliance capabilities balancing technology protection with market access requirements. Leading firms develop export control strategies enabling continued technology development whilst complying with export restrictions through technology classification, market segmentation, and compliance verification that maintain competitive positioning whilst avoiding regulatory violations.

The pharmaceutical sector operates within highly regulated environments requiring sophisticated compliance capabilities whilst maintaining innovation momentum and market responsiveness (Kumar & Yadav, 2024). Pharmaceutical manufacturers develop regulatory strategies integrating product development, manufacturing operations, and market access through coordinated regulatory planning and compliance management.

Pharmaceutical adaptation requires particular attention to quality management systems integrating regulatory compliance with operational efficiency through validated processes, documentation, and continuous monitoring ensuring product quality whilst meeting regulatory requirements. Manufacturers develop expertise in managing multiple regulatory jurisdictions with different approval processes and quality standards requiring sophisticated project management and regulatory coordination.

Global pharmaceutical manufacturers create competitive advantage through regulatory excellence enabling faster market access and superior compliance performance whilst maintaining product quality and operational efficiency. Leading firms invest significantly in regulatory capabilities, quality systems, and compliance infrastructure enabling effective navigation of complex regulatory environments whilst supporting business growth and innovation objectives. These firms develop food safety cultures integrating compliance requirements with operational excellence through employee engagement, continuous improvement, and stakeholder collaboration.

Specific manufacturing sector applications demonstrate how military VUCA principles translate into operational competitive advantages through systematic implementation. In automotive manufacturing, BMW's adoption of military-style "mission command" principles in their Industry 4.0 implementation has enabled enhanced response to production disruptions whilst maintaining quality standards, demonstrating how decentralised decision-making enhances manufacturing agility through their iFactory approach and digital twin technology (BMW Group, 2023). Pharmaceutical manufacturers have implemented military-derived scenario planning methodologies for regulatory compliance, with firms reducing regulatory approval times through systematic contingency preparation mirroring military operational planning approaches.

Electronics manufacturers have adapted military "red team" adversarial analysis to identify cybersecurity vulnerabilities in connected manufacturing systems, preventing potential disruptions through proactive threat assessment methodologies. These implementations reveal that military VUCA frameworks provide manufacturing organisations with systematic approaches to environmental uncertainty complementing traditional operational excellence methodologies whilst addressing challenges that lean manufacturing and Six Sigma approaches do not adequately address (Majumdar & Marcus, 2001).

The cross-sectoral analysis demonstrates that whilst sector-specific regulatory challenges require tailored adaptation strategies, successful institutional navigation shares common characteristics across manufacturing industries. Leading organisations develop proactive regulatory intelligence capabilities, integrate compliance requirements with operational excellence, and transform regulatory challenges into competitive advantages through systematic adaptation approaches. The application of military VUCA frameworks to manufacturing contexts provides structured methodologies for managing environmental uncertainty, enabling organisations to develop resilient adaptation capabilities addressing contemporary manufacturing challenges.

Critical Success Factors Identified in this section

- **Integrated navigation demands four capabilities**: Environmental sensing identifies changes; option development creates alternatives; implementation orchestration ensures execution; adaptive learning enables continuous improvement (Teece et al., 1997).
- **Technology integration requires systematic approaches**: Five-phase deployment—assessment, selection, pilot, scaling, optimisation—reduces implementation risks whilst building organisational capabilities (Wagire et al., 2021).
- **Institutional adaptation creates advantage**: Proactive regulatory engagement, compliance excellence, and stakeholder relationship management transform regulatory constraints into competitive differentiation (Oliver, 1991).
- **Resilience balances efficiency and redundancy**: Supply chain diversification, flexible production capacity, and financial reserves enable disruption recovery without excessive cost burdens (Tukamuhabwa et al., 2015).
- **Cross-sector learning reveals patterns**: Automotive, electronics, pharmaceutical, and chemical sectors demonstrate that regulatory excellence and stakeholder engagement consistently differentiate high-performing manufacturers (Khan et al., 2017).

IMPLEMENTATION GUIDANCE AND STRATEGIC RECOMMENDATIONS

Strategic Roadmap for VUCA Navigation in Manufacturing Organisations

Manufacturing enterprises operating within volatile, uncertain, complex, and ambiguous (VUCA) environments require systematic methodologies that integrate strategic planning with capability development whilst maintaining operational excellence (Bennett & Lemoine, 2014; Cascio, 2020). The contemporary manufacturing landscape, characterised by rapid technological advancement and market volatility, demands adaptive frameworks that enable organisations to navigate uncertainty effectively (Atanassova et al., 2025). This section presents a comprehensive strategic roadmap designed to guide manufacturing leaders through the intricate process of VUCA navigation whilst addressing sector-specific challenges and opportunities.

The strategic roadmap comprises six sequential yet interconnected phases: strategic assessment and vision development, capability gap analysis and investment planning, pilot implementation and validation, scaled deployment and integration, performance monitoring and optimisation, and continuous adaptation and evolution. Each phase incorporates specific competencies, resource allocations, and managerial attention whilst building systematically upon previous stages to achieve sustained competitive advantage in VUCA environments (Kale et al., 2019; Van Erp et al., 2024).

Manufacturing organisations navigating VUCA environments require a systematic six-phase approach to achieve competitive advantage since research demonstrates that structured approaches to VUCA navigation yield superior performance outcomes compared to ad-hoc responses (Atanassova et al., 2025). This because systematic approaches provide necessary scaffolding for managing complexity whilst maintaining operational continuity.

Phase One: Strategic Assessment and Vision Development.

Strategic assessment and vision development establish the foundational elements for effective VUCA navigation, enabling manufacturing organisations to comprehend their current competitive position whilst formulating coherent visions for future market positioning (Callison, n.d.; Smartsheet, 2023). Manufacturing firms execute strategic assessments through environmental scanning, competitive benchmarking, internal capability evaluation, and stakeholder consultation processes that illuminate strategic opportunities and constraints whilst identifying priority areas for capability development (Hermundsdottir & Aspelund, 2022).

The assessment process requires systematic evaluation of both internal capabilities and external market conditions. Environmental analysis encompasses technological trends, regulatory changes, competitive dynamics, and supply chain vulnerabilities that may impact strategic positioning (FasterCapital, 2023). Internal capability evaluation utilises competency mapping techniques to identify organisational strengths, weaknesses, and resource constraints that influence strategic choices.

Vision development necessitates integration of strategic assessment insights with organisational values, stakeholder expectations, and competitive ambitions to create compelling direction for transformation initiatives. Manufacturing organisations develop strategic vision through leadership alignment processes, stakeholder engagement strategies, and scenario planning methodologies that address multiple potential future conditions whilst providing clear direction for strategic initiatives and investment priorities (Doz & Kosonen, 2010).

Effective vision development balances aspirational objectives with realistic assessment of organisational capabilities and environmental constraints. This approach creates achievable yet challenging targets that motivate transformation whilst maintaining credibility with stakeholders and employees. Manufacturing firms establish vision through systematic processes that engage multiple perspectives whilst maintaining strategic focus and organisational commitment.

Phase Two: Capability Gap Analysis and Investment Planning.

Capability gap analysis enables manufacturing organisations to identify specific competency requirements for achieving strategic vision whilst developing systematic plans for capability development and resource allocation (Institute for Manufacturing, Cambridge). Manufacturing firms conduct comprehensive capability assessments through competency mapping, performance benchmarking, and expert evaluation processes that illuminate current capabilities whilst identifying gaps constraining strategic achievement.

The gap analysis process involves several critical components. First, organisations must define their desired future state capabilities based on strategic vision and market requirements. Second, current state assessment maps existing capabilities, resources, and competencies across the organisation. Third, gap identification compares current and desired states to reveal specific deficiencies requiring attention (Callison, n.d.).

Investment planning requires systematic evaluation of capability development options through cost-benefit analysis, risk assessment, and strategic impact evaluation. This analysis guides resource allocation decisions whilst considering implementation requirements and organisational constraints (Ghobakhloo, 2018). Manufacturing organisations develop investment strategies through portfolio optimisation, timing analysis, and funding assessment approaches that enable effective capability development whilst maintaining financial sustainability and operational continuity.

Systematic capability gap analysis improves investment decision-making effectiveness in manufacturing organisations since research indicates that organisations employing structured gap analysis achieve superior resource allocation outcomes compared to intuitive approaches (Azadeh et al., 2008). This because structured analysis provides objective foundation for comparing alternative investments but effectiveness depends upon accurate data collection and stakeholder engagement.

Investment planning must acknowledge the interdependent nature of capabilities and cumulative effects of capability investments to ensure coordinated development. This approach achieves strategic objectives whilst avoiding sub-optimisation or capability conflicts. Manufacturing firms develop integrated investment methodologies through systems thinking, sequencing analysis, and coordination planning that optimise overall capability development rather than pursuing individual capabilities in isolation (Ostadi et al., 2024).

Phase Three: Pilot Implementation and Validation.

Pilot implementation and validation provide opportunities for manufacturing organisations to test strategic initiatives within controlled environments whilst building implementation capabilities and organisational confidence. Manufacturing firms establish pilot programmes through limited scope implementation, performance measurement, and learning capture processes that demonstrate strategic value whilst identifying implementation challenges and refinement opportunities (Bachinger et al., 2021).

The pilot validation process requires systematic performance measurement, stakeholder feedback collection, and lessons learned documentation. These activities inform decisions regarding scaled deployment whilst building organisational understanding of effective implementation approaches. Manufacturing organisations develop validation criteria through success metrics definition, evaluation frameworks, and decision protocols that enable objective assessment of pilot results whilst maintaining strategic momentum and organisational commitment.

Pilot implementation must balance risk management with learning objectives to enable meaningful testing whilst avoiding excessive risk exposure. Such exposure may compromise organisational performance or stakeholder confidence. Manufacturing firms develop pilot strategies through scope definition, risk mitigation planning, and learning design that maximise learning value whilst managing implementation risks effectively.

The validation phase employs multiple assessment methodologies to ensure comprehensive evaluation. Quantitative metrics measure operational performance improvements, cost reductions, and quality enhancements. Qualitative assessments capture stakeholder feedback, cultural impacts, and organisational learning outcomes (Augmentir, 2024). This mixed-methods approach provides holistic understanding of pilot programme effectiveness.

Phase Four: Scaled Deployment and Integration.

Scaled deployment and integration encompass systematic rollout of validated strategic initiatives across manufacturing operations whilst maintaining operational continuity and performance standards. Manufacturing firms develop deployment strategies through phased implementation, change management protocols, and performance monitoring systems that enable smooth transition whilst capturing strategic benefits and building organisational capabilities (Digital Leadership, 2024).

Deployment success requires careful coordination of technical implementation, organisational change, and stakeholder engagement activities. These coordinated efforts ensure strategic initiatives achieve intended benefits whilst maintaining operational effectiveness and employee engagement levels. Manufacturing organisations establish deployment capabilities through project management excellence, change leadership development, and performance management systems that enable effective strategic implementation across complex organisational structures.

The deployment process addresses multiple dimensions simultaneously. Technical deployment involves system integration, process modification, and technology adoption. Organisational deployment encompasses training programmes, role redefinition, and culture change initiatives. Stakeholder deployment includes communication strategies, resistance management, and support system establishment (Cosa & Torelli, 2024).

Deployment planning must consider resource requirements, timing constraints, and interdependencies between strategic initiatives to ensure coordinated implementation. This coordination avoids resource conflicts whilst maintaining strategic momentum and organisational focus. Manufacturing firms develop deployment approaches through resource planning, scheduling optimisation, and coordination mechanisms that enable effective strategic implementation whilst managing organisational complexity.

Phase Five: Performance Monitoring and Optimisation.

Performance monitoring and optimisation ensure manufacturing organisations capture maximum value from strategic investments whilst identifying improvement opportunities and adaptation requirements. Manufacturing firms establish monitoring systems through performance metrics specification, tracking mechanisms implementation, and reporting processes development that provide visibility into strategic performance whilst enabling timely intervention when performance falls short of expectations (Neely et al., 2005).

The monitoring framework incorporates multiple performance dimensions. Operational metrics assess efficiency, quality, and productivity improvements. Financial metrics evaluate return on investment, cost reduction achievements, and revenue enhancement. Strategic metrics measure competitive positioning, market share development, and innovation outcomes (Gomes et al., 2004).

Optimisation requires systematic analysis of performance data, improvement identification, and enhancement implementation processes that maximise strategic value whilst adapting to changing conditions and emerging opportunities. Manufacturing organisations develop optimisation capabilities through data analytics deployment, improvement processes establishment, and enhancement implementation protocols that enable sustained value creation whilst maintaining strategic focus and organisational alignment.

Monitoring and optimisation must balance performance accountability with learning orientation to enable both performance achievement and strategic adaptation. This balance accommodates environmental condition changes and organisational capability evolution. Manufacturing firms develop balanced approaches through performance management systems, learning system implementation, and adaptation mechanisms that optimise current performance whilst building future capabilities (Alokshe et al., 2025).

Phase Six: Continuous Adaptation and Evolution.

Continuous adaptation and evolution ensure manufacturing organisations maintain strategic relevance and competitive advantage as environmental conditions change and new opportunities emerge. Manufacturing firms establish adaptation processes through environmental monitoring, strategic review protocols, and capability evolution mechanisms that enable strategic renewal whilst maintaining organisational coherence and stakeholder confidence (Bachinger et al., 2021).

The adaptation process requires systematic evaluation of strategic assumptions, environmental changes, and performance outcomes. This evaluation informs strategic adjustments whilst maintaining commitment to core strategic objectives and organisational values. Manufacturing organisations develop adaptation capabilities through strategic sensing systems, evaluation processes, and change management protocols that enable effective strategic evolution whilst preserving organisational strengths and stakeholder relationships.

Continuous adaptation capabilities are essential for sustained competitive advantage in VUCA environments since research demonstrates that organisations with strong adaptation capabilities outperform static competitors in volatile environments (Atanassova et al., 2025).This because environmental volatility requires organisational flexibility to maintain relevance. Multiple studies across manufacturing sectors confirm the positive relationship between adaptive capacity and performance outcomes but effectiveness depends upon organisational learning capabilities and leadership commitment.

Adaptation encompasses multiple organisational levels. Strategic adaptation involves vision refinement, objective modification, and investment reallocation. Operational adaptation includes process improvement, technology adoption, and capability enhancement. Cultural adaptation addresses mindset changes, behaviour modification, and value system evolution (VUCA World, 2025).

The evolution process recognises that adaptation is not merely reactive but proactive. Forward-looking organisations anticipate environmental changes and prepare adaptive responses before conditions necessitate change. This proactive approach provides competitive advantage through early mover benefits and reduced adaptation costs.

Critical Success Factors and Implementation Considerations.

Successful implementation of this strategic roadmap requires several critical success factors. Leadership commitment provides necessary resources and organisational support. Change management capabilities facilitate smooth transitions between phases. Learning orientation enables continuous improvement and adaptation. Stakeholder engagement ensures broad-based support and commitment (Harvard Business School, 2024).

The roadmap implementation must acknowledge organisational context and industry-specific requirements. Manufacturing organisations vary in size, complexity, technological sophistication, and market positioning. Successful implementation requires customisation of the general framework to specific organisational circumstances whilst maintaining core methodological principles.

Risk management throughout the implementation process addresses potential challenges and obstacles. Common risks include resource constraints, resistance to change, technological failures, and market disruptions. Effective risk management involves early identification, mitigation planning, and contingency preparation (TechTarget, 2015).

This strategic roadmap provides manufacturing organisations with a structured approach to navigating VUCA environments effectively. The six-phase methodology addresses the complexity and uncertainty characteristic of contemporary manufacturing whilst providing practical guidance for implementation. Success requires commitment to systematic implementation, continuous learning, and adaptive refinement based on experience and changing conditions.

The framework's strength lies in its recognition that VUCA navigation is not a one-time activity but an ongoing organisational capability. Manufacturing organisations that develop systematic approaches to uncertainty management whilst maintaining operational excellence will achieve sustained competitive advantage in increasingly volatile business environments.

Technology Implementation Best Practices

Successful deployment of artificial intelligence, automation, and robotics technologies within manufacturing operations necessitates systematic methodologies that maximise value creation whilst managing implementation risks and organisational disruption (Chen, Zhang, & Wang, 2024; IBM, 2024). Contemporary manufacturing environments demand sophisticated approaches to technology adoption that balance innovation opportunities with operational continuity requirements (Kayande & Kediya, 2025). This section presents evidence-based best practices derived from successful technology implementation initiatives across diverse manufacturing contexts, providing practical guidance for manufacturing leaders seeking to capture technology benefits whilst avoiding common implementation pitfalls.

Technology implementation best practices encompass eight critical dimensions that collectively ensure successful adoption outcomes: strategic alignment and objective clarity, readiness assessment, stakeholder engagement and change management, phased implementation and risk management, integration planning and systems coordination, performance measurement and value validation, continuous improvement and optimisation, and knowledge capture and organisational learning (Lawrence, 2024; Margherita & Braccini, 2022).

Manufacturing organisations require systematic eight-dimensional approaches to achieve successful AI, automation, and robotics implementation since research demonstrates that structured implementation frameworks yield superior outcomes compared to ad-hoc adoption approaches (Joglekar, Parker, & Srai, 2024; Cheng et al., 2024). Systematic frameworks provide necessary scaffolding for managing technological complexity whilst maintaining operational excellence. Multiple empirical studies across manufacturing sectors confirm the relationship between structured approaches and implementation success rates.

The contemporary manufacturing landscape, characterised by increasing technological sophistication and market volatility, demands adaptive frameworks that enable organisations to harness advanced technologies effectively (Schwarz, 2024; Sajid, 2024). Successful technology implementation transcends mere technical deployment, encompassing organisational transformation, capability development, and strategic alignment processes that collectively determine long-term success outcomes.

Dimension One: Strategic Alignment and Objective Clarity.

Strategic alignment demands more than stakeholder workshops and vision statements—it requires leadership practices that translate strategic intent into operational reality. Large-scale research shows that digital transformations are significantly more likely to exceed expectations when senior leaders make them a top priority and commit time and resources, with expectation-beating performance 1.5 times more likely under such conditions (Bughin, Deakin, & O'Beirne, 2019). This time investment focuses not on micromanagement but on obstacle removal, decision acceleration, and political navigation across organisational silos. Consider how Festo's transformation to Industry 4.0 manufacturing required their chief digital officer to personally mediate 17 interdepartmental conflicts about data ownership, system access rights, and performance metric definitions—negotiations that would have stalled implementation for months without direct leadership intervention (Festo, 2017). The practical implication challenges conventional delegation models: manufacturing leaders cannot simply approve digital transformation budgets and await results. Effective implementation requires establishing what Kotter (2012) terms a "guiding coalition"—a cross-functional team with sufficient authority to make binding decisions about technology selection, process modification, and resource allocation without requiring constant executive approval. Toyota addresses this challenge by empowering a Chief Engineer (shusa) with cross-functional responsibility and using hoshin kanri to align strategy and execution, rather than a universal "sponsor/driver" pairing (Morgan & Liker, 2006; Shook, 2009). The Chief Engineer removes organisational barriers whilst operational teams manage technical execution, preventing the common failure mode where senior leaders lack implementation knowledge whilst operational managers lack political authority. Manufacturing organisations navigating VUCA environments require similar structural mechanisms that bridge strategic intent and operational execution, ensuring that alignment extends beyond planning documents into daily decision-making.

The alignment process requires systematic integration of technology capabilities with business objectives. Strategic analysis encompasses market positioning considerations, competitive advantage opportunities, and operational efficiency requirements that collectively inform technology selection and implementation priorities (IBM, 2024). This comprehensive approach ensures technology investments align with broader organisational goals whilst addressing specific operational challenges.

Objective clarity necessitates specific, measurable, achievable, relevant, and time-bound (SMART) goals that enable progress tracking whilst providing motivation and direction for implementation teams. Manufacturing organisations develop clear objectives through stakeholder consultation, business case development, and metric definition processes that ensure technology initiatives address priority business needs whilst delivering quantifiable benefits justifying continued investment and organisational commitment (Lawrence, 2024).

SMART objectives significantly improve technology implementation success rates in manufacturing environments. Empirical studies demonstrate that organisations with clearly defined implementation objectives achieve higher success rates than those with vague goals (Cheng et al., 2024). Clear objectives provide necessary direction and accountability mechanisms for complex technology initiatives. Effectiveness depends upon stakeholder alignment and regular objective refinement based on implementation experience.

Strategic alignment must consider both direct benefits from technology implementation and indirect effects on organisational capabilities, competitive positioning, and strategic flexibility. These indirect effects may provide longer-term value beyond immediate operational improvements. Manufacturing firms develop value propositions through strategic analysis, benefit quantification, and impact assessment processes that capture full technology value whilst building organisational understanding and commitment (Hinterhuber, 2017).

Dimension Two: Readiness Assessment.

Readiness assessment enables manufacturing organisations to understand their current capabilities whilst identifying gaps that must be addressed before successful technology implementation. Manufacturing firms conduct readiness assessments through technical infrastructure evaluation, organisational capability analysis, and resource availability assessment that illuminate implementation requirements whilst identifying potential obstacles and mitigation strategies (Margherita & Braccini, 2022).

The assessment process encompasses multiple organisational dimensions. Technical readiness evaluation examines data infrastructure, systems integration capabilities, cybersecurity frameworks, and network capacity that enable effective technology deployment whilst maintaining operational security and performance (Safar, Hidayat, & Chen, 2023). Manufacturing organisations evaluate technical readiness through infrastructure audits, systems analysis, and capacity assessment processes that identify upgrade requirements whilst ensuring adequate technical foundations for successful implementation.

Organisational readiness includes leadership commitment, change management capabilities, employee skills, and cultural factors that influence technology adoption success whilst affecting long-term sustainability of technology benefits.

Manufacturing firms assess organisational readiness through capability evaluation, cultural assessment, and stakeholder analysis processes that identify development requirements whilst building implementation support and organisational confidence (Chen, Zhang, & Wang, 2024).

The readiness assessment must acknowledge the interdependent nature of technical and organisational factors. Technical capabilities without corresponding organisational readiness often result in implementation failures, whilst strong organisational readiness cannot overcome fundamental technical deficiencies (Singh et al., 2023). Successful assessment processes evaluate both dimensions systematically whilst identifying specific actions required to address identified gaps.

Dimension Three: Stakeholder Engagement and Change Management.

Stakeholder engagement and change management ensure that technology implementation receives adequate support whilst addressing resistance and building organisational capabilities for effective technology utilisation. Manufacturing firms develop stakeholder engagement strategies through communication planning, participation design, and feedback mechanisms that build understanding whilst addressing concerns and building commitment to technology initiatives (Panorama Consulting Group, 2024; Margherita & Braccini, 2022).

Effective stakeholder engagement requires systematic identification and analysis of all affected parties. Primary stakeholders include direct users, operational managers, and senior leadership, whilst secondary stakeholders encompass support functions, suppliers, and customers potentially impacted by technology changes (Rittenhouse, 2015). Each stakeholder group requires tailored engagement approaches that address specific concerns whilst building support for implementation initiatives.

Change management requires systematic approaches to addressing organisational impacts of technology implementation through training programmes, communication strategies, and support systems that enable smooth transition whilst maintaining employee engagement and organisational performance (Errida & Lotfi, 2021; Saghafian et al., 2021). Manufacturing organisations develop change management capabilities through leadership development, communication excellence, and support systems that enable effective organisational transformation whilst preserving valuable aspects of existing operations and culture.

Systematic stakeholder engagement reduces technology implementation resistance and improves adoption rates. Studies demonstrate that organisations employing structured stakeholder engagement achieve higher implementation success rates (Margherita & Braccini, 2022; Cheng et al., 2024). Engagement processes build understanding and commitment whilst addressing legitimate concerns. Multiple change management studies across manufacturing contexts support the stakeholder engagement-success relationship. Success depends upon authentic engagement rather than superficial consultation processes.

Effective change management balances technology promotion with realistic assessment of implementation challenges to maintain credibility whilst building enthusiasm and commitment for technology initiatives. Manufacturing firms develop balanced communication approaches that acknowledge implementation challenges whilst highlighting technology benefits and providing adequate support for employees navigating technological change (Errida & Lotfi, 2021).

Dimension Four: Phased Implementation and Risk Management.

Phased implementation and risk management enable manufacturing organisations to capture technology benefits whilst managing implementation risks through systematic rollout strategies that enable learning capture and course correction (Joglekar, Parker, & Srai, 2024; DealHub, 2024). Manufacturing firms develop phased approaches through scope definition, sequencing analysis, and milestone planning that enable controlled implementation whilst building organisational capabilities and confidence.

The phased approach provides multiple advantages over 'big bang' implementations. Risk reduction occurs through limited initial exposure, allowing organisations to identify and address challenges before full-scale deployment. Learning opportunities emerge at each phase, enabling continuous improvement and adaptation of implementation strategies (Zhou & Zhang, 2025; Makhanya, 2023). These advantages become particularly important for complex technology implementations involving multiple organisational units or sophisticated technical integrations.

Risk management requires identification of potential implementation challenges through risk assessment, mitigation planning, and contingency preparation that enable effective response to implementation difficulties whilst maintaining project momentum and organisational confidence. Manufacturing organisations develop risk management capabilities through systematic risk identification, mitigation strategy development, and contingency planning addressing technical, organisational, and external risks that may affect implementation success (Safar, Hidayat, & Chen, 2023; Margherita & Braccini, 2022).

Phased implementation must balance risk management with implementation momentum to enable meaningful progress whilst avoiding excessive caution that may prevent technology benefits realisation or organisational learning. Manufacturing firms develop balanced approaches through milestone definition, success criteria establishment, and decision protocols that enable systematic progress whilst maintaining appropriate risk management and organisational protection (Joglekar, Parker, & Srai, 2024).

Phased implementation approaches significantly reduce technology implementation risks whilst enabling organisational learning. Empirical evidence demonstrates lower failure rates for phased versus 'big bang' technology implementations (DealHub, 2024; Joglekar, Parker, & Srai, 2024). Phased approaches enable risk identification and mitigation whilst building implementation capabilities progressively. Success requires careful phase planning and adequate resource allocation across implementation stages.

Dimension Five: Integration Planning and Systems Coordination.

Integration planning and systems coordination ensure that new technologies integrate effectively with existing manufacturing systems whilst maintaining operational continuity and performance standards. Manufacturing firms develop integration strategies through systems analysis, interface design, and coordination planning that enable seamless technology deployment whilst avoiding disruption to critical operations and business processes (Azadeh et al., 2007; Frohlich & Westbrook, 2001).

The integration process encompasses technical and organisational dimensions. Technical integration involves data flows, systems interfaces, and operational procedures that connect new technologies with existing infrastructure. Organisational integration addresses process changes, role modifications, and workflow adaptations required to accommodate new technological capabilities (Margherita & Braccini, 2022).

Systems coordination requires careful attention to data flows, process integration, and organisational interfaces that connect new technologies with existing operations whilst maintaining system reliability and performance effectiveness. Manufacturing organisations develop coordination capabilities through systems engineering, project management excellence, and technical expertise that enable effective integration whilst maintaining operational excellence and competitive positioning (Bititci, Carrie, & McDevitt, 1997; Xu, Lu, Vogel-Heuser, & Wang, 2021).

Integration planning must consider both immediate implementation requirements and long-term system evolution needs. Short-term planning addresses immediate compatibility and operational continuity requirements, whilst long-term planning ensures technology investments support future organisational growth and development (Bititci, Carrie, & McDevitt, 1997). This dual perspective enables organisations to capture immediate benefits whilst building foundations for continued technological advancement.

Integration planning must consider both technical and organisational integration requirements to ensure that new technologies enhance rather than disrupt existing capabilities whilst building organisational competencies for future technology development and deployment. Manufacturing firms develop integration approaches through holistic planning, stakeholder coordination, and performance management that optimise overall system performance rather than focusing solely on new technology functionality (Cheng et al., 2024).

Dimension Six: Performance Measurement and Value Validation.

Performance measurement and value validation enable manufacturing organisations to demonstrate technology benefits whilst identifying improvement opportunities and building organisational confidence in technology investments. Manufacturing firms establish measurement systems through metric definition, data collection, and analysis processes that provide visibility into technology performance whilst enabling evidence-based decision-making about continued investment and enhancement opportunities (Azadeh et al., 2007; Bititci, Carrie, & McDevitt, 1997).

The measurement framework must encompass multiple performance dimensions. Operational metrics assess efficiency improvements, quality enhancements, and productivity gains directly attributable to technology implementation. Financial metrics evaluate return on investment, cost reduction achievements, and revenue enhancement outcomes. Strategic metrics measure competitive positioning improvements, innovation capability development, and organisational learning outcomes (Cheng et al., 2024).

Value validation requires systematic comparison of actual technology benefits with projected outcomes through quantitative analysis and qualitative assessment that demonstrate return on investment whilst identifying factors that enhance or constrain technology value creation. Manufacturing organisations develop validation capabilities through data analytics, benchmarking processes, and assessment frameworks that enable objective evaluation of technology benefits whilst building organisational understanding of effective technology deployment and utilisation (Azadeh et al., 2007).

Performance measurement must balance short-term operational metrics with longer-term strategic benefits to capture full technology value whilst maintaining focus on immediate implementation success and organisational performance. Manufacturing firms develop balanced measurement approaches through metric portfolios, reporting systems, and analysis processes that illuminate both immediate and strategic technology benefits whilst supporting continued improvement and optimisation efforts (Bititci, Carrie, & McDevitt, 1997).

Comprehensive performance measurement systems significantly improve technology implementation outcomes and organisational learning. Research demonstrates that organisations with systematic measurement approaches achieve better technology adoption results (Cheng et al., 2024; Azadeh et al., 2007). Measurement systems provide visibility and accountability mechanisms essential for effective technology management. Multiple studies across manufacturing contexts confirm the measurement-performance relationship.

Dimension Seven: Continuous Improvement and Optimisation.

Continuous improvement and optimisation ensure that manufacturing organisations maximise value from technology investments whilst adapting to changing conditions and emerging opportunities that may enhance technology benefits or require implementation adjustments. Manufacturing firms establish improvement processes through performance monitoring, opportunity identification, and enhancement implementation that enable sustained value creation whilst building organisational capabilities for future technology development and deployment (Bessant, Caffyn, & Gallagher, 2001; Tupa et al., 2017).

The improvement process requires systematic analysis of technology performance data, user feedback, and operational outcomes. Performance analysis identifies areas where technology delivers expected benefits and areas requiring attention or enhancement. User feedback provides insights into practical implementation challenges and opportunities for improving technology usability and effectiveness (Cheng et al., 2024).

Optimisation requires systematic analysis of technology performance, user feedback, and improvement opportunities that guide enhancement investments whilst maintaining focus on value creation and organisational objectives. Manufacturing organisations develop optimisation capabilities through data analytics, user engagement processes, and improvement implementation frameworks that enable sustained technology benefits whilst building organisational competencies for effective technology management and utilisation (Tupa et al., 2017).

The optimisation process must balance stability with continuous improvement to ensure reliable technology operations whilst capturing emerging enhancement opportunities. Excessive optimisation efforts may create system instability, whilst insufficient optimisation fails to capture available value improvements (Bessant, Caffyn, & Gallagher, 2001). Successful optimisation approaches establish systematic review cycles that enable planned improvements whilst maintaining operational stability.

Dimension Eight: Knowledge Capture and Organisational Learning.

Knowledge capture and organisational learning ensure that manufacturing organisations build capabilities for future technology implementation whilst preserving valuable insights and experiences that enhance organisational competencies for navigating technological change. Manufacturing firms establish learning processes through documentation, reflection, and knowledge sharing that capture implementation lessons whilst building organisational wisdom about effective technology deployment and management (Ardito, Messeni Petruzzelli, & Albino, 2018; Tupa et al., 2017).

The learning process encompasses multiple knowledge categories. Technical knowledge includes implementation procedures, integration approaches, and troubleshooting solutions. Organisational knowledge encompasses change management insights, stakeholder engagement lessons, and cultural adaptation strategies. Strategic knowledge includes investment decision criteria, technology selection processes, and value assessment frameworks (Cheng et al., 2024; Castaldi & Giarratana, 2015).

Learning requires systematic reflection on implementation experiences, success factors, and improvement opportunities that inform future technology initiatives whilst building organisational confidence and capabilities for managing technological transformation. Manufacturing organisations develop learning capabilities through knowledge management systems, best practice identification processes, and capability development programmes that enhance future technology implementation whilst preserving valuable organisational knowledge and experience (Tupa et al., 2017; Alegre & Chiva, 2008).

Systematic knowledge capture and organisational learning significantly improve future technology implementation success rates. Studies demonstrate that organisations with strong learning capabilities achieve consistently better technology adoption outcomes over time (Ardito, Messeni Petruzzelli, & Albino, 2018; Tupa et al., 2017). Organisational learning enables continuous capability development and avoids repeated implementation errors. Knowledge management research across manufacturing contexts supports the learning-performance relationship. Learning effectiveness depends upon systematic documentation and active knowledge sharing processes.

The knowledge capture process must balance systematic documentation with practical accessibility to ensure captured knowledge influences future implementation efforts. Overly complex documentation systems may discourage use, whilst insufficient documentation fails to preserve critical implementation insights (Castaldi & Giarratana, 2015). Effective knowledge management systems balance comprehensiveness with usability to maximise learning value.

Integration and Implementation Considerations.

Successful implementation of these eight dimensions requires coordinated attention to multiple organisational factors. Leadership commitment provides necessary resources and organisational support for sustained implementation efforts. Technical infrastructure ensures adequate foundations for technology deployment and operation. Organisational culture influences acceptance and adoption of technological changes (Lawrence, 2024; Kestria, 2024).

The implementation process must acknowledge industry-specific requirements and organisational contexts that influence technology adoption success. Manufacturing organisations vary in size, technological sophistication, market positioning, and resource availability. Successful implementation requires customisation of general best practices to specific organisational circumstances whilst maintaining core methodological principles (IBM, 2024; Schwarz, 2024).

Risk factors throughout implementation include technical failures, organisational resistance, resource constraints, and external market changes. Effective risk management addresses these factors through systematic identification, mitigation planning, and contingency preparation. Regular risk assessment and mitigation adjustment enable proactive response to emerging challenges whilst maintaining implementation momentum (Safar, Hidayat, & Chen, 2023; Margherita & Braccini, 2022).

Success Factors and Performance Outcomes.

Implementation success depends upon several critical factors working in combination. Technical competence ensures adequate foundation for technology deployment and operation. Organisational readiness provides necessary capabilities for managing change and adopting new operational approaches. Strategic alignment ensures technology investments support broader organisational objectives and competitive positioning (Sukathong et al., 2021; Chen, Zhang, & Wang, 2024).

Performance outcomes from successful technology implementation include operational improvements, strategic advantages, and organisational capability development. Operational improvements encompass efficiency gains, quality enhancements, and cost reductions directly attributable to technology adoption. Strategic advantages include competitive positioning improvements, innovation capability enhancement, and market responsiveness increases (IBM, 2024; Lawrence, 2024).

Organisational capability development represents longer-term benefits that extend beyond immediate technology impacts. These capabilities include enhanced change management competencies, improved technology assessment skills, and strengthened innovation capacities that support continued organisational development and technological advancement (Cheng et al., 2024; Sajid, 2024).

Technology implementation best practices provide manufacturing organisations with systematic approaches to capturing technology benefits whilst managing implementation risks and organisational disruption. The eight-dimensional framework addresses critical aspects of successful technology adoption whilst recognising the interconnected nature of technical, organisational, and strategic factors that collectively determine implementation outcomes.

Future technology implementations will require continued evolution of these best practices as artificial intelligence, automation, and robotics technologies advance in sophistication and capability. Manufacturing organisations must maintain adaptive approaches to technology implementation whilst preserving core methodological principles that support successful adoption outcomes.

Manufacturing organisations adopting systematic technology implementation best practices achieve superior outcomes compared to ad-hoc approaches. Comprehensive evidence across multiple manufacturing contexts demonstrates the effectiveness of structured implementation methodologies (Lawrence, 2024; IBM, 2024; Cheng et al., 2024). Systematic approaches address critical success factors whilst managing implementation risks effectively. Multiple empirical studies and industry case studies support the structured approach-success relationship.

The framework's strength lies in its recognition that technology implementation represents organisational transformation rather than mere technical deployment. Manufacturing organisations that develop systematic approaches to technology implementation whilst maintaining focus on organisational capability development will achieve sustained competitive advantage in increasingly technology-dependent manufacturing environments.

Implementation Priorities Identified in this section

- **Strategic roadmap provides structure**: Six-phase implementation—vision development, gap analysis, pilot validation, scaled deployment, performance monitoring, continuous adaptation—ensures systematic capability building (Santos & Martinho, 2020).
- **Technology success demands eight dimensions**: Strategic alignment, readiness assessment, stakeholder engagement, phased implementation, systems integration, performance measurement, continuous improvement, and organisational learning determine outcomes (Margherita & Braccini, 2022).
- **Leadership commitment proves decisive**: Senior executive involvement in obstacle removal, resource allocation, and cross-functional coordination increases implementation success probability by 1.5 times (Bughin et al., 2019).
- **Change management addresses resistance**: Systematic stakeholder engagement, training programmes, and communication strategies reduce resistance and improve adoption rates significantly (Errida & Lotfi, 2021).

- **Learning captures value**: Post-implementation reviews, best practice identification, and knowledge sharing enable organisations to improve continuously and avoid repeated implementation errors (Ardito et al., 2018).

CONCLUSIONS AND FUTURE DIRECTIONS

Synthesis of Key Findings

This examination reveals a paradox confronting manufacturing leadership: the frameworks, technologies, and practices designed to navigate VUCA conditions often amplify the very "volatility, uncertainty, complexity, and ambiguity" they aim to resolve. Digital transformation initiatives intended to increase agility can create technological brittleness—systems so optimised for current conditions that they resist adaptation when conditions change (Tushman & O'Reilly, 2002). Supply chain diversification strategies meant to reduce uncertainty can multiply coordination complexity across geographically dispersed suppliers, each operating under different regulatory regimes and quality standards (Shih, 2020). Sustainability programmes designed to address ESG pressures introduce measurement ambiguity, as organisations struggle to verify Scope 3 emissions from supplier networks or calculate the carbon impact of circular economy initiatives (Bjørn et al., 2021). The critical insight emerging from this analysis suggests that effective VUCA navigation requires not merely implementing frameworks but developing meta-capabilities—the ability to recognise when established approaches create new problems requiring different solutions. Manufacturing organisations demonstrate such meta-capabilities through systematic experimentation, rapid learning from failures, and willingness to abandon practices that no longer serve strategic objectives even when those practices required substantial investment (Reeves et al., 2016). The theoretical contributions of this chapter matter because they reframe VUCA navigation from problem-solving to capability-building, from seeking stability to developing adaptive capacity, from managing change to institutionalising continuous transformation.

Evidence demonstrates that VUCA conditions in manufacturing have intensified considerably over the past five years. Quantitative indicators reveal unprecedented levels of volatility, uncertainty, complexity, and ambiguity across key performance dimensions (Federal Reserve Board, 2025). Manufacturing capacity utilisation volatility has increased by 34% compared to pre-2020 baselines, whilst supply chain disruption frequency has risen from 41% to 73% of manufacturers experiencing significant interruptions (Business Continuity Institute & Resilience360, 2020). These empirical findings confirm that VUCA represents a fundamental shift in manufacturing operating environments rather than temporary disruption that will return to previous stability levels.

Table 1. VUCA exposure analysis of the most sophisticated manufacturing economies in the OECD (Data compiled from OECD (2024), World Bank (2024), and McKinsey Global Institute (2024). Author estimates based on comparative economic analysis)

Country	Volatility (0-100)	Uncertainty 0-100)	Complexity (0-100)	Ambiguity (0-100)	Grounds for the exposure estimate
South Korea	80	75	95	80	Semiconductor cycle volatility + China-US strategic ambiguity
United Kingdom	80	85	85	80	Post-Brexit uncertainty + complex new trade arrangements
Ireland	75	70	85	80	FDI dependence + multi-jurisdictional policy complexity
Japan	75	65	90	70	Energy dependency + supply chain complexity
Netherlands	70	65	90	75	Global trade hub exposure + EU governance complexity
France	70	75	80	75	Aerospace volatility + EU-national policy balancing
Germany	70	60	95	65	Industry 4.0 complexity + post-Ukraine energy transition
Sweden	65	55	85	70	Nordic countries balancing green transition with global market forces
Finland	65	60	80	70	Nordic countries balancing green transition with global market forces
United States	60	55	85	70	Large domestic market provides buffer despite technological complexity
Austria	60	55	80	65	Germanic integration with moderate exposure levels
Switzerland	55	50	80	60	Lowest exposure due to specialization
Canada	60	55	70	60	Lowest exposure due to stable frameworks

Note. V = Volatility (0-100); U = Uncertainty (0-100); C = Complexity (0-100); A = Ambiguity (0-100). Scores represent relative exposure levels based on analysis of manufacturing sector indicators, trade dependencies, and regulatory environments.

The comparative analysis in Table 1 reveals significant variation in VUCA exposure across advanced manufacturing economies. South Korea and the United Kingdom face particularly acute challenges across multiple dimensions, whilst Switzerland and Canada demonstrate relatively lower exposure levels. The complexity dimension scores consistently high across all economies, reflecting the universal challenge of managing interconnected global manufacturing networks (Caldara & Iacoviello, 2022; McKinsey Global Institute, 2024)

Theoretical analysis reveals that traditional strategic management approaches prove insufficient for navigating VUCA conditions effectively. Integration of dynamic capabilities theory, institutional theory, and strategic management perspectives becomes essential to address the multifaceted challenges confronting contemporary manufacturing organisations. The analysis demonstrates that Mintzberg's five Ps framework provides valuable insights for understanding strategic multiplicity in VUCA contexts, where strategy as plan, ploy, pattern, position, and perspective must operate simultaneously to address environmental complexity and uncertainty (Mintzberg, 1987).

Dynamic capabilities emerge as essential requirements for VUCA navigation, with sensing, seizing, and transforming capabilities enabling manufacturing organisations to identify environmental changes, capture value from emerging opportunities, and adapt organisational systems to maintain competitive advantage (Teece, 2007; Teece et al., 1997). However, analysis reveals that dynamic capabilities alone prove insufficient without complementary institutional adaptation strategies that enable manufacturing firms to navigate complex regulatory environments whilst leveraging institutional changes as sources of competitive advantage.

Technology integration represents both opportunity and challenge within VUCA manufacturing environments. Artificial intelligence, automation, and robotics technologies provide capabilities for enhancing operational efficiency, supply chain resilience, and decision-making effectiveness whilst creating new sources of complexity and vulnerability requiring careful management. Successful technology implementation requires systematic approaches that integrate strategic planning, organisational change management, and performance monitoring whilst maintaining operational continuity and competitive positioning.

Case study analysis reveals common success factors across diverse manufacturing contexts, including senior leadership commitment, cross-functional collaboration, phased implementation approaches, and systematic change management addressing both technical and organisational dimensions of technology integration. These findings demonstrate that technology success depends more upon implementation approach than technology selection, with organisational capabilities and management processes proving more critical than specific technology features or capabilities.

Primary research evidence from manufacturing practitioners reveals specific applications of military-derived VUCA principles extending beyond published case studies. Interviews with senior manufacturing executives indicate systematic adoption of military-style “war gaming” exercises for supply chain disruption planning, with 78% of surveyed companies conducting quarterly scenario-based crisis simulations adapted from military contingency planning (Manufacturing Leadership Council, 2024). Manufacturing organisations increasingly employ military-derived "red team" exercises to stress-test production systems, with participants noting that military adversarial thinking helps identify vulnerabilities in manufacturing processes that traditional quality assurance methods miss. Furthermore, primary observations of Industry 4.0 implementations reveal manufacturers adapting military "mission command" principles, where front-line operators receive strategic intent but tactical flexibility in responding to production disruptions—a direct adaptation of military decentralised command structures to manufacturing contexts (Holmström & Partanen. 2014).

Institutional adaptation strategies prove essential for manufacturing organisations operating across multiple regulatory jurisdictions with evolving compliance requirements and conflicting institutional pressures. Analysis reveals that successful institutional navigation requires proactive engagement with regulatory development processes whilst developing compliance strategies creating competitive advantage rather than simply meeting minimum requirements. Manufacturing organisations excelling at institutional adaptation develop sophisticated stakeholder engagement capabilities whilst building regulatory intelligence and compliance excellence enabling superior performance within complex institutional environments.

Risk management and resilience development emerge as critical capabilities for VUCA navigation, requiring approaches extending beyond traditional risk assessment and mitigation to encompass adaptive capacity building and organisational transformation capabilities. Analysis demonstrates that resilience requires balance between efficiency optimisation and redundancy development, with successful manufacturing organisations creating flexible systems that maintain operational effectiveness whilst enabling rapid adaptation to disruption scenarios.

Theoretical Contributions and Implications

This analysis makes several significant contributions to strategic management and operations management literature whilst advancing theoretical understanding of VUCA navigation in manufacturing contexts. The integrated framework developed synthesises multiple theoretical perspectives to provide guidance for manufacturing leaders whilst addressing gaps in existing literature that treat VUCA dimensions, dynamic capabilities, institutional adaptation, and technology integration as separate rather than interconnected phenomena.

The first theoretical contribution involves demonstrating the interconnected nature of VUCA dimensions in manufacturing contexts, extending Bennett and Lemoine's (2014) conceptual framework by showing how volatility, uncertainty, complexity, and ambiguity create cascading effects that amplify challenges through dimensional interactions. This contribution advances VUCA theory by moving beyond dimensional independence assumptions to illuminate interaction effects that create compound challenges requiring integrated rather than sequential responses.

The second contribution encompasses integration of dynamic capabilities theory with institutional theory to explain how manufacturing organisations develop capabilities for navigating complex institutional environments whilst maintaining operational effectiveness and competitive positioning (Teece, 2007). This integration addresses limitations in existing literature that examine dynamic capabilities and institutional adaptation as separate phenomena whilst providing theoretical foundation for understanding their interaction effects in manufacturing contexts.

The third theoretical contribution involves extending technology adoption and implementation literature by demonstrating how VUCA conditions affect technology integration processes whilst showing how technology deployment can both mitigate and amplify VUCA dimensions depending upon implementation approaches and organisational capabilities. This contribution advances technology management theory by incorporating environmental turbulence considerations into technology adoption models whilst providing guidance for technology implementation within uncertain environments.

The fourth contribution encompasses development of integrated frameworks that connect strategic management, operations management, and technology management perspectives within approaches to VUCA navigation. This integration addresses fragmentation in existing literature whilst providing theoretical foundation for understanding how different management domains interact within complex organisational systems confronting environmental turbulence.

These theoretical contributions have important implications for future research directions and management practice. The integrated nature of VUCA dimensions suggests that research examining individual dimensions in isolation may miss critical interaction effects that determine organisational performance in turbulent environments (Bennett & Lemoine, 2014). Future research should examine VUCA interactions more systematically whilst developing measurement approaches that capture compound effects rather than treating dimensions independently.

Practical Implications for Manufacturing Leaders

The findings provide several critical implications for manufacturing leaders seeking to navigate VUCA environments effectively whilst building competitive advantage through strategic technology integration and institutional adaptation. These implications address both immediate operational challenges and longer-term strategic positioning requirements that enable sustained competitive advantage in turbulent environments.

Manufacturing leaders must develop sophisticated environmental sensing capabilities that enable early identification of emerging opportunities and threats across technological, market, regulatory, and geopolitical domains (Teece, 2007). This requires investment in environmental scanning systems, expert networks, and analytical capabilities that provide awareness of environmental changes whilst enabling proactive response to emerging challenges and opportunities.

Effective sensing requires integration across multiple organisational functions and external partnerships to capture diverse perspectives on environmental conditions whilst avoiding blind spots that may prevent identification of critical changes. Manufacturing leaders should establish cross-functional sensing teams whilst developing external partnerships that enhance environmental awareness and early warning capabilities.

Strategic option development emerges as essential capability for managing uncertainty and complexity within VUCA environments (Mintzberg, 1987). Manufacturing leaders must develop systematic approaches to creating and evaluating alternative responses to environmental changes whilst maintaining strategic flexibility that enables rapid adaptation to unexpected developments. This requires investment in scenario planning capabilities, strategic analysis competencies, and decision-making processes that enable effective option evaluation and selection.

Technology integration decisions require careful balance between value creation opportunities and implementation risks, with successful integration depending more upon implementation approach than technology selection. Manufacturing leaders should develop technology assessment capabilities whilst building implementation competencies that enable effective technology deployment without compromising operational continuity or competitive positioning..

Future Research Directions

Despite widespread adoption, significant research gaps remain in VUCA measurement and validation within manufacturing contexts. The literature reveals contested origins and attribution disputes, with some sources incorrectly crediting Bennis and Nanus (1985) as VUCA creators rather than recognising their conceptual influence on military development of the acronym (Cuban, 2010). More critically, standardised metrics for VUCA dimensions in manufacturing remain underdeveloped, with volatility and uncertainty having clear statistical measures whilst complexity and ambiguity metrics lack standardisation across studies (Giones & Brem, 2022). Kristjánsdóttir's (2019) development of a “VUCA meter” diagnostic questionnaire for manufacturing projects represents promising progress, demonstrating that whilst single aggregate VUCA scores prove less useful, graphical representation of relative component severity provides accurate "mapping" of operational challenges. However, the literature remains dominated by conceptual models and prescriptive advice rather than rigorous quantitative studies linking VUCA-oriented strategies to tangible performance outcomes (Troise et al., 2022). This empirical gap limits both theoretical advancement and practical guidance for manufacturing leaders seeking to operationalise military-derived VUCA frameworks within contemporary production systems.

This analysis identifies several promising directions for future research that would advance understanding of VUCA navigation in manufacturing contexts whilst addressing limitations and gaps in current knowledge. These research directions encompass both theoretical development and empirical investigation that would enhance both scholarly understanding and practical guidance for manufacturing leaders confronting contemporary challenges.

Longitudinal studies examining VUCA navigation over extended time periods would provide valuable insights into how organisational capabilities evolve whilst identifying factors that enable sustained competitive advantage in turbulent environments. Such studies should examine how dynamic capabilities, institutional adaptation strategies, and technology integration approaches change over time whilst investigating their long-term effects on organisational performance and competitive positioning.

Concluding Remarks

This analysis of strategic navigation in VUCA manufacturing environments demonstrates that contemporary manufacturing organisations face unprecedented challenges requiring sophisticated strategic approaches that integrate multiple theoretical perspectives and practical capabilities. The evidence presented confirms that VUCA conditions represent fundamental shifts in manufacturing operating environments rather than temporary disruptions, requiring sustained investment in dynamic capabilities, institutional adaptation strategies, and technology integration competencies.

The integrated framework developed provides both theoretical contributions to strategic management literature and practical guidance for manufacturing leaders seeking to transform environmental challenges into competitive advantages. The synthesis of dynamic capabilities theory, institutional theory, and strategic management perspectives within VUCA navigation frameworks addresses gaps in existing literature whilst providing actionable insights for manufacturing practitioners confronting complex environmental challenges.

Manufacturing organisations that implement the frameworks and recommendations presented in this analysis will be better positioned to navigate contemporary challenges whilst capturing opportunities for competitive advantage through superior VUCA navigation capabilities. The investment in dynamic capabilities, institutional adaptation strategies, and technology integration competencies outlined will enable manufacturing firms to transform environmental turbulence from constraint into competitive advantage whilst building sustainable competitive positioning for future success.

REFERENCES

Abdallah, Y. O., Shehab, E., & Al-Ashaab, A. (2021). *Towards managing digital transformation in manufacturing industry: Theoretical framework*. Cranfield University Repository. https://dspace.lib.cranfield.ac.uk/server/api/core/bitstreams/0107b34c-0128-4d88-ab16-107fed663cdc/content

Abdullah, F. M., Al-Ahmari, A. M., & Anwar, S. (2023). Analyzing interdependencies among influencing factors in smart manufacturing. *Sustainability (Basel)*, *15*(4), 3864. DOI: 10.3390/su15043864

Acquah, I. S. K., Essel, D., Baah, C., Agyabeng-Mensah, Y., & Afum, E. (2021). Investigating the efficacy of isomorphic pressures on the adoption of green manufacturing practices and its influence on organizational legitimacy and financial performance. *Journal of Manufacturing Technology Management*, *32*(7), 1399–1420. DOI: 10.1108/JMTM-10-2020-0404

Adams, C. A., & Abhayawansa, S. (2022). Connecting the COVID-19 pandemic, environmental, social and governance (ESG) investing and calls for "harmonisation" of sustainability reporting. *Critical Perspectives on Accounting*, *82*, 102309. DOI: 10.1016/j.cpa.2021.102309

Adler, P. S., Goldoftas, B., & Levine, D. I. (1999). Flexibility versus efficiency? A case study of model changeovers in the Toyota Production System. *Organization Science*, *10*(1), 43–68. https://www.jstor.org/stable/2640387. DOI: 10.1287/orsc.10.1.43

Ahmed, S., Abd Aziz, N., Haque, R., & Rahman, M. T. (2024). Digital transformation in Malaysian manufacturing: A study of change sensing and seizing capabilities. *Cogent Business & Management*, *11*(1), 2392046. DOI: 10.1080/23311975.2024.2392046

Al-Talib, M., Melhem, W. Y., Anosike, A. I., Garza-Reyes, J. A., Nadeem, S. P., & Kumar, A. (2020). Achieving resilience in the supply chain by applying IoT technology. *Procedia CIRP*, *91*, 752–757. DOI: 10.1016/j.procir.2020.02.231

Alegre, J., & Chiva, R. (2008). Assessing the impact of organizational learning capability on product innovation performance: An empirical test. *Technovation*, *28*(6), 315–326. DOI: 10.1016/j.technovation.2007.09.003

Alokshe, M. M. F., Adedokun, M. W., & Iyiola, K. (2025). Advanced manufacturing technologies, strategic agility, business network and sustained competitive performance: An empirical evidence from an emerging economy. *Humanities & Social Sciences Communications*, *12*(1), 391. https://www.nature.com/articles/s41599-025-04671-9. DOI: 10.1057/s41599-025-04671-9

Andrianandrianina Johanesa, T. V., Equeter, L., & Mahmoudi, S. A. (2024). Survey on AI applications for product quality control and predictive maintenance in industry 4.0. *Electronics (Basel), 13*(5), 976. DOI: 10.3390/electronics13050976

Appelbaum, S. H., & Goransson, L. (1997). Transformational and adaptive learning within the learning organization: A framework for research and application. *The Learning Organization, 4*(3), 115–128. DOI: 10.1108/09696479710182803

Aral, K. D., Giambona, E., & Lopez-Aliouchkin, R. (2020). *The COVID-19 pandemic: Supply chain disruption, wealth effects, and corporate responses.* SSRN *Electronic Journal.* https://doi.org/DOI: 10.2139/ssrn.3686744

Ardito, L., Messeni Petruzzelli, A., & Albino, V. (2018). Approaches for organizational learning: A literature review. *SAGE Open, 8*(3), 1–20. DOI: 10.1177/2158244018794224 PMID: 32983596

Ashena, M., Laal Khezri, H., & Shahpari, G. (2024). Investigation into the dynamic relationships between global economic uncertainty and price volatilities of commodities, raw materials, and energy. *Applied Economic Analysis, 32*(94), 23–40. DOI: 10.1108/AEA-06-2023-0207

Atanassova, I., Bednar, P., Khan, H., & Khan, Z. (2025). Managing the VUCA environment: The dynamic role of organizational learning and strategic agility in B2B versus B2C firms. *Industrial Marketing Management, 125*, 12–28. DOI: 10.1016/j.indmarman.2024.12.008

Augmentir. (2024, June 1). *Performance management in manufacturing: Best practices and examples.* https://www.augmentir.com/blog/performance-management-in-manufacturing-best-practices-and-examples

Axios. (2021, May 19). *Colonial Pipeline CEO says company paid hacker group $4.4 million.* https://www.axios.com/2021/05/19/colonial-pipeline-darkside-4-million-ransomware

Axios. (2024, February 20). *Ransomware hackers focus on North America, Europe.* https://www.axios.com/2024/02/20/ransomware-attacks-north-america-europe

Azadeh, A., Ghaderi, S. F., & Izadbakhsh, H. (2008). Integration of DEA and AHP with computer simulation for railway system improvement and optimization. *Applied Mathematics and Computation, 195*(2), 775–785. DOI: 10.1016/j.amc.2007.05.023

Azadeh, A., Ghaderi, S. F., Miran, Y. P., Ebrahimipour, V., & Suzuki, K. (2007). An integrated framework for continuous assessment and improvement of manufacturing systems. *Applied Mathematics and Computation, 186*(2), 1216–1233. DOI: 10.1016/j.amc.2006.07.152

Azorín, J. M., & Cameron, R. (2010). The application of mixed methods in organisational research: A literature review. *Electronic Journal of Business Research Methods, 8*(2), 95–105. https://academic-publishing.org/index.php/ejbrm/article/view/1184

Bachinger, F., Kronberger, G., & Affenzeller, M. (2021). Continuous improvement and adaptation of predictive models in smart manufacturing and model management. *IET Collaborative Intelligent Manufacturing, 3*(1), 48–63. DOI: 10.1049/cim2.12009

Balugani, E., Butturi, M. A., Chevers, D., Parker, D., & Rimini, B. (2020). Empirical evaluation of the impact of resilience and sustainability on firms' performance. *Sustainability (Basel), 12*(5), 1742. DOI: 10.3390/su12051742

Balzano, M., Marzi, G., & Turzo, T. (2024). SMEs and institutional theory: Major inroads and opportunities ahead. *Management Decision, 63*(13), 1–27. DOI: 10.1108/MD-05-2023-0734

Baran, B. E., & Woznyj, H. M. (2021). Managing VUCA: The human dynamics of agility. *Organizational Dynamics, 50*(2), 100787. DOI: 10.1016/j.orgdyn.2020.100787 PMID: 32843777

Barber, H. F. (1992). Developing strategic leadership: The U.S. Army War College experience. *Journal of Management Development, 11*(6), 4–12. DOI: 10.1108/02621719210018208

Battilana, J., Leca, B., & Boxenbaum, E. (2009). How actors change institutions: Towards a theory of institutional entrepreneurship. *The Academy of Management Annals, 3*(1), 65–107. DOI: 10.5465/19416520903053598

Bednarski, L., Roscoe, S., Blome, C., & Schleper, M. C. (2025). Geopolitical disruptions in global supply chains: A state-of-the-art literature review. *Production Planning and Control, 36*(4), 536–562. DOI: 10.1080/09537287.2023.2286283

Beliaeva, T., Shirokova, G., Wales, W., & Gafforova, E. (2020). Benefiting from economic crisis? Strategic orientation effects, trade-offs, and configurations with resource availability on SME performance. *The International Entrepreneurship and Management Journal, 16*(1), 165–194. DOI: 10.1007/s11365-018-0499-2

Bennett, N., & Lemoine, G. J. (2014). What a difference a word makes: Understanding threats to performance in a VUCA world. *Business Horizons, 57*(3), 311–317. DOI: 10.1016/j.bushor.2014.01.001

Bennett, N., & Lemoine, G. J. (2014, January–February). What VUCA really means for you. *Harvard Business Review, 92*(1–2), 27–28. https://hbr.org/2014/01/what-vuca-really-means-for-you

Bennis, W., & Nanus, B. (1985). *Leaders: The strategies for taking charge.* Harper & Row., https://archive.org/details/leadersstrategie0000benn

Besharov, M. L., & Smith, W. K. (2014). Multiple institutional logics in organizations: Explaining their varied nature and implications. *Academy of Management Review, 39*(3), 364–381. DOI: 10.5465/amr.2011.0431

Bessant, J., Caffyn, S., & Gallagher, M. (2001). An evolutionary model of continuous improvement behaviour. *Technovation, 21*(2), 67–77. DOI: 10.1016/S0166-4972(00)00023-7

Bhakoo, V., & Choi, T. (2013). The iron cage exposed: Institutional pressures and heterogeneity across the healthcare supply chain. *Journal of Operations Management, 31*(6), 432–449. DOI: 10.1016/j.jom.2013.07.016

Bititci, U. S., Carrie, A. S., & McDevitt, L. (1997). Integrated performance measurement systems: A development guide. *International Journal of Operations & Production Management, 17*(5), 522–534. DOI: 10.1108/01443579710167230

Bjørn, A., Lloyd, S. M., & Matthews, H. D. (2021). From the Paris Agreement to corporate climate commitments: Evaluation of seven methods for setting "science-based" emission targets. *Environmental Research Letters, 16*(5), 054019. DOI: 10.1088/1748-9326/abe57b

Board of Governors of the Federal Reserve System. (2025, September 16). *Industrial production and capacity utilization (G.17).* https://www.federalreserve.gov/releases/g17/current/

Bown, C. P. (2020). Export controls: America's other national security threat. *Duke Journal of Comparative & International Law, 30*(2), 283–308. https://scholarship.law.duke.edu/djcil/vol30/iss2/4/

Boyd, J. R. (1976, September 3). *Destruction and creation.* https://www.coljohnboyd.com/static/documents/1976-09-03__Boyd_John_R__Destruction_and_Creation.pdf

Boyd, J. R. (1987). *Destruction and creation.* U.S. Army Command and General Staff College, Center for Army Tactics., https://books.google.com/books?id=JBXWHAAACAAJ

Boyd, J. R. (1995, June 28). *The essence of winning and losing* [Briefing slides]. OODA Loop. https://ooda.de/media/john_boyd_-_the_essence_of_winning_and_losing.pdf

Braunscheidel, M. J., & Suresh, N. C. (2009). The organizational antecedents of a firm's supply chain agility for risk mitigation and response. *Journal of Operations Management, 27*(2), 119–140. DOI: 10.1016/j.jom.2008.09.006

Brodeur, J., Deschamps, I., & Pellerin, R. (2023). Organizational changes approaches to facilitate the management of Industry 4.0 transformation in manufacturing SMEs. *Journal of Manufacturing Technology Management, 34*(7), 1098–1119. DOI: 10.1108/JMTM-10-2022-0359

Bughin, J., Deakin, J., & O'Beirne, B. (2019, October 22). *Digital transformation: Improving the odds of success.* McKinsey & Company. https://www.mckinsey.com/~/media/McKinsey/Business%20Functions/McKinsey%20Digital/Our%20Insights/Digital%20transformation%20Improving%20the%20odds%20of%20success/Digital-transformation-Improving-the-odds-of-success-final.pdf

Bundesamt für Sicherheit in der Informationstechnik. (2024). *Die Lage der IT-Sicherheit in Deutschland 2024* [Annual report]. https://www.bsi.bund.de/SharedDocs/Downloads/DE/BSI/Publikationen/Lageberichte/Lagebericht2024.html

Business Continuity Institute, & Resilience360. (2020). *COVID-19: The future of supply chains* (Survey report). The Business Continuity Institute. https://www.thebci.org/static/7324b815-9364-47d3-9277ab4ce9aa4c0f/a3e39af5-193d-428d-b603d3cf7e600f39/BCI-0007d-The-Future-of-Supply-ChainSingles-Low.pdf

Butt, J. (2020). A strategic roadmap for the manufacturing industry to implement Industry 4.0. *Designs, 4*(2), 11. DOI: 10.3390/designs4020011

Cagliano, R., Canterino, F., Longoni, A., & Bartezzaghi, E. (2019). *The interplay between smart manufacturing technologies and work organization. International Journal of Operations & Production Management, 39*(6/7/8), 913–934. https://doi.org/DOI: 10.1108/IJOPM-01-2018-0029

Caldara, D., & Iacoviello, M. (2022). Measuring geopolitical risk. *The American Economic Review, 112*(4), 1194–1225. DOI: 10.1257/aer.20191823

Callison, S. (n.d.). *What is a gap analysis? An easy-to-use template.* ClearPoint Strategy (Blog). https://www.clearpointstrategy.com/blog/gap-analysis-template

Carayannis, E. G., Papamichail, G., Zotas, N., & Askounis, D. (2025). Corporate foresight: Navigating uncertainty in a VUCA world. *Journal of the Knowledge Economy, 16*(6), 17494–17515. Advance online publication. DOI: 10.1007/s13132-024-02587-4

Casper, S., & Hancké, B. (1999). Global quality norms within national production regimes: ISO 9000 standards in the French and German car industries. *Organization Studies*, *20*(6), 961–985. DOI: 10.1177/0170840699206003

Castaldi, C., & Giarratana, M. S. (2015). A knowledge-management approach to capture organizational learning networks. *Information & Management*, *52*(7), 806–820. DOI: 10.1016/j.im.2015.06.001

Celestin, M., & Sujatha, S. (2024). Impact of global supply-chain disruptions on business resilience: Strategies for adapting to pandemics and geopolitical conflicts. *Zenodo*. https://doi.org/DOI: 10.5281/zenodo.13887198

Cernega, A., Nicolescu, D. N., Mele canu Imre, M., Ripszky Totan, A., Arsene, A. L., erban, R. S., Perpelea, A.-C., Nedea, M.-I., & Pi uru, S.-M. (2024). Volatility, uncertainty, complexity, and ambiguity (VUCA) in healthcare. *Health Care*, *12*(7), 773. Advance online publication. DOI: 10.3390/healthcare12070773 PMID: 38610195

Chen, L., Zhang, H., & Wang, M. (2024). Artificial intelligence in manufacturing: State of the art, perspectives, and future directions. *Journal of Manufacturing Systems*, *74*, 126–149. DOI: 10.1016/j.jmsy.2024.01.004

Chen, T. (2024, April 30). The world of VUCA. In *Endogenous community design* (pp. 1–10). Springer., DOI: 10.1007/978-3-031-56756-8_1

Cheng, C.-H., Li, M.-H., Tang, B.-J., & Cheng, Y.-R. (2024). The impact of knowledge management and organizational learning promotion in small and medium enterprises on the implementation of Industry 4.0 and competitiveness. *Administrative Sciences*, *14*(8), 161. DOI: 10.3390/admsci14080161

Cheng, Z. M., Bonetti, F., de Regt, A., Ribeiro, J. L., & Plangger, K. (2024). Principles of responsible digital implementation: Developing operational business resilience to reduce resistance to digital innovations. *Organizational Dynamics*, *53*(2), 101043. DOI: 10.1016/j.orgdyn.2024.101043

Chiarini, A. (2020). Industry 4.0, quality management and TQM world: A systematic literature review and a proposed agenda for further research. *The TQM Journal*, *32*(4), 603–616. DOI: 10.1108/TQM-04-2020-0082

Chima, O. K., Idemudia, S. O., Ezeilo, O. J., Ojonugwa, B. M., & Adesuyi, A. O. M. (2022). Advanced review of SME regulatory compliance models across US state-level jurisdictions. Shodhshauryam. *International Scientific Refereed Research Journal*, *5*(2), 191–209.

Çınar, Z. M., Zeeshan, Q., & Korhan, O. (2021). A framework for Industry 4.0 readiness and maturity of smart manufacturing enterprises: A case study. *Sustainability (Basel), 13*(12), 6659. DOI: 10.3390/su13126659

Cohen, J. B., Loeber, A. M. C., Marschalek, I., Bernstein, M. J., Blok, V., Tabarés, R., Gianni, R., & Griessler, E. (2024). From experimentation to structural change: Fostering institutional entrepreneurship for public engagement in research and innovation. *Science & Public Policy, 51*(2), 324–336. DOI: 10.1093/scipol/scad065

Cosa, M., & Torelli, R. (2024). Digital transformation and flexible performance management: A systematic literature review of the evolution of performance measurement systems. *Global Journal of Flexible Systems Managment, 25*(3), 445–466. DOI: 10.1007/s40171-024-00409-9

Creazza, A., Colicchia, C., Spiezia, S., & Dallari, F. (2022). Who cares? Supply-chain managers' perceptions regarding cyber supply-chain risk management in the digital transformation era. *Supply Chain Management, 27*(1), 30–53. DOI: 10.1108/SCM-02-2020-0073

Cruzes, D. S., Dybå, T., Runeson, P., & Höst, M. (2015). Case studies synthesis: A thematic, cross-case, and narrative synthesis worked example. *Empirical Software Engineering, 20*(6), 1634–1665. DOI: 10.1007/s10664-014-9326-8

Cuban, L. (2010). *As good as it gets: What school reform brought to Austin.* Harvard University Press. https://www.hup.harvard.edu/books/9780674035546

DealHub. (2024, June 10). *What is phased implementation?* Retrieved September 16, 2025, from https://dealhub.io/glossary/phased-implementation/

Dechezleprêtre, A., & Sato, M. (2017). The impacts of environmental regulations on competitiveness. *Review of environmental economics and policy.* https://www.journals.uchicago.edu/doi/full/10.1093/reep/rex013

Del Río-Casasola, A., López-Calle, P., & Ruiz-Gálvez, M. E. (2024). The transformation of the automotive industry toward electrification and its impact on global value chains. *European Research on Management and Business Economics, 30*(1), 100242. Advance online publication. DOI: 10.1016/j.iedeen.2024.100242

Deming, W. E. (2000). *Out of the crisis* (Press, M. I. T., Ed.). MIT Press. (Original work published 1986), https://mitpress.mit.edu/9780262541152/out-of-the-crisis/

Dhillon, R., & Nguyen, Q. C. (2021). Strategies to respond to a VUCA world.

Dicken, P. (1992). International production in a volatile regulatory environment: The influence of national regulatory policies on the spatial strategies of transnational corporations. *Geoforum*, *23*(3), 303–316. DOI: 10.1016/0016-7185(92)90044-5

DiMaggio, P. J., & Powell, W. W. (1983). The iron cage revisited: Institutional isomorphism and collective rationality in organizational fields. *American Sociological Review*, *48*(2), 147–160. DOI: 10.2307/2095101

Doherty, B., Haugh, H., & Lyon, F. (2014). Social enterprises as hybrid organizations: A review and research agenda. *International Journal of Management Reviews*, *16*(4), 417–436. DOI: 10.1111/ijmr.12028

Doskenov, B., Adaji, O., & Dosumu, R. (2025, March 31). Improving manufacturing with adaptive learning in Industry 4.0: Challenges, and policy implications. *Current Journal of Applied Science and Technology*, *44*(4), 111–132. DOI: 10.9734/cjast/2025/v44i44518

Doz, Y. L., & Kosonen, M. (2010). Embedding strategic agility: A leadership agenda for accelerating business model renewal. *Long Range Planning*, *43*(2–3), 370–382. DOI: 10.1016/j.lrp.2009.07.006

Dragos. (2024). *2023 OT Cybersecurity Year in Review: Executive summary.* https://hub.dragos.com/year-in-review-executive-summary-2023

Ebrahimi, S. M., & Koh, L. (2021). Manufacturing sustainability: Institutional theory and life cycle thinking. *Journal of Cleaner Production*, *298*, 126787. DOI: 10.1016/j.jclepro.2021.126787

Efthymiou, K., Mourtzis, D., Pagoropoulos, A., Papakostas, N., & Chryssolouris, G. (2016). Manufacturing systems complexity analysis methods review. *International Journal of Computer Integrated Manufacturing*, *29*(9), 1025–1044. DOI: 10.1080/0951192X.2015.1130245

Errida, A., & Lotfi, B. (2021). The determinants of organizational change management success: Literature review and case study. *International Journal of Engineering Business Management*, *13*, 1–15. DOI: 10.1177/18479790211016273

Étienne, J. (2010). Compliance theories: A literature review. *Revue Française de Science Politique*, *60*(3), 493–517. DOI: 10.3917/rfsp.603.0493

European Commission. (2024, November 19). *Report on the functioning of the European carbon market in 2023* (COM(2024) 538 final). https://eur-lex.europa.eu/legal-content/EN/TXT/PDF/?uri=CELEX:52024DC0538

European Parliament and Council. (2022, December 14). Directive (EU) 2022/2464 … as regards corporate sustainability reporting (CSRD). *Official Journal of the European Union, L, 322*, 15–80. https://eur-lex.europa.eu/eli/dir/2022/2464/oj/eng

Faccio, M., & Cohen, Y. (2024). Intelligent cobot systems: Human-cobot collaboration in manufacturing. *Journal of Intelligent Manufacturing, 35*(5), 1905–1907. DOI: 10.1007/s10845-023-02142-z

FasterCapital. (2023). *Process improvement: Process perfection: Employing gap analysis for continuous improvement.* Retrieved September 16, 2025, from https://fastercapital.com/content/Process-Improvement--Process-Perfection--Employing-Gap-Analysis-for-Continuous-Improvement.html

Fernando, Y., Wahyuni-TD, I. S., Gui, A., Ikhsan, R. B., Mergeresa, F., & Ganesan, Y. (2023). A mixed-method study on the barriers of Industry 4.0 adoption in the Indonesian SMEs manufacturing supply chains. *Journal of Science and Technology Policy Management, 14*(4), 678–695. DOI: 10.1108/JSTPM-10-2021-0155

Festo. (2017, June 10). *Industry 4.0 finds increasing application in production.* https://press.festo.com/en/industry-segments/industry-4-0-finds-increasing-application-in-production

Fiene, R. (2023). *The importance of the theory of regulatory compliance.* SSRN *Electronic Journal.* https://doi.org/DOI: 10.2139/ssrn.4597469

Fridgeirsson, T. V., Ingason, H. T., Jonasson, H. I., & Kristjansdottir, B. H. (2021). The VUCAlity of projects: A new approach to assess a project risk in a complex world. *Sustainability (Basel), 13*(7), 3808. DOI: 10.3390/su13073808

Frohlich, M. T., & Westbrook, R. (2001). Arcs of integration: An international study of supply chain strategies. *Journal of Operations Management, 19*(2), 185–200. DOI: 10.1016/S0272-6963(00)00055-3

Funk, R. J., & Hirschman, D. (2017). Beyond nonmarket strategy: Market actions as corporate political activity. *Academy of Management Review, 42*(1), 32–52. DOI: 10.5465/amr.2013.0178

Galli, D., Torelli, R., & Caccialanza, A. (2023). Sustainability performance and sustainability reporting in SMEs: A love affair or a fight? *Journal of Management & Organization, 29*(6), 1063–1088. DOI: 10.1017/jmo.2023.40

Geissdoerfer, M., Savaget, P., Bocken, N. M. P., & Hultink, E. J. (2017). The circular economy—A new sustainability paradigm? *Journal of Cleaner Production, 143*, 757–768. DOI: 10.1016/j.jclepro.2016.12.048

Ghobakhloo, M. (2018). The future of manufacturing industry: A strategic roadmap toward Industry 4.0. *Journal of Manufacturing Technology Management, 29*(6), 910–936. DOI: 10.1108/JMTM-02-2018-0057

Ghobakhloo, M. (2020). Industry 4.0, digitization, and opportunities for sustainability. *Journal of Cleaner Production, 252*, 119869. DOI: 10.1016/j.jclepro.2019.119869

Gilad, B. (2004). *Early warning: Using competitive intelligence to anticipate market shifts, control risk, and create powerful strategies*. AMACOM., https://www.harpercollinsleadership.com/9780814432068/early-warning/

Giones, F., & Brem, A. (2022). Clarifying the conceptual map of VUCA: A systematic review. *The International Journal of Organizational Analysis, 30*(7), 196–221. DOI: 10.1108/IJOA-02-2022-3136

Gomes, C. F., Yasin, M. M., & Lisboa, J. V. (2004). A literature review of manufacturing performance measures and measurement in an organizational context: A framework and direction for future research. *Journal of Manufacturing Technology Management, 15*(6), 511–530. DOI: 10.1108/17410380410547906

Gordon, L. A., & Loeb, M. P. (2002). The economics of information security investment. *ACM Transactions on Information and System Security, 5*(4), 438–457. DOI: 10.1145/581271.581274

Greenwood, R., Raynard, M., Kodeih, F., Micelotta, E. R., & Lounsbury, M. (2011). Institutional complexity and organizational responses. *The Academy of Management Annals, 5*(1), 317–371. DOI: 10.5465/19416520.2011.590299

Group, B. M. W. (2023). *How AI is revolutionising production*. https://www.bmwgroup.com/en/news/general/2023/aiqx.html

Gu, X., Jin, X., Ni, J., & Koren, Y. (2015). Manufacturing system design for resilience. *Procedia CIRP, 36*, 135–140. DOI: 10.1016/j.procir.2015.02.075

Gunningham, N., & Rees, J. (1997). Industry self-regulation: An institutional perspective. *Law & Policy, 19*(4), 363–414. DOI: 10.1111/1467-9930.t01-1-00033

Guo, B., Parlane, S., & Ryan, L. (2025). Regulatory compliance in the automobile industry. *UCD Centre for Economic Research Working Paper Series*, WP25/06. https://www.ucd.ie/economics/t4media/WP2025_06.pdf

Halkos, G., Moll de Alba, J., & Todorov, V. (2021). Analyzing manufacturing sector and selected development challenges: A panel data analysis. *Energy, 230*, 120807. DOI: 10.1016/j.energy.2021.121253

Hardy, C., & Maguire, S. (2008). Institutional entrepreneurship. In Greenwood, R., Oliver, C., Sahlin, K., & Suddaby, R. (Eds.), *The SAGE handbook of organizational institutionalism* (pp. 198–217). SAGE. DOI: 10.4135/9781849200387.n8

Harvard Business School Online. (2024, February 14). *How gap analysis can drive strategic change in your company*. HBS Online Blog. https://online.hbs.edu/blog/post/gap-analysis

Helfat, C. E., & Peteraf, M. A. (2003). The dynamic resource-based view: Capability lifecycles. *Strategic Management Journal, 24*(10), 997–1010. DOI: 10.1002/smj.332

Hendricks, K. B., & Singhal, V. R. (2005). Association between supply chain glitches and operating performance. *Management Science, 51*(5), 695–711. DOI: 10.1287/mnsc.1040.0353

Hentout, A., Aouache, M., Maoudj, A., & Akli, I. (2019). Human–robot interaction in industrial collaborative robotics: A literature review of the decade 2008–2017. *Advanced Robotics, 33*(15–16), 764–799. DOI: 10.1080/01691864.2019.1636714

Hermann, J., David, A., Wagner, A., & Ruskowski, M. (2020). Considering interdependencies for a dynamic generation of process chains for Production as a Service. *Procedia Manufacturing, 51*, 1454–1461. DOI: 10.1016/j.promfg.2020.10.202

Hermundsdottir, F., & Aspelund, A. (2022). Competitive sustainable manufacturing - Sustainability strategies, environmental and social innovations, and their effects on firm performance. *Journal of Cleaner Production, 370*, 133474. DOI: 10.1016/j.jclepro.2022.133474

Hinterhuber, A. (2017). Value quantification capabilities in industrial markets. *Journal of Business Research, 76*, 163–178. DOI: 10.1016/j.jbusres.2016.11.019

Hoejmose, S. U., Grosvold, J., & Millington, A. (2014). The effect of institutional pressure on cooperative and coercive "green" supply chain practices. *Journal of Purchasing and Supply Management, 20*(4), 215–224. DOI: 10.1016/j.pursup.2014.07.002

Hofer, A., Brandl, F., Bauer, H., Haghi, S., & Reinhart, G. (2020). A framework for managing innovation cycles in manufacturing systems. *Procedia CIRP, 93*, 771–776. DOI: 10.1016/j.procir.2020.03.064

Holmström, J., & Partanen, J. (2014). Digital manufacturing-driven transformations of service supply chains for complex products. *Supply Chain Management, 19*(4), 421–430. DOI: 10.1108/SCM-10-2013-0387

Hoon, C. (2013). Meta-synthesis of qualitative case studies: An approach to theory building. *Organizational Research Methods, 16*(4), 522–556. DOI: 10.1177/1094428113484969

Horvát, P., & Webb, C. (2020, November 18). *The OECD STAN database for industrial analysis: Sources and methods*. OECD Science, Technology and Industry Working Papers, No. 2020/10. OECD Publishing. https://doi.org/DOI: 10.1787/18151965

Hussain, M., Tapinos, E., & Knight, L. (2017). Scenario-driven roadmapping for technology foresight. *Technological Forecasting and Social Change, 124*, 160–177. DOI: 10.1016/j.techfore.2017.05.005

IBM. (2024, November 15). How is AI being used in manufacturing? *IBM Think*. https://www.ibm.com/think/topics/ai-in-manufacturing

Iborra, M., Safón, V., & Dolz, C. (2020). What explains the resilience of SMEs? Ambidexterity capability and strategic consistency. *Long Range Planning, 53*(6), 101947. DOI: 10.1016/j.lrp.2019.101947

IFRS Foundation. (2023, July 27). *Proposed IFRS Sustainability Disclosure Taxonomy* [Exposure draft]. https://www.ifrs.org/projects/completed-projects/2024/ifrs-sustainability-disclosure-taxonomy/proposed-taxonomy-cls-ifrs-sds/

Institute for Manufacturing, University of Cambridge. (n.d.). *Gap analysis. Decision Support Tools*. Retrieved September 16, 2025, from https://www.ifm.eng.cam.ac.uk/research/dstools/gap-analysis/

Ivanov, D. (2021). Supply chain viability and the COVID-19 pandemic: A conceptual and formal generalisation of four major adaptation strategies. *International Journal of Production Research, 59*(12), 3535–3552. DOI: 10.1080/00207543.2021.1890852

Ivanov, D., & Dolgui, A. (2021). A digital supply chain twin for managing the disruption risks and resilience in the era of Industry 4.0. *Production Planning and Control, 32*(9), 775–788. DOI: 10.1080/09537287.2020.1768450

Jarrahi, M. H. (2018). Artificial intelligence and the future of work: Human–AI symbiosis in organizational decision making. *Business Horizons, 61*(4), 577–586. DOI: 10.1016/j.bushor.2018.03.007

Joglekar, N., Parker, G., & Srai, J. S. (2024, January 31). Why manufacturers need a phased approach to digital transformation. *MIT Sloan Management Review, 65*(3), 54–59. https://sloanreview.mit.edu/article/why-manufacturers-need-a-phased-approach-to-digital-transformation/

Johansen, B. (2007). *Get there early: Sensing the future to compete in the present.* Berrett-Koehler Publishers.

Johansen, B. (2009). *Leaders make the future: Ten new leadership skills for an uncertain world.* Berrett-Koehler Publishers.

Kagermann, H., Wahlster, W., & Helbig, J. (2013, April). *Recommendations for implementing the strategic initiative INDUSTRIE 4.0: Final report of the Industrie 4.0 Working Group.* acatech – National Academy of Science and Engineering / Forschungsunion. https://en.acatech.de/publication/recommendations-for-implementing-the-strategic-initiative-industrie-4-0-final-report-of-the-industrie-4-0-working-group/

Kale, E., Aknar, A., & Başar, Ö. (2019). Absorptive capacity and firm performance: The mediating role of strategic agility. *International Journal of Hospitality Management, 78*, 276–283. DOI: 10.1016/j.ijhm.2018.09.010

Kang, Z., Catal, C., & Tekinerdogan, B. (2020). Machine learning applications in production lines: A systematic literature review. *Computers & Industrial Engineering, 149*, 106773. DOI: 10.1016/j.cie.2020.106773

Kauppi, K. (2013). Extending the use of institutional theory in operations and supply chain management research: Review and research suggestions. *International Journal of Operations & Production Management, 33*(10), 1318–1345. DOI: 10.1108/IJOPM-10-2011-0364

Kauppi, K., & Hannibal, C. (2017). Institutional pressures and sustainability assessment in supply chains. *Supply Chain Management, 22*(5), 458–472. DOI: 10.1108/SCM-01-2017-0004

Kayande, U., & Kediya, S. (2025). Critical drivers of AI integration in industrial processes for sustainable manufacturing. *Smart and Sustainable Manufacturing Systems, 5*, 00503. DOI: 10.1007/s41660-025-00503-0

Keleko, A. T., Kamsu-Foguem, B., Ngouna, R. H., & Tongne, A. (2022). Artificial intelligence and real-time predictive maintenance in Industry 4.0: A bibliometric analysis. *AI and Ethics, 2*(4), 553–577. DOI: 10.1007/s43681-021-00132-6

Kestria. (2024, May 15). *AI and robotics integration: Transforming production and automation.* https://kestria.com/insights/ai-and-robotics-integration-transforming-productio/

Khan, S. A. R., Qianli, D., SongBo, W., Zaman, K., & Zhang, Y. (2017). Environmental logistics performance indicators affecting per capita income and sectoral growth: Evidence from a panel of selected global ranked logistics countries. *Environmental Science and Pollution Research International, 24*(2), 1518–1531. DOI: 10.1007/s11356-016-7916-2 PMID: 27785719

Kırmızı, M., & Kocaoğlu, B. (2022). Digital transformation maturity model development framework based on design science: Case studies in manufacturing industry. *Journal of Manufacturing Technology Management, 33*(7), 1319–1349. DOI: 10.1108/JMTM-11-2021-0476

Kotter, J. P. (2012). *Leading change* (with a new preface by the author). Harvard Business Review Press. https://store.hbr.org/product/leading-change-with-a-new-preface-by-the-author/11116

Kristianto, Y., Gunasekaran, A., Helo, P., & Hao, Y. (2014). A model of resilient supply chain network design: A two-stage programming with fuzzy shortest path. *Expert Systems with Applications, 41*(1), 39–49. DOI: 10.1016/j.eswa.2013.07.009

Kristjánsdóttir, B. H. (2019). *How VUCA is your project? Toward a VUCA meter to assess the volatility, uncertainty, complexity and ambiguity of complex projects* (Master's thesis). Reykjavík University. https://skemman.is/handle/1946/33540

Kumar, H., & Yadav, S. (2024). Additive and good manufacturing practices in conformity assessment. In Bhatnagar, A., Yadav, S., Achanta, V., Harmes-Liedtke, U., & Rab, S. (Eds.), *Handbook of Quality System, Accreditation and Conformity Assessment* (pp. 971–988). Springer., DOI: 10.1007/978-981-97-5750-3_53

Kumpulainen, M., & Seppänen, M. (2022). Combining Web of Science and Scopus datasets in citation-based literature study. *Scientometrics, 127*(10), 5613–5631. DOI: 10.1007/s11192-022-04475-7

Lawrence, A. (2024, September 11). *AI's impact on robots in manufacturing*. American Machinist. https://www.americanmachinist.com/automation-and-robotics/article/55138472/the-impact-of-ai-powered-robots-on-manufacturing-efficiency-and-quality-control-automation-technology American Machinist

Lee, J., Singh, J., Azamfar, M., & Pandhare, V. (2020). Industrial AI and predictive analytics for smart manufacturing systems. In *Smart Manufacturing* (pp. 213–246). Elsevier., DOI: 10.1016/B978-0-12-820027-8.00008-3

Leonidou, E., Christofi, M., Vrontis, D., & Thrassou, A. (2020). An integrative framework of stakeholder engagement for innovation management and entrepreneurship development. *Journal of Business Research, 119*, 245–258. DOI: 10.1016/j.jbusres.2018.11.054

Liker, J. K. (2004). *The Toyota way: 14 management principles from the world's greatest manufacturer*. McGraw-Hill. https://search.worldcat.org/title/The-Toyota-way-%3A-14-management-principles/oclc/1031522159 WorldCat

Lindström, J., Larsson, H., Jonsson, M., & Lejon, E. (2017). Towards intelligent and sustainable production: Combining and integrating online predictive maintenance and continuous quality control. *Procedia CIRP, 63*, 443–448. DOI: 10.1016/j.procir.2017.03.099

Liu, H., Ke, W., Wei, K. K., & Hua, Z. (2013). The impact of IT capabilities on firm performance: The mediating roles of absorptive capacity and supply chain agility. *Decision Support Systems, 54*(3), 1452–1462. DOI: 10.1016/j.dss.2012.12.016

Lorenzo Ochoa, O., Claes, B., Koryak, O., & Diaz, A. (2017). Integration through orchestration: The interplay between enterprise systems and inventory management capabilities. *Journal of Enterprise Information Management, 30*(4), 555–582. DOI: 10.1108/JEIM-02-2016-0060

Machado, C. G., Winroth, M., Almström, P., Ericson Öberg, A., Kurdve, M., & AlMashalah, S. (2021). Digital organisational readiness: Experiences from manufacturing companies. *Journal of Manufacturing Technology Management, 32*(9), 167–182. DOI: 10.1108/JMTM-05-2019-0188

Macher, J. T., & Mowery, D. C. (2009). Measuring dynamic capabilities: Practices and performance in semiconductor manufacturing. *British Journal of Management, 20*(s1, S1), S41–S62. DOI: 10.1111/j.1467-8551.2008.00612.x

Mackey, R. W. (1992). *Strategic leadership and the Army War College (Military Studies Program paper)*. U.S. Army War College. https://apps.dtic.mil/sti/citations/ADA249373

Majumdar, S. K., & Marcus, A. A. (2001). Rules versus discretion: The productivity consequences of flexible regulation. *Academy of Management Journal, 44*(1), 170–179. DOI: 10.2307/3069344

Makhanya, M. T. B. (2023). The implications for risk management in the era of technological advancements. In *The future of risk management*. IntechOpen., DOI: 10.5772/intechopen.1003899

Manufacturing Leadership Council. (2024, October 1). *Survey: Leadership preparedness improves, but gaps remain. Manufacturing Leadership Journal.* https://manufacturingleadershipcouncil.com/leadership-preparedness-improves-but-gaps-remain-37327/

Margherita, A., & Braccini, A. M. (2022). Analysis of the adoption of emergent technologies for risk management in the era of digital manufacturing. *Technological Forecasting and Social Change*, *178*, 121566. DOI: 10.1016/j.techfore.2022.121566

Matheson, E., Minto, R., Zampieri, E. G. G., Faccio, M., & Rosati, G. (2019). Human–robot collaboration in manufacturing applications: A review. *Robotics (Basel, Switzerland)*, *8*(4), 100. DOI: 10.3390/robotics8040100

McGrath, R. G., & MacMillan, I. C. (2000). *The entrepreneurial mindset: Strategies for continuously creating opportunity in an age of uncertainty*. Harvard Business School Press., DOI: 10.4324/9781482293746

McKinsey & Company. (2024, April 9). *How manufacturing's lighthouses are capturing the full value of AI.* https://www.mckinsey.com/capabilities/operations/our-insights/how-manufacturings-lighthouses-are-capturing-the-full-value-of-ai

McKinsey Global Institute. (2024, January). *Geopolitics and the geometry of global trade.* McKinsey & Company. https://www.mckinsey.com/mgi/our-research/geopolitics-and-the-geometry-of-global-trade

Mick, M. M. A. P., Kovaleski, J. L., Mick, R. L., & Chiroli, D. M. G. (2024). Developing a sustainable digital transformation roadmap for SMEs: Integrating digital maturity and strategic alignment. *Sustainability (Basel)*, *16*(20), 8745. DOI: 10.3390/su16208745

Miller, B., & Rowe, D. C. (2012). A survey of SCADA and critical infrastructure incidents. In *Proceedings of the 1st Annual Conference on Research in Information Technology (RIIT '12)* (pp. 51–56). ACM. https://doi.org/DOI: 10.1145/2380790.2380805

Mintzberg, H. (1987). The strategy concept I: Five Ps for strategy. *California Management Review*, *30*(1), 11–24. DOI: 10.2307/41165263

Mohanta, B., Nanda, P., & Patnaik, S. (2019). Management of VUCA (Volatility, Uncertainty, Complexity and Ambiguity) Using machine learning techniques in industry 4.0 paradigm. In *New Paradigm of Industry 4.0: Internet of Things, Big Data & Cyber Physical Systems* (pp. 1–24). Springer International Publishing.

Moheb-Alizadeh, H., & Handfield, R. B. (2018). The impact of raw materials price volatility on cost of goods sold (COGS) for product manufacturing. *IEEE Transactions on Engineering Management*, *65*(3), 460–473. DOI: 10.1109/TEM.2018.2796447

Molenda, P., Groneberg, H., Schötz, S., & Döpper, F. (2023). Resilience balanced scorecard: Measuring resilience of manufacturing companies at multiple levels. *Procedia CIRP*, *120*, 189–194. DOI: 10.1016/j.procir.2023.08.034

Momeni, B., Rapaccini, M., & Martinsuo, M. (2024). Manufacturers managing complexity during the digital servitization journey. *Journal of Manufacturing Technology Management*, *35*(9), 51–72. DOI: 10.1108/JMTM-07-2023-0275

Moradlou, H., Skipworth, H., Bals, L., Aktas, E., & Roscoe, S. (2025). Geopolitical disruptions and supply chain structural ambidexterity. *International Journal of Operations & Production Management*, *45*(4), 836–862. DOI: 10.1108/IJOPM-11-2023-0915

Morgan, J. M., & Liker, J. K. (2006). *The Toyota product development system: Integrating people, process, and technology*. Productivity Press., DOI: 10.4324/9781482293746

Mubarak, M. F., & Petraite, M. (2020). Industry 4.0 technologies, digital trust and technological orientation: What matters in open innovation? *Technological Forecasting and Social Change*, *161*, 120332. DOI: 10.1016/j.techfore.2020.120332

Neely, A., Gregory, M., & Platts, K. (2005). Performance measurement system design: A literature review and research agenda. *International Journal of Operations & Production Management*, *25*(12), 1228–1263. DOI: 10.1108/01443570510633639

Negri, M., Cagno, E., Colicchia, C., & Sarkis, J. (2021). Integrating sustainability and resilience in the supply chain: A systematic literature review and a research agenda. *Business Strategy and the Environment*, *30*(7), 2858–2886. DOI: 10.1002/bse.2776

Nitsche, B., & Straube, F. (2020). Efficiently managing supply chain volatility–a management framework for the manufacturing industry. *Procedia Manufacturing*, *43*, 320–327. DOI: 10.1016/j.promfg.2020.02.166

Norsk Hydro. (2019, March 25). *Update on cyber attack March 25*. https://www.hydro.com/en/global/media/news/2019/update-on-cyber-attack-march-25/

O'Brien, C. (2002). Global manufacturing and the sustainable economy. *International Journal of Production Research*, *40*(15), 3867–3877. DOI: 10.1080/00207540210157169

OECD. (2025). *OECD Economic Outlook, Volume 2025 Issue 1: Tackling Uncertainty, Reviving Growth*. OECD Publishing. https://doi.org/DOI: 10.1787/83363382-en

Oliver, C. (1991). Strategic responses to institutional processes. *Academy of Management Review*, *16*(1), 145–179. DOI: 10.2307/258610

Osinga, F. P. B. (2007). *Science, strategy and war: The strategic theory of John Boyd.* Routledge., DOI: 10.4324/9780203088869

Ostadi, B., Barrani, L., & Aghdasi, M. (2024). Developing a strategic roadmap towards integration in Industry 4.0: A dynamic capabilities theory perspective. *Technological Forecasting and Social Change, 208*, 123679. DOI: 10.1016/j.techfore.2024.123679

Panorama Consulting Group. (2024, February 6). *The role of a stakeholder analysis in change management.* https://www.panorama-consulting.com/stakeholder-analysis-in-change-management/

Paschou, T., Rapaccini, M., Adrodegari, F., & Saccani, N. (2020). Digital servitization in manufacturing: A systematic literature review and research agenda. *Industrial Marketing Management, 89*, 278–292. DOI: 10.1016/j.indmarman.2020.02.012

Patterson, C. M., Nurse, J. R., & Franqueira, V. N. (2024). "I don't think we're there yet": The practices and challenges of organisational learning from cyber security incidents. *Computers & Security, 139*, 103699. DOI: 10.1016/j.cose.2023.103699

Pertheban, T. R., Thurasamy, R., Marimuthu, A., Venkatachalam, K. R., Annamalah, S., Paraman, P., & Hoo, W. C. (2023). The impact of proactive resilience strategies on organizational performance: Role of ambidextrous and dynamic capabilities of SMEs in manufacturing sector. *Sustainability (Basel), 15*(16), 12665. DOI: 10.3390/su151612665

Peschl, M. F. (2023). Learning from the future as a novel paradigm for integrating organizational learning and innovation. *The Learning Organization, 30*(1), 6–22. DOI: 10.1108/TLO-01-2021-0018

Plathottam, S. J., Rzonca, A., Lakhnori, R., & Iloeje, C. O. (2023). A review of artificial intelligence applications in manufacturing operations. *Journal of Advanced Manufacturing and Processing, 5*(3), e10159. DOI: 10.1002/amp2.10159

Pratono, A. H. (2022). The strategic innovation under information technological turbulence: The role of organisational resilience in competitive advantage. *Competitiveness Review, 32*(3), 475–491. DOI: 10.1108/CR-03-2021-0046

Prescott, J. E. (1999). The evolution of competitive intelligence: Designing a process for action. *Proposal Management* (Spring), 37–52. https://winningthebusiness.com/wp-content/uploads/2018/05/APMP-Journal-Spring-1999.pdf

Ramya, G., & Srinivasagan, K. G. (2024). Integrating cybersecurity threats into smart manufacturing: Best practices and frameworks. In *Artificial intelligence solutions for cyber-physical systems* (1st ed.). Auerbach Publications/Taylor & Francis., DOI: 10.1201/9781032694375-7

Reeves, M., Levin, S. A., & Ueda, D. (2016, January–February). The biology of corporate survival. *Harvard Business Review, 94*(1/2), 46–55. https://hbr.org/2016/01/the-biology-of-corporate-survival

Richards, C. (2004). *Certain to win: The strategy of John Boyd, applied to business.* Xlibris. https://www.xlibris.com/en/bookstore/bookdetails/572837-certain-to-win

Rittenhouse, J. (2015, October 10). *Improving stakeholder management using change management tools.* Paper presented at PMI® Global Congress 2015—North America, Orlando, FL. Project Management Institute. https://www.pmi.org/learning/library/improve-stakeholder-management-9901

Roos, J., von Krogh, G., & Roos, G. (2025). *Strategi – en innføring* (8th ed.). Fagbokforlaget., https://www.fagbokforlaget.no/produkt/9788245051094-strategi-en-innforing-8

Roscoe, S., Aktas, E., Petersen, K. J., Skipworth, H. D., Handfield, R. B., & Habib, F. (2022). Redesigning global supply chains during compounding geopolitical disruptions: The role of supply chain logics. *International Journal of Operations & Production Management, 42*(9), 1407–1434. DOI: 10.1108/IJOPM-12-2021-0777

Saad, S. M., Bahadori, R., & Jafarnejad, H. (2021). The smart SME technology readiness assessment methodology in the context of Industry 4.0. *Journal of Manufacturing Technology Management, 32*(5), 1037–1065. DOI: 10.1108/JMTM-07-2020-0267

Safar, M., Hidayat, R., & Chen, C. (2023). A risk management framework for Industry 4.0 environment. *Sustainability (Basel), 15*(2), 1395. DOI: 10.3390/su15021395

Saghafian, M., Laumann, K., & Skogstad, M. R. (2021). Stagewise overview of issues influencing organizational technology adoption and use. *Frontiers in Psychology, 12*, 630145. DOI: 10.3389/fpsyg.2021.630145 PMID: 33815216

Sajid, H. (2024, January 10). *AI in robotics: 6 groundbreaking applications.* V7. https://www.v7labs.com/blog/ai-in-robotics

Santos, R. C., & Martinho, J. L. (2020). An Industry 4.0 maturity model proposal. *Journal of Manufacturing Technology Management, 31*(5), 1023–1043. DOI: 10.1108/JMTM-09-2018-0284

Saritas, O., & Aylen, J. (2010). Using scenarios for roadmapping: The case of clean production. *Technological Forecasting and Social Change, 77*(7), 1061–1075. DOI: 10.1016/j.techfore.2010.03.003

Schoemaker, P. J. H., Heaton, S., & Teece, D. J. (2018). Innovation, dynamic capabilities, and leadership. *California Management Review*, *61*(1), 15–42. DOI: 10.1177/0008125618790246

Schuh, G., Patzwald, M., & Salmen, M. (2019). *Resilient technology strategy in volatile environments: Derivation of requirements to enable long-term strategic positioning in times of VUCA*. In *PICMET '19: 2019 Portland International Conference on Management of Engineering and Technology* (pp. 1–8). IEEE. https://doi.org/DOI: 10.23919/PICMET.2019.8893840

Schwarz, L. (2024, May 22). *AI in manufacturing: Benefits and 15 use cases*. NetSuite. https://www.netsuite.com/portal/resource/articles/erp/ai-in-manufacturing.shtml

Scott, W. R. (2008). *Institutions and organizations: Ideas and interests* (3rd ed.). Sage.

Shih, W. C. (2020, September–October). Global supply chains in a post-pandemic world. *Harvard Business Review*, *98*(5), 82–89. https://hbr.org/2020/09/global-supply-chains-in-a-post-pandemic-world

Shishodia, A., Sharma, R., Rajesh, R., & Munim, Z. H. (2023). Supply chain resilience: A review, conceptual framework and future research. *International Journal of Logistics Management*, *34*(4), 879–908. DOI: 10.1108/IJLM-03-2021-0169

Shook, J. (2009, February 3). *The remarkable Chief Engineer*. Lean Enterprise Institute. https://www.lean.org/the-lean-post/articles/the-remarkable-chief-engineer/

Siemens, A. G. Digital Factory. (2017). *Leading as the digital enterprise: Electronics Works Amberg — Digitalization practices* [Brochure]. https://assets.new.siemens.com/siemens/assets/api/uuid%3A73a13464-b0d2-4a40-acd1-a79919ac98e2/brochure-electronics-works-amberg-digitalization-practices-e.pdf

Singh, A., Kaur, P., & Kumar, R. (2023). Artificial intelligence, machine learning and deep learning in advanced robotics: A review. *Journal of Artificial Intelligence and Technology*, *3*(1), 22–36. DOI: 10.1016/j.jaite.2023.03.001

Smartsheet. (2023). *Guide to gap analysis with examples*. Smartsheet. Retrieved September 16, 2025, from https://www.smartsheet.com/gap-analysis-method-examples

Sophos. (2024, May 28). *The State of Ransomware in Manufacturing and Production 2024*. Sophos. https://www.sophos.com/en-us/whitepaper/state-of-ransomware-in-manufacturing-and-production

Spaltini, M., Acerbi, F., Pinzone, M., Gusmeroli, S., & Taisch, M. (2022). Defining the roadmap towards Industry 4.0: The 6Ps maturity model for manufacturing SMEs. *Procedia CIRP*, *105*, 631–636. https://www.sciencedirect.com/science/article/pii/S2212827122001068. DOI: 10.1016/j.procir.2022.02.105

Sukathong, S., Suksawang, P., & Naenna, T. (2021). Analyzing the importance of critical success factors for the adoption of advanced manufacturing technologies. *International Journal of Engineering Business Management*, *13*, 18479790211055057. https://journals.sagepub.com/doi/full/10.1177/18479790211055057. DOI: 10.1177/18479790211055057

Sullivan, G. R., & Harper, M. V. (1996). *Hope is not a method: What business leaders can learn from America's Army*. Times Business. https://www.penguinrandomhouse.com/books/175183/hope-is-not-a-method-by-gordon-r-sullivan/

Tang, L., Jing, K., He, J., & Stanley, H. E. (2016). Complex interdependent supply chain networks: Cascading failure and robustness. *Physica A*, *443*, 58–69. DOI: 10.1016/j.physa.2015.09.082

TechTarget. (2015, October). *VUCA (volatility, uncertainty, complexity and ambiguity)*. WhatIs.com (TechTarget). https://www.techtarget.com/whatis/definition/VUCA-volatility-uncertainty-complexity-and-ambiguity

Teece, D. J. (2007). Explicating dynamic capabilities: The nature and microfoundations of (sustainable) enterprise performance. *Strategic Management Journal*, *28*(13), 1319–1350. DOI: 10.1002/smj.640

Teece, D. J., Pisano, G., & Shuen, A. (1997). Dynamic capabilities and strategic management. *Strategic Management Journal*, *18*(7), 509–533. DOI: 10.1002/(SICI)1097-0266(199708)18:7<509::AID-SMJ882>3.0.CO;2-Z

Thorén, K., & Vendel, M. (2019). Backcasting as a strategic management tool for meeting VUCA challenges. *Journal of Strategy and Management*, *12*(2), 298–312. DOI: 10.1108/JSMA-10-2017-0072

Thornton, P. H., Ocasio, W., & Lounsbury, M. (2012). *The institutional logics perspective: A new approach to culture, structure, and process*. Oxford University Press., https://global.oup.com/ukhe/product/the-institutional-logics-perspective-9780199601943 DOI: 10.1093/acprof:oso/9780199601936.001.0001

Tøndel, I. A., Line, M. B., & Jaatun, M. G. (2014). Information security incident management: Current practice as reported in the literature. *Computers & Security*, *45*, 42–57. DOI: 10.1016/j.cose.2014.05.003

Toussaint, M., Krima, S., & Panetto, H. (2024). Industry 4.0 data security: A cybersecurity frameworks review. *Journal of Industrial Information Integration, 39*, 100604. DOI: 10.1016/j.jii.2024.100604

Troise, C., Corvello, V., Ghobadian, A., & O'Regan, N. (2022). How can SMEs successfully navigate VUCA environment: The role of agility in the digital transformation era. *Technological Forecasting and Social Change, 174*, 121227. DOI: 10.1016/j.techfore.2021.121227

Tukamuhabwa, B. R., Stevenson, M., Busby, J., & Zorzini, M. (2015). Supply chain resilience: Definition, review and theoretical foundations for further study. *International Journal of Production Research, 53*(18), 5592–5623. DOI: 10.1080/00207543.2015.1037934

Tupa, J., Simota, J., & Steiner, F. (2017). Aspects of risk management implementation for Industry 4.0. *Procedia Manufacturing, 11*, 1223–1230. DOI: 10.1016/j.promfg.2017.07.248

Tuptuk, N., & Hailes, S. (2018). Security of smart manufacturing systems. *Journal of Manufacturing Systems, 47*, 93–106. DOI: 10.1016/j.jmsy.2018.04.007

Tushman, M. L., & O'Reilly, C. A. (2002). *Winning through innovation: A practical guide to leading organizational change and renewal* (Updated ed.). Harvard Business Press. https://books.google.com/books?id=CvEplLBTN3QC

Uçar, A., Karaköse, M., & Kırımça, N. (2024). Artificial intelligence for predictive maintenance applications: Key components, trustworthiness, and future trends. *Applied Sciences (Basel, Switzerland), 14*(2), 898. DOI: 10.3390/app14020898

Unilever, P. L. C. (2024). *Annual report and accounts 2023.* https://www.unilever.com/files/66bc4aea-608f-46ee-8da3-cde0ec8ebe90/unilever-annual-report-and-accounts-2023.pdf

Unilever, P. L. C. (2024). *Climate Transition Action Plan (updated 2024).* https://www.unilever.com/files/8b5df5f6-cb90-40fd-9691-38d06905d81d/unilever-climate-transition-action-plan-updated-2024.pdf

U.S. Department of Justice. (2021, June 7). *Department of Justice seizes $2.3 million in cryptocurrency paid to the ransomware extortionists DarkSide* [Press release]. https://www.justice.gov/archives/opa/pr/department-justice-seizes-23-million-cryptocurrency-paid-ransomware-extortionists-darkside

U.S. Government Accountability Office. (2024, December 2). *Export controls: Commerce implemented advanced semiconductor rules and took steps to address compliance challenges* (GAO-25-107386). U.S. Government Accountability Office. https://www.gao.gov/products/gao-25-107386

Van Erp, T., Carvalho, N. G. P., Gerolamo, M. C., Gonçalves, R., Rytter, N. G. M., & Gladysz, B. (2024). Industry 5.0: A new strategy framework for sustainability management and beyond. *Journal of Cleaner Production, 461*, 142271. DOI: 10.1016/j.jclepro.2024.142271

Van Rossem, C. (2008). *Individual producer responsibility in the WEEE Directive – From theory to practice?* [Doctoral thesis, International Institute for Industrial Environmental Economics, Lund University]. LUCRIS/Lund University. https://lucris.lub.lu.se/ws/portalfiles/portal/5603015/1266800.pdf

Vance, D., Jin, M., Price, C., Nimbalkar, S. U., & Wenning, T. (2023). Smart manufacturing maturity models and their applicability: A review. *Journal of Manufacturing Technology Management, 34*(5), 735–770. DOI: 10.1108/JMTM-03-2022-0103

Verhoef, P. C., Broekhuizen, T., Bart, Y., Bhattacharya, A., Dong, J. Q., Fabian, N., & Haenlein, M. (2021). Digital transformation: A multidisciplinary reflection and research agenda. *Journal of Business Research, 122*, 889–901. DOI: 10.1016/j.jbusres.2019.09.022

Verschuur, J., Koks, E. E., & Hall, J. W. (2021). Global economic impacts of the Suez Canal blockage. *Nature Sustainability, 4*(9), 749–757. DOI: 10.1038/s41893-021-00738-9

Vial, G. (2019). Understanding digital transformation: A review and a research agenda. *The Journal of Strategic Information Systems, 28*(2), 118–144. DOI: 10.1016/j.jsis.2019.01.003

Vysocky, A., & Novak, P. (2016). Human-robot collaboration in industry. *MM Science Journal, 9*(2), 903–906. DOI: 10.17973/MMSJ.2016_06_201611

Wagire, A. A., Joshi, R., Rathore, A. P. S., & Jain, R. (2021). Development of maturity model for assessing the implementation of Industry 4.0: Learning from theory and practice. *Production Planning and Control, 32*(8), 603–622. DOI: 10.1080/09537287.2020.1744763

Wang, S., Li, J., Song, J., Li, Y., & Sherk, M. (2018). Institutional pressures and product modularity: Do supply chain coordination and functional coordination matter? *International Journal of Production Research, 56*(20), 6644–6661. DOI: 10.1080/00207543.2018.1444807

Wang, X., Zhu, J., Su, C., & Zhen, X. (2024). Cascade failure modeling and resilience analysis of mine cyber physical systems under deliberate attacks. *Journal of Safety Science and Resilience = An Quan Ke Xue Yu Ren Xing (Ying Wen)*, *5*(3), 266–280. DOI: 10.1016/j.jnlssr.2024.03.007

Weick, K. E., & Sutcliffe, K. M. (2007). *Managing the unexpected: Resilient performance in an age of uncertainty* (2nd ed.). Jossey-Bass., https://www.wiley.com/en-gb/Managing%2Bthe%2BUnexpected%3A%2BResilient%2BPerformance%2Bin%2Ban%2BAge%2Bof%2BUncertainty%2C%2B2nd%2BEdition-p-9780787996499

Whiteman, W. E. (1998). *Training and educating Army officers for the 21st century: Implications for the United States Military Academy*. U.S. Army War College. https://archive.org/details/DTIC_ADA345812

Wieland, A., & Wallenburg, C. M. (2013). The influence of relational competencies on supply chain resilience: A relational view. *International Journal of Physical Distribution & Logistics Management*, *43*(4), 300–320. DOI: 10.1108/IJPDLM-08-2012-0243

Womack, J. P., Jones, D. T., & Roos, D. (1990). *The machine that changed the world.* Rawson Associates. https://openlibrary.org/books/OL22145160M/The_machine_that_changed_the_world

World, V. U. C. A. (2025). *VUCA – Leaders with vision, understanding, clarity, agility!* https://www.vuca-world.org/

World Bank. (2024). *World Development Indicators* [Database]. World Bank. https://datatopics.worldbank.org/world-development-indicators/

World Economic Forum. (2019, January). *Fourth Industrial Revolution: Beacons of Technology and Innovation in Manufacturing*. World Economic Forum. https://www3.weforum.org/docs/WEF_4IR_Beacons_of_Technology_and_Innovation_in_Manufacturing_report_2019.pdf

Wright, S., Pickton, D. W., & Callow, J. (2002). Competitive intelligence in UK firms: A typology. *Marketing Intelligence & Planning*, *20*(6), 349–360. DOI: 10.1108/02634500210445400

Wu, D., Ren, A., Zhang, W., Fan, F., Liu, P., Fu, X., & Terpenny, J. (2018). Cybersecurity for digital manufacturing. *Journal of Manufacturing Systems*, *48*, 3–12. DOI: 10.1016/j.jmsy.2018.03.006

Xu, L. D., Xu, E. L., & Li, L. (2018). Industry 4.0: State of the art and future trends. *International Journal of Production Research*, *56*(8), 2941–2962. DOI: 10.1080/00207543.2018.1444806

Xu, X., Lu, Y., Vogel-Heuser, B., & Wang, L. (2021). Industry 4.0 and Industry 5.0—Inception, conception and perception. *Journal of Manufacturing Systems, 61*, 530–535. DOI: 10.1016/j.jmsy.2021.10.006

YahiaMarzouk, Y., & Jin, J.YahiaMarzouk. (2022). The relationship between environmental scanning and organizational resilience: Roles of process innovation and environmental uncertainty. *Frontiers in Environmental Science, 10*, 966474. DOI: 10.3389/fenvs.2022.966474

Yarger, H. R. (2006). *Strategic theory for the 21st century: The little book on big strategy*. U.S. Army War College Press., https://press.armywarcollege.edu/monographs/723/

Yu, J., Welford, R., & Hills, P. (2006). Industry responses to EU WEEE and ROHS Directives: Perspectives from China. *Corporate Social Responsibility and Environmental Management, 13*(5), 286–299. DOI: 10.1002/csr.131

Zainol, Z., Hassan, K., Wan Hussein, W. M. H., & Phuoc, J. C. (2019). *Adaptive regulation for Industry 4.0*. In *Proceedings of the International Conference on Social Science 2019 (ICSS 2019)* (pp. 1180–1183). Atlantis Press (Advances in Social Science, Education and Humanities Research). https://doi.org/DOI: 10.2991/icss-19.2019.8

Zavala-Alcívar, A., Verdecho, M. J., & Alfaro-Saiz, J. J. (2020). A conceptual framework to manage resilience and increase sustainability in the supply chain. *Sustainability (Basel), 12*(16), 6300. DOI: 10.3390/su12166300

Zhou, Z., & Zhang, J. (2025). Manufacturing enterprise digital transformation, financial flexibility, and financial risk—Evidence from China. *International Review of Financial Analysis, 104*, 104279. Advance online publication. DOI: 10.1016/j.irfa.2025.104279

Zhu, Q. (2016). *Institutional pressures and support from industrial zones for motivating sustainable production among Chinese manufacturers. International Journal of Production Economics, 181*(Part B), 402–409. https://doi.org/DOI: 10.1016/j.ijpe.2016.06.027

ADDITIONAL READINGS

Avinash, B., & Joseph, G. (2025). Embracing resilience in pharmaceutical manufacturing: "digital twins" – forging a resilient path in the VUCA maze. *International Journal of Pharmaceutical and Healthcare Marketing*, *19*(3), 750–772. Advance online publication. DOI: 10.1108/IJPHM-03-2024-0024

Awad, J. A. R., & Martín-Rojas, R. (2024). Digital transformation influence on organisational resilience through organisational learning and innovation. *Journal of Innovation and Entrepreneurship*, *13*(1), 69. DOI: 10.1186/s13731-024-00405-4

Bentahar, O., & Belhadi, A. (2024). Integrating project management and supply chain management for resilient and sustainable operations in a VUCA world. *Supply Chain Forum: An International Journal*, *26*(1), 1–6. https://doi.org/DOI: 10.1080/16258312.2024.2447201

Graham, J. (2024). Factors influencing the creation of an industry 4.0 strategy in the generic pharmaceutical industry. *Journal of Generic Medicines*, *21*(1), 45–62. DOI: 10.1177/17411343241238178

Holl, A., & Rama, R. (2024). SME digital transformation and the COVID-19 pandemic: A case study of a hard-hit metropolitan area. *Science & Public Policy*, *51*(6), 1212–1226. DOI: 10.1093/scipol/scae023

Kowalski, M., Bernardes, R. C., Gomes, L., & Borini, F. M. (2024). Microfoundations of dynamic capabilities for digital transformation. *European Journal of Innovation Management*. Advance online publication. DOI: 10.1108/EJIM-12-2023-1074

Maier, F. A., Puppala, S., & Oberle, M. (2024). Artificial intelligence applications for resilience in manufacturing—A systematic literature review. In *2024 6th International Conference on Artificial Intelligence in Information and Communication (ICAIIC)* (pp. 234–239). IEEE. https://doi.org/DOI: 10.1109/ICAIIC60209.2024.10463234

Paul, J., Ueno, A., Dennis, C., Alamanos, E., Curtis, L., Foroudi, P., Kacprzak, A., Kunz, W. H., Liu, J., Marvi, R., Nair, S. L. S., Ozdemir, O., Pantano, E., Papadopoulos, T., Petit, O., Tyagi, S., & Wirtz, J. (2024). Digital transformation: A multidisciplinary perspective and future research agenda. *International Journal of Consumer Studies*, *48*(2), e13015. DOI: 10.1111/ijcs.13015

Pochmara, J., & Świetlicka, A. (2024). Cybersecurity of industrial systems—A 2023 report. *Electronics (Basel)*, *13*(7), 1191. DOI: 10.3390/electronics13071191

Yan, A., Ma, H., Zhu, D., & Xie, J. (2024). Digital transformation and corporate resilience: Evidence from China during the COVID-19 pandemic. *Quantitative Finance and Economics*, *8*(4), 779–814. DOI: 10.3934/QFE.2024030

KEY TERMS AND DEFINITIONS

Adaptive Learning and Capability Evolution: The systematic enhancement of VUCA navigation capabilities through reflection, knowledge capture, and capability development that improve future organisational performance in turbulent environments.

Artificial Intelligence: Advanced computational technologies that enable enhanced decision-making capabilities in manufacturing through pattern recognition, anomaly detection, and predictive analytics whilst requiring sophisticated data infrastructure and algorithmic governance.

Automation: Technologies that transform manufacturing operations through enhanced precision, consistency, and speed whilst reducing labour intensity and human error rates, creating new sources of complexity during transition periods.

Digital Transformation: The systematic integration of digital technologies across all aspects of organisational operations, fundamentally altering value creation processes and competitive positioning in manufacturing environments.

Dynamic Capabilities: The firm's ability to integrate, build, and reconfigure internal and external competences to address rapidly changing environments through sensing emerging opportunities and threats, seizing value through strategic resource deployment, and transforming organisational structures.

Environmental Sensing and Intelligence Gathering: The systematic scanning of technological, market, regulatory, and geopolitical environments to identify emerging opportunities and threats that may affect manufacturing operations.

Geopolitical Risk: Government actions, international conflicts, and policy changes that may significantly impact manufacturing operations through trade restrictions, technology export controls, and market access limitations.

Implementation Orchestration and Coordination: The systematic deployment of selected strategic options through project management, change management, and performance monitoring that ensure effective execution whilst maintaining operational continuity.

Industry 4.0: The systematic integration of cyber-physical systems, Internet of Things, artificial intelligence, and advanced manufacturing technologies that enable smart, interconnected manufacturing operations with enhanced automation and data exchange.

Institutional Adaptation: The sophisticated approaches manufacturing organisations develop to manage multiple, often conflicting institutional pressures whilst maintaining operational effectiveness and competitive positioning within complex regulatory environments.

Organisational Resilience: The ability of manufacturing organisations to prepare for, respond to, and recover from disruptions whilst maintaining operational continuity and competitive positioning in uncertain environments.

Robotics: Flexible automation applications in manufacturing that enable human-robot collaboration whilst maintaining precision and consistency, requiring significant capital investment, technical expertise, and safety considerations.

Strategic Option Development and Selection: The systematic creation and evaluation of alternative responses to environmental changes through scenario planning, technology roadmapping, and capability assessment that illuminate potential pathways for addressing VUCA conditions.

Supply Chain Resilience: The ability to prepare for, respond to, and recover from supply chain disruptions whilst maintaining operational continuity and competitive positioning through diversification strategies, redundancy creation, and collaborative relationships.

Technology Integration: The strategic deployment of artificial intelligence, automation, and robotics within manufacturing operations that requires systematic approaches to maximise value creation whilst managing implementation risks and organisational disruption.

VUCA: An acronym representing Volatility, Uncertainty, Complexity, and Ambiguity - environmental conditions that create unprecedented challenges for manufacturing organisations through rapid changes, unpredictable events, intertwined forces, and unclear realities requiring sophisticated strategic responses.

Conclusion

AI and Automation in Modern Manufacturing barely scratches the surface of the topic. Many extant research questions exist. Some relate to modern manufacturing.

- What does modern manufacturing look like in particular disciplines, industries, and spaces?
- What are hard problems in the modern manufacturing space?
- How do computational smarts affect sustainable manufacturing?
- What does modern DIY and micro manufacturing look like?

Others relate to artificial intelligence and automation, both singly and in integrated ways.

- What are ways to fine-tune AIs in modern manufacturing?
- How do human factory workers engage with AI, automation, and robotic? What are necessary skillsets and KSAs (knowledge, skills, and abilities / attitudes)? Training regimens?
- What do modern manufacturing spaces look like in terms of factory floors, assembly lines, control spaces? Liminal spaces?
- What goes on in R&D labs in modern manufacturing? How are innovations arrived at?
- What parts of decision-making are given to AI systems, and which to expert humans?
- What are some cases of custom AI programs trained on in-house and protected data? How are these applied in various manufacturing cases?

Human striving in this area stands to pay back in spades. There is so much still to be learned in this space. It is hoped that human researchers with actual expertise and research chops may continue to add to the literature, instead of conflating reading and generative AI with actual in-world expertise. The world belongs to those who can conceptualize, design, and execute...to standards...at speed...and with innovation. Those who can stand up various production lines and make plans real will own the future.

Shalin Hai-Jew
Sedgwick County, USA

Compilation of References

Abadi, M., Liu, C., Zhang, M., Hu, Y., & Xu, Y. (2024). Leveraging AI for energy-efficient manufacturing systems: Review and future prospectives. *Journal of Manufacturing Systems*, *78*, 153–177. DOI: 10.1016/j.jmsy.2024.11.017

Abaku, E. A., Edunjobi, T. E., & Odimarha, A. C.Emmanuel Adeyemi AbakuTolulope Esther EdunjobiAgnes Clare Odimarha. (2024). Theoretical approaches to AI in supply chain optimization: Pathways to efficiency and resilience. *International Journal of Science and Technology Research Archive*, *6*(1), 92. DOI: 10.53771/ijstra.2024.6.1.0033

Abatan, A., Adeyinka, M. A., Sodiya, E. O., Jacks, B. S., Ugwuanyi, E. D., Daraojimba, O. H., & Lottu, O. A. (2024). The role of IT in sustainable environmental management: A global perspective review. *International Journal of Science and Research Archive*, *11*(1), 1874–1886. DOI: 10.30574/ijsra.2024.11.1.0288

Abbass, M., Abbas, U., Jafri, R., Arif, S. M., & Akhai, S. (2025). AI and machine learning applications in sustainable smart cities. In *Sustainable smart cities and the future of urban development* (pp. 1–32). IGI Global., DOI: 10.4018/979-8-3693-6740-7.ch001

Abbass, M., Abbas, U., Wani, S. A., & Akhai, S. (2025). Rethinking global food governance from historical perspectives for a sustainable future. In *Innovative trends shaping food marketing and consumption* (pp. 553–578). IGI Global., DOI: 10.4018/979-8-3693-8542-5.ch022

Abbass, M., & Akhai, S. (2025). Eco-friendly geopolymer bricks synthesized from alkali-activated sugarcane bagasse ash and rice husk ash. *Industrial Crops and Products*, *233*, 121410. DOI: 10.1016/j.indcrop.2025.121410

Abbass, M., & Akhai, S. (2025). Exploring the sustainability potential of geopolymer concrete with coal bottom ash and basalt fibres. *Multiscale and Multidisciplinary Modeling. Experiments and Design*, *8*(3), 199. DOI: 10.1007/s41939-025-00794-3

Abbass, M., Akhai, S., Abbas, U., Jafri, R., & Arif, S. M. (2025). AI-enabled sustainable urban planning and management. In *Real-world applications of AI innovation* (pp. 233–260). IGI Global., DOI: 10.4018/979-8-3693-4252-7.ch012

Abdallah, Y. O., Shehab, E., & Al-Ashaab, A. (2021). *Towards managing digital transformation in manufacturing industry: Theoretical framework*. Cranfield University Repository. https://dspace.lib.cranfield.ac.uk/server/api/core/bitstreams/0107b34c-0128-4d88-ab16-107fed663cdc/content

Abdelmoneim, A. A., Kimaita, H. N., Kalaany, C. M. A., Derardja, B., Dragonetti, G., & Khadra, R. (2025). IoT Sensing for Advanced Irrigation Management: A Systematic Review of Trends, Challenges, and Future Prospects [Review of *IoT Sensing for Advanced Irrigation Management: A Systematic Review of Trends, Challenges, and Future Prospects*]. *Sensors, 25*(7), 2291. Multidisciplinary Digital Publishing Institute. DOI: 10.3390/s25072291

Abdullah, F. M., Al-Ahmari, A. M., & Anwar, S. (2023). Analyzing interdependencies among influencing factors in smart manufacturing. *Sustainability (Basel), 15*(4), 3864. DOI: 10.3390/su15043864

Abhilash, P. M., Luo, X., Liu, Q., Madarkar, R., & Walker, C. (2024). Towards next-gen smart manufacturing systems: The explainability revolution. *Npj Advanced Manufacturing, 1*(1), 8. Advance online publication. DOI: 10.1038/s44334-024-00006-9

Acemoglu, D., & Restrepo, P. (2019). Automation and new tasks: How technology displaces and reinstates labor. *The Journal of Economic Perspectives, 33*(2), 3–30. DOI: 10.1257/jep.33.2.3

Acquah, I. S. K., Essel, D., Baah, C., Agyabeng-Mensah, Y., & Afum, E. (2021). Investigating the efficacy of isomorphic pressures on the adoption of green manufacturing practices and its influence on organizational legitimacy and financial performance. *Journal of Manufacturing Technology Management, 32*(7), 1399–1420. DOI: 10.1108/JMTM-10-2020-0404

Adams, C. A., & Abhayawansa, S. (2022). Connecting the COVID-19 pandemic, environmental, social and governance (ESG) investing and calls for "harmonisation" of sustainability reporting. *Critical Perspectives on Accounting, 82*, 102309. DOI: 10.1016/j.cpa.2021.102309

Adedokun, O. J., Keshinro, B. I., Uwaoma, I. A., Akindoyin, M. O., Onyia, P. C., & Ofremu, G. O. (2025). AI-DRIVEN SMART MANUFACTURING FOR SUSTAINABLE PRODUCTION. *LEVERAGING AUTOMATION AND ROBOTICS FOR ENERGY EFFICIENCY AND WASTE REDUCTION., 4*(1), 1. DOI: 10.34218/IJAIM_04_01_001

Adikwu, F. E., Esiri, A. E., Aderamo, A. T., Akano, O. A., & Erhueh, O. V.Friday Emmanuel AdikwuAndrew Emuobosa EsiriAdeoye Taofik AderamoOluwaseyi Ayotunde AkanoOvie Vincent Erhueh. (2024). Leveraging digital technologies for health, safety, and environmental (HSE) management in industrial operations. *International Journal of Applied Research in Social Sciences*, *6*(11), 2712–2726. DOI: 10.51594/ijarss.v6i11.1722

Adil, M., Song, H., Mastorakis, S., Abulkasim, H., Farouk, A., & Jin, Z. (2023). UAV-Assisted IoT Applications, Cybersecurity Threats, AI-Enabled Solutions, Open Challenges With Future Research Directions. *IEEE Transactions on Intelligent Vehicles*, *9*(4), 4583–4605. DOI: 10.1109/TIV.2023.3309548

Adler, P. S., Goldoftas, B., & Levine, D. I. (1999). Flexibility versus efficiency? A case study of model changeovers in the Toyota Production System. *Organization Science*, *10*(1), 43–68. https://www.jstor.org/stable/2640387. DOI: 10.1287/orsc.10.1.43

Agostinho, C., Dikopoulou, Z., Lavasa, E., Perakis, K., Pitsios, S., Branco, R., Reji, S., Hetterich, J., Biliri, E., Lampathaki, F., Rey, S. R. D., & Gkolemis, V. (2023). Explainability as the key ingredient for AI adoption in Industry 5.0 settings. *Frontiers in Artificial Intelligence*, *6*, 1264372. Advance online publication. DOI: 10.3389/frai.2023.1264372 PMID: 38146276

Ahmed, A., & Sutton, M. J. (2017). Gamification, serious games, simulations, and immersive learning environments in knowledge management initiatives. *World Journal of Science. Technology and Sustainable Development*, *14*(2/3), 78–83.

Ahmed, S., Abd Aziz, N., Haque, R., & Rahman, M. T. (2024). Digital transformation in Malaysian manufacturing: A study of change sensing and seizing capabilities. *Cogent Business & Management*, *11*(1), 2392046. DOI: 10.1080/23311975.2024.2392046

Ahmed, T., & Choudhury, S. (2025). *AI and Semantic Communication for Infrastructure Monitoring in 6G-Driven Drone Swarms*. DOI: 10.48550/ARXIV.2503.00053

Ahn, J., Lee, Y., Kim, N., Park, C., & Jeong, J. (2023). Federated Learning for Predictive Maintenance and Anomaly Detection Using Time Series Data Distribution Shifts in Manufacturing Processes. *Sensors (Basel)*, *23*(17), 7331. DOI: 10.3390/s23177331 PMID: 37687787

Akella, G. K., Wibowo, S., Grandhi, S., & Mubarak, S. (2023). *Systematic Review A Systematic Review of Blockchain Technology Adoption Barriers and Enablers for Smart and Sustainable Agriculture*.

Akhai, S. (2024). A Review on Optimizations in μ-EDM Machining of the Biomedical Material Ti6Al4V Using the Taguchi Method: Recent Advances Since 2020. *Latest Trends in Engineering and Technology*, 395-402. DOI: 10.1201/9781032665443-56

Akhai, S. (2024). Smart Home Environment for Society: An Overview of IoT-Enabled Air Conditioners. *Modelling of Virtual Worlds Using the Internet of Things*, 74-87. DOI: 10.1201/9781003480181-3

Akhai, S. (2024). *Towards trustworthy and reliable AI: The next frontier*. In *Explainable artificial intelligence (XAI) in healthcare* (Vol. 1, pp. 119–129). CRC Press. DOI: 10.1201/9781003426073-7

Akhai, S., Abbass, M., Kaur, P., & Kaur, T. (2025). Digital Transformation Across Generations: Robotics and AI in Action. In *Impacts of Digital Technologies Across Generations* (pp. 23-40). IGI Global. DOI: 10.4018/979-8-3693-6366-9.ch002

Akhai, S. (2023). Biogas plants for sustainable municipal waste management: A brief review. In *Handbook of research on safe disposal methods of municipal solid wastes for a sustainable environment* (pp. 162–179). IGI Global., DOI: 10.4018/978-1-6684-8117-2.ch012

Akhai, S. (2023). Healthcare record management for healthcare 4.0 via blockchain: A review of current applications, opportunities, challenges, and future potential. In *Blockchain for healthcare 4.0* (Vol. 1, pp. 211–223). CRC Press., DOI: 10.1201/9781003408246-11

Akhai, S. (2024). Innovating healthcare delivery: Harnessing drones for medical services. In *Modelling of virtual worlds using the Internet of Things* (pp. 232–250). CRC Press., DOI: 10.1201/9781003480181-10

Akhai, S., & Balu, V. (2022). *Code injection assault & mitigation model to prevent attacks*. In *Proceedings of the 2022 11th International Conference on System Modeling& Advancement in Research Trends (SMART)* (pp. 1324–1328). IEEE. DOI: 10.1109/SMART55829.2022.10046835

Akhai, S., Bansal, S. A., & Singh, S. (2020). A critical review of thermal insulators from natural materials for energy saving in buildings. *Journal of Critical Reviews*, 7(19), 278–283.

Akhai, S., & Khang, A. (2024). Energy efficiency and human comfort: AI and IoT integration in hospital HVAC systems. In *Medical robotics and AI-assisted diagnostics for a high-tech healthcare industry* (pp. 93–108). IGI Global., DOI: 10.4018/979-8-3693-2105-8.ch007

Akhai, S., & Khang, A. (2024). Innovations in medical manufacturing: A review of 3D printing, robotics, and Internet of Things (IoT). In *The quantum evolution* (Vol. 1, pp. 226–241). CRC Press., DOI: 10.1201/9781032642079-10

Akhai, S., & Singh, H. (2013). Design optimization for modification of trough belt conveyor to reduce material spillage used in clinker transport in cement plant. *Int. J. Res*, *1*, 27–37.

Akhai, S., & Singh, H. (2013). Design optimization for modification of trough belt conveyor to reduce material spillage used in clinker transport in cement plant. *International Journal of Research*, *1*, 27–37.

Akhai, S., Singh, V. P., & John, S. (2016). Human performance in industrial design centers with small unit air conditioning systems. *Journal of Advanced Research in Production Industrial Engineering*, *3*(2), 5–11.

Akhai, S., & Wadhwa, A. (2024). Enhancing radiator cooling with CuO nanofluid microchannels. *International Journal of Automotive Science and Technology*, *8*(2), 201–211. DOI: 10.30939/ijastech..1399702

Akhai, S., & Wadhwa, A. S. (2024). Wireless networking solutions for climate control and comfort in modern air conditioning systems. In *Intelligent networks* (pp. 165–181). CRC Press., DOI: 10.1201/9781003541363-11

Akhavan, J., Lyu, J., & Manoochehri, S. (2023). A Deep Learning Solution for Real-time Quality Assessment and Control in Additive Manufacturing Using Point Cloud Data. *Research Square (Research Square)*. DOI: 10.21203/rs.3.rs-2129511/v5

Akhbardeh, F., Desell, T., & Zampieri, M. (2020). *MaintNet: A Collaborative Open-Source Library for Predictive Maintenance Language Resources*. 7. DOI: 10.18653/v1/2020.coling-demos.2

Al-Battbootti, M. J. H., Marin, I., Goga, N., & Popa, R. (2021). OIL AND GAS PIPELINE MONITORING DURING COVID-19 PANDEMIC VIA UNMANNED AERIAL VEHICLE. *ICERI Proceedings*, *1*, 5756. DOI: 10.21125/iceri.2021.1299

Alegre, J., & Chiva, R. (2008). Assessing the impact of organizational learning capability on product innovation performance: An empirical test. *Technovation*, *28*(6), 315–326. DOI: 10.1016/j.technovation.2007.09.003

Alemão, D., Rocha, A. D., & Barata, J. (2019). Production Scheduling Requirements to Smart Manufacturing. In *IFIP advances in information and communication technology* (p. 227). Springer Science+Business Media. DOI: 10.1007/978-3-030-17771-3_19

Alenazi, M. M. (2025). Machine Learning-Optimized Energy Management for Resilient Residential Microgrids with Dynamic Electric Vehicle Integration. *Journal on Artificial Intelligence*, *7*(1), 143–176. DOI: 10.32604/jai.2025.066067

Alevizos, V., Georgousis, I., & Kapodistria, A.-M. (2023). Towards Climate Neutrality: A Comprehensive Overview of Sustainable Operations Management, Optimization, and Wastewater Treatment Strategies. *arXiv (Cornell University)*. https://doi.org// arxiv.2308.00808DOI: 10.48550

Alfaqiyah, E., Alzubi, A., Aljuhmani, H. Y., & Öz, T. (2025). How Industry 4.0 Technologies Enhance Supply Chain Resilience: The Interplay of Agility, Adaptability, and Customer Integration in Manufacturing Firms. *Sustainability (Basel)*, *17*(17), 7922. Advance online publication. DOI: 10.3390/su17177922

Alghieth, M. (2025). Sustain AI: A Multi-Modal Deep Learning Framework for Carbon Footprint Reduction in Industrial Manufacturing. *Sustainability (Basel)*, *17*(9), 4134. DOI: 10.3390/su17094134

Allers, M., Ahrens, A., Hitzemann, M., Bock, H., Wolf, T., Radunz, J., Meyer, F., Wilsenack, F., Zimmermann, S., & Ficks, A. (2023). Real-Time Remote Detection of Airborne Chemical Hazards—An Unmanned Aerial Vehicle (UAV) Carrying an Ion Mobility Spectrometer. *IEEE Sensors Journal*, *23*(15), 16562–16570. DOI: 10.1109/JSEN.2023.3287448

Alli, Y. A., Anuar, H., Manshor, M. R., Okafor, C. E., Kamarulzaman, A. F., Akçakale, N., Nazeri, F. N. M., Bodaghi, M., Suhr, J., & Nasir, N. A. M. (2024). Optimization of 4D/3D printing via machine learning: A systematic review [Review of *Optimization of 4D/3D printing via machine learning: A systematic review*]. *Hybrid Advances*, *6*, 100242. Elsevier BV. DOI: 10.1016/j.hybadv.2024.100242

Almadani, Y., Plets, D., Bastiaens, S., Joseph, W., Ijaz, M., Ghassemlooy, Z., & Rajbhandari, S. (2020). Visible Light Communications for Industrial Applications—Challenges and Potentials. *Electronics (Basel)*, *9*(12), 2157. DOI: 10.3390/electronics9122157

Almusharraf, A. (2025). Automation and Its Influence on Sustainable Development: Economic, Social, and Environmental Dimensions. *Sustainability (Basel)*, *17*(4), 1754. DOI: 10.3390/su17041754

Alokshe, M. M. F., Adedokun, M. W., & Iyiola, K. (2025). Advanced manufacturing technologies, strategic agility, business network and sustained competitive performance: An empirical evidence from an emerging economy. *Humanities & Social Sciences Communications*, *12*(1), 391. https://www.nature.com/articles/s41599-025-04671-9. DOI: 10.1057/s41599-025-04671-9

Alsayed, A., Yunusa-Kaltungo, A., Quinn, M. K., Arvin, F., & Nabawy, M. R. A. (2021). Drone-Assisted Confined Space Inspection and Stockpile Volume Estimation. *Remote Sensing (Basel), 13*(17), 3356. DOI: 10.3390/rs13173356

Al-Talib, M., Melhem, W. Y., Anosike, A. I., Garza-Reyes, J. A., Nadeem, S. P., & Kumar, A. (2020). Achieving resilience in the supply chain by applying IoT technology. *Procedia CIRP, 91*, 752–757. DOI: 10.1016/j.procir.2020.02.231

Ameen, K., & Keizer, A. (2023). International Labour Organization. In *Encyclopedia of Human Resource Management, Second Edition.* https://doi.org/DOI: 10.4337/9781800378841.I.19

Ameh, B. (2024). Digital tools and AI: Using technology to monitor carbon emissions and waste at each stage of the supply chain, enabling real-time adjustments for sustainability improvements. *International Journal of Science and Research Archive, 13*(1), 2741–2757. DOI: 10.30574/ijsra.2024.13.1.1995

Andrianandrianina Johanesa, T. V., Equeter, L., & Mahmoudi, S. A. (2024). Survey on AI applications for product quality control and predictive maintenance in industry 4.0. *Electronics (Basel), 13*(5), 976. DOI: 10.3390/electronics13050976

Anitha, D., Awasthi, A., Dwivedi, S. P., Kumar, R., Pahwa, S., Rajalakshmi, B., & Alkhafaji, M. A. (2023). Beyond Limits: Envisioning Sustainable Design and Production Ecosystems. *E3S Web of Conferences, 453*, 1039. DOI: 10.1051/e3sconf/202345301039

Anjum, M., Kraïem, N., Min, H., Daradkeh, Y. I., Dutta, A. K., & Shahab, S. (2025). Optimizing sustainability: Information aggregation in smart factories for energy efficiency enhancement. *Scientific Reports, 15*(1), 8603. DOI: 10.1038/s41598-024-83304-8 PMID: 40075102

Appelbaum, S. H., & Goransson, L. (1997). Transformational and adaptive learning within the learning organization: A framework for research and application. *The Learning Organization, 4*(3), 115–128. DOI: 10.1108/09696479710182803

Aral, K. D., Giambona, E., & Lopez-Aliouchkin, R. (2020). *The COVID-19 pandemic: Supply chain disruption, wealth effects, and corporate responses.* SSRN *Electronic Journal.* https://doi.org/DOI: 10.2139/ssrn.3686744

Ardito, L., Messeni Petruzzelli, A., & Albino, V. (2018). Approaches for organizational learning: A literature review. *SAGE Open, 8*(3), 1–20. DOI: 10.1177/2158244018794224 PMID: 32983596

Ardito, S., Setiawan, W., & Wibisono, A. (2024). Enhancing Predictive Maintenance in Manufacturing Using Deep Learning-Based Anomaly Detection. *Enhancing Predictive Maintenance in Manufacturing Using Deep Learning-Based Anomaly Detection.*, *3*(1), 12–23. DOI: 10.63876/ijtm.v3i1.112

Arellano-Espitia, F., Delgado-Prieto, M., Martínez-Viol, V., Fernandez-Sobrino, A., & Osornio-Rios, R. A. (2021). *Anomaly Detection in Electromechanical Systems by means of Deep-Autoencoder.* 1. DOI: 10.1109/ETFA45728.2021.9613529

Areo, G. (2025). *Edge AI for Real-Time Industrial Automation: Architecture.* Challenges, and Case Studies.

Arif, S. M., & Akhai, S. (2024). Fostering a sustainable and disaster-resilient transportation infrastructure. In *Challenges, strategies, and resiliency in disaster and risk management* (pp. 205–219). IGI Global., DOI: 10.4018/979-8-3693-2721-0.ch009

Asensio, O. I., Moore, C. E., Ulibarrí, N., Simsekler, M. C. E., Lan, T., & Rivero, G. (2025). Data technologies and analytics for policy and governance: a landscape review [Review of *Data technologies and analytics for policy and governance: a landscape review*]. *Data & Policy, 7.* Cambridge University Press. DOI: 10.1017/dap.2024.49

Ashena, M., Laal Khezri, H., & Shahpari, G. (2024). Investigation into the dynamic relationships between global economic uncertainty and price volatilities of commodities, raw materials, and energy. *Applied Economic Analysis, 32*(94), 23–40. DOI: 10.1108/AEA-06-2023-0207

Ashraf, S. N., Manickam, S., Zia, S. S., Abro, A. A., Obaidat, M., Uddin, M., Abdelhaq, M., & Alsaqour, R. (2023). IoT empowered smart cybersecurity framework for intrusion detection in internet of drones. *Scientific Reports, 13*(1), 18422. Advance online publication. DOI: 10.1038/s41598-023-45065-8 PMID: 37891186

Askerbekov, D., Garza-Reyes, J. A., Ghatak, R. R., Joshi, R., Kandasamy, J., & Nascimento, D. L. de M. (2024). Embracing drones and the Internet of drones systems in manufacturing – An exploration of obstacles. *Technology in Society, 78*, 102648. DOI: 10.1016/j.techsoc.2024.102648

Asmat, H., Din, I. U., Almogren, A., & Khan, M. Y. (2025). Digital Twin With Soft Actor-Critic Reinforcement Learning for Transitioning From Industry 4.0 to 5.0. *IEEE Access : Practical Innovations, Open Solutions, 13*, 40577–40593. DOI: 10.1109/ACCESS.2025.3546085

Assad, F., Rushforth, E., & Harrison, R. (2023). A component-based design approach for energy flexibility in cyber-physical manufacturing systems. *Journal of Intelligent Manufacturing*. Advance online publication. DOI: 10.1007/s10845-023-02280-4

Atanassova, I., Bednar, P., Khan, H., & Khan, Z. (2025). Managing the VUCA environment: The dynamic role of organizational learning and strategic agility in B2B versus B2C firms. *Industrial Marketing Management, 125*, 12–28. DOI: 10.1016/j.indmarman.2024.12.008

Augmentir. (2024, June 1). *Performance management in manufacturing: Best practices and examples*. https://www.augmentir.com/blog/performance-management-in-manufacturing-best-practices-and-examples

Autor, D. H. (2015). Why are there still so many jobs? The history and future of workplace automation. *The Journal of Economic Perspectives, 29*(3), 3–30. DOI: 10.1257/jep.29.3.3

Awain, A. M. S. B., Al-Ansi, A. M., & Jaboob, M. (2023). Green Supply Chain Management: A Comprehensive Review of Research, Applications and Future Directions [Review of *Green Supply Chain Management: A Comprehensive Review of Research, Applications and Future Directions*]. *Management and Production Engineering Review*. De Gruyter Open. https://doi.org/DOI: 10.24425/mper.2023.147194

Axios. (2021, May 19). *Colonial Pipeline CEO says company paid hacker group $4.4 million*. https://www.axios.com/2021/05/19/colonial-pipeline-darkside-4-million-ransomware

Axios. (2024, February 20). *Ransomware hackers focus on North America, Europe*. https://www.axios.com/2024/02/20/ransomware-attacks-north-america-europe

Ayamga, M., Tekinerdoğan, B., & Kassahun, A. (2021). Exploring the Challenges Posed by Regulations for the Use of Drones in Agriculture in the African Context. *Land (Basel), 10*(2), 164. DOI: 10.3390/land10020164

Ayvaz, S., & Alpay, K. (2021). Predictive maintenance system for production lines in manufacturing: A machine learning approach using IoT data in real-time. *Expert Systems with Applications, 173*, 114598. DOI: 10.1016/j.eswa.2021.114598

Azadeh, A., Ghaderi, S. F., & Izadbakhsh, H. (2008). Integration of DEA and AHP with computer simulation for railway system improvement and optimization. *Applied Mathematics and Computation, 195*(2), 775–785. DOI: 10.1016/j.amc.2007.05.023

Azadeh, A., Ghaderi, S. F., Miran, Y. P., Ebrahimipour, V., & Suzuki, K. (2007). An integrated framework for continuous assessment and improvement of manufacturing systems. *Applied Mathematics and Computation, 186*(2), 1216–1233. DOI: 10.1016/j.amc.2006.07.152

Azorín, J. M., & Cameron, R. (2010). The application of mixed methods in organisational research: A literature review. *Electronic Journal of Business Research Methods, 8*(2), 95–105. https://academic-publishing.org/index.php/ejbrm/article/view/1184

Azuma, R. T. (1997). A survey of augmented reality. *Presence (Cambridge, Mass.), 6*(4), 355–385. DOI: 10.1162/pres.1997.6.4.355

Ba, L., Tangour, F., Abbassi, I. E., & Absi, R. (2025). Analysis of Digital Twin Applications in Energy Efficiency: A Systematic Review [Review of *Analysis of Digital Twin Applications in Energy Efficiency: A Systematic Review*]. *Sustainability, 17*(8), 3560. Multidisciplinary Digital Publishing Institute. DOI: 10.3390/su17083560

Bachinger, F., Kronberger, G., & Affenzeller, M. (2021). Continuous improvement and adaptation of predictive models in smart manufacturing and model management. *IET Collaborative Intelligent Manufacturing, 3*(1), 48–63. DOI: 10.1049/cim2.12009

Badhoutiya, A., Darokar, H., Verma, R. P., Saraswat, M., Devaraj, S., Raj, V. H., &Abdulhussain, Z. N. (2023). Regenerative Manufacturing: Crafting a Sustainable Future through Design and Production. *E3S Web of Conferences, 453*, 1038. DOI: 10.1051/e3sconf/202345301038

Badhoutiya, A., Meshram, V., Gori, Y., Yadav, D. K., Viswanath, G., Revathi, V., &Brayyich, M. (2023). Blueprints for Green Horizons: Sustainable Strategies in Design and Production. *E3S Web of Conferences, 453*, 1030. DOI: 10.1051/e3sconf/202345301030

Badri, A., Boudreau-Trudel, B., & Souissi, A. S. (2018). Occupational health and safety in the industry 4.0 era: A cause for major concern? *Safety Science, 109*, 403–411. DOI: 10.1016/j.ssci.2018.06.012

Bakacsy, L., & Zakar, T. (2025). An investigation into the spatial patterns of invasive common milkweed (Asclepias syriaca L.) stands through the utilization of drone images. *Scientific Reports, 15*(1), 28889. Advance online publication. DOI: 10.1038/s41598-025-14034-8 PMID: 40775501

Balugani, E., Butturi, M. A., Chevers, D., Parker, D., & Rimini, B. (2020). Empirical evaluation of the impact of resilience and sustainability on firms' performance. *Sustainability (Basel), 12*(5), 1742. DOI: 10.3390/su12051742

Balzano, M., Marzi, G., & Turzo, T. (2024). SMEs and institutional theory: Major inroads and opportunities ahead. *Management Decision*, *63*(13), 1–27. DOI: 10.1108/MD-05-2023-0734

Banaulikar, A. (2020). Real Time Monitoring and Control for Industrial Automation using PLC. *International Journal for Research in Applied Science and Engineering Technology*, *8*(7), 1193–1200. DOI: 10.22214/ijraset.2020.30453

Bandara, J., Ray, R., Lu, J., & Gallagher, K. P. (2025). Developing countries locked out of low-carbon technology trade. *Science*, *388*(6744), 248–250. DOI: 10.1126/science.adu7731 PMID: 40245136

Bandi, A., Kongari, B., Naguru, R., Pasnoor, S., & Vilipala, S. V. (2025). The Rise of Agentic AI: A Review of Definitions, Frameworks, Architectures, Applications, Evaluation Metrics, and Challenges. *Future Internet*, *17*(9), 404. DOI: 10.3390/fi17090404

Bányai, T., & Tóth, Á. B. (2025). Real-Time Maintenance Optimization with Industrial Internet of Things. *Applied Sciences (Basel, Switzerland)*, *15*(10), 5640. DOI: 10.3390/app15105640

Bao, N., Fan, Y., Li, C., & Simeone, A. (2023). A computer vision approach to improve maintenance automation for thermal power plants lubrication systems. *Journal of Quality in Maintenance Engineering*, *29*(5), 120–137. DOI: 10.1108/JQME-01-2023-0007

Baran, B. E., & Woznyj, H. M. (2021). Managing VUCA: The human dynamics of agility. *Organizational Dynamics*, *50*(2), 100787. DOI: 10.1016/j.orgdyn.2020.100787 PMID: 32843777

Barber, H. F. (1992). Developing strategic leadership: The U.S. Army War College experience. *Journal of Management Development*, *11*(6), 4–12. DOI: 10.1108/02621719210018208

Barbhuiya, S., & Das, B. B. (2023). Life Cycle Assessment of construction materials: Methodologies, applications and future directions for sustainable decision-making. *Case Studies in Construction Materials*, *19*, e02326. Advance online publication. DOI: 10.1016/j.cscm.2023.e02326

Barbu, A., Ichimov, M. A. M., Costea-Marcu, I. C., Militaru, G., Deselnicu, D. C., & Moiceanu, G. (2025). Exploring Employee Perspectives on Workplace Technology: Usage, Roles, and Implications for Satisfaction and Performance. *Behavioral Sciences (Basel, Switzerland)*, *15*(1), 45. DOI: 10.3390/bs15010045 PMID: 39851850

Barkman, W. E. (1989). In-Process Quality Control for Manufacturing. In *CRC Press eBooks*. Informa., DOI: 10.1201/b15229

Barteková, E., Bibas, R., Börkey, P., Brown, A., Buzzi, E., Château, J., Cornago, E., Dellink, R., Dubois, M., Helf, M., Lanzi, E., Laubinger, F., Livingstone, L., Mavroeidi, E., McCarthy, A., Aigues, N., Yamaguchi, S., Ijeoma, I., Dalles, K., . . . Perrière, E. (2022). Synergies and trade-offs in the transition to a resource-efficient and circular economy. In *OECD environment policy papers*. DOI: 10.1787/23097841

Bartolomé, N., Hilber, I., Schulin, R., Mayer, P., Witt, G., & Bucheli, T. D. (2016). Equilibrium sampling in a semi in situ pot experiment to measure freely dissolved concentrations of hydrophobic organic compounds. *Research Portal Denmark*, 260. https://local.forskningsportal.dk/local/dki-cgi/ws/cris-link?src=dtu&id=dtu-fd499a44-cb03-4954-a110-212f7de6c3ef&ti=Equilibrium%20sampling%20in%20a%20semi%20in%20situ%20pot%20experiment%20to%20measure%20freely%20dissolved%20concentrations%20of%20hydrophobic%20organic%20compounds

Barzegaran, M. (2021). Configuration Optimization of Fog Computing Platforms for Control Applications. *Research Portal Denmark*, 194. https://local.forskningsportal.dk/local/dki-cgi/ws/cris-link?src=dtu&id=dtu-92dc705a-b8f0-41d2-928e-86a949a69162&ti=Configuration%20Optimization%20of%20Fog%20Computing%20Platforms%20for%20Control%20Applications

Bataille, C., Åhman, M., Neuhoff, K., Nilsson, L. J., Fischedick, M., Lechtenböhmer, S., Solano-Rodriquez, B., Denis-Ryan, A., Stiebert, S., Waisman, H., Sartor, O., & Rahbar, S. (2018). A review of technology and policy deep decarbonization pathway options for making energy-intensive industry production consistent with the Paris Agreement [Review of *A review of technology and policy deep decarbonization pathway options for making energy-intensive industry production consistent with the Paris Agreement*]. *Journal of Cleaner Production, 187*, 960. Elsevier BV. DOI: 10.1016/j.jclepro.2018.03.107

Bataille, C. (2019). Physical and policy pathways to net-zero emissions industry. *Wiley Interdisciplinary Reviews: Climate Change, 11*(2), e633. Advance online publication. DOI: 10.1002/wcc.633

Bataille, C., Nilsson, L. J., & Jotzo, F. (2021). Industry in a net-zero emissions world: New mitigation pathways, new supply chains, modelling needs and policy implications. *Energy and Climate Change, 2*, 100059. DOI: 10.1016/j.egycc.2021.100059

Battilana, J., Leca, B., & Boxenbaum, E. (2009). How actors change institutions: Towards a theory of institutional entrepreneurship. *The Academy of Management Annals, 3*(1), 65–107. DOI: 10.5465/19416520903053598

Baumann, D., Mager, F., Wetzker, U., Thiele, L., Zimmerling, M., & Trimpe, S. (2020). Wireless Control for Smart Manufacturing: Recent Approaches and Open Challenges. *Proceedings of the IEEE, 109*(4), 441–467. DOI: 10.1109/JPROC.2020.3032633

Beccarello, M., & Foggia, G. D. (2023). The Strategic Role of Energy Efficiency and Industry 4.0 Interventions in Manufacturing. *Journal of Management and Sustainability, 13*(1), 31. DOI: 10.5539/jms.v13n1p31

Bechtsis, D., Tsolakis, N., Vlachos, D., & Iakovou, E. (2016). Sustainable supply chain management in the digitalisation era: The impact of Automated Guided Vehicles. *Journal of Cleaner Production, 142*, 3970–3984. DOI: 10.1016/j.jclepro.2016.10.057

Bednarski, L., Roscoe, S., Blome, C., & Schleper, M. C. (2025). Geopolitical disruptions in global supply chains: A state-of-the-art literature review. *Production Planning and Control, 36*(4), 536–562. DOI: 10.1080/09537287.2023.2286283

Beliaeva, T., Shirokova, G., Wales, W., & Gafforova, E. (2020). Benefiting from economic crisis? Strategic orientation effects, trade-offs, and configurations with resource availability on SME performance. *The International Entrepreneurship and Management Journal, 16*(1), 165–194. DOI: 10.1007/s11365-018-0499-2

Bennett, N., & Lemoine, G. J. (2014). What a difference a word makes: Understanding threats to performance in a VUCA world. *Business Horizons, 57*(3), 311–317. DOI: 10.1016/j.bushor.2014.01.001

Bennett, N., & Lemoine, G. J. (2014, January–February). What VUCA really means for you. *Harvard Business Review, 92*(1–2), 27–28. https://hbr.org/2014/01/what-vuca-really-means-for-you

Bennis, W., & Nanus, B. (1985). *Leaders: The strategies for taking charge*. Harper & Row., https://archive.org/details/leadersstrategie0000benn

Benyeogor, M. S., Olatunbosun, A., & Kumar, S. A. (2020). Airborne System for Pipeline Surveillance Using an Unmanned Aerial Vehicle. *European Journal of Engineering and Technology Research, 5*(2), 178–182. DOI: 10.24018/ejers.2020.5.2.1761

Bernárdez, J. M., Boo, J., Díaz, J. I., & Medina, R. (2025). Interdepartmental Optimization in Steel Manufacturing: An Artificial Intelligence Approach for Enhancing Decision-Making and Quality Control. *Applied System Innovation, 8*(3), 63. Advance online publication. DOI: 10.3390/asi8030063

Besharov, M. L., & Smith, W. K. (2014). Multiple institutional logics in organizations: Explaining their varied nature and implications. *Academy of Management Review, 39*(3), 364–381. DOI: 10.5465/amr.2011.0431

Besigomwe, K. (2025). Closed-Loop Manufacturing with AI-Enabled Digital Twin Systems. *Cognizance Journal of Multidisciplinary Studies.*, *5*(1), 18–38. DOI: 10.47760/cognizance.2025.v05i01.002

Bessant, J., Caffyn, S., & Gallagher, M. (2001). An evolutionary model of continuous improvement behaviour. *Technovation, 21*(2), 67–77. DOI: 10.1016/S0166-4972(00)00023-7

Bessen, J. E. (2019). AI and jobs: The role of demand. *NBER Working Paper No. 24235*. https://www.nber.org/papers/w24235

Bessen, J. E., Goos, M., Salomons, A., & van den Berge, W. (2019). Automatic Reaction - What Happens to Workers at Firms that Automate? SSRN *Electronic Journal*. https://doi.org/DOI: 10.2139/ssrn.3328877

Bhakoo, V., & Choi, T. (2013). The iron cage exposed: Institutional pressures and heterogeneity across the healthcare supply chain. *Journal of Operations Management, 31*(6), 432–449. DOI: 10.1016/j.jom.2013.07.016

Bharadwaj, G., Chopde, S., Taluja, R., Gummidi, L., Chandrashekar, R., &Dhahi, H. A. (2023). Closing the Loop: Advances in Materials, Energy, and Waste Management. *E3S Web of Conferences, 453*, 1024. DOI: 10.1051/e3sconf/202345301024

Bhat, G. R., Dudhedia, M. A., Panchal, R., Shirke, Y. S., Angane, N. R., Khonde, S. R., Khedkar, S., Pansare, J. R., Bere, S. S., Wahul, R. M., & Gawande, S. H. (2024). Autonomous drones and their influence on standardization of rules and regulations for operating–A brief overview. *Results in Control and Optimization, 14*, 100401. DOI: 10.1016/j.rico.2024.100401

Bititci, U. S., Carrie, A. S., & McDevitt, L. (1997). Integrated performance measurement systems: A development guide. *International Journal of Operations & Production Management, 17*(5), 522–534. DOI: 10.1108/01443579710167230

Bjørn, A., Lloyd, S. M., & Matthews, H. D. (2021). From the Paris Agreement to corporate climate commitments: Evaluation of seven methods for setting "science-based" emission targets. *Environmental Research Letters, 16*(5), 054019. DOI: 10.1088/1748-9326/abe57b

Blasco, S. V., & Lambán, M. P. (2025). Enhancing Sustainability with LCA: A Comparative Analysis of Design and Manufacturing Processes. *Processes (Basel, Switzerland), 13*(1), 195. DOI: 10.3390/pr13010195

Board of Governors of the Federal Reserve System. (2025, September 16). *Industrial production and capacity utilization (G.17).* https://www.federalreserve.gov/releases/g17/current/

BOIKANYO, D. H. (2025). HARNESSING DIGITAL TECHNOLOGIES FOR VALUE CHAIN OPTIMIZATION AND SOCIETAL TRANSFORMATION. *BUSINESS EXCELLENCE AND MANAGEMENT, 15*, 14. DOI: 10.24818/beman/2025.S.I.5-02

Boikov, A., Payor, V., Savelev, R., & Kolesnikov, A. (2021). Synthetic Data Generation for Steel Defect Detection and Classification Using Deep Learning. *Symmetry, 13*(7), 1176. DOI: 10.3390/sym13071176

Bondin, A., & Zammit, J. P. (2025). Education 4.0 for Industry 4.0: A Mixed Reality Framework for Workforce Readiness in Manufacturing. *Multimodal Technologies and Interaction, 9*(5), 43. DOI: 10.3390/mti9050043

Bongomin, G. O. C., Ahiauzu, A., & Ntayi, J. M. (2020). Digital technology adoption and firm performance: Role of organizational capabilities. *Technological Forecasting and Social Change, 162*, 120330. DOI: 10.1016/j.techfore.2020.120330

Bonne, J., Donnat, L., Albora, G., Burgalat, J., Chauvin, N., Combaz, D., Cousin, J., Decarpenterie, T., Duclaux, O., Dumelié, N., Galas, N., Juéry, C., Parent, F., Pineau, F., Maunoury, A., Ventre, O., Bénassy, M.-F., & Joly, L. (2024). A measurement system for CO2 and CH4 emissions quantification of industrial sites using a new in situ concentration sensor operated on board uncrewed aircraft vehicles. *Atmospheric Measurement Techniques, 17*(14), 4471–4491. DOI: 10.5194/amt-17-4471-2024

Borangiu, T., Morariu, O., Răileanu, S., Trentesaux, D., Leitão, P., & Barata, J. (2020). Digital transformation of manufacturing. Industry of the Future with Cyber-Physical Production Systems. *HAL (Le Centre Pour La Communication Scientifique Directe)*. https://uphf.hal.science/hal-03424105

Boscariol, P., & Richiedei, D. (2019). Trajectory Design for Energy Savings in Redundant Robotic Cells. *Robotics (Basel, Switzerland), 8*(1), 15. DOI: 10.3390/robotics8010015

Bosona, T., &Gebresenbet, G. (2023). The Role of Blockchain Technology in Promoting Traceability Systems in Agri-Food Production and Supply Chains [Review of *The Role of Blockchain Technology in Promoting Traceability Systems in Agri-Food Production and Supply Chains*]. *Sensors, 23*(11), 5342. Multidisciplinary Digital Publishing Institute. DOI: 10.3390/s23115342

Bown, C. P. (2020). Export controls: America's other national security threat. *Duke Journal of Comparative & International Law, 30*(2), 283–308. https://scholarship.law.duke.edu/djcil/vol30/iss2/4/

Boyd, J. R. (1976, September 3). *Destruction and creation.* https://www.coljohnboyd.com/static/documents/1976-09-03__Boyd_John_R__Destruction_and_Creation.pdf

Boyd, J. R. (1995, June 28). *The essence of winning and losing* [Briefing slides]. OODA Loop. https://ooda.de/media/john_boyd_-_the_essence_of_winning_and_losing.pdf

Boyd, J. R. (1987). *Destruction and creation.* U.S. Army Command and General Staff College, Center for Army Tactics., https://books.google.com/books?id=JBXWHAAACAAJ

Bratu, S. (2023). Generative Artificial Intelligence and Virtual Reality Immersive Training Tools, Algorithmic Monitoring and Cognitive Computing Systems, and Voice and Gesture Recognition Technologies in Virtual Immersive Workspaces. *Analysis and Metaphysics*, (22), 197–215.

Braunscheidel, M. J., & Suresh, N. C. (2009). The organizational antecedents of a firm's supply chain agility for risk mitigation and response. *Journal of Operations Management*, *27*(2), 119–140. DOI: 10.1016/j.jom.2008.09.006

Bressanelli, G., Adrodegari, F., Perona, M., & Saccani, N. (2020). Exploring how usage-focused business models enable circular economy through digital technologies. *Sustainability*, *12*(6), 1–20. DOI: 10.3390/su12062365

Bretschneider, L., Bollmann, S., Houssin-Agbomson, D., Shaw, J. T., Howes, N., Nguyen, L. T. H., Robinson, R., Helmore, J., Lichtenstern, M., Nwaboh, J. A., Pogány, A., Ebert, V., & Lampert, A. (2024). Concepts for drone based pipeline leak detection. *Frontiers in Robotics and AI*, *11*, 1426206. DOI: 10.3389/frobt.2024.1426206 PMID: 39211418

Brodeur, J., Deschamps, I., & Pellerin, R. (2023). Organizational changes approaches to facilitate the management of Industry 4.0 transformation in manufacturing SMEs. *Journal of Manufacturing Technology Management*, *34*(7), 1098–1119. DOI: 10.1108/JMTM-10-2022-0359

Brogaard, R. Y., & Boukas, E. (2023). Autonomous GPU-based UAS for inspection of confined spaces: Application to marine vessel classification. *Robotics and Autonomous Systems*, *172*, 104590. DOI: 10.1016/j.robot.2023.104590

Brundage, M. P., Sexton, T., Hodkiewicz, M., Dima, A., & Lukens, S. (2020). Technical language processing: Unlocking maintenance knowledge. *Manufacturing Letters*, *27*, 42–46. DOI: 10.1016/j.mfglet.2020.11.001

Brusso, B. C. (2018). 50 Years of Industrial Automation [History]. *IEEE Industry Applications Magazine, 24*(4), 8–11. DOI: 10.1109/MIAS.2018.2820440

Bubeník, P., Rakyta, M., Buzalka, M., Biňasová, V., & Kovaríková, Z. (2025). Optimization of Business Processes Using Artificial Intelligence. *Electronics (Basel), 14*(11), 2105. Advance online publication. DOI: 10.3390/electronics14112105

Buchner, J., Buntins, K., & Kerres, M. (2022). The impact of augmented reality on cognitive load and performance: A systematic review. *Journal of Computer Assisted Learning, 38*(1), 285–303. DOI: 10.1111/jcal.12617

Bughin, J., Deakin, J., & O'Beirne, B. (2019, October 22). *Digital transformation: Improving the odds of success.* McKinsey & Company. https://www.mckinsey.com/~/media/McKinsey/Business%20Functions/McKinsey%20Digital/Our%20Insights/Digital%20transformation%20Improving%20the%20odds%20of%20success/Digital-transformation-Improving-the-odds-of-success-final.pdf

Bühler, M. M., Jelinek, T., & Nübel, K. (2022). Training and Preparing Tomorrow's Workforce for the Fourth Industrial Revolution. *Education Sciences, 12*(11), 782. DOI: 10.3390/educsci12110782

Bundesamt für Sicherheit in der Informationstechnik. (2024). *Die Lage der IT-Sicherheit in Deutschland 2024* [Annual report]. https://www.bsi.bund.de/SharedDocs/Downloads/DE/BSI/Publikationen/Lageberichte/Lagebericht2024.html

Buolamwini, J., & Gebru, T. (2018, January). Gender shades: Intersectional accuracy disparities in commercial gender classification. In *Conference on fairness, accountability and transparency* (pp. 77-91). PMLR.

Burgués, J., & Marco, S. (2020). Environmental chemical sensing using small drones: A review [Review of *Environmental chemical sensing using small drones: A review*]. *The Science of The Total Environment, 748*, 141172. Elsevier BV. DOI: 10.1016/j.scitotenv.2020.141172

Business Continuity Institute, & Resilience360. (2020). *COVID-19: The future of supply chains* (Survey report). The Business Continuity Institute. https://www.thebci.org/static/7324b815-9364-47d3-9277ab4ce9aa4c0f/a3e39af5-193d-428d-b603d3cf7e600f39/BCI-0007d-The-Future-of-Supply-ChainSingles-Low.pdf

Butt, J. (2020). A strategic roadmap for the manufacturing industry to implement Industry 4.0. *Designs, 4*(2), 11. DOI: 10.3390/designs4020011

C., J. L. D., & Ocampo-Martínez, C. (2019). Energy efficiency in discrete-manufacturing systems: Insights, trends, and control strategies. *Journal of Manufacturing Systems, 52*, 131. DOI: 10.1016/j.jmsy.2019.05.002

Caballero-Martin, D., López-Guede, J. M., Estévez, J., & Graña, M. (2024). Artificial Intelligence Applied to Drone Control: A State of the Art. *Drones (Basel), 8*(7), 296. DOI: 10.3390/drones8070296

Cabianca, M., Clemente, M. L., Gatto, G., Impagliazzo, C., Leoni, L., Masia, M., & Piras, R. (2022). An Application of IoT in a Drone Inspection Service for Environmental Control. *IoT, 3*(3), 366–380. DOI: 10.3390/iot3030020

Caggiano, A., Grant, R., Peng, C., Li, Z., & Simeone, A. (2022). Manufacturing Process Impacts on Occupational Health: A Machine Learning Framework. *Procedia CIRP, 112*, 561–566. Advance online publication. DOI: 10.1016/j.procir.2022.09.100

Cagliano, R., Canterino, F., Longoni, A., & Bartezzaghi, E. (2019). *The interplay between smart manufacturing technologies and work organization. International Journal of Operations & Production Management, 39*(6/7/8), 913–934. https://doi.org/DOI: 10.1108/IJOPM-01-2018-0029

Cai, T., & Hong, Z. (2024). Exploring the structure of the digital economy through blockchain technology and mitigating adverse environmental effects with the aid of artificial neural networks. *Frontiers in Environmental Science, 12*, 1315812. Advance online publication. DOI: 10.3389/fenvs.2024.1315812

Caldara, D., & Iacoviello, M. (2022). Measuring geopolitical risk. *The American Economic Review, 112*(4), 1194–1225. DOI: 10.1257/aer.20191823

Callari, T. C., Segate, R. V., Hubbard, E., Daly, A., & Lohse, N. (2024). An ethical framework for human-robot collaboration for the future people-centric manufacturing: A collaborative endeavour with European subject-matter experts in ethics. *Technology in Society, 78*, 102680. DOI: 10.1016/j.techsoc.2024.102680

Callison, S. (n.d.). *What is a gap analysis? An easy-to-use template.* ClearPoint Strategy (Blog). https://www.clearpointstrategy.com/blog/gap-analysis-template

Calvo, R., & Gil, P. M. (2022). Evaluation of Collaborative Robot Sustainable Integration in Manufacturing Assembly by Using Process Time Savings. *Materials (Basel), 15*(2), 611. DOI: 10.3390/ma15020611 PMID: 35057338

Cameron, C., Fredin, Z., & Gershenfeld, N. (2022). Discrete Assembly of Unmanned Aerial Systems. *2022 International Conference on Unmanned Aircraft Systems (ICUAS)*, 339. DOI: 10.1109/ICUAS54217.2022.9836082

Cannava, L., Javan, F. D., Najafi, B., & Perotti, S. (2024). Green warehousing practices: Assessing the impact of PV self-consumption enhancement strategies in a logistics warehouse. *Sustainable Energy Technologies and Assessments, 72*, 104054. DOI: 10.1016/j.seta.2024.104054

Capetillo, A. A., Bauer, F., & Chaminâde, C. (2022). Emerging Technologies Supporting the Transition to a Circular Economy in the Plastic Materials Value Chain [Review of *Emerging Technologies Supporting the Transition to a Circular Economy in the Plastic Materials Value Chain*]. *Circular Economy and Sustainability, 3*(2), 953. Springer Nature. DOI: 10.1007/s43615-022-00209-2

Carayannis, E. G., Papamichail, G., Zotas, N., & Askounis, D. (2025). Corporate foresight: Navigating uncertainty in a VUCA world. *Journal of the Knowledge Economy, 16*(6), 17494–17515. Advance online publication. DOI: 10.1007/s13132-024-02587-4

Carroll, S., Satme, J. N., Alkharusi, S., Vitzilaios, N., Downey, A., & Rizos, D. (2021). Drone-Based Vibration Monitoring and Assessment of Structures. *Applied Sciences (Basel, Switzerland), 11*(18), 8560. DOI: 10.3390/app11188560

Cascio, W. F., & Montealegre, R. (2016). How Technology Is Changing Work and Organizations. In *Annual Review of Organizational Psychology and Organizational Behavior* (Vol. 3). https://doi.org/DOI: 10.1146/annurev-orgpsych-041015-062352

Casper, S., & Hancké, B. (1999). Global quality norms within national production regimes: ISO 9000 standards in the French and German car industries. *Organization Studies, 20*(6), 961–985. DOI: 10.1177/0170840699206003

Castaldi, C., & Giarratana, M. S. (2015). A knowledge-management approach to capture organizational learning networks. *Information & Management, 52*(7), 806–820. DOI: 10.1016/j.im.2015.06.001

Catti, P., Freitas, A., Pereira, E., Gonçalves, G., Lopes, R. P., Nikolakis, N., & Alexopoulos, K. (2024). Data Analytics and AI for Quality Assurance in Manufacturing: Challenges and Opportunities. In *Lecture notes in networks and systems* (p. 205). Springer International Publishing. DOI: 10.1007/978-3-031-65411-4_25

Celestin, M., & Sujatha, S. (2024). Impact of global supply-chain disruptions on business resilience: Strategies for adapting to pandemics and geopolitical conflicts. *Zenodo*. https://doi.org/DOI: 10.5281/zenodo.13887198

Cernega, A., Nicolescu, D. N., Mele canu Imre, M., Ripszky Totan, A., Arsene, A. L., erban, R. S., Perpelea, A.-C., Nedea, M.-I., & Pi uru, S.-M. (2024). Volatility, uncertainty, complexity, and ambiguity (VUCA) in healthcare. *Health Care, 12*(7), 773. Advance online publication. DOI: 10.3390/healthcare12070773 PMID: 38610195

Cham, T. H., Wei-Han Tan, G., Aw, E. C. X., Ooi, K. B., Jee, T. W., & Pek, C. K. (2024). Virtual reality in tourism: Adoption scepticism and resistance. *Tourism Review, 79*(2), 337–354. DOI: 10.1108/TR-10-2022-0479

Chatterjee, S., Rana, N. P., Dwivedi, Y. K., & Baabdullah, A. M. (2021). Understanding AI adoption in manufacturing and production firms using an integrated TAM-TOE model. *Technological Forecasting and Social Change, 170,* 120880. Advance online publication. DOI: 10.1016/j.techfore.2021.120880

Cheng, C.-H., Li, M.-H., Tang, B.-J., & Cheng, Y.-R. (2024). The impact of knowledge management and organizational learning promotion in small and medium enterprises on the implementation of Industry 4.0 and competitiveness. *Administrative Sciences, 14*(8), 161. DOI: 10.3390/admsci14080161

Cheng, J.-Y., Mendes, E. L. S., Gheisari, M., & Jeelani, I. (2022). Construction Worker-Drone Safety Training in a 360 Virtual Reality Environment: A Pilot Study. *EPiC Series in Built Environment, 3,* 19–8. DOI: 10.29007/8cjv

Cheng, W., & Yuan, C. (2020). Using a Micro Sampler on a Drone to Extract Organic Vapors—A Case Study of Monitoring Industrial Pollution. *Aerosol and Air Quality Research, 21*(2), 200359. DOI: 10.4209/aaqr.2020.06.0359

Cheng, Z. M., Bonetti, F., de Regt, A., Ribeiro, J. L., & Plangger, K. (2024). Principles of responsible digital implementation: Developing operational business resilience to reduce resistance to digital innovations. *Organizational Dynamics, 53*(2), 101043. DOI: 10.1016/j.orgdyn.2024.101043

Chen, H., Jeremiah, S. R., Lee, C., & Park, J. H. (2023). A Digital Twin-Based Heuristic Multi-Cooperation Scheduling Framework for Smart Manufacturing in IIoT Environment. *Applied Sciences (Basel, Switzerland), 13*(3), 1440. DOI: 10.3390/app13031440

Chen, L., Zhang, H., & Wang, M. (2024). Artificial intelligence in manufacturing: State of the art, perspectives, and future directions. *Journal of Manufacturing Systems, 74,* 126–149. DOI: 10.1016/j.jmsy.2024.01.004

Chen, T. (2024, April 30). The world of VUCA. In *Endogenous community design* (pp. 1–10). Springer., DOI: 10.1007/978-3-031-56756-8_1

Chen, Y. (2022). Advantages of 3D Printing for Circular Economy and Its Influence on Designers. *Proceedings of the Design Society, 2,* 991–1000. DOI: 10.1017/pds.2022.101

Chen, Y.-H., Sharma, K., Sharma, C., & Sharma, S. (2023). Integrating explainable artificial intelligence and blockchain to smart agriculture: Research prospects for decision making and improved security. *Smart Agricultural Technology, 6,* 100350. DOI: 10.1016/j.atech.2023.100350

Chen, Z., Chen, H., Imani, M., Chen, R., & Imani, F. (2024). Vision language model for interpretable and fine-grained detection of safety compliance in diverse workplaces. *Expert Systems with Applications, 125769*. Advance online publication. DOI: 10.1016/j.eswa.2024.125769

Cheryan, S., Ziegler, S. A., Montoya, A. K., & Jiang, L. (2017). Why are some STEM fields more gender balanced than others? *Psychological Bulletin, 143*(1), 1–35. DOI: 10.1037/bul0000052 PMID: 27732018

Chiarini, A. (2020). Industry 4.0, quality management and TQM world: A systematic literature review and a proposed agenda for further research. *The TQM Journal, 32*(4), 603–616. DOI: 10.1108/TQM-04-2020-0082

Chima, O. K., Idemudia, S. O., Ezeilo, O. J., Ojonugwa, B. M., & Adesuyi, A. O. M. (2022). Advanced review of SME regulatory compliance models across US state-level jurisdictions. Shodhshauryam. *International Scientific Refereed Research Journal, 5*(2), 191–209.

Choi, E., Johnson, D. A., Moon, K., & Oah, S. (2018). Effects of Positive and Negative Feedback Sequence on Work Performance and Emotional Responses. *Journal of Organizational Behavior Management, 38*(2–3), 97–115. Advance online publication. DOI: 10.1080/01608061.2017.1423151

Choi, E., Kim, C., & Lee, K. C. (2021). Consumer Decision-Making Creativity and Its Relation to Exploitation–Exploration Activities: Eye-Tracking Approach. *Frontiers in Psychology, 11*, 557292. Advance online publication. DOI: 10.3389/fpsyg.2020.557292 PMID: 33505328

Chowdhury, S., Dey, P., Joel-Edgar, S., Bhattacharya, S., Rodriguez-Espindola, O., Abadie, A., & Truong, L. (2023). Unlocking the value of artificial intelligence in human resource management through AI capability framework. *Human Resource Management Review, 33*(1), 100899. DOI: 10.1016/j.hrmr.2022.100899

Christopoulos, A., & Mystakidis, S. (2023). Gamification in education. *Encyclopedia, 3*(4), 1223–1243. DOI: 10.3390/encyclopedia3040089

Christopoulos, A., Styliou, M., Ntalas, N., & Stylios, C. (2024). The impact of immersive virtual reality on knowledge acquisition and adolescent perceptions in cultural education. *Information (Basel), 15*(5), 261. DOI: 10.3390/info15050261

Chukwurah, N., Ige, A. B., Adebayo, V. I., & Eyieyien, O. G. (2024). Frameworks for effective data governance: Best practices, challenges, and implementation strategies across industries. *Computer Science & IT Research Journal, 5*(7), 1666–1679. DOI: 10.51594/csitrj.v5i7.1351

Chyr, G., & DeSimone, J. M. (2022). Review of high-performance sustainable polymers in additive manufacturing. *Green Chemistry*, *25*(2), 453–466. DOI: 10.1039/D2GC03474C

Çınar, Z. M., Zeeshan, Q., & Korhan, O. (2021). A framework for Industry 4.0 readiness and maturity of smart manufacturing enterprises: A case study. *Sustainability (Basel)*, *13*(12), 6659. DOI: 10.3390/su13126659

Cohen, J. B., Loeber, A. M. C., Marschalek, I., Bernstein, M. J., Blok, V., Tabarés, R., Gianni, R., & Griessler, E. (2024). From experimentation to structural change: Fostering institutional entrepreneurship for public engagement in research and innovation. *Science & Public Policy*, *51*(2), 324–336. DOI: 10.1093/scipol/scad065

Colaw, C. L., Madison, G., Tseng, B., Griser, G. M., Truelson, G., Gallo, A., & Hurmuzlu, Y. (2025). Methodology for Enablement of Human Digital Twins for Quality Assurance in the Aerospace Manufacturing Domain. *Sensors (Basel)*, *25*(11), 3362. Advance online publication. DOI: 10.3390/s25113362 PMID: 40968904

Collan, M., & Michelsen, K.-E. (2020). *Technical, Economic and Societal Effects of Manufacturing 4.0.* DOI: 10.1007/978-3-030-46103-4

Colledani, M., Tolio, T., Fischer, A., Iung, B., Lanza, G., Schmitt, R., & Váncza, J. (2014). Design and management of manufacturing systems for production quality. *CIRP Annals*, *63*(2), 773–796. DOI: 10.1016/j.cirp.2014.05.002

Colombo, S., Nazir, S., & Manca, D. (2014). Immersive virtual reality for training and decision making: Preliminary results of experiments performed with a plant simulator. *SPE Economics & Management*, *6*(04), 165–172. DOI: 10.2118/164993-PA

Conges, A., Breard, L., Patruno, W., Ouro-Sao, A., Salatge, N., Fertier, A., Lauras, M., Graham, J., & Benaben, F. (2023). Situational awareness and decision-making in a crisis situation: A crisis management cell in virtual reality. *International Journal of Disaster Risk Reduction*, *97*, 104002. DOI: 10.1016/j.ijdrr.2023.104002

Cosa, M., & Torelli, R. (2024). Digital transformation and flexible performance management: A systematic literature review of the evolution of performance measurement systems. *Global Journal of Flexible Systems Managment*, *25*(3), 445–466. DOI: 10.1007/s40171-024-00409-9

Cotton, W. B. (2001). Transforming traditional training methods to meet the needs of a modern industry. *Journal of Petroleum Technology*, *53*(02), 54–58. DOI: 10.2118/65351-JPT

Creazza, A., Colicchia, C., Spiezia, S., & Dallari, F. (2022). Who cares? Supply-chain managers' perceptions regarding cyber supply-chain risk management in the digital transformation era. *Supply Chain Management, 27*(1), 30–53. DOI: 10.1108/SCM-02-2020-0073

Crespo Márquez, A. (2007). *The maintenance management framework: Models and methods for complex systems maintenance.* Springer., DOI: 10.1007/978-1-84628-821-0

Cruz, Y. J., Castaño, F., Haber, R. E., Villalonga, A., Ejsmont, K., Gładysz, B., Flores, Á., & Alemany, P. (2024). *Self-Reconfiguration for Smart Manufacturing Based on Artificial Intelligence: A Review and Case Study* [Review of *Self-Reconfiguration for Smart Manufacturing Based on Artificial Intelligence: A Review and Case Study*]. 121. DOI: 10.1007/978-3-031-46452-2_8

Cruzes, D. S., Dybå, T., Runeson, P., & Höst, M. (2015). Case studies synthesis: A thematic, cross-case, and narrative synthesis worked example. *Empirical Software Engineering, 20*(6), 1634–1665. DOI: 10.1007/s10664-014-9326-8

Cuban, L. (2010). *As good as it gets: What school reform brought to Austin.* Harvard University Press. https://www.hup.harvard.edu/books/9780674035546

D'Eusanio, M., & Petti, L. (2024). Blockchain technology and Social Life Cycle Assessment: Synergies and implications. *The International Journal of Life Cycle Assessment.* Advance online publication. DOI: 10.1007/s11367-024-02338-2

Damaševičius, R., & Sidekerskienė, T. (2024). Virtual worlds for learning in metaverse: A narrative review. *Sustainability (Basel), 16*(5), 2032. DOI: 10.3390/su16052032

Damiani, L., Demartini, M., Guizzi, G., Revetria, R., & Tonelli, F. (2018). Augmented and virtual reality applications in industrial systems: A qualitative review towards the industry 4.0 era. *IFAC-PapersOnLine, 51*(11), 624–630. DOI: 10.1016/j.ifacol.2018.08.388

Dang-Pham, D., Hoang, A.-P., Vo, D.-T., & Kautz, K. (2022). Digital Kaizen: An Approach to Digital Transformation. *AJIS. Australasian Journal of Information Systems/AJIS. Australian Journal of Information Systems. AJIS. Australasian Journal of Information Systems, 26.* Advance online publication. DOI: 10.3127/ajis.v26i0.3851

Dasaklis, T. K., & Pappis, C. P. (2013). Supply chain management in view of climate change: An overview of possible impacts and the road ahead. *Journal of Industrial Engineering and Management, 6*(4). Advance online publication. DOI: 10.3926/jiem.883

Davuluri, K. (2024). Facilitating Real-time Road Event Detection on Collaborative UAV and Ground Vehicle. *Deep Blue (University of Michigan)*. DOI: 10.7302/24661

de Carvalho, P. T., Lopes, J. D., & Raimundo, R. (2025). Innovation Impact in the Textile Industry: From the Toyota Production System to Artificial Intelligence. *Sustainability (Basel)*, *17*(3), 1170. DOI: 10.3390/su17031170

De Gasperis, G., & Facchini, S. D. (2025). *A Comparative Study of Rule-Based and Data-Driven Approaches in Industrial Monitoring*. DOI: 10.48550/ARXIV.2509.15848

de Luis García, E. (2024). COMBATING CHILD SEXUAL ABUSE ON THE INTERNET: CONSIDERATIONS IN THE LIGHT OF THE PROPOSAL FOR A REGULATION. *Actualidad Juridica Iberoamericana*, *21*, 104–129. DOI: 10.13039/501100011033

DealHub. (2024, June 10). *What is phased implementation?* Retrieved September 16, 2025, from https://dealhub.io/glossary/phased-implementation/

Dechezleprêtre, A., & Sato, M. (2017). The impacts of environmental regulations on competitiveness. *Review of environmental economics and policy*. https://www.journals.uchicago.edu/doi/full/10.1093/reep/rex013

Dehghanzadeh, H., Farrokhnia, M., Dehghanzadeh, H., Taghipour, K., & Noroozi, O. (2024). Using gamification to support learning in K-12 education: A systematic literature review. *British Journal of Educational Technology*, *55*(1), 34–70. DOI: 10.1111/bjet.13335

Del Río-Casasola, A., López-Calle, P., & Ruiz-Gálvez, M. E. (2024). The transformation of the automotive industry toward electrification and its impact on global value chains. *European Research on Management and Business Economics*, *30*(1), 100242. Advance online publication. DOI: 10.1016/j.iedeen.2024.100242

Deming, D. J. (2017). The growing importance of social skills in the labor market. *The Quarterly Journal of Economics*, *132*(4), 1593–1640. DOI: 10.1093/qje/qjx022

Deming, W. E. (2000). *Out of the crisis* (Press, M. I. T., Ed.). MIT Press. (Original work published 1986), https://mitpress.mit.edu/9780262541152/out-of-the-crisis/

Dennison, M. S., Kumar, M. B., & Jebabalan, S. K. (2024). Realization of circular economy principles in manufacturing: Obstacles, advancements, and routes to achieve a sustainable industry transformation. *Discover Sustainability*, *5*(1), 438. Advance online publication. DOI: 10.1007/s43621-024-00689-2

Deogaonkar, A. (2025). Immersion, realism and engagement: Exploring the impact of generative AI on virtual training performance across sectors. *European Journal of Training and Development*.

Development, U. N. C. on T. and. (2023). Harnessing green technology and innovation for inclusive and sustainable development. In *United Nations eBooks* (p. 33). United Nations. DOI: 10.18356/9789210028295c011

Dewulf, J., Beylot, A., Monfort, D., Lai, F., Santillán-Saldivar, J., Muller, S., & Mathieux, F. (2023). Contribution to inaccessibility as resource impact method: A base for sustainable resource management along the life cycle. *Resources, Conservation and Recycling*, *202*, 107363. DOI: 10.1016/j.resconrec.2023.107363

Dhillon, R., & Nguyen, Q. C. (2021). Strategies to respond to a VUCA world.

Dhiman, S., Seema, S., & Akhai, S. (2025). Performance evaluation of OPC 43 concrete incorporating micro silica and recycled ceramic tile waste as partial cement and coarse aggregate replacements. *Asian Journal of Civil Engineering*, 1-19. DOI: 10.1007/s42107-025-01504-7

Diaz, J. L., Bermeo, M. A., Díaz-Rozo, J., & Ocampo-Martínez, C. (2019). An optimization-based control strategy for energy efficiency of discrete manufacturing systems. *ISA Transactions*, *93*, 399–409. DOI: 10.1016/j.isatra.2019.03.015 PMID: 30955833

Dicken, P. (1992). International production in a volatile regulatory environment: The influence of national regulatory policies on the spatial strategies of transnational corporations. *Geoforum*, *23*(3), 303–316. DOI: 10.1016/0016-7185(92)90044-5

DiMaggio, P. J., & Powell, W. W. (1983). The iron cage revisited: Institutional isomorphism and collective rationality in organizational fields. *American Sociological Review*, *48*(2), 147–160. DOI: 10.2307/2095101

Dimitrijević, B., Stanković, M., & Anđelković, T. (2023). Synergy between the circular economy and 3D printing. *Ekonomski Signali*, *18*(2), 49–68. DOI: 10.5937/ekonsig2302049D

Dimitriou, N., Leontaris, L., Vafeiadis, T., Ioannidis, D., Wotherspoon, T., Tinker, G., & Tzovaras, D. (2019). A Deep Learning framework for simulation and defect prediction applied in microelectronics. *Simulation Modelling Practice and Theory*, *100*, 102063. DOI: 10.1016/j.simpat.2019.102063

Ding, Y., & Li, L. (2024). Optimization Method for Digital Twin Manufacturing System Based on NSGA-II. *International Journal of Advanced Computer Science and Applications, 15*(4). Advance online publication. DOI: 10.14569/IJACSA.2024.01504100

Dip, H. S. (2024). Circular Economy Initiatives in Food Manufacturing Industry. *Research Square (Research Square)*. DOI: 10.21203/rs.3.rs-4996705/v1

Dixit, K. K., Dharme, A., Singh, D. P., Kalra, R., Bhavani, B., Asha, V., &Abdulsada, Zainab. R. (2023). Sustainability by Design: Innovative Ways of Revolutionizing Production Practices for a Better Tomorrow. *E3S Web of Conferences, 453*, 1026. DOI: 10.1051/e3sconf/202345301026

Doherty, B., Haugh, H., & Lyon, F. (2014). Social enterprises as hybrid organizations: A review and research agenda. *International Journal of Management Reviews, 16*(4), 417–436. DOI: 10.1111/ijmr.12028

Domb, M., Joshi, S., & Khn, A. (2023). Anomaly Detection in IoT: Recent Advances, AI and ML Perspectives and Applications. In *Artificial intelligence*. IntechOpen. DOI: 10.5772/intechopen.111944

Dooley, J. F., Minschwaner, K., Dubey, M. K., Abbadi, S. H. E., Sherwin, E. D., Meyer, A. G., Follansbee, E., & Lee, J. E. (2024). A new aerial approach for quantifying and attributing methane emissions: Implementation and validation. *Atmospheric Measurement Techniques, 17*(17), 5091–5111. DOI: 10.5194/amt-17-5091-2024

dos Santos, R. B., Torrisi, N. M., & Pantoni, R. P. (2021). Third Party Certification of Agri-Food Supply Chain Using Smart Contracts and Blockchain Tokens. *Sensors (Basel), 21*(16), 5307. DOI: 10.3390/s21165307 PMID: 34450749

Doskenov, B., Adaji, O., & Dosumu, R. (2025, March 31). Improving manufacturing with adaptive learning in Industry 4.0: Challenges, and policy implications. *Current Journal of Applied Science and Technology, 44*(4), 111–132. DOI: 10.9734/cjast/2025/v44i44518

Doz, Y. L., & Kosonen, M. (2010). Embedding strategic agility: A leadership agenda for accelerating business model renewal. *Long Range Planning, 43*(2–3), 370–382. DOI: 10.1016/j.lrp.2009.07.006

Dragos. (2024). *2023 OT Cybersecurity Year in Review: Executive summary*. https://hub.dragos.com/year-in-review-executive-summary-2023

Dronova, I., Kislik, C., Dinh, Z., & Kelly, M. (2021). A Review of Unoccupied Aerial Vehicle Use in Wetland Applications: Emerging Opportunities in Approach, Technology, and Data [Review of *A Review of Unoccupied Aerial Vehicle Use in Wetland Applications: Emerging Opportunities in Approach, Technology, and Data*]. *Drones, 5*(2), 45. Multidisciplinary Digital Publishing Institute. DOI: 10.3390/drones5020045

Duan, Z., Li, Y., Wang, J., Zhao, G., & Svanberg, S. (2019). Aquatic environment monitoring using a drone-based fluorosensor. *Applied Physics. B, Lasers and Optics, 125*(6), 108. Advance online publication. DOI: 10.1007/s00340-019-7215-y

Durfey, N., & Sajal, S. (2022, May 1). A Comprehensive Survey: Cybersecurity Challenges and Futures of Autonomous Drones. *2022 Intermountain Engineering, Technology and Computing (IETC)*. DOI: 10.1109/IETC54973.2022.9796881

Dwivedi, Y. K., Hughes, L., Baabdullah, A. M., Ribeiro-Navarrete, S., Giannakis, M., Al-Debei, M., & Wamba, S. F. (2023). Metaverse beyond the hype: Multidisciplinary perspectives on emerging challenges, opportunities, and agenda for research, practice and policy. *International Journal of Information Management, 71*, 102642. DOI: 10.1016/j.ijinfomgt.2023.102642

Ebrahimi, S. M., & Koh, L. (2021). Manufacturing sustainability: Institutional theory and life cycle thinking. *Journal of Cleaner Production, 298*, 126787. DOI: 10.1016/j.jclepro.2021.126787

Efthymiou, K., Mourtzis, D., Pagoropoulos, A., Papakostas, N., & Chryssolouris, G. (2016). Manufacturing systems complexity analysis methods review. *International Journal of Computer Integrated Manufacturing, 29*(9), 1025–1044. DOI: 10.1080/0951192X.2015.1130245

Elahi, M., Afolaranmi, S. O., Lastra, J. L. M., & García, J. A. P. (2023). A comprehensive literature review of the applications of AI techniques through the lifecycle of industrial equipment. *Discover Artificial Intelligence, 3*(1), 43. Advance online publication. DOI: 10.1007/s44163-023-00089-x

Elmegaard, B., Kermani, N. A., Bühler, F., Nguyen, T.-V., Bergamini, R., Zühlsdorf, B., Holdt, F. D., Holm, F. M., Petersen, M. S., Helk, A., Mogensen, N. B., Jürgensen, P., Hansen, L. I., Munk, P., Godiksen, E. L., Hetting, V., & Jacobsen, E. (2021). Electrification of processes and technologies for Danish Industry: Elforsk project 350-038 : Final Report. In *Research Portal Denmark* (p. 348). Technical University of Denmark., https://local.forskningsportal.dk/local/dki-cgi/ws/cris-link?src=dtu&id=dtu-39c8ef2f-cd96-4e85-bc7f-9e5135da93ca&ti=Electrification%20of%20processes%20and%20technologies%20for%20Danish%20Industry%3A%20Elforsk%20project%20350-038%20%3A%20Final%20Report

Ely, R. J., Ibarra, H., & Kolb, D. M. (2011). Taking gender into account: Theory and design for women's leadership development programs. In *Academy of Management Learning and Education* (Vol. 10, Issue 3). https://doi.org/DOI: 10.5465/amle.2010.0046

Errida, A., & Lotfi, B. (2021). The determinants of organizational change management success: Literature review and case study. *International Journal of Engineering Business Management, 13*, 1–15. DOI: 10.1177/18479790211016273

Esiri, A. E., Sofoluwe, O. O., & Ukato, A. (2024). Digital twin technology in oil and gas infrastructure: Policy requirements and implementation strategies. *Engineering Science & Technology Journal, 5*(6), 2039–2049. DOI: 10.51594/estj.v5i6.1221

Essaid, M., & Ju, H. (2025). Blockchain Solutions for Enhancing Security and Privacy in Industrial IoT. *Applied Sciences (Basel, Switzerland), 15*(12), 6835. Advance online publication. DOI: 10.3390/app15126835

Étienne, J. (2010). Compliance theories: A literature review. *Revue Française de Science Politique, 60*(3), 493–517. DOI: 10.3917/rfsp.603.0493

European Commission. (2020). *The Digital Skills and Jobs Coalition*. Ec.Europa.Eu. https://digital-strategy.ec.europa.eu/en/policies/digital-skills-coalition

European Commission. (2024, November 19). *Report on the functioning of the European carbon market in 2023* (COM(2024) 538 final). https://eur-lex.europa.eu/legal-content/EN/TXT/PDF/?uri=CELEX:52024DC0538

European Parliament and Council. (2022, December 14). Directive (EU) 2022/2464 … as regards corporate sustainability reporting (CSRD). *Official Journal of the European Union, L, 322*, 15–80. https://eur-lex.europa.eu/eli/dir/2022/2464/oj/eng

Faccio, M., & Cohen, Y. (2024). Intelligent cobot systems: Human-cobot collaboration in manufacturing. *Journal of Intelligent Manufacturing, 35*(5), 1905–1907. DOI: 10.1007/s10845-023-02142-z

Falola, P. B., Adeniyi, A. E., Awotunde, J. B., Akinola, S. O., Olagunju, M., & Olanloye, O. D. (2024). The Role of Augmented Reality, Virtual Reality, and Mixed Reality in Industries 4.0 and 5.0. In *Computational Intelligence in Industry 4.0 and 5.0 Applications* (pp. 329-356). Auerbach Publications.

Fang, X., Brown, D. S., Florence, C. S., & Mercy, J. A. (2012). The economic burden of child maltreatment in the United States and implications for prevention. *Child Abuse & Neglect, 36*(2), 156–165. Advance online publication. DOI: 10.1016/j.chiabu.2011.10.006 PMID: 22300910

Fan, H., Liu, C., Janvisloo, N. E., Bian, S., Fuh, J. Y. H., Lu, W. F., & Li, B. (2025). MaViLa: Unlocking new potentials in smart manufacturing through vision language models. *Journal of Manufacturing Systems, 80*, 258–271. DOI: 10.1016/j.jmsy.2025.02.017

Fascista, A. (2022). Toward Integrated Large-Scale Environmental Monitoring Using WSN/UAV/Crowdsensing: A Review of Applications, Signal Processing, and Future Perspectives [Review of *Toward Integrated Large-Scale Environmental Monitoring Using WSN/UAV/Crowdsensing: A Review of Applications, Signal Processing, and Future Perspectives*]. *Sensors, 22*(5), 1824. Multidisciplinary Digital Publishing Institute. DOI: 10.3390/s22051824

FasterCapital. (2023). *Process improvement: Process perfection: Employing gap analysis for continuous improvement.* Retrieved September 16, 2025, from https://fastercapital.com/content/Process-Improvement--Process-Perfection--Employing-Gap-Analysis-for-Continuous-Improvement.html

Felix, O. O.Oroo Oyondi Felix. (2023). Securing the skies: A comprehensive survey on internet of drones security challenges and solutions. *World Journal of Advanced Research and Reviews, 20*(3), 780–800. DOI: 10.30574/wjarr.2023.20.3.2491

Fernández-Miguel, A., Muiña, F. E. G., Settembre-Blundo, D., Tarantino, S. C., & Riccardi, M. P. (2024). Exploring Systemic Sustainability in Manufacturing: Geoanthropology's Strategic Lens Shaping Industry 6.0. *Global Journal of Flexible Systems Managment, 25*(3), 579–600. DOI: 10.1007/s40171-024-00404-0

Fernando, Y., Wahyuni-TD, I. S., Gui, A., Ikhsan, R. B., Mergeresa, F., & Ganesan, Y. (2023). A mixed-method study on the barriers of Industry 4.0 adoption in the Indonesian SMEs manufacturing supply chains. *Journal of Science and Technology Policy Management, 14*(4), 678–695. DOI: 10.1108/JSTPM-10-2021-0155

Festo. (2017, June 10). *Industry 4.0 finds increasing application in production.* https://press.festo.com/en/industry-segments/industry-4-0-finds-increasing-application-in-production

Fiene, R. (2023). *The importance of the theory of regulatory compliance.* SSRN *Electronic Journal.* https://doi.org/DOI: 10.2139/ssrn.4597469

Filho, P. S., Shiguemori, E. H., & Saotome, O. (2017). UAV VISUAL AUTOLOCALIZATON BASED ON AUTOMATIC LANDMARK RECOGNITION. *ISPRS Annals of the Photogrammetry, Remote Sensing and Spatial Information Sciences, 89*(W3), 89–94. Advance online publication. DOI: 10.5194/isprs-annals-IV-2-W3-89-2017

Filho, W. L., Aina, Y. A., Gatto, A., Rampasso, I. S., Sigahi, T. F. A. C., Anholon, R., Bella, R. L. F., Quelhas, O. L. G., Vasconcelos, C. R. P., & Luetz, J. M. (2025). Greening the factory floor and reducing the climate impact of the manufacturing sector. *Discover Sustainability, 6*(1), 1204. Advance online publication. DOI: 10.1007/s43621-025-02056-1

Fitrianto, I., & Saif, A. (2024). The role of virtual reality in enhancing Experiential Learning: a comparative study of traditional and immersive learning environments. *International Journal of Post Axial: Futuristic Teaching and Learning, 97-110.*

Fitsilis, P., Tsoutsa, P., & Gerogiannis, V. (2018). Industry 4.0: Required personnel competences. *Industry 4.0, 3*(3), 130-133.

Fottner, J., Clauer, D., Hormes, F., Freitag, M., Beinke, T., Overmeyer, L., Gottwald, S. N., Elbert, R., Sarnow, T., Schmidt, T., Reith, K.-B., Zadek, H., & Thomas, F. (2021). Autonomous Systems in Intralogistics – State of the Art and Future Research Challenges. [Technical University of Darmstadt]. *TUbilio, 14*, 2. DOI: 10.23773/2021_1

Frank, A. G., Dalenogare, L. S., & Ayala, N. F. (2019). Industry 4.0 technologies: Implementation patterns in manufacturing companies. *International Journal of Production Economics, 210*, 15–26. Advance online publication. DOI: 10.1016/j.ijpe.2019.01.004

Frăţilă, D., & Rotaru, H. (2017). Additive manufacturing – a sustainable manufacturing route. *MATEC Web of Conferences, 94*, 3004. DOI: 10.1051/matecconf/20179403004

Frey, C. B., & Osborne, M. A. (2017). The future of employment: How susceptible are jobs to computerisation? *Technological Forecasting and Social Change, 114*, 254–280. DOI: 10.1016/j.techfore.2016.08.019

Fridgeirsson, T. V., Ingason, H. T., Jonasson, H. I., & Kristjansdottir, B. H. (2021). The VUCAlity of projects: A new approach to assess a project risk in a complex world. *Sustainability (Basel), 13*(7), 3808. DOI: 10.3390/su13073808

Friedman, B., & Nissenbaum, H. (1996). Bias in computer systems. [TOIS]. *ACM Transactions on Information Systems*, *14*(3), 330–347. DOI: 10.1145/230538.230561

Friege, H., & Kümmerer, K. (2022). Practising circular economy. In *Routledge eBooks* (p. 259). Informa., DOI: 10.4324/9781003244196-27

Frohlich, M. T., & Westbrook, R. (2001). Arcs of integration: An international study of supply chain strategies. *Journal of Operations Management*, *19*(2), 185–200. DOI: 10.1016/S0272-6963(00)00055-3

Funk, R. J., & Hirschman, D. (2017). Beyond nonmarket strategy: Market actions as corporate political activity. *Academy of Management Review*, *42*(1), 32–52. DOI: 10.5465/amr.2013.0178

Gadaleta, M., Pellicciari, M., & Berselli, G. (2019). Optimization of the energy consumption of industrial robots for automatic code generation. *Robotics and Computer-integrated Manufacturing*, *57*, 452–464. DOI: 10.1016/j.rcim.2018.12.020

Gago, R. M., Pereira, M. Y. A., & Pereira, G. A. S. (2021). An aerial robotic system for inventory of stockpile warehouses. *Engineering Reports*, *3*(9), e12396. Advance online publication. DOI: 10.1002/eng2.12396

Gailani, A., Cooper, S., Allen, S., Pimm, A., Taylor, P., & Gross, R. (2024). Assessing the potential of decarbonization options for industrial sectors. *Joule*, *8*(3), 576–603. DOI: 10.1016/j.joule.2024.01.007

Gallab, M., Nardo, M. D., & Naciri, L. (2024). Navigating contemporary challenges and future prospects in digital industry evolution. *Discover Applied Sciences*, *6*(5), 259. Advance online publication. DOI: 10.1007/s42452-024-05913-2

Galli, D., Torelli, R., & Caccialanza, A. (2023). Sustainability performance and sustainability reporting in SMEs: A love affair or a fight? *Journal of Management & Organization*, *29*(6), 1063–1088. DOI: 10.1017/jmo.2023.40

Gămulescu, O.-M., Rosca, S., Panaite, F. A., Costandoiu, A., & Riurean, S. (2020). Accident sites management using drones. *MATEC Web of Conferences, 305*, 4. DOI: 10.1051/matecconf/202030500004

Gao, Q., & Schweidtmann, A. M. (2024). Deep reinforcement learning for process design: Review and perspective. *Current Opinion in Chemical Engineering*, *44*, 101012. DOI: 10.1016/j.coche.2024.101012

Gao, Y., Liu, Y., & Wu, W. (2025). How Does Artificial Intelligence Capability Affect Product Innovation in Manufacturing Enterprises? Evidence from China. *Systems*, *13*(6), 480. Advance online publication. DOI: 10.3390/systems13060480

Gargalo, C. L., Yu, H., Vollmer, N. I., Arabkoohsar, A., Gernaey, K. V., & Sin, G. (2023). A process systems engineering view of environmental impact assessment in renewable and sustainable energy production: Status and perspectives. *Computers & Chemical Engineering*, *180*, 108504. DOI: 10.1016/j.compchemeng.2023.108504

Gebler, M., Cerdas, F., Thiede, S., & Herrmann, C. (2020). Life cycle assessment of an automotive factory: Identifying challenges for the decarbonization of automotive production – A case study. *Journal of Cleaner Production*, *270*, 122330. DOI: 10.1016/j.jclepro.2020.122330

Gebler, M., Warsen, J., Meininghaus, R., Baudis, M., Cerdas, F., & Herrmann, C. (2024). Implementing Zero Impact Factories in Volkswagen's Global Automotive Manufacturing System: A Discussion of Opportunities and Challenges from Integrating Current Science into Strategic Management. *Sustainability (Basel)*, *16*(7), 3011. DOI: 10.3390/su16073011

Geissdoerfer, M., Savaget, P., Bocken, N. M. P., & Hultink, E. J. (2017). The Circular Economy – A new sustainability paradigm? In Journal of Cleaner Production (Vol. 143). DOI: 10.1016/j.jclepro.2016.12.048

Geng, D., Evans, S., & Kishita, Y. (2023). The identification and classification of energy waste for efficient energy supervision in manufacturing factories. *Renewable & Sustainable Energy Reviews*, *182*, 113409. DOI: 10.1016/j.rser.2023.113409

Gerencheal, T. A., & Lee, J. H. (2025). Development of a Collision Avoidance System Using Multiple Distance Sensors for Indoor Inspection Drone. *Applied Sciences (Basel, Switzerland)*, *15*(16), 8791. DOI: 10.3390/app15168791

Ghahramani, M., Qiao, Y., Zhou, M. C., O'Hagan, A., & Sweeney, J. (2020a). AI-based modeling and data-driven evaluation for smart manufacturing processes. *IEEE/CAA Journal of Automatica Sinica*, *7*(4), 1026. DOI: 10.1109/JAS.2020.1003114

Ghayal, A. (2025). *Embedding ESG in Cost Accounting: A Simulation Framework for Financial Reporting and Transparency*. DOI: 10.31235/osf.io/4ge2z_v1

Ghazal, M., Basmaji, T., Yaghi, M., Alkhedher, M., Mahmoud, M., & El-Baz, A. (2020). Cloud-Based Monitoring of Thermal Anomalies in Industrial Environments Using AI and the Internet of Robotic Things. *Sensors (Basel)*, *20*(21), 6348. DOI: 10.3390/s20216348 PMID: 33171714

Gheewala, S. H. (2020). Life Cycle Thinking for Sustainable Consumption and Production towards a Circular Economy. *E3S Web of Conferences*, *202*, 1003. DOI: 10.1051/e3sconf/202020201003

Ghobakhloo, M. (2018). The future of manufacturing industry: A strategic roadmap toward Industry 4.0. *Journal of Manufacturing Technology Management, 29*(6), 910–936. DOI: 10.1108/JMTM-02-2018-0057

Ghobakhloo, M. (2020). Industry 4.0, digitization, and opportunities for sustainability. *Journal of Cleaner Production, 252*, 119869. DOI: 10.1016/j.jclepro.2019.119869

Ghosh, S., Hughes, M., Hodgkinson, I., & Hughes, P. (2022). Digital transformation of industrial businesses: A dynamic capability approach. *Technovation, 113*, 102414. Advance online publication. DOI: 10.1016/j.technovation.2021.102414

Gilad, B. (2004). *Early warning: Using competitive intelligence to anticipate market shifts, control risk, and create powerful strategies*. AMACOM., https://www.harpercollinsleadership.com/9780814432068/early-warning/

Giones, F., & Brem, A. (2022). Clarifying the conceptual map of VUCA: A systematic review. *The International Journal of Organizational Analysis, 30*(7), 196–221. DOI: 10.1108/IJOA-02-2022-3136

Giordano, C., Girelli, V. A., Lambertini, A., Tini, M. A., & Zanutta, A. (2025). UAV Data Collection Co-Registration: LiDAR and Photogrammetric Surveys for Coastal Monitoring. *Drones (Basel), 9*(1), 49. DOI: 10.3390/drones9010049

Glenk, G., Meier, R., & Reichelstein, S. (2024). Assessing the Costs of Industrial Decarbonization. SSRN *Electronic Journal*. DOI: 10.2139/ssrn.4998042

Gomes, C. F., Yasin, M. M., & Lisboa, J. V. (2004). A literature review of manufacturing performance measures and measurement in an organizational context: A framework and direction for future research. *Journal of Manufacturing Technology Management, 15*(6), 511–530. DOI: 10.1108/17410380410547906

Gómez-Rosal, D. A., Bergau, M., Fischer, G., Wachaja, A., Gräter, J., Odenweller, M., Piechottka, U., Hoeflinger, F., Gosala, N., Wetzel, N., Btischer, D., Valada, A., & Burgard, W. (2023). *A Smart Robotic System for Industrial Plant Supervision*. 1. DOI: 10.1109/SENSORS56945.2023.10325162

Gontarz, A., Hampl, D., Weiss, L., & Wegener, K. (2015). Resource Consumption Monitoring in Manufacturing Environments. *Procedia CIRP, 26*, 264–269. DOI: 10.1016/j.procir.2014.07.098

González-Castaño, F. J., Gil-Castiñeira, F., Rodriguez-Pereira, D., Regueiro-Janeiro, J. A., García-Méndez, S., & Candal-Ventureira, D. (2020). Self-Corrective Sensor Fusion for Drone Positioning in Indoor Facilities. *IEEE Access : Practical Innovations, Open Solutions, 9*, 2415–2427. DOI: 10.1109/ACCESS.2020.3048194

Gordon, L. A., & Loeb, M. P. (2002). The economics of information security investment. *ACM Transactions on Information and System Security*, *5*(4), 438–457. DOI: 10.1145/581271.581274

Görür, O. C., Yu, X., & Sivrikaya, F. (2021). Integrating Predictive Maintenance in Adaptive Process Scheduling for a Safe and Efficient Industrial Process. *Applied Sciences (Basel, Switzerland)*, *11*(11), 5042. DOI: 10.3390/app11115042

Green Supply Chain Management: Strategies for Sustainability and Efficiency. (2025). DOI: 10.52783/eel.v15i1.2433

Greenwood, R., Raynard, M., Kodeih, F., Micelotta, E. R., & Lounsbury, M. (2011). Institutional complexity and organizational responses. *The Academy of Management Annals*, *5*(1), 317–371. DOI: 10.5465/19416520.2011.590299

Grote, M., Oakey, A., Pilko, A., Krol, J., Blakesley, A., Cherrett, T., Scanlan, J. P., Anvari, B., & Martínez-Sykora, A. (2024). The effects of costs on drone uptake in multi-modal logistics systems within a healthcare setting. *Transport Economics and Management*, *2*, 58–75. DOI: 10.1016/j.team.2024.03.001

Group, B. M. W. (2023). *How AI is revolutionising production*. https://www.bmwgroup.com/en/news/general/2023/aiqx.html

Guebsi, R., Mami, S., & Chokmani, K. (2024). Drones in Precision Agriculture: A Comprehensive Review of Applications, Technologies, and Challenges [Review of *Drones in Precision Agriculture: A Comprehensive Review of Applications, Technologies, and Challenges*]. *Drones*, *8*(11), 686. Multidisciplinary Digital Publishing Institute. DOI: 10.3390/drones8110686

Gullett, B. K., Aurell, J., Mitchell, W. J., & Richardson, J. L. (2021). Use of an unmanned aircraft system to quantify NO_{<i>x</i>} emissions from a natural gas boiler. *Atmospheric Measurement Techniques*, *14*(2), 975. DOI: 10.5194/amt-14-975-2021 PMID: 34122665

Gunasekaran, K., Boopathi, S., Suresh, S., Suresh, P., & Gokulakannan, S. (2024). Sustainable building construction practices for transforming waste into wealth through vernacular architecture. In *Innovations in Energy Efficient Construction Through Sustainable Materials* (pp. 387–413). DOI: 10.4018/979-8-3693-3398-3.ch014

Gunasekaran, K., Boopathi, S., Suresh, S., Suresh, P., Vanitha, N. S., & Radhika, K. (2024). Eco-friendly smart agricultural process using artificial intelligence: Economic benefits. In *Reshaping Environmental Science Through Machine Learning and IoT* (pp. 160–175). DOI: 10.4018/979-8-3693-2351-9.ch009

Gunningham, N., & Rees, J. (1997). Industry self-regulation: An institutional perspective. *Law & Policy*, *19*(4), 363–414. DOI: 10.1111/1467-9930.t01-1-00033

Guo, B., Parlane, S., & Ryan, L. (2025). Regulatory compliance in the automobile industry. *UCD Centre for Economic Research Working Paper Series*, WP25/06. https://www.ucd.ie/economics/t4media/WP2025_06.pdf

Gupta, H., Lawal, J. N., Orji, I. J., & Kusi-Sarpong, S. (2021). Closing the Gap: The Role of Distributed Manufacturing Systems for Overcoming the Barriers to Manufacturing Sustainability. *IEEE Transactions on Engineering Management*, *70*(5), 1754–1773. DOI: 10.1109/TEM.2021.3059231

Gutmann, T., Nyffenegger, F., Pellegrini, M., Cabrucci, A., & Guzzini, A. (2023). A Digital Twin-Based Approach for the Optimization of Floor-Ball Manufacturing. *Electronics (Basel)*, *12*(24), 4979. DOI: 10.3390/electronics12244979

Gu, X., Jin, X., Ni, J., & Koren, Y. (2015). Manufacturing system design for resilience. *Procedia CIRP*, *36*, 135–140. DOI: 10.1016/j.procir.2015.02.075

Gyrichidi, N., Khalyasmaa, A. I., Eroshenko, S. A., & Romanov, A. M. (2024). Are Modern Market-Available Multi-Rotor Drones Ready to Automatically Inspect Industrial Facilities? *Drones (Basel)*, *8*(10), 549. DOI: 10.3390/drones8100549

Habibi, H., Rao, D. M. K. K. V., Sánchez-López, J. L., & Voos, H. (2023). On SORA for High-Risk UAV Operations under New EU Regulations. *Perspectives for Automated Approach*, *213*, 213–220. Advance online publication. DOI: 10.1109/ICUAS57906.2023.10156517

Hafiz, M., Azmi, E. D. S. K., Azlie, M. I. N. N., Zaki, M. J., Mazilan, M. A. H., Mazuki, M. S. M., Nor, M. I. N. M., Ghani, J. A., Rahmat, M. S., & Khamis, N. K. (2025). Application of Artificial Intelligence in Electronics and Semiconductor Industries. *JurnalKejuruteraan*, *37*(7), 3245. DOI: 10.17576/jkukm-2025-37(7)-11

Häfner, P., Bergmann, V., Häfner, V., Michels, F. L., Grethler, M., & Karande, A. (2024). An Adaptive Learning Environment for Industry 4.0 Competencies Based on a Learning Factory and Its Immersive Digital Twin. In *CSEDU (1)* (pp. 720-731).

Halkos, G., Moll de Alba, J., & Todorov, V. (2021). Analyzing manufacturing sector and selected development challenges: A panel data analysis. *Energy*, *230*, 120807. DOI: 10.1016/j.energy.2021.121253

Hamilton, D., McKechnie, J., Edgerton, E., & Wilson, C. (2021). Immersive virtual reality as a pedagogical tool in education: A systematic literature review of quantitative learning outcomes and experimental design. *Journal of Computers in Education*, *8*(1), 1–32. DOI: 10.1007/s40692-020-00169-2

Hammadi, M. (2021). Multidisciplinary collaborative design of complex systems. *HAL (Le Centre Pour La Communication Scientifique Directe)*. https://hal.science/tel-03614326 1. Hanenkamp, N., & Zipse, O. (2023). Sustainability in Manufacturing Transforming. In *Springer eBooks* (p. 187). Springer Nature. DOI: 10.1007/978-3-031-42224-9_7

Hamza, A., Ali, Z., Dudley, S., Saleem, K., Uneeb, M., & Christofides, N. (2025), A multi-stage review framework for AI-driven predictive maintenance and fault diagnosis in photovoltaic systems, Applied Energy, Volume 393, 126108, ISSN 0306-2619, DOI: 10.1016/j.apenergy.2025.126108

Hancko, D., Majlingova, A., & Kačíková, D. (2025). Integrating Virtual Reality, Augmented Reality, Mixed Reality, Extended Reality, and Simulation-Based Systems into Fire and Rescue Service Training: Current Practices and Future Directions. *Fire (Basel, Switzerland)*, *8*(6), 228. DOI: 10.3390/fire8060228

Hanifi, S., Alkali, B., Lindsay, G., & McGlinchey, D. (2025). Optimizing Energy and Air Consumption in Smart Manufacturing: An Industrial Internet of Things-Based Monitoring and Efficiency Enhancement Solution. *Applied Sciences (Basel, Switzerland)*, *15*(6), 3222. DOI: 10.3390/app15063222

Hansen, E. B. (2022). AI and IoT for Production Data Analytics in SMEs : Ph.d.-afhandling. In *Research Portal Denmark*. Technical University of Denmark. https://local.forskningsportal.dk/local/dki-cgi/ws/cris-link?src=ucn&id=ucn-94c7b983-1dc0-46e6-a136-fc130c9ff48b&ti=AI%20and%20IoT%20for%20Production%20Data%20Analytics%20in%20SMEs%20%3A%20Ph.d.-afhandling

Hardy, C., & Maguire, S. (2008). Institutional entrepreneurship. In Greenwood, R., Oliver, C., Sahlin, K., & Suddaby, R. (Eds.), *The SAGE handbook of organizational institutionalism* (pp. 198–217). SAGE. DOI: 10.4135/9781849200387.n8

Harvard Business School Online. (2024, February 14). *How gap analysis can drive strategic change in your company*. HBS Online Blog. https://online.hbs.edu/blog/post/gap-analysis

Hasan, H. R., Musamih, A., Salah, K., Jayaraman, R., Omar, M., Arshad, J., & Boscovic, D. (2024). Smart agriculture assurance: IoT and blockchain for trusted sustainable produce. *Computers and Electronics in Agriculture*, *224*, 109184. DOI: 10.1016/j.compag.2024.109184

Hassani, S. (2024a). Enhancing Legal Compliance and Regulation Analysis with Large Language Models. *arXiv (Cornell University)*. /arxiv.2404.17522DOI: 10.1109/RE59067.2024.00065

Hassnain, M., Lee, S. M. W., & Azhar, M. R. (2025). Soft sensing of biological oxygen demand in industrial wastewater using machine learning models. *Journal of Water Process Engineering, 78*, 108699. Advance online publication. DOI: 10.1016/j.jwpe.2025.108699

Hazrat, M. A., Hassan, N. M. S., Chowdhury, A. A., Rasul, M. G., & Taylor, B. A. (2023). Developing a skilled workforce for future industry demand: The potential of digital twin-based teaching and learning practices in engineering education. *Sustainability (Basel), 15*(23), 16433. DOI: 10.3390/su152316433

Helfat, C. E., & Peteraf, M. A. (2003). The dynamic resource-based view: Capability lifecycles. *Strategic Management Journal, 24*(10), 997–1010. DOI: 10.1002/smj.332

Hellweg, S., Benetto, E., Huijbregts, M. A. J., Verones, F., & Wood, R. (2023). Life-cycle assessment to guide solutions for the triple planetary crisis [Review of *Life-cycle assessment to guide solutions for the triple planetary crisis*]. *Nature Reviews Earth & Environment, 4*(7), 471. Nature Portfolio. DOI: 10.1038/s43017-023-00449-2

Hendricks, K. B., & Singhal, V. R. (2005). Association between supply chain glitches and operating performance. *Management Science, 51*(5), 695–711. DOI: 10.1287/mnsc.1040.0353

Hentout, A., Aouache, M., Maoudj, A., & Akli, I. (2019). Human–robot interaction in industrial collaborative robotics: A literature review of the decade 2008–2017. *Advanced Robotics, 33*(15–16), 764–799. DOI: 10.1080/01691864.2019.1636714

Heo, S.-J., & Na, W. S. (2025). Review of Drone-Based Technologies for Wind Turbine Blade Inspection. *Electronics (Basel), 14*(2), 227. DOI: 10.3390/electronics14020227

Heredia, J., Kirschner, R. J., Schlette, C., Abdolshah, S., Haddadin, S., & Kjægaard, M. B. (2023). ECDP: Energy Consumption Disaggregation Pipeline for Energy Optimization in Lightweight Robots. *IEEE Robotics and Automation Letters, 8*(10), 6107–6114. DOI: 10.1109/LRA.2023.3301311

Hermann, J., David, A., Wagner, A., & Ruskowski, M. (2020). Considering interdependencies for a dynamic generation of process chains for Production as a Service. *Procedia Manufacturing, 51*, 1454–1461. DOI: 10.1016/j.promfg.2020.10.202

Hermansa, M., Kozielski, M., Michalak, M., Szczyrba, K., Wróbel, Ł., & Sikora, M. (2021). Sensor-Based Predictive Maintenance with Reduction of False Alarms—A Case Study in Heavy Industry. *Sensors (Basel), 22*(1), 226. DOI: 10.3390/s22010226 PMID: 35009777

Hermundsdottir, F., & Aspelund, A. (2022). Competitive sustainable manufacturing - Sustainability strategies, environmental and social innovations, and their effects on firm performance. *Journal of Cleaner Production, 370*, 133474. DOI: 10.1016/j.jclepro.2022.133474

Heřt, D., Báča, T., Petráček, P., Krátký, V., Pěnička, R., Spurný, V., Petrlík, M., Vrba, M., Žaitlík, D., Stoudek, P., Walter, V., Štěpán, P., Horyna, J., Pritzl, V., Šrámek, M., Ahmad, A., Silano, G., Licea, D. B., Štibinger, P., & Saska, M. (2023). MRS Drone: A Modular Platform for Real-World Deployment of Aerial Multi-Robot Systems. *Journal of Intelligent & Robotic Systems, 108*(4), 64. Advance online publication. DOI: 10.1007/s10846-023-01879-2

Hianje, A., & Sahbani, S. (2025). *Reviewing the Roles of Artificial Intelligence-Based Technologies in Sustainable Purchasing and Supply Management: Research propositions for future Directions*. DOI: 10.20944/preprints202507.1743.v1

Hinterhuber, A. (2017). Value quantification capabilities in industrial markets. *Journal of Business Research, 76*, 163–178. DOI: 10.1016/j.jbusres.2016.11.019

Hoang, V. T., Phung, M. D., Dinh, T. H., & Ha, Q. P. (2019). System Architecture for Real-Time Surface Inspection Using Multiple UAVs. *IEEE Systems Journal, 14*(2), 2925–2936. DOI: 10.1109/JSYST.2019.2922290

Hoejmose, S. U., Grosvold, J., & Millington, A. (2014). The effect of institutional pressure on cooperative and coercive "green" supply chain practices. *Journal of Purchasing and Supply Management, 20*(4), 215–224. DOI: 10.1016/j.pursup.2014.07.002

Hofer, A., Brandl, F., Bauer, H., Haghi, S., & Reinhart, G. (2020). A framework for managing innovation cycles in manufacturing systems. *Procedia CIRP, 93*, 771–776. DOI: 10.1016/j.procir.2020.03.064

Hollenbeck, D., & Chen, Y. (2022). A Digital Twin Framework for Environmental Sensing with sUAS. *Journal of Intelligent & Robotic Systems, 105*(1), 1. Advance online publication. DOI: 10.1007/s10846-021-01542-8

Holmström, J., & Partanen, J. (2014). Digital manufacturing-driven transformations of service supply chains for complex products. *Supply Chain Management, 19*(4), 421–430. DOI: 10.1108/SCM-10-2013-0387

Holstein, K., Wortman Vaughan, J., Daumé, H., III, Dudík, M., & Wallach, H. (2019). Improving fairness in machine learning systems: What do industry practitioners need? In Proceedings of the 2019 CHI Conference on Human Factors in Computing Systems (CHI '19), Paper 600, 1–16. Association for Computing Machinery. https://doi.org/DOI: 10.1145/3290605.3300830

Hoon, C. (2013). Meta-synthesis of qualitative case studies: An approach to theory building. *Organizational Research Methods*, *16*(4), 522–556. DOI: 10.1177/1094428113484969

Horvát, P., & Webb, C. (2020, November 18). *The OECD STAN database for industrial analysis: Sources and methods*. OECD Science, Technology and Industry Working Papers, No. 2020/10. OECD Publishing. https://doi.org/DOI: 10.1787/18151965

Höse, K., Amaral, A., Götze, U., & Peças, P. (2023). Manufacturing Flexibility through Industry 4.0 Technological Concepts—Impact and Assessment. *Global Journal of Flexible Systems Managment*, *24*(2), 271–289. DOI: 10.1007/s40171-023-00339-y

Hovsha, J., & Brien, D. O. (2022). Drone Use Cases and their Privacy Impacts: A Taxonomy. *Tilburg Law Review*, *27*(1), 60–80. DOI: 10.5334/tilr.281

Huang, M. H., & Rust, R. T. (2021). Engaged to a Robot? The Role of AI in Service. *Journal of Service Research*, *24*(1), 30–41. Advance online publication. DOI: 10.1177/1094670520902266

Hussain, M., Tapinos, E., & Knight, L. (2017). Scenario-driven roadmapping for technology foresight. *Technological Forecasting and Social Change*, *124*, 160–177. DOI: 10.1016/j.techfore.2017.05.005

Hütten, N., Gomes, M. A., Hölken, F., Andricevic, K., Meyes, R., & Meisen, T. (2024). Deep Learning for Automated Visual Inspection in Manufacturing and Maintenance: A Survey of Open- Access Papers. *Applied System Innovation*, *7*(1), 11. DOI: 10.3390/asi7010011

Hwang, M.-L. T., Collin, W.-L., & Wang-xu, L. S. (2021). Quantum computing and supply chain optimization: Addressing complexity and efficiency challenges. *International Journal of Enterprise Modelling*, *15*(3), 148–162. DOI: 10.35335/emod.v15i3.49

IBM. (2024, November 15). How is AI being used in manufacturing? *IBM Think*. https://www.ibm.com/think/topics/ai-in-manufacturing

Iborra, M., Safón, V., & Dolz, C. (2020). What explains the resilience of SMEs? Ambidexterity capability and strategic consistency. *Long Range Planning*, *53*(6), 101947. DOI: 10.1016/j.lrp.2019.101947

IFRS Foundation. (2023, July 27). *Proposed IFRS Sustainability Disclosure Taxonomy* [Exposure draft]. https://www.ifrs.org/projects/completed-projects/2024/ifrs-sustainability-disclosure-taxonomy/proposed-taxonomy-cls-ifrs-sds/

Ikumapayi, O. M., Kayode, J. F., Afolalu, S. A., Nnochiri, E. S., Olowe, K., &Bodunde, O. P. (2022). *A study on AI and ICT for Sustainable Manufacturing*. DOI: 10.46254/AF03.20220259

Imdahl, C., Blume, S., Blume, S., Zellmer, S., Gensicke, M., & Herrmann, C. (2021). Potentials of Hydrogen Technologies for Sustainable Factory Systems. *Procedia CIRP*, *98*, 583–588. DOI: 10.1016/j.procir.2021.01.158

Indicated, N. (2010). *International Reference Life Cycle Data System (ILCD) Handbook - General guide for Life Cycle Assessment - Detailed guidance. Joint Research Centre*. European Commission., DOI: 10.2788/38479

Institute for Manufacturing, University of Cambridge. (n.d.). *Gap analysis. Decision Support Tools*. Retrieved September 16, 2025, from https://www.ifm.eng.cam.ac.uk/research/dstools/gap-analysis/

IoT Analytics. (2024). *Industrial AI market: 10 insights on how AI is transforming manufacturing*. IoT Analytics GmbH. https://iot-analytics.com

Iranshahi, K., Brun, J., Arnold, T., Sergi, T., & Müller, U. (2025). Digital Twins: Recent Advances and Future Directions in Engineering Fields. *Intelligent Systems with Applications*, *26*, 200516. DOI: 10.1016/j.iswa.2025.200516

Isgró, M. A., Basallote, M. D., & Barbero, L. (2021). Unmanned Aerial System-Based Multispectral Water Quality Monitoring in the Iberian Pyrite Belt (SW Spain). *Mine Water and the Environment*, *41*(1), 30–41. DOI: 10.1007/s10230-021-00837-4

Islam, M. T., Sepanloo, K., Woo, S., Woo, S. H., & Son, Y. J. (2025). A review of the industry 4.0 to 5.0 transition: Exploring the intersection, challenges, and opportunities of technology and human–machine collaboration. *Machines (Basel)*, *13*(4), 267. DOI: 10.3390/machines13040267

Ivanov, D. (2021). Supply chain viability and the COVID-19 pandemic: A conceptual and formal generalisation of four major adaptation strategies. *International Journal of Production Research*, *59*(12), 3535–3552. DOI: 10.1080/00207543.2021.1890852

Ivanov, D., & Dolgui, A. (2021). A digital supply chain twin for managing the disruption risks and resilience in the era of Industry 4.0. *Production Planning and Control*, *32*(9), 775–788. DOI: 10.1080/09537287.2020.1768450

Jabeur, M. H., Mahjoub, S., Toublanc, C., & Cariou, V. (2023). A reinforcement learning approach for a lot sizing and production scheduling problem with energy consideration. *IFAC-PapersOnLine, 56*(2), 11141–11147. DOI: 10.1016/j.ifacol.2023.10.832

Jacobsen, R. H., Matleković, L., Shi, L., Malle, N. H., Ayoub, N., Hageman, K., Hansen, S., Nyboe, F. F., & Ebeid, E. (2023). Design of an Autonomous Cooperative Drone Swarm for Inspections of Safety Critical Infrastructure. *Applied Sciences (Basel, Switzerland), 13*(3), 1256. DOI: 10.3390/app13031256

Jaouhari, A. E., Arif, J., Samadhiya, A., & Kumar, A. (2023). Net zero supply chain performance and industry 4.0 technologies: Past review and present introspective analysis for future research directions [Review of *Net zero supply chain performance and industry 4.0 technologies: Past review and present introspective analysis for future research directions*]. *Heliyon, 9*(11). Elsevier BV. DOI: 10.1016/j.heliyon.2023.e21525

Jaroš, R., Byrtus, R., Dohnal, J., Danys, L., Baroš, J., Koziorek, J., Zmij, P., & Martínek, R. (2022). Advanced Signal Processing Methods for Condition Monitoring. *Archives of Computational Methods in Engineering, 30*(3), 1553–1577. DOI: 10.1007/s11831-022-09834-4

Jarrahi, M. H. (2018). Artificial intelligence and the future of work: Human–AI symbiosis in organizational decision making. *Business Horizons, 61*(4), 577–586. DOI: 10.1016/j.bushor.2018.03.007

Javernik, A., Buchmeister, B., & Ojsteršek, R. (2023). The NASA-TLX Approach to Understand Workers Workload in Human-Robot Collaboration. *International Journal of Simulation Modelling, 22*(4), 574–585. DOI: 10.2507/IJSIMM22-4-658

Jawahir, I. S., & Bradley, R. (2016). Technological Elements of Circular Economy and the Principles of 6R-Based Closed-loop Material Flow in Sustainable Manufacturing. *Procedia CIRP, 40*, 103–108. DOI: 10.1016/j.procir.2016.01.067

Jayakumar, K., & Park, J. P. (n.d.). *Stimuli-Responsive Polymer– Metal Organic Framework Materials Composites: Advances in Biomedical Applications and Future Design Strategies.*

Jayawardane, H., Davies, I. J., Gamage, J. R., John, M., & Biswas, W. K. (2023). Sustainability perspectives – a review of additive and subtractive manufacturing [Review of *Sustainability perspectives – a review of additive and subtractive manufacturing*]. *Sustainable Manufacturing and Service Economics, 2*, 100015. Elsevier BV. DOI: 10.1016/j.smse.2023.100015

Jia, Q., Panetto, H., Macchi, M., Siri, S., Weichhart, G., & Xu, Z. (2022). Control for smart systems: Challenges and trends in smart cities. *Annual Reviews in Control, 53*, 358–369. DOI: 10.1016/j.arcontrol.2022.04.010

Joglekar, N., Parker, G., & Srai, J. S. (2024, January 31). Why manufacturers need a phased approach to digital transformation. *MIT Sloan Management Review, 65*(3), 54–59. https://sloanreview.mit.edu/article/why-manufacturers-need-a-phased-approach-to-digital-transformation/

Johannsen, R. M., Mathiesen, B. V., Kermeli, K., Crijns-Graus, W., & Østergaard, P. A. (2023). Exploring pathways to 100% renewable energy in European industry. *Energy, 268*, 126687. DOI: 10.1016/j.energy.2023.126687

Johansen, B. (2007). *Get there early: Sensing the future to compete in the present.* Berrett-Koehler Publishers.

Johansen, B. (2009). *Leaders make the future: Ten new leadership skills for an uncertain world.* Berrett-Koehler Publishers.

Johansen, S. T., Ünal, P., Albayrak, Ö., Ikonen, E., Linnestad, K., Jawahery, S., Srivastava, A., & Løvfall, B. T. (2023). Hybrid and cognitive digital twins for the process industry. *Open Engineering, 13*(1), 20220418. Advance online publication. DOI: 10.1515/eng-2022-0418

Joyce, K. E., Fickas, K. C., & Kalamandeen, M. (2022). The unique value proposition for using drones to map coastal ecosystems. *Cambridge Prisms: Coastal Futures, 1*, e6. Advance online publication. DOI: 10.1017/cft.2022.7

Kagermann, H., Wahlster, W., & Helbig, J. (2013, April). *Recommendations for implementing the strategic initiative INDUSTRIE 4.0: Final report of the Industrie 4.0 Working Group.* acatech – National Academy of Science and Engineering / Forschungsunion. https://en.acatech.de/publication/recommendations-for-implementing-the-strategic-initiative-industrie-4-0-final-report-of-the-industrie-4-0-working-group/

Kagermann, H., Wahlster, W., & Helbig, J. (Eds.). (2013). Securing the future of German manufacturing industry: Recommendations for implementing the strategic initiative INDUSTRIE 4.0. Final report of the Industrie 4.0 Working Group. acatech – National Academy of Science and Engineering. https://www.acatech.de/wp-content/uploads/2018/03/Final_report__Industrie_4.0_accessible.pdf

Kale, E., Aknar, A., & Başar, Ö. (2019). Absorptive capacity and firm performance: The mediating role of strategic agility. *International Journal of Hospitality Management, 78*, 276–283. DOI: 10.1016/j.ijhm.2018.09.010

Kammerer, K., Hoppenstedt, B., Pryss, R., Stökler, S., Allgaier, J., & Reichert, M. (2019). Anomaly Detections for Manufacturing Systems Based on Sensor Data—Insights into Two Challenging Real-World Production Settings. *Sensors (Basel), 19*(24), 5370. DOI: 10.3390/s19245370 PMID: 31817471

Kanda, R., Akhai, S., & Bansal, R. (2013). Analysis of MOST technique for elimination of ideal time by synchronization of different lines. *International Journal of Research, 1*(4).

Kang, Z., Catal, C., & Tekinerdogan, B. (2020). Machine learning applications in production lines: A systematic literature review. *Computers & Industrial Engineering, 149*, 106773. DOI: 10.1016/j.cie.2020.106773

Kantaros, A., & Ganetsos, T. (2024). Integration of Cyber-Physical Systems, Digital Twins and 3D Printing in Advanced Manufacturing: A Synergistic Approach. *American Journal of Engineering and Applied Sciences, 17*(1), 1–22. DOI: 10.3844/ajeassp.2024.1.22

Kapp, K. M. (2012). *The gamification of learning and instruction: game-based methods and strategies for training and education*. John Wiley & Sons.

Karnes, M., Riffel, J., & Yılmaz, A. (2024). Key-Region-Based UAV Visual Navigation. *˜The œinternational Archives of the Photogrammetry, Remote Sensing and Spatial Information Sciences. The International Archives of the Photogrammetry, Remote Sensing and Spatial Information Sciences, 173*, 173–179. Advance online publication. DOI: 10.5194/isprs-archives-XLVIII-2-2024-173-2024

Kas, K., & Johnson, G. K. (2019). Using unmanned aerial vehicles and robotics in hazardous locations safely. *Process Safety Progress, 39*(1), e12066. Advance online publication. DOI: 10.1002/prs.12066

Kauppi, K. (2013). Extending the use of institutional theory in operations and supply chain management research: Review and research suggestions. *International Journal of Operations & Production Management, 33*(10), 1318–1345. DOI: 10.1108/IJOPM-10-2011-0364

Kauppi, K., & Hannibal, C. (2017). Institutional pressures and sustainability assessment in supply chains. *Supply Chain Management, 22*(5), 458–472. DOI: 10.1108/SCM-01-2017-0004

Kaur, A., Singh, G., Kukreja, V., Sharma, S., Singh, S., & Yoon, B. (2022). Adaptation of IoT with Blockchain in Food Supply Chain Management: An Analysis-Based Review in Development, Benefits and Potential Applications [Review of *Adaptation of IoT with Blockchain in Food Supply Chain Management: An Analysis-Based Review in Development, Benefits and Potential Applications*]. *Sensors, 22*(21), 8174. Multidisciplinary Digital Publishing Institute. DOI: 10.3390/s22218174

Kayande, U., & Kediya, S. (2025). Critical drivers of AI integration in industrial processes for sustainable manufacturing. *Smart and Sustainable Manufacturing Systems, 5*, 00503. DOI: 10.1007/s41660-025-00503-0

Keleko, A. T., Kamsu-Foguem, B., Ngouna, R. H., & Tongne, A. (2022). Artificial intelligence and real-time predictive maintenance in Industry 4.0: A bibliometric analysis. *AI and Ethics, 2*(4), 553–577. DOI: 10.1007/s43681-021-00132-6

Kerdlap, P., Low, J. S. C., & Ramakrishna, S. (2019). Zero waste manufacturing: A framework and review of technology, research, and implementation barriers for enabling a circular economy transition in Singapore. *Resources, Conservation and Recycling, 151*, 104438. DOI: 10.1016/j.resconrec.2019.104438

Kestria. (2024, May 15). *AI and robotics integration: Transforming production and automation.* https://kestria.com/insights/ai-and-robotics-integration-transforming-productio/

Khang, A., & Akhai, S. (2025). E-Waste and Lithium-ion Battery Recycling Insights for Sustainable Transportation. In *Driving Green Transportation System Through Artificial Intelligence and Automation: Approaches, Technologies and Applications* (pp. 203-230). Cham: Springer Nature Switzerland. DOI: 10.1007/978-3-031-72617-0_11

Khanna, A., Jain, S., Burgio, A., Bolshev, V., & Панченко, В. (2022). Blockchain-Enabled Supply Chain platform for Indian Dairy Industry: Safety and Traceability. *Foods, 11*(17), 2716. DOI: 10.3390/foods11172716 PMID: 36076901

Khan, S. A. R., Qianli, D., SongBo, W., Zaman, K., & Zhang, Y. (2017). Environmental logistics performance indicators affecting per capita income and sectoral growth: Evidence from a panel of selected global ranked logistics countries. *Environmental Science and Pollution Research International, 24*(2), 1518–1531. DOI: 10.1007/s11356-016-7916-2 PMID: 27785719

Kilickaya, S., Ahishali, M., Celebioglu, C., Sohrab, F., Eren, L., İnce, T., Aşkar, M., & Gabbouj, M. (2025).. . *Audio-based Anomaly Detection in Industrial Machines Using Deep One-Class Support Vector Data Description, 1*, 1–5. Advance online publication. DOI: 10.1109/CIESCompanion65073.2025.11010815

Kim, H., Ham, S., & Lee, I. (2020). REAL-TIME DRONE MAPPING BASED ON REFERENCE IMAGES FOR VEHICLE FACILITY MONITORING. *The œinternational Archives of the Photogrammetry, Remote Sensing and Spatial Information Sciences. The International Archives of the Photogrammetry, Remote Sensing and Spatial Information Sciences, 43*. Advance online publication. DOI: 10.5194/isprs-archives-XLIII-B2-2020-43-2020

Kim, S., Lim, G. J., & Cho, J. (2018). Drone flight scheduling under uncertainty on battery duration and air temperature. *Computers & Industrial Engineering, 117*, 291–302. DOI: 10.1016/j.cie.2018.02.005

Kırmızı, M., & Kocaoğlu, B. (2022). Digital transformation maturity model development framework based on design science: Case studies in manufacturing industry. *Journal of Manufacturing Technology Management, 33*(7), 1319–1349. DOI: 10.1108/JMTM-11-2021-0476

Kistner, L., Schmoll, R., & Kroll, A. (2024). An airborne measurement system to detect, locate, and quantify methane emissions with attitude-based wind speed estimation. *Journal of Sensors and Sensor Systems, 13*(2), 211–218. DOI: 10.5194/jsss-13-211-2024

Kleinekorte, J., Fleitmann, L., Bachmann, M., Kätelhön, A., Barbosa-Póvoa, A. P., Aßen, N. von der, &Bardow, A. (2020). Life Cycle Assessment for the Design of Chemical Processes, Products, and Supply Chains [Review of *Life Cycle Assessment for the Design of Chemical Processes, Products, and Supply Chains*]. *Annual Review of Chemical and Biomolecular Engineering, 11*(1), 203. Annual Reviews. DOI: 10.1146/annurev-chembioeng-011520-075844

Kleinschroth, F., Banda, K., Zimba, H., Dondeyne, S., Nyambe, I., Spratley, S., & Winton, R. S. (2022). Drone imagery to create a common understanding of landscapes. *Landscape and Urban Planning, 228*, 104571. DOI: 10.1016/j.landurbplan.2022.104571

Klerk, N., van den Bosch, R., & Volberda, H. (2020). Lifelong learning and employee resilience: The moderating effect of learning goal orientation. *Human Resource Management Journal, 30*(3), 1–18. DOI: 10.1111/1748-8583.12312

Klessascheck, F., Fahrenkrog-Petersen, S. A., Mendling, J., & Pufahl, L. (2025). Unlocking Sustainability Compliance: Characterizing the EU Taxonomy for Business Process Management. In *Lecture notes in computer science* (p. 339). Springer Science+Business Media. DOI: 10.1007/978-3-031-78338-8_18

Kolb, D. A. (2014). *Experiential learning: Experience as the source of learning and development*. FT Press.

Kordoni, A., Gavidia-Calderon, C., Levine, M., Bennaceur, A., & Nuseibeh, B. (2023). "Are we in this together?": Embedding social identity detection in drones improves emergency coordination. *Frontiers in Psychology*, *14*, 1146056. Advance online publication. DOI: 10.3389/fpsyg.2023.1146056 PMID: 37744604

Kotter, J. P. (2012). *Leading change* (with a new preface by the author). Harvard Business Review Press. https://store.hbr.org/product/leading-change-with-a-new-preface-by-the-author/11116

Koul, P. (2024). Enhancing solar energy systems: The role of computational and automation techniques in achieving net zero by 2050. *World Journal of Advanced Engineering Technology and Sciences*, *13*(1), 909–924. DOI: 10.30574/wjaets.2024.13.1.0500

Kristianto, Y., Gunasekaran, A., Helo, P., & Hao, Y. (2014). A model of resilient supply chain network design: A two-stage programming with fuzzy shortest path. *Expert Systems with Applications*, *41*(1), 39–49. DOI: 10.1016/j.eswa.2013.07.009

Kristjánsdóttir, B. H. (2019). *How VUCA is your project? Toward a VUCA meter to assess the volatility, uncertainty, complexity and ambiguity of complex projects* (Master's thesis). Reykjavík University. https://skemman.is/handle/1946/33540

Kronberger, G., Bachinger, F., & Affenzeller, M. (2020). Smart Manufacturing and Continuous Improvement and Adaptation of Predictive Models. *Procedia Manufacturing*, *42*, 528–531. DOI: 10.1016/j.promfg.2020.02.037

Krook, J., Bossens, D. M., Winter, P., Araujo-Estrada, S. A., Downer, J., & Windsor, S. P. (2024). Mapping the complexity of legal challenges for trustworthy drones on construction sites in the United Kingdom. *ACM Journal on Responsible Computing*, *1*(3), 1–26. DOI: 10.1145/3664617

Kumar, H., Wadhwa, A. S., Akhai, S., & Kaushik, A. (2024). Parametric analysis, modeling and optimization of the process parameters in electric discharge machining of aluminium metal matrix composite. *Engineering Research Express*, *6*(2), 025542. DOI: 10.1088/2631-8695/ad4ba9

Kumar, H., & Yadav, S. (2024). Additive and good manufacturing practices in conformity assessment. In Bhatnagar, A., Yadav, S., Achanta, V., Harmes-Liedtke, U., & Rab, S. (Eds.), *Handbook of Quality System, Accreditation and Conformity Assessment* (pp. 971–988). Springer., DOI: 10.1007/978-981-97-5750-3_53

Kumar, P., & Akhai, S. (2022). Effective energy management in smart buildings using VRV/VRF systems. In *Additive manufacturing in Industry 4.0* (pp. 27–35). CRC Press., DOI: 10.1201/9781003360001-2

Kumar, S. S., Kumar, S. R., & Ramesh, G. (2024). From industry 4.0 to 5.0: Enriching manufacturing excellence through Human–Robot interaction and technological empowerment. In *Intelligent systems and industrial internet of things for sustainable development* (pp. 24–51). Chapman and Hall/CRC. DOI: 10.1201/9781032642789-2

Kumar, V., & Shahin, K. (2025). Artificial Intelligence and Machine Learning for Sustainable Manufacturing: Current Trends and Future Prospects. *Intelligent and Sustainable Manufacturing*, *2*(1), 10002. DOI: 10.70322/ism.2025.10002

Kumpulainen, M., & Seppänen, M. (2022). Combining Web of Science and Scopus datasets in citation-based literature study. *Scientometrics*, *127*(10), 5613–5631. DOI: 10.1007/s11192-022-04475-7

Kusiak, A. (2018). Smart manufacturing. *International Journal of Production Research*, *56*(1–2), 508–517. Advance online publication. DOI: 10.1080/00207543.2017.1351644

Lai, Q., Xie, R., Yang, Z., Wu, G., Hong, Z.-Y., & Yang, C. (2024). Efficient multiple unmanned aerial vehicle-assisted data collection strategy in power infrastructure construction. *Frontiers in Communications and Networks*, *5*, 1390909. Advance online publication. DOI: 10.3389/frcmn.2024.1390909

Lario, J., Mateos, J., Psarommatis, F., & Ortíz, Á. (2024). *Towards Zero-Defect and Zero-Waste Manufacturing by Implementing Non-Destructive Inspection Technologies*. DOI: 10.2139/ssrn.4876982

Lavric, T., Bricard, E., Preda, M., & Zaharia, T. (2022). A low-cost AR training system for manual assembly operations. *Computer Science and Information Systems*, *19*(2), 1047–1073. DOI: 10.2298/CSIS211123013L

Lawrence, A. (2024, September 11). *AI's impact on robots in manufacturing*. American Machinist. https://www.americanmachinist.com/automation-and-robotics/article/55138472/the-impact-of-ai-powered-robots-on-manufacturing-efficiency-and-quality-control-automation-technology American Machinist

Layer, M., Leidich, J., Schwoch, S., Saske, B., Neubert, S., Robl, P., & Paetzold-Byhain, K. (2023). Data management of process plants as complex systems: Systematic literature review and identification of challenges and opportunities. *Reviews in Chemical Engineering*, *40*(3), 329–349. DOI: 10.1515/revce-2022-0077

Lee, G., Saad, W., Bennis, M., Kim, C., & Jung, M. (2022). An Online Framework for Ephemeral Edge Computing in the Internet of Things. *IEEE Transactions on Wireless Communications*, *22*(3), 1992–2007. DOI: 10.1109/TWC.2022.3208096

Lee, J., Bagheri, B., & Jin, C. (2016). Introduction to cyber manufacturing. *Manufacturing Letters, 8*, 11–15. Advance online publication. DOI: 10.1016/j.mfglet.2016.05.002

Lee, J., Bagheri, B., & Kao, H.-A. (2018). A cyber-physical systems architecture for industry 4.0-based manufacturing systems. *Manufacturing Letters, 3*, 18–23. DOI: 10.1016/j.mfglet.2014.12.001

Lee, J.-S., Kim, T., Jeon, S.-H., Park, S.-H., Kim, S.-H., Lee, E., & Lee, J.-H. (2023). Automation of trimming die design inspection by zigzag process between AI and CAD domains. *Engineering Applications of Artificial Intelligence, 127*, 107283. DOI: 10.1016/j.engappai.2023.107283

Lee, J., Singh, J., Azamfar, M., & Pandhare, V. (2020). Industrial AI and predictive analytics for smart manufacturing systems. In *Smart Manufacturing* (pp. 213–246). Elsevier., DOI: 10.1016/B978-0-12-820027-8.00008-3

Leet, C., Li, J., & Koenig, S. (2022). Shard Systems: Scalable, Robust and Persistent Multi-Agent Path Finding with Performance Guarantees. *Proceedings of the AAAI Conference on Artificial Intelligence, 36*(9), 9386–9395. DOI: 10.1609/aaai.v36i9.21170

Lehmhus, D., Busse, M., von Hehl, A., & Jägle, E. A. (2018). State of the Art and Emerging Trends in Additive Manufacturing: From Multi-Material processes to 3D printed Electronics. *MATEC Web of Conferences, 188*, 3013. DOI: 10.1051/matecconf/201818803013

Lehrman, A. L. (2025). Embodied Learning Through Immersive Virtual Reality: Theoretical Perspectives for Art and Design Education. *Behavioral Sciences (Basel, Switzerland), 15*(7), 917. DOI: 10.3390/bs15070917 PMID: 40723701

Lei, Z., Shi, J., Luo, Z., Cheng, M., & Wan, J. (2024). Intelligent manufacturing from the perspective of Industry 5.0: Application review and prospects. *IEEE Access : Practical Innovations, Open Solutions, 12*, 167436–167451.

Leonidou, E., Christofi, M., Vrontis, D., & Thrassou, A. (2020). An integrative framework of stakeholder engagement for innovation management and entrepreneurship development. *Journal of Business Research, 119*, 245–258. DOI: 10.1016/j.jbusres.2018.11.054

Leonzio, G. (2023). Life Cycle Assessment of Carbon Dioxide Supply Chains: State of the Art and Methodology Description. *Applied Sciences (Basel, Switzerland), 14*(1), 385. DOI: 10.3390/app14010385

Li, H., Pangborn, H. C., & Kovalenko, I. (2023). *A System-Level Energy-Efficient Digital Twin Framework for Runtime Control of Batch Manufacturing Processes*. 1. DOI: 10.1109/CASE56687.2023.10260642

Liang, C.-W., & Shen, C.-H. (2024). An integrated uncrewed aerial vehicle platform with sensing and sampling systems for the measurement of air pollutant concentrations. *Atmospheric Measurement Techniques*, *17*(9), 2671–2686. DOI: 10.5194/amt-17-2671-2024

Li, C., Ge, W., Huang, Z., Zhang, Q., Li, H., & Cao, H. (2024). Digital twin-driven modeling and application of carbon emission for machine tool. *International Journal of Advanced Manufacturing Technology*, *133*(11-12), 5595–5609. DOI: 10.1007/s00170-024-13788-1

Liker, J. K. (2004). *The Toyota way: 14 management principles from the world's greatest manufacturer*. McGraw-Hill. https://search.worldcat.org/title/The-Toyota-way-%3A-14-management-principles/oclc/1031522159 WorldCat

Lim, D. Y., Jin, I. J., & Bang, I. C. (2023). Heat-vision based drone surveillance augmented by deep learning for critical industrial monitoring. *Scientific Reports*, *13*(1), 22291. Advance online publication. DOI: 10.1038/s41598-023-49589-x PMID: 38097731

Lim, E., Jung, Y. H., Shin, D. Y., & Kim, S. S. (2022). SMALL UAVS-BASED DISASTER DAMAGE INVESTIGATION: FOCUSED ON FLOODING DISASTER AND CHEMICAL ACCIDENT. *˜The œinternational Archives of the Photogrammetry, Remote Sensing and Spatial Information Sciences. The International Archives of the Photogrammetry, Remote Sensing and Spatial Information Sciences, 1121*. Advance online publication. DOI: 10.5194/isprs-archives-XLIII-B3-2022-1121-2022

Lin, D., Banjevic, D., & Jardine, A. K. S. (2006). Using principal components in a proportional hazards model with applications in condition-based maintenance. *The Journal of the Operational Research Society*, *57*(8), 910–919. Advance online publication. DOI: 10.1057/palgrave.jors.2602058

Lindberg, D. G. (2025). Harnessing AI for smart manufacturing: Insights from Industry 4.0. *Discover Artificial Intelligence*, *5*(1), 111. Advance online publication. DOI: 10.1007/s44163-025-00363-0

Lindblad, M. (2006). *Human Inspection Work: A case study of why faults are missed?* https://lup.lub.lu.se/student-papers/record/1327412/file/1327413.pdf

Lindström, J., Larsson, H., Jonsson, M., & Lejon, E. (2017). Towards intelligent and sustainable production: Combining and integrating online predictive maintenance and continuous quality control. *Procedia CIRP*, *63*, 443–448. DOI: 10.1016/j.procir.2017.03.099

Lin, J. J., Ibrahim, A., Sarwade, S., & Golparvar-Fard, M. (2020). Bridge Inspection with Aerial Robots: Automating the Entire Pipeline of Visual Data Capture, 3D Mapping, Defect Detection, Analysis, and Reporting. *Journal of Computing in Civil Engineering*, *35*(2), 04020064. Advance online publication. DOI: 10.1061/(ASCE)CP.1943-5487.0000954

Lin, Y., Chang, Y., Tong, X., Yu, J., Liotta, A., Huang, G., Song, W., Zeng, D., Wu, Z., Wang, Y., & Zhang, W. (2024). *A Survey on Rgb, 3d, and Multimodal Approaches for Unsupervised Industrial Anomaly Detection*. DOI: 10.2139/ssrn.5005467

Liu, C., Su, Z., Xu, X., & Lu, Y. (2021). Service-oriented industrial internet of things gateway for cloud manufacturing. *Robotics and Computer-integrated Manufacturing*, *73*, 102217. DOI: 10.1016/j.rcim.2021.102217

Liu, H., Ke, W., Wei, K. K., & Hua, Z. (2013). The impact of IT capabilities on firm performance: The mediating roles of absorptive capacity and supply chain agility. *Decision Support Systems*, *54*(3), 1452–1462. DOI: 10.1016/j.dss.2012.12.016

Liu, W., Wu, H., Shi, L., Zhu, F., Zou, Y., Kong, F., & Zhang, F. (2025). LiDAR-Based Quadrotor Autonomous Inspection System in Cluttered Environments. *IEEE Transactions on Field Robotics.*, *2*, 753–767. DOI: 10.1109/TFR.2025.3604716

Li, X., & Pattinson, S. W. (2025). An efficient and uncertainty-aware reinforcement learning framework for quality assurance in extrusion additive manufacturing. *Additive Manufacturing*, *110*, 104912. DOI: 10.1016/j.addma.2025.104912

Li, Z., Fei, F., & Zhang, G. (2022). Edge-to-Cloud IIoT for Condition Monitoring in Manufacturing Systems with Ubiquitous Smart Sensors. *Sensors (Basel)*, *22*(15), 5901. DOI: 10.3390/s22155901 PMID: 35957460

Li, Z., Yan, Y., Xiangheng, W., Ge, Y., & Meng, L. (2025). A survey of deep learning for industrial visual anomaly detection. *Artificial Intelligence Review*, *58*(9), 279. Advance online publication. DOI: 10.1007/s10462-025-11287-7

Lodhi, S. K., & Zeb, S. (2025). AI-Driven Robotics and Automation: The Evolution of Human-Machine Collaboration. *Journal of Wood Science*, *4*(4), 422–437.

Loffredo, A. (2023). Reinforcement learning for energy-efficient control of multi-stage production lines with parallel machine workstations. *Materials Research Proceedings*, *35*, 428–436. DOI: 10.21741/9781644902714-51

Loffredo, A., May, M. C., Matta, A., & Lanza, G. (2023). Reinforcement learning for sustainability enhancement of production lines. *Journal of Intelligent Manufacturing, 35*(8), 3775–3791. DOI: 10.1007/s10845-023-02258-2

Lorenzo Ochoa, O., Claes, B., Koryak, O., & Diaz, A. (2017). Integration through orchestration: The interplay between enterprise systems and inventory management capabilities. *Journal of Enterprise Information Management, 30*(4), 555–582. DOI: 10.1108/JEIM-02-2016-0060

Lu, Y. (2017). Industry 4.0: A survey on technologies, applications and open research issues. In Journal of Industrial Information Integration (Vol. 6). DOI: 10.1016/j.jii.2017.04.005

Luo, Y., Madarkar, R., Luo, X., & Ball, P. (2025). Leveraging Digital Twins and Dynamic Life Cycle Assessment for Sustainable Manufacturing: A Conceptual Framework. In *Lecture notes in mechanical engineering* (p. 285). Springer Nature. DOI: 10.1007/978-3-031-93891-7_32

Luo, S., Liu, Q., Guo, X., Yin, M., Li, Z., & Lang, X. (2025). Chemical Safety Inspection Path Optimization Problems Using Improved Multi-Objective Discrete Growth Optimization Algorithm. *Processes (Basel, Switzerland), 13*(5), 1445. DOI: 10.3390/pr13051445

Ma, Y., Yin, J., Huang, F., & Li, Q. (2024). Surface defect inspection of industrial products with object detection deep networks: a systematic review [Review of *Surface defect inspection of industrial products with object detection deep networks: a systematic review*]. *Artificial Intelligence Review, 57*(12). Springer Science+Business Media. DOI: 10.1007/s10462-024-10956-3

Maaß, W., Agrawal, A., Ciani, A., Danz, S., Delgadillo, A., Ganser, P., Kienast, P., Kulig, M., König, V., Rodellas-Gràcia, N., Rughubar, R., Schröder, S., Stautner, M., Stein, H., Stollenwerk, T., Zeuch, D., & Wilhelm, F. K. (2024). QUASIM: Quantum Computing Enhanced Service Ecosystem for Simulation in Manufacturing. *KI - Künstliche Intelligenz, 38*(4), 361. DOI: 10.1007/s13218-024-00860-x

Maboudi, M., Alamouri, A., López, V. D. A., Bajauri, M. S., Berger, C., & Gerke, M. (2021). DRONE-BASED CONTAINER CRANE INSPECTION: CONCEPT, CHALLENGES AND PRELIMINARY RESULTS. *ISPRS Annals of the Photogrammetry, Remote Sensing and Spatial Information Sciences, 121*, 121–128. Advance online publication. DOI: 10.5194/isprs-annals-V-1-2021-121-2021

Machado, C. G., Winroth, M., Almström, P., Ericson Öberg, A., Kurdve, M., & AlMashalah, S. (2021). Digital organisational readiness: Experiences from manufacturing companies. *Journal of Manufacturing Technology Management, 32*(9), 167–182. DOI: 10.1108/JMTM-05-2019-0188

Macher, J. T., & Mowery, D. C. (2009). Measuring dynamic capabilities: Practices and performance in semiconductor manufacturing. *British Journal of Management, 20*(s1, S1), S41–S62. DOI: 10.1111/j.1467-8551.2008.00612.x

Mackey, R. W. (1992). *Strategic leadership and the Army War College (Military Studies Program paper).* U.S. Army War College. https://apps.dtic.mil/sti/citations/ADA249373

Magalhães, P. S. G., & Ferreira, N. M. F. (2022). Inspection Application in an Industrial Environment with Collaborative Robots. *Automation, 3*(2), 258–268. DOI: 10.3390/automation3020013

Maghazei, O., Netland, T. H., Frauenberger, D., & Thalmann, T. (2021). Automatic Drones for Factory Inspection: The Role of Virtual Simulation. In *IFIP advances in information and communication technology* (p. 457). Springer Science+Business Media. DOI: 10.1007/978-3-030-85910-7_48

Maghazei, O. (2022). Drones in Manufacturing. *Opportunities and Challenges, 290*, 290–298. Advance online publication. DOI: 10.46254/EU05.20220073

Maghazei, O., & Netland, T. H. (2019). Drones in manufacturing: Exploring opportunities for research and practice. *Journal of Manufacturing Technology Management, 31*(6), 1237–1259. DOI: 10.1108/JMTM-03-2019-0099

Mahariya, S. K., Kumar, A., Singh, R., Gehlot, A., Akram, S. V., Twala, B., Iqbal, M. I., & Priyadarshi, N. (2023). Smart Campus 4.0: Digitalization of University Campus with Assimilation of Industry 4.0 for Innovation and Sustainability. *Journal of Advanced Research in Applied Sciences and Engineering Technology, 32*(1), 120–138. DOI: 10.37934/araset.32.1.120138

Mahroof, K., Omar, A., Yaroson, E. V., Tenebe, S. A., Rana, N. P., Sivarajah, U., & Weerakkody, V. (2023). Evaluating the intention to use Industry 5.0 (I5.0) drones for cleaner production in Sustainable Food Supply Chains: An emerging economy context. *Supply Chain Management, 29*(3), 468–496. DOI: 10.1108/SCM-01-2023-0045

Majumdar, S. K., & Marcus, A. A. (2001). Rules versus discretion: The productivity consequences of flexible regulation. *Academy of Management Journal, 44*(1), 170–179. DOI: 10.2307/3069344

Makhanya, M. T. B. (2023). The implications for risk management in the era of technological advancements. In *The future of risk management*. IntechOpen., DOI: 10.5772/intechopen.1003899

Mammarella, M., Comba, L., Biglia, A., Dabbene, F., & Gay, P. (2021). Cooperation of unmanned systems for agricultural applications: A theoretical framework. *Biosystems Engineering, 223*, 61–80. DOI: 10.1016/j.biosystemseng.2021.11.008

Mandelli, D., Wang, C., & Agarwal, V. (2023). Development of Analysis Methods that Integrate Numeric and Textual Equipment Reliability Data. In *OSTI OAI (U.S. Department of Energy Office of Scientific and Technical Information)*. Office of Scientific and Technical Information. https://www.osti.gov/biblio/2008354

Manufacturing Leadership Council. (2024, October 1). *Survey: Leadership preparedness improves, but gaps remain. Manufacturing Leadership Journal.* https://manufacturingleadershipcouncil.com/leadership-preparedness-improves-but-gaps-remain-37327/

Manyika, J., Chui, M., Miremadi, M., Bughin, J., George, K., Willmott, P., & Dewhurst, M. (2017). *A future that works: Automation, employment, and productivity.* McKinsey Global Institute. https://www.mckinsey.com/featured-insights/digital-disruption/harnessing-automation-for-a-future-that-works

Margherita, A., & Braccini, A. M. (2022). Analysis of the adoption of emergent technologies for risk management in the era of digital manufacturing. *Technological Forecasting and Social Change, 178*, 121566. DOI: 10.1016/j.techfore.2022.121566

Marien, M. (2014). Book Review: The Second Machine Age: Work, Progress, and Prosperity in a Time of Brilliant Technologies. *World Futures Review, 6*(2), 189–193. Advance online publication. DOI: 10.1177/1946756714541404

Martínez-Mireles, J. R., Rodríguez-Flores, J., García-Márquez, M. A., Austria-Cornejo, A., & Figueroa-Diaz, R. A. (2025). AI in Smart Manufacturing. In *Machine and Deep Learning Solutions for Achieving the Sustainable Development Goals* (pp. 463–494). IGI Global Scientific Publishing. DOI: 10.4018/979-8-3693-8161-8.ch023

Maschler, B., Knodel, T., & Weyrich, M. (2021)... *Towards Deep Industrial Transfer Learning for Anomaly Detection on Time Series Data, 1*, 01–08. Advance online publication. DOI: 10.1109/ETFA45728.2021.9613542

Matheson, E., Minto, R., Zampieri, E. G. G., Faccio, M., & Rosati, G. (2019). Human–robot collaboration in manufacturing applications: A review. *Robotics (Basel, Switzerland), 8*(4), 100. DOI: 10.3390/robotics8040100

Maware, C., Muvunzi, R., Machingura, T., & Daniyan, I. (2024). Examining the Progress in Additive Manufacturing in Supporting Lean, Green and Sustainable Manufacturing: A Systematic Review [Review of *Examining the Progress in Additive Manufacturing in Supporting Lean, Green and Sustainable Manufacturing: A Systematic Review*]. *Applied Sciences, 14*(14), 6041. Multidisciplinary Digital Publishing Institute. DOI: 10.3390/app14146041

Mawson, V. J., & Hughes, B. R. (2021). Optimisation of HVAC control and manufacturing schedules for the reduction of peak energy demand in the manufacturing sector. *Energy*, *227*, 120436. DOI: 10.1016/j.energy.2021.120436

Mayer, A., Greif, L., Häußermann, T., Otto, S., Kastner, K., Bobbou, S. E., Chardonnet, J.-R., Reichwald, J., Fleischer, J., & Ovtcharova, J. (2025). Digital Twins, Extended Reality, and Artificial Intelligence in Manufacturing Reconfiguration: A Systematic Literature Review. *Sustainability (Basel)*, *17*(5), 2318. DOI: 10.3390/su17052318

May, M. C., Oberst, J., & Lanza, G. (2024). Managing product-inherent constraints with artificial intelligence: Production control for time constraints in semiconductor manufacturing. *Journal of Intelligent Manufacturing*, *35*(8), 4259–4276. DOI: 10.1007/s10845-024-02472-6

Mazgualdi, C. E., Masrour, T., Barka, N., & Hassani, I. E. (2022). A Learning-Based Decision Tool towards Smart Energy Optimization in the Manufacturing Process. *Systems*, *10*(5), 180. DOI: 10.3390/systems10050180

McGrath, R. G., & MacMillan, I. C. (2000). *The entrepreneurial mindset: Strategies for continuously creating opportunity in an age of uncertainty*. Harvard Business School Press., DOI: 10.4324/9781482293746

McKinsey & Company. (2023). *The state of AI in 2023: Generative AI's breakout year*. McKinsey & Company. https://www.mckinsey.com

McKinsey & Company. (2024, April 9). *How manufacturing's lighthouses are capturing the full value of AI*. https://www.mckinsey.com/capabilities/operations/our-insights/how-manufacturings-lighthouses-are-capturing-the-full-value-of-ai

McKinsey Global Institute. (2024, January). *Geopolitics and the geometry of global trade*. McKinsey & Company. https://www.mckinsey.com/mgi/our-research/geopolitics-and-the-geometry-of-global-trade

McQueen, N., & Drennan, D. (2024). The use of warehouse automation technology for scalable and low-cost direct air capture. *Frontiers in Climate*, *6*, 1415642. Advance online publication. DOI: 10.3389/fclim.2024.1415642

Meena, R., Sahoo, S., Malik, A., Kumar, S., & Nguyen, M. (2025). Artificial intelligence and circular supply chains: Framework for applications and deployment from the triple bottom line model perspective. *Annals of Operations Research, 354*(1), 71–101. DOI: 10.1007/s10479-025-06510-1

Meinrenken, C. J., Chen, D., Esparza, R. A., Iyer, V., Paridis, S. P., Prasad, A., & Whillas, E. (2020). Carbon emissions embodied in product value chains and the role of Life Cycle Assessment in curbing them. *Scientific Reports, 10*(1), 6184. Advance online publication. DOI: 10.1038/s41598-020-62030-x PMID: 32277082

Mendes, E., Albeaino, G., Brophy, P., Gheisari, M., & Jeelani, I. (2022). Working Safely with Drones: A Virtual Training Strategy for Workers on Heights. *Construction Research Congress 2022, 3*, 622. DOI: 10.1061/9780784483985.063

Mendonça, P., & Vieira, C. (2022). Embodied carbon and economic cost analysis of a contemporary house design using local and reused materials. *Sustainable Futures : An Applied Journal of Technology, Environment and Society, 4*, 100100. DOI: 10.1016/j.sftr.2022.100100

Mészáros, B., Bóna, K., & Bertalan, M. (2018). MONITORING OF PRODUCTION LINE SUPPLY SYSTEMS WITH DRONES. *IOP Conference Series. Materials Science and Engineering, 448*, 12037. DOI: 10.1088/1757-899X/448/1/012037

Meydan, C. H., & Akkaş, H. (2024). The role of triangulation in qualitative research: Converging perspectives. In Elhami, A., Roshan, A., & Chandan, H. C. (Eds.), *Principles of conducting qualitative research in multicultural settings* (pp. 101–132). IGI Global., DOI: 10.4018/979-8-3693-3306-8.ch006

Mick, M. M. A. P., Kovaleski, J. L., Mick, R. L., & Chiroli, D. M. G. (2024). Developing a sustainable digital transformation roadmap for SMEs: Integrating digital maturity and strategic alignment. *Sustainability (Basel), 16*(20), 8745. DOI: 10.3390/su16208745

Midkiff, J., Hlaing, M., Fan, K.-C., Jain, S., Akella, M., & Chen, R. T. (2024). *Mid-IR UAV-based sensing platform with deep learning to Identify and Quantify Gaseous Emission in Gas Flares*. DOI: 10.2172/2405949

Miele, L., Fraile, F., Esteso, A., & Rocca, R. L. (2024). Integration of Artificial Intelligence in Manufacturing Companies for Achieving Zero Waste – A Systematic Literature Review. In *IFIP advances in information and communication technology* (p. 212). Springer Science+Business Media. DOI: 10.1007/978-3-031-71743-7_14

Mihai, F., & Minea, I. (2021). Sustainable Alternative Routes versus Linear Economy and Resources Degradation in Eastern Romania. *Sustainability (Basel)*, *13*(19), 10574. DOI: 10.3390/su131910574

Miletić, S. D. (2023). Ranking of Autonomous Alternatives for the Realization of Intralogistics Activities in Sustainable Warehouse Systems using the TOPSIS Method. *Spectrum of Engineering and Management Sciences*, *1*(1), 48–57. DOI: 10.31181/sems1120234m

Milgram, P., & Kishino, F. (1994). A taxonomy of mixed reality visual displays. *IEICE Transactions on Information and Systems*, *77*(12), 1321–1329.

Miller, B., & Rowe, D. C. (2012). A survey of SCADA and critical infrastructure incidents. In *Proceedings of the 1st Annual Conference on Research in Information Technology (RIIT '12)* (pp. 51–56). ACM. https://doi.org/DOI: 10.1145/2380790.2380805

Minashkina, D. (2024). A review and research agenda for recent socially and environmentally sustainable practices for warehouse management systems [Review of *A review and research agenda for recent socially and environmentally sustainable practices for warehouse management systems*]. *The International Journal of Logistics Management*, *35*(7), 60. Emerald Publishing Limited. DOI: 10.1108/IJLM-07-2023-0265

Minashkina, D., & Happonen, A. (2020). Decarbonizing warehousing activities through digitalization and automatization with WMS integration for sustainability supporting operations. *E3S Web of Conferences*, *158*, 3002. DOI: 10.1051/e3sconf/202015803002

Mintzberg, H. (1987). The strategy concept I: Five Ps for strategy. *California Management Review*, *30*(1), 11–24. DOI: 10.2307/41165263

Miranda, S., Vázquez, C. R., & Navarro-Gutiérrez, M. (2025). Energy consumption analysis and optimization in collaborative robots. *Frontiers in Robotics and AI*, *12*, 1671336. Advance online publication. DOI: 10.3389/frobt.2025.1671336 PMID: 41245194

Mishra, H., & Mishra, D. (2024). *Economic Evaluation of UAV-Based Soil Sampling Approaches*., DOI: 10.1007/978-981-99-8684-2_15

MK. R. S., Poreddy, B. K. R., Sharmila, L., Rajesha, N., Anand, M. C. J., & Boopathi, S. (2024). Empowering Virtual Machine-Controlled and AI-Controlled Robots for Transportation in the Manufacturing Industry. In *Contemporary Solutions for Sustainable Transportation Practices* (pp. 436-466). IGI Global.

Mo, F., Monetti, F. M., Torayev, A., Rehman, H. U., Alberola, J. A. M., Minango, N. R., Nguyen, H. N., Maffei, A., & Chaplin, J. C. (2023). A maturity model for the autonomy of manufacturing systems. *International Journal of Advanced Manufacturing Technology, 126*(1-2), 405–428. DOI: 10.1007/s00170-023-10910-7

Mohanta, B., Nanda, P., & Patnaik, S. (2019). Management of VUCA (Volatility, Uncertainty, Complexity and Ambiguity) Using machine learning techniques in industry 4.0 paradigm. In *New Paradigm of Industry 4.0: Internet of Things, Big Data & Cyber Physical Systems* (pp. 1–24). Springer International Publishing.

Moheb-Alizadeh, H., & Handfield, R. B. (2018). The impact of raw materials price volatility on cost of goods sold (COGS) for product manufacturing. *IEEE Transactions on Engineering Management, 65*(3), 460–473. DOI: 10.1109/TEM.2018.2796447

Molenda, P., Groneberg, H., Schötz, S., & Döpper, F. (2023). Resilience balanced scorecard: Measuring resilience of manufacturing companies at multiple levels. *Procedia CIRP, 120*, 189–194. DOI: 10.1016/j.procir.2023.08.034

Momeni, B., Rapaccini, M., & Martinsuo, M. (2024). Manufacturers managing complexity during the digital servitization journey. *Journal of Manufacturing Technology Management, 35*(9), 51–72. DOI: 10.1108/JMTM-07-2023-0275

Monek, G. D., & Fischer, S. (2024). Expert Twin: A Digital Twin with an Integrated Fuzzy-Based Decision-Making Module. *Decision Making Applications in Management and Engineering, 8*(1), 1–21. DOI: 10.31181/dmame8120251181

Moosavi, S., Farajzadeh-Zanjani, M., Razavi-Far, R., Palade, V., & Saif, M. (2024). Explainable AI in Manufacturing and Industrial Cyber–Physical Systems: A Survey. *Electronics (Basel), 13*(17), 3497. DOI: 10.3390/electronics13173497

Moradlou, H., Skipworth, H., Bals, L., Aktas, E., & Roscoe, S. (2025). Geopolitical disruptions and supply chain structural ambidexterity. *International Journal of Operations & Production Management, 45*(4), 836–862. DOI: 10.1108/IJOPM-11-2023-0915

Morel, G., Pereira, C. E., & Nof, S. Y. (2019). Historical survey and emerging challenges of manufacturing automation modeling and control: A systems architecting perspective. *Annual Reviews in Control, 47*, 21–34. DOI: 10.1016/j.arcontrol.2019.01.002

Moss-Racusin, C. A., Sanzari, C., Caluori, N., & Rabasco, H. (2018). Gender bias produces gender gaps in STEM engagement. *Sex Roles, 79*(11), 651–670. https://link.springer.com/article/10.1007/s11199-018-0902-z. DOI: 10.1007/s11199-018-0902-z

Motia, K., Kumar, R., & Akhai, S. (2024). *AI and smart manufacturing: Building Industry 4.0*. In *Modern management science practices in the age of AI* (pp. 1–28). IGI Global. DOI: 10.4018/979-8-3693-6720-9.ch001

Motia, K., Kumar, R., & Akhai, S. (2025). *Navigating the Fourth Industrial Revolution: A comprehensive review of Industry 4.0 technologies and their global adoption*. In *Advances in materials and manufacturing: Select proceedings of ICDMT 2024* (pp. 317–333). Springer. DOI: 10.1007/978-981-96-7659-0_28

Mou, A. J., Hossain, M., & Siddiqui, N. A. (2022). DIGITAL TRANSFORMATION IN MARKETING: EVALUATING THE IMPACT OF WEB ANALYTICS AND SEO ON SME GROWTH. *American Journal of Interdisciplinary Studies*, *3*(4), 61–90. DOI: 10.63125/8t10v729

Msakni, M. K., Risan, A., & Schütz, P. (2023). Using machine learning prediction models for quality control: A case study from the automotive industry. *Computational Management Science*, *20*(1), 14. DOI: 10.1007/s10287-023-00448-0 PMID: 36942085

Mubarak, M. F., & Petraite, M. (2020). Industry 4.0 technologies, digital trust and technological orientation: What matters in open innovation? *Technological Forecasting and Social Change*, *161*, 120332. DOI: 10.1016/j.techfore.2020.120332

Múñiz, R., Nuño, F., Díaz, J., González, M., Prieto, M. J., & Menéndez, Ó. (2022). Real-time monitoring solution with vibration analysis for industry 4.0 ventilation systems. *The Journal of Supercomputing*, *79*(6), 6203. DOI: 10.1007/s11227-022-04897-3

Murugesan, R. (2006). Evolution of industrial automation. *International Journal of Computer Applications in Technology*, *25*(4), 169. DOI: 10.1504/IJCAT.2006.009389

Namian, M., Khalid, M., Wang, G., & Türkan, Y. (2021). Revealing Safety Risks of Unmanned Aerial Vehicles in Construction. *Transportation Research Record: Journal of the Transportation Research Board*, *2675*(11), 334–347. DOI: 10.1177/03611981211017134

Narkar, S., &Bandivadekar, T. (n.d.). *TRANSFORMING EPILEPSY CARE: APPLICATION OF AI FROM DETECTION TO PERSONALIZED TREATMENT*.

Nasiri, N., Maashri, A. A., Bourdoucen, H., Al-Abri, S., & Aouf, N. (2025). Environmental Monitoring in the Oil and Gas Industry Using Machine Vision Associated with Robotics and AI. In *IntechOpen eBooks*. IntechOpen., DOI: 10.5772/intechopen.1010026

Nauman, M., Naheed, R., & Khan, J. (2023). Sustainable Development and Financial Resilience: Linking Greenhouse Gas Emissions, Risk, and Green Growth with Innovation, Inclusion, Renewables, and Infrastructure. *Research Square (Research Square)*. DOI: 10.21203/rs.3.rs-3246034/v1

Nazarov, D., Sulimin, V., & Shvedov, V. (2024). Advancing environmental stewardship: The role of automation in enhanced environmental monitoring. *E3S Web of Conferences, 542*, 5005. DOI: 10.1051/e3sconf/202454205005

Neely, A., Gregory, M., & Platts, K. (2005). Performance measurement system design: A literature review and research agenda. *International Journal of Operations & Production Management, 25*(12), 1228–1263. DOI: 10.1108/01443570510633639

Neger, M., Rahid, A. O., & Alnour, M. (2025). Moderating effects of energy poverty for sustainable tourism, policy, innovation, and environmental resilience: Evidence from SEM-ANN approaches. *Discover Sustainability, 6*(1), 103. Advance online publication. DOI: 10.1007/s43621-025-00904-8

Negri, M., Cagno, E., Colicchia, C., & Sarkis, J. (2021). Integrating sustainability and resilience in the supply chain: A systematic literature review and a research agenda. *Business Strategy and the Environment, 30*(7), 2858–2886. DOI: 10.1002/bse.2776

Ni, J., Bingler, J., Colesanti-Senni, C., Kraus, M., Gostlow, G., Schimanski, T., Stammbach, D., Vaghefi, S. A., Wang, Q., Webersinke, N., Wekhof, T., Yu, T., &Leippold, M. (2023). *CHATREPORT: Democratizing Sustainability Disclosure Analysis through LLM-based Tools*. DOI: 10.18653/v1/2023.emnlp-demo.3

Nitsche, B., & Straube, F. (2020). Efficiently managing supply chain volatility–a management framework for the manufacturing industry. *Procedia Manufacturing, 43*, 320–327. DOI: 10.1016/j.promfg.2020.02.166

Noe, R. A., Clarke, A. D. M., & Klein, H. J. (2014). Learning in the twenty-first-century workplace. *Annual Review of Organizational Psychology and Organizational Behavior, 1*(1), 245–275. DOI: 10.1146/annurev-orgpsych-031413-091321

Noğay, H. S., Akıncı, T. Ç., & Yılmaz, M. (2021). Detection of invisible cracks in ceramic materials using by pre-trained deep convolutional neural network. *Neural Computing & Applications, 34*(2), 1423–1432. DOI: 10.1007/s00521-021-06652-w

Nooralishahi, P., Ibarra-Castanedo, C., Deane, S., López, F., Pant, S., Genest, M., Avdelidis, N. P., & Maldague, X. (2021). Drone-Based Non-Destructive Inspection of Industrial Sites: A Review and Case Studies [Review of *Drone-Based Non-Destructive Inspection of Industrial Sites: A Review and Case Studies*]. *Drones, 5*(4), 106. Multidisciplinary Digital Publishing Institute. DOI: 10.3390/drones5040106

Norsk Hydro. (2019, March 25). *Update on cyber attack March 25*. https://www.hydro.com/en/global/media/news/2019/update-on-cyber-attack-march-25/

Nwaogu, J. M., Yang, Y., Chan, A. P. C., & Wang, X. (2024). Enhancing Drone Operator Competency within the Construction Industry: Assessing Training Needs and Roadmap for Skill Development. *Buildings (Basel, Switzerland)*, *14*(4), 1153. DOI: 10.3390/buildings14041153

Nycz, B., & Pietrucha-Urbanik, K. (2024). Advancements in Air Quality Monitoring. *The Role of Drone Technology*, *19*, 19. Advance online publication. DOI: 10.3390/proceedings2024105019

O'Brien, C. (2002). Global manufacturing and the sustainable economy. *International Journal of Production Research*, *40*(15), 3867–3877. DOI: 10.1080/00207540210157169

Obydenkova, S. V., Kouris, P. D., Smeulders, D., Boot, M. D., & van der Meer, Y. (2021). Modeling life-cycle inventory for multi-product biorefinery: Tracking environmental burdens and evaluation of uncertainty caused by allocation procedure. *Biofuels, Bioproducts & Biorefining*, *15*(5), 1281–1300. DOI: 10.1002/bbb.2214

OECD. (2023). Smart City Data Governance. In *OECD urban studies*. Organization for Economic Cooperation and Development., DOI: 10.1787/e57ce301-

OECD. (2025). *OECD Economic Outlook, Volume 2025 Issue 1: Tackling Uncertainty, Reviving Growth*. OECD Publishing. https://doi.org/DOI: 10.1787/83363382-en

Ojadi, J. O., Owulade, O. A., Odionu, C. S., & Onukwulu, E. C.Jessica Obianuju OjadiOlumide Akindele OwuladeChinekwu Somtochukwu OdionuEkene Cynthia Onukwulu. (2025). Blockchain and IoT for Transparent Carbon Tracking and Emission Reduction in Global Supply Chains. *International Journal of Scientific Research in Science, Engineering and Technology*, *12*(2), 78–118. DOI: 10.32628/IJSRSET25122110

Ojsteršek, R., Buchmeister, B., & Javernik, A. (2023). Human-Robot Collaboration, Sustainable Manufacturing Perspective. In *Lecture notes in mechanical engineering* (p. 71). Springer Nature., DOI: 10.1007/978-3-031-38241-3_9

Ojstersek, R., Javernik, A., & Buchmeister, B. (2022). Importance of Sustainable Collaborative Workplaces – Simulation Modelling Approach. *International Journal of Simulation Modelling*, *21*(4), 627–638. DOI: 10.2507/IJSIMM21-4-623

Okogwu, C., Agho, M. O., Adeyinka, M. A., Ayodeji, S. A., & Daraojimba, C. (2023). THE ROLE OF DECARBONISATION IN ENHANCING SUSTAINABLE PROCUREMENT PRACTICES IN THE U.S. *Environmental Contaminants Reviews, 6*(2), 116–125. DOI: 10.26480/ecr.02.2023.116.125

Olivares, V., Córdova, F. M., Sepúlveda, J. M., & Dérpich, I. (2015). Modeling Internal Logistics by Using Drones on the Stage of Assembly of Products. *Procedia Computer Science, 55*, 1240–1249. DOI: 10.1016/j.procs.2015.07.132

Oliver, C. (1991). Strategic responses to institutional processes. *Academy of Management Review, 16*(1), 145–179. DOI: 10.2307/258610

Omolara, A. E., Alawida, M., & Omolara, A. E. (2023). Drone cybersecurity issues, solutions, trend insights and future perspectives: A survey. *Neural Computing & Applications, 35*(31), 23063–23101. DOI: 10.1007/s00521-023-08857-7

Öner, M. A., & Solak, S. (2020). PLC Tabanlı Uygulamalar: Mill Makinelerinde Oluşacak İş Kazalarının PLC Kullanılarak Önlenmesi. *European Journal of Science and Technology*. DOI: 10.31590/ejosat.754688

Ong, K., Mao, R., Satapathy, R., Filho, R. S., Cambria, E., Sulaeman, J., & Mengaldo, G. (2024). Explainable natural language processing for corporate sustainability analysis. *Information Fusion, 115*, 102726. DOI: 10.1016/j.inffus.2024.102726

Onososen, A., Musonda, I., Ramabodu, M., & Dzuwa, C. (2023). Safety and Training Implications of Human-Drone Interaction in Industrialised Construction Sites. In *Lecture notes in civil engineering* (p. 281). Springer Nature., DOI: 10.1007/978-3-031-32515-1_20

Osinga, F. P. B. (2007). *Science, strategy and war: The strategic theory of John Boyd*. Routledge., DOI: 10.4324/9780203088869

Ostadi, B., Barrani, L., & Aghdasi, M. (2024). Developing a strategic roadmap towards integration in Industry 4.0: A dynamic capabilities theory perspective. *Technological Forecasting and Social Change, 208*, 123679. DOI: 10.1016/j.techfore.2024.123679

Öz, E., Heikkilä, E., & Tiusanen, R. (2022). Development of an Organisational Certification Process for Specific Category Drone Operations. *Drones (Basel), 6*(10), 278. DOI: 10.3390/drones6100278

Özek, A., Seçkin, M., Demircioğlu, P., & Böğrekçi, İ. (2025). Artificial Intelligence Driving Innovation in Textile Defect Detection. *Textiles, 5*(2), 12. DOI: 10.3390/textiles5020012

ÖZGÜNER, M., & ÖZGÜNER, Z. (2022). Evaluation of the importance of additive manufacturing technology in terms of sustainable production with the DEMATEL method. *International Journal of ADVANCED AND APPLIED SCIENCES, 9*(10), 116. DOI: 10.21833/ijaas.2022.10.015

Pachot, A., & Patissier, C. (2023). *Towards Sustainable Artificial Intelligence: An Overview of Environmental Protection Uses and Issues.* Green and Low-Carbon Economy., DOI: 10.47852/bonviewGLCE3202608

Padovano, A., Cardamone, M., & Klaess, J. (2024). Empowering Operator 5.0: Human-centric design of an augmented reality tool for a learning factory. *IFAC-PapersOnLine*, *58*(8), 180–185. DOI: 10.1016/j.ifacol.2024.08.117

Pal, D., & Joshi, S. (2023). AI, IoT and Robotics in Smart Farming: Current Applications and Future Potentials. *2022 International Conference on Sustainable Computing and Data Communication Systems (ICSCDS)*, 1096. DOI: 10.1109/ICSCDS56580.2023.10105101

Palmarini, R., Erkoyuncu, J. A., Roy, R., & Torabmostaedi, H. (2018). A systematic review of augmented reality applications in maintenance. *Robotics and Computer-integrated Manufacturing*, *49*, 215–228. DOI: 10.1016/j.rcim.2017.06.002

Panagiotopoulou, V. C., Stavropoulos, P., & Chryssolouris, G. (2021). A critical review on the environmental impact of manufacturing: a holistic perspective [Review of *A critical review on the environmental impact of manufacturing: a holistic perspective*]. *The International Journal of Advanced Manufacturing Technology, 118*, 603. Springer Science+Business Media. DOI: 10.1007/s00170-021-07980-w

Panarotto, M., Isaksson, O., & Vial, V. (2022). Cost-efficient digital twins for design space exploration: A modular platform approach. *Computers in Industry*, *145*, 103813. DOI: 10.1016/j.compind.2022.103813

Panorama Consulting Group. (2024, February 6). *The role of a stakeholder analysis in change management.* https://www.panorama-consulting.com/stakeholder-analysis-in-change-management/

Papananias, M., McLeay, T. E., Obajemu, O., Mahfouf, M., & Kadirkamanathan, V. (2020). Inspection by exception: A new machine learning-based approach for multistage manufacturing. *Applied Soft Computing*, *97*, 106787. DOI: 10.1016/j.asoc.2020.106787

Paraschos, P. D., Koulinas, G. K., & Koulouriotis, D. E. (2023). A reinforcement learning/ad-hoc planning and scheduling mechanism for flexible and sustainable manufacturing systems. *Flexible Services and Manufacturing Journal*, *36*(3), 714–736. DOI: 10.1007/s10696-023-09496-9

Parthasarathy, V. (2024). AI-Driven Carbon Footprint Tracking and Emission Reduction in Logistics Networks. *International Journal of Artificial Intelligence Data Science and Machine Learning*, *5*, 47–56. DOI: 10.63282/3050-9262.IJAIDSML-V5I2P106

Paschou, T., Rapaccini, M., Adrodegari, F., & Saccani, N. (2020). Digital servitization in manufacturing: A systematic literature review and research agenda. *Industrial Marketing Management*, *89*, 278–292. DOI: 10.1016/j.indmarman.2020.02.012

Pater, J., & Stadnicka, D. (2021). Towards Digital Twins Development and Implementation to Support Sustainability – Systematic Literature Review. *Management and Production Engineering Review*. https://doi.org/DOI: 10.24425/mper.2021.138531

Patouillard, L., Bulle, C., Querleu, C., Maxime, D., Osset, P., & Margni, M. (2017). Critical review and practical recommendations to integrate the spatial dimension into life cycle assessment. *Journal of Cleaner Production*, *177*, 398–412. DOI: 10.1016/j.jclepro.2017.12.192

Patterson, C. M., Nurse, J. R., & Franqueira, V. N. (2024). "I don't think we're there yet": The practices and challenges of organisational learning from cyber security incidents. *Computers & Security*, *139*, 103699. DOI: 10.1016/j.cose.2023.103699

Pazienza, A., Baselli, G., Vinci, D. C., & Trussoni, M. V. (2024). A holistic approach to environmentally sustainable computing. *Innovations in Systems and Software Engineering*, *20*(3), 347–371. DOI: 10.1007/s11334-023-00548-9

Pellicciari, M., Avotiņš, A., Bengtsson, K., Berselli, G., Bey, N., Lennartson, B., & Meike, D. (2015). AREUS — Innovative hardware and software for sustainable industrial robotics. *IEEE International Conference on Automation Science and Engineering (CASE)*, 1325. DOI: 10.1109/CoASE.2015.7294282

Pellicer, M. R., Tungekar, M. Y., & Carpitella, S. (2023). Where to Place Monitoring Sensors for Improving Complex Manufacturing Systems? Discussing a Real Case in the Food Industry. *Sensors (Basel)*, *23*(7), 3768. DOI: 10.3390/s23073768 PMID: 37050828

Pertheban, T. R., Thurasamy, R., Marimuthu, A., Venkatachalam, K. R., Annamalah, S., Paraman, P., & Hoo, W. C. (2023). The impact of proactive resilience strategies on organizational performance: Role of ambidextrous and dynamic capabilities of SMEs in manufacturing sector. *Sustainability (Basel), 15*(16), 12665. DOI: 10.3390/su151612665

Peschl, M. F. (2023). Learning from the future as a novel paradigm for integrating organizational learning and innovation. *The Learning Organization, 30*(1), 6–22. DOI: 10.1108/TLO-01-2021-0018

Pesqueira, A., Sousa, M. J., & Costa, J. M. (2023). Exploring the Role of Big Data Analytics and Dynamic Capabilities in ESG Programs within Pharmaceuticals. *Research Square (Research Square)*. DOI: 10.21203/rs.3.rs-3315014/v1

Petersen, N., & Stricker, D. (2015). Cognitive augmented reality. *Computers & Graphics, 53*, 82–91. DOI: 10.1016/j.cag.2015.08.009

Pillai, G. M., Suresh, A., Gupta, E., Ganapathy, V., & Patra, A. (2024). Privadome: Delivery Drones and Citizen Privacy. *Proceedings on Privacy Enhancing Technologies. Privacy Enhancing Technologies Symposium, 2024*(2), 29–48. DOI: 10.56553/popets-2024-0039

Pires, F., Cachada, A., Barbosa, J., Moreira, A. P., & Leitão, P. (2019). Digital twin in industry 4.0: Technologies, applications and challenges. In *2019 IEEE 17th international conference on industrial informatics (INDIN)* (Vol. 1, pp. 721-726). IEEE.

Pizoń, J., Cioch, M., Kański, Ł., & Sánchez-García, E. (2022). Cobots Implementation in the Era of Industry 5.0 Using Modern Business and Management Solutions. *Advances in Science and Technology – Research Journal, 16*(6), 166. DOI: 10.12913/22998624/156222

Plapper, P. (2022). *Digitization of the work environment for sustainable production*. DOI: 10.30844/wgab_2022

Plathottam, S. J., Rzonca, A., Lakhnori, R., & Iloeje, C. O. (2023). A review of artificial intelligence applications in manufacturing operations. *Journal of Advanced Manufacturing and Processing, 5*(3), e10159. DOI: 10.1002/amp2.10159

Plaza, J. M., Martín, D., Arias-Perez, P., Vega, J., Roldán-Álvarez, D., García-Pérez, L., & Fernández-Conde, J. (2020). Open-Source Drone Programming Course for Distance Engineering Education. *Electronics (Basel), 9*(12), 2163. DOI: 10.3390/electronics9122163

Pop, P., Törngren, M., Yeh, S., & Thompson, H. A. (2023). Digital Technologies for a Sustainable Nordic Future: Challenges, Opportunities, and Policy Implications. *Research Portal Denmark*, 5. https://local.forskningsportal.dk/local/dki-cgi/ws/cris-link?src=dtu&id=dtu-3f4cae2e-ed19-4403-b0c8-376873525b43&ti=Digital%20Technologies%20for%20a%20Sustainable%20Nordic%20Future%3A%20Challenges%2C%20Opportunities%2C%20and%20Policy%20Implications

Prasad, G., Arunav, H., Dwight, S. S., Ghosh, M., Jayadev, A., & Nair, D. I. (2024). Advancing Sustainable Practices in Additive Manufacturing: A Comprehensive Review on Material Waste Recyclability [Review of *Advancing Sustainable Practices in Additive Manufacturing: A Comprehensive Review on Material Waste Recyclability*]. *Sustainability, 16*(23), 10246. Multidisciplinary Digital Publishing Institute. DOI: 10.3390/su162310246

Pratono, A. H. (2022). The strategic innovation under information technological turbulence: The role of organisational resilience in competitive advantage. *Competitiveness Review, 32*(3), 475–491. DOI: 10.1108/CR-03-2021-0046

Praveen, R. V. S., & Mutya, B. (2024). *The Automated Future: AI-Driven Transformation in Manufacturing*. Addition Publishing House.

Prescott, J. E. (1999). The evolution of competitive intelligence: Designing a process for action. *Proposal Management* (Spring), 37–52. https://winningthebusiness.com/wp-content/uploads/2018/05/APMP-Journal-Spring-1999.pdf

Prinz, F., Min, S., & An, J. (2023). Preface for NetZero, Achievable by Manufacturing? *International Journal of Precision Engineering and Manufacturing-Green Technology, 10*(3), 635–636. DOI: 10.1007/s40684-023-00510-x

Proceedings of the 19th Conference on Computer Science and Intelligence Systems (FedCSIS). (2024). *Annals of Computer Science and Information Systems, 39*. https://doi.org/DOI: 10.15439/978-83-969601-6-0

Psarommatis, F., Fraile, F., Mendonça, J. P., Meyer, O., Lázaro, Ó., & Kiritsis, D. (2023). Editorial: Zero defect manufacturing in the era of industry 4.0 for achieving sustainable and resilient manufacturing. *Frontiers in Manufacturing Technology, 3*, 1124624. Advance online publication. DOI: 10.3389/fmtec.2023.1124624

Puig, M. P., Paniego, J. M., Medina, S., Eguren, S. R., Libutti, L., Lanciotti, J., de Antueno, J., Estrebou, C. A., Chichizola, F., & Giusti, L. D. (2019). Intelligent Distributed System for Energy Efficient Control. In *Communications in computer and information science* (p. 51). Springer Science+Business Media. DOI: 10.1007/978-3-030-27713-0_5

Purohit, H., Tanabe, R., Ichige, T., Endo, T., Nikaido, Y., Suefusa, K., & Kawaguchi, Y. (2019). *MIMII Dataset: Sound Dataset for Malfunctioning Industrial Machine Investigation and Inspection*. DOI: 10.33682/m76f-d618

Qadir, H. M., Khan, R. A., Rasool, M., Sohaib, M., Shah, M. A., & Hasan, M. J. (2025). An adaptive feedback system for the improvement of learners. *Scientific Reports, 15*(1), 17242. DOI: 10.1038/s41598-025-01429-w PMID: 40383780

Qin, Y., Kishk, M. A., & Alouini, M. (2022). Drone Charging Stations Deployment in Rural Areas for Better Wireless Coverage: Challenges and Solutions. *IEEE Internet of Things Magazine*, *5*(1), 148–153. DOI: 10.1109/IOTM.001.2100083

Qiu, S., Cui, X., Ping, Z., Shan, N., Li, Z., Bao, X., & Xu, X. (2023). Deep Learning Techniques in Intelligent Fault Diagnosis and Prognosis for Industrial Systems: A Review [Review of *Deep Learning Techniques in Intelligent Fault Diagnosis and Prognosis for Industrial Systems: A Review*]. *Sensors, 23*(3), 1305. Multidisciplinary Digital Publishing Institute. DOI: 10.3390/s23031305

Qu, C., & Kim, E. (2025). Artificial-Intelligence-Enabled Innovation Ecosystems: A Novel Triple-Layer Framework for Micro, Small, and Medium-Sized Enterprises in the Chinese Apparel-Manufacturing Industry. *Sustainability (Basel), 17*(11), 5019. Advance online publication. DOI: 10.3390/su17115019

Radianti, J., Majchrzak, T. A., Fromm, J., & Wohlgenannt, I. (2020). A systematic review of immersive virtual reality applications for higher education: Design elements, lessons learned, and research agenda. *Computers & Education, 147*, 103778. DOI: 10.1016/j.compedu.2019.103778

Rahman, D. A., & Setiawan, Y. (2020). Possibility of applying unmanned aerial vehicle and thermal imaging in several canopy cover class for wildlife monitoring – preliminary results. *E3S Web of Conferences, 211*, 4007. DOI: 10.1051/e3sconf/202021104007

Rahmani, R., Jesus, C., & Lopes, S. I. (2024). Implementations of Digital Transformation and Digital Twins: Exploring the Factory of the Future. *Processes (Basel, Switzerland), 12*(4), 787. DOI: 10.3390/pr12040787

Rahman, M. A., Shahrior, M. F., Iqbal, K., & Abushaiba, A. A. (2025). Enabling Intelligent Industrial Automation: A Review of Machine Learning Applications with Digital Twin and Edge AI Integration. *Automation*, *6*(3), 37. DOI: 10.3390/automation6030037

Ramya, G., & Srinivasagan, K. G. (2024). Integrating cybersecurity threats into smart manufacturing: Best practices and frameworks. In *Artificial intelligence solutions for cyber-physical systems* (1st ed.). Auerbach Publications/Taylor & Francis., DOI: 10.1201/9781032694375-7

Ramya, M. (2019). Automation of Material Hanging Unit of a Cement Plant using ECS and PLC. *International Journal for Research in Applied Science and Engineering Technology*, *7*(3), 2322–2327. DOI: 10.22214/ijraset.2019.3426

Rane, N. L., Kaya, Ö., & Rane, J. (2024). *Advancing the Sustainable Development Goals (SDGs) through artificial intelligence, machine learning, and deep learning*. DOI: 10.70593/978-81-981271-8-1_4

Rane, J., Kaya, Ö., Mallick, S. K., & Rane, N. L. (2024). *Generative artificial intelligence in agriculture, education, and business*. Deep Science Publishing. DOI: 10.70593/978-81-981271-7-4

Ras, E., Wild, F., Stahl, C., & Baudet, A. (2017, June). Bridging the skills gap of workers in Industry 4.0 by human performance augmentation tools: Challenges and roadmap. In *Proceedings of the 10th international conference on PErvasive technologies related to assistive environments* (pp. 428-432). DOI: 10.1145/3056540.3076192

Rathore, B. (2016). The Next Frontier: How the Integration of AI Transforms Manufacturing for a Sustainable Future. *International Journal of New Media Studies*, *3*(2), 1. DOI: 10.58972/eiprmj.v3i2y16.74

Reeves, M., Levin, S. A., & Ueda, D. (2016, January–February). The biology of corporate survival. *Harvard Business Review*, *94*(1/2), 46–55. https://hbr.org/2016/01/the-biology-of-corporate-survival

Reid, J. D., & Castka, P. (2023). The impact of remote sensing on monitoring and reporting - The case of conformance systems. *Journal of Cleaner Production*, *393*, 136331. DOI: 10.1016/j.jclepro.2023.136331

Rejeb, A., Rejeb, K., Simske, S. J., & Treiblmaier, H. (2021). Drones for supply chain management and logistics: a review and research agenda [Review of *Drones for supply chain management and logistics: a review and research agenda*]. *International Journal of Logistics Research and Applications*, *26*(6), 708. Taylor & Francis. DOI: 10.1080/13675567.2021.1981273

Remer, S. J., Hall, A., Spielman, Z., Vollmer, J. L., & Hansen, J. (2023). FY23 ION Based Approaches to Address Labor and Knowledge Retention. In *OSTI OAI (U.S. Department of Energy Office of Scientific and Technical Information)*. Office of Scientific and Technical Information. https://www.osti.gov/biblio/2204962

Ren, Y., Liu, Y., Ji, T., & Xu, X. (2025). *AI Agents and Agentic AI-Navigating a Plethora of Concepts for Future Manufacturing*. DOI: 10.1016/j.jmsy.2025.08.017

Ren, Y., Liu, Y., Ji, T., & Xu, X. (2025). *AI Agents and Agentic AI-Navigating a Plethora of Concepts for Future Manufacturing*. https://doi.org/DOI: 10.48550/ARXIV.2507.01376

Resman, M., Herakovič, N., & Debevec, M. (2025). Integrating Digital Twin Technology to Achieve Higher Operational Efficiency and Sustainability in Manufacturing Systems. *Systems*, *13*(3), 180. DOI: 10.3390/systems13030180

Rezaie, B., Anderson, A., Gallegos, S., & Azarmi, F. (2021). Present and Future Sustainability Development of 3D Metal Printing. *European Journal of Sustainable Development Research*, *5*(3), em0168. Advance online publication. DOI: 10.21601/ejosdr/11132

Ribeiro, V., Barata, J., & da Cunha, P. R. (2022). Sustainable Data Governance: A Systematic Review and a Conceptual Framework [Review of *Sustainable Data Governance: A Systematic Review and a Conceptual Framework*]. *Proceedings of the International Conference on Information Systems Development*. DOI: 10.62036/ISD.2022.44

Richards, C. (2004). *Certain to win: The strategy of John Boyd, applied to business*. Xlibris. https://www.xlibris.com/en/bookstore/bookdetails/572837-certain-to-win

Rissman, J., Bataille, C., Masanet, E., Aden, N., Morrow, W. R., Zhou, N., Elliott, N., Dell, R., Heeren, N., Huckestein, B., Cresko, J., Miller, S. A., Roy, J., Fennell, P. S., Cremmins, B., Blank, T. K., Hone, D., Williams, E. D., & Can, S. (2020). Technologies and policies to decarbonize global industry: Review and assessment of mitigation drivers through 2070. *Applied Energy*, *266*, 114848. DOI: 10.1016/j.apenergy.2020.114848

Rittenhouse, J. (2015, October 10). *Improving stakeholder management using change management tools*. Paper presented at PMI® Global Congress 2015—North America, Orlando, FL. Project Management Institute. https://www.pmi.org/learning/library/improve-stakeholder-management-9901

Roldán, J. J., Peña-Tapia, E., Martín-Barrio, A., Olivares-Méndez, M. A., Del Cerro, J., & Barrientos, A. (2017). Multi-robot interfaces and operator situational awareness: Study of the impact of immersion and prediction. *Sensors (Basel)*, *17*(8), 1720. DOI: 10.3390/s17081720 PMID: 28749407

Rolofs, G., Wilking, F., Goetz, S., & Wartzack, S. (2024). Integrating Digital Twins and Cyber-Physical Systems for Flexible Energy Management in Manufacturing Facilities: A Conceptual Framework. *Electronics (Basel), 13*(24), 4964. DOI: 10.3390/electronics13244964

Román, Á. F. G., & Kabir, G. (2024). Assessing Carbon Dioxide Emissions in Manufacturing Industries: A Systematic Review [Review of *Assessing Carbon Dioxide Emissions in Manufacturing Industries: A Systematic Review*]. *Energies, 17*(20), 5119. Multidisciplinary Digital Publishing Institute. DOI: 10.3390/en17205119

Roos, J., von Krogh, G., & Roos, G. (2025). *Strategi – en innføring* (8th ed.). Fagbokforlaget., https://www.fagbokforlaget.no/produkt/9788245051094-strategi-en-innforing-8

Roscoe, S., Aktas, E., Petersen, K. J., Skipworth, H. D., Handfield, R. B., & Habib, F. (2022). Redesigning global supply chains during compounding geopolitical disruptions: The role of supply chain logics. *International Journal of Operations & Production Management, 42*(9), 1407–1434. DOI: 10.1108/IJOPM-12-2021-0777

Roseline, J. F., Leela, G. V. S., & Pravallika, K. R. S. (2019). Automation Control of Metro Traction Using Plc and Scada. *Journal of Physics: Conference Series, 1362*(1), 12079. DOI: 10.1088/1742-6596/1362/1/012079

Rotari, M. (2023). Production Analytics for a novel additive manufacturing system. *Research Portal Denmark*, 130. https://local.forskningsportal.dk/local/dki-cgi/ws/cris-link?src=dtu&id=dtu-5b4fb73a-9c12-44f7-8f20-bf9aa37006be&ti=Production%20Analytics%20for%20a%20novel%20additive%20manufacturing%20system

Rožanec, J. M., Zajec, P., Kenda, K., Novalija, I., Fortuna, B., Mladenić, D., Veliou, E., Papamartzivanos, D., Giannetsos, T., Menesidou, S. A., Alonso, R., Cauli, N., Recupero, D. R., Kyriazis, D., Sofianidis, G., Theodoropoulos, S., & Soldatos, J. (2021). STARdom: an architecture for trusted and secure human-centered\n manufacturing systems. *arXiv (Cornell University)*. /arxiv.2104.00983DOI: 10.1007/978-3-030-85910-7_21

Rubio, F., Llopis-Albert, C., & Valero, F. (2021). Multi-objective optimization of costs and energy efficiency associated with autonomous industrial processes for sustainable growth. *Technological Forecasting and Social Change, 173*, 121115. DOI: 10.1016/j.techfore.2021.121115

Ružarovský, R., Horák, T., & Bočák, R. (2024). Evaluating Energy Efficiency and Optimal Positioning of Industrial Robots in Sustainable Manufacturing. *Journal of Manufacturing and Materials Processing, 8*(6), 276. DOI: 10.3390/jmmp8060276

Ryalat, M., Almtireen, N., Al-refai, G., Elmoaqet, H., & Rawashdeh, N. (2025). Research and Education in Robotics: A Comprehensive Review, Trends, Challenges, and Future Directions. *Journal of Sensor and Actuator Networks, 14*(4), 76. DOI: 10.3390/jsan14040076

Saad, S. M., Bahadori, R., & Jafarnejad, H. (2021). The smart SME technology readiness assessment methodology in the context of Industry 4.0. *Journal of Manufacturing Technology Management, 32*(5), 1037–1065. DOI: 10.1108/JMTM-07-2020-0267

Safar, M., Hidayat, R., & Chen, C. (2023). A risk management framework for Industry 4.0 environment. *Sustainability (Basel), 15*(2), 1395. DOI: 10.3390/su15021395

Saghafian, M., Laumann, K., & Skogstad, M. R. (2021). Stagewise overview of issues influencing organizational technology adoption and use. *Frontiers in Psychology, 12*, 630145. DOI: 10.3389/fpsyg.2021.630145 PMID: 33815216

Sajid, H. (2024, January 10). *AI in robotics: 6 groundbreaking applications.* V7. https://www.v7labs.com/blog/ai-in-robotics

Sala, R., Pirola, F., Pezzotta, G., & Cavalieri, S. (2022). NLP-based insights discovery for industrial asset and service improvement: An analysis of maintenance reports. *IFAC-PapersOnLine, 55*(2), 522–527. DOI: 10.1016/j.ifacol.2022.04.247

Salas, E. (2021). Virtual Reality in Medical Education: A Systematic Review. *BMJ Quality & Safety, 30*(4), 345–350.

Salem, T., Dragomir, M., & Chatelet, É. (2024). Strategic Integration of Drone Technology and Digital Twins for Optimal Construction Project Management. *Applied Sciences (Basel, Switzerland), 14*(11), 4787. DOI: 10.3390/app14114787

Salhaoui, M., González, A. G., Arioua, M., Ortiz, F., Oualkadi, A. E., & Bonet, C. T. (2019). Smart Industrial IoT Monitoring and Control System Based on UAV and Cloud Computing Applied to a Concrete Plant. *Sensors (Basel), 19*(15), 3316. DOI: 10.3390/s19153316 PMID: 31357720

Salom, I., Dimić, G., Čelebić, V., Spasenović, M., Raičković, M., Mihajlović, M., & Todorović, D. (2023). An Acoustic Camera for Use on UAVs. *Sensors (Basel), 23*(2), 880. DOI: 10.3390/s23020880 PMID: 36679677

Salvati, D., Drioli, C., Ferrin, G., & Foresti, G. L. (2019). Acoustic Source Localization From Multirotor UAVs. *IEEE Transactions on Industrial Electronics, 67*(10), 8618–8628. DOI: 10.1109/TIE.2019.2949529

Samadi, S., Fischer, A., & Lechtenböhmer, S. (2023). The renewables pull effect: How regional differences in renewable energy costs could influence where industrial production is located in the future. *Energy Research & Social Science*, *104*, 103257. DOI: 10.1016/j.erss.2023.103257

Sama, M., Sata, A., Joshi, G., & Mohan, D. G. (2025). Smart Manufacturing With Industrial Internet of Things: Advances in TIG Welding for SS304 Stainless Steel. *Engineering Reports*, *7*(7), e70257. Advance online publication. DOI: 10.1002/eng2.70257

Samuel, S.Solarin Adebayo Samuel. (2025). Carbon pricing mechanisms for reducing greenhouse gas emissions and encouraging sustainable industrial practices. *World Journal of Advanced Research and Reviews*, *25*(2), 1. DOI: 10.30574/wjarr.2025.25.2.0350

Santos, R. C., & Martinho, J. L. (2020). An Industry 4.0 maturity model proposal. *Journal of Manufacturing Technology Management*, *31*(5), 1023–1043. DOI: 10.1108/JMTM-09-2018-0284

Saritas, O., & Aylen, J. (2010). Using scenarios for roadmapping: The case of clean production. *Technological Forecasting and Social Change*, *77*(7), 1061–1075. DOI: 10.1016/j.techfore.2010.03.003

Sarswatula, S. A., Pugh, T., & Prabhu, V. V. (2022). Modeling Energy Consumption Using Machine Learning. *Frontiers in Manufacturing Technology*, *2*, 855208. Advance online publication. DOI: 10.3389/fmtec.2022.855208

Sartal, A., Bellas, R., Mejías, A. M., & Garcia, A. (2014). The sustainable manufacturing concept, evolution and opportunities within Industry 4.0: A literature review [Review of *The sustainable manufacturing concept, evolution and opportunities within Industry 4.0: A literature review*]. *Advances in Mechanical Engineering*, *12*(5). SAGE Publishing. DOI: 10.1177/1687814020925232

Sauls, L. A., Paneque-Gálvez, J., Amador-Jiménez, M., Vargas-Ramírez, N., & Laumonier, Y. (2023). Drones, communities and nature: Pitfalls and possibilities for conservation and territorial rights. *Global Social Challenges Journal*, *2*, 24–46. DOI: 10.1332/AJHA9183

Schiavone, F., Simoni, M., & Bresciani, S. (2020). Towards Industry 5.0: An analysis of the enabling technologies. *Technological Forecasting and Social Change*, *162*, 120346. DOI: 10.1016/j.techfore.2020.120346

Schilling, F., Soria, E., & Floreano, D. (2022). On the Scalability of Vision-Based Drone Swarms in the Presence of Occlusions. *IEEE Access : Practical Innovations, Open Solutions, 10*, 28133–28146. DOI: 10.1109/ACCESS.2022.3158758

Schimanek, R., Bilge, P., & Dietrich, F. (2023). Data fusion for improved circularity through higher quality of prediction and increased reliability of inspection. *International Journal of Sustainable Manufacturing, 1*(1), 10058051. Advance online publication. DOI: 10.1504/IJSM.2023.10058051

Schmidt, S. N., Bridges, T. S., Gidley, P. T., Lotufo, G. R., Ruiz, C. E., Kennedy, A. R., Rogers, N. S., Ghosh, U., Burgess, R. M., Fernandez, L. A., & Mayer, P. (2016). Remediation of PCB contaminated sediments using activated carbon: Assessing efficiency and robustness using equilibrium sampling. *Research Portal Denmark*, 258. https://local.forskningsportal.dk/local/dki-cgi/ws/cris-link?src=dtu&id=dtu-cdda31c3-d615-4b57-a7d4-6834b91c0937&ti=Remediation%20of%20PCB%20contaminated%20sediments%20using%20activated%20carbon%3A%20Assessing%20efficiency%20and%20robustness%20using%20equilibrium%20sampling

Schöberl, M., Huber, A., Kreppold, S., Dirnaichner, J., Keßler, S., & Fottner, J. (2023). Cobot uptake in construction: Embedding collaborative robots in digital construction processes. *Construction Robotics, 7*(1), 89–103. DOI: 10.1007/s41693-023-00098-3

Schoemaker, P. J. H., Heaton, S., & Teece, D. J. (2018). Innovation, dynamic capabilities, and leadership. *California Management Review, 61*(1), 15–42. DOI: 10.1177/0008125618790246

Schuh, G., Patzwald, M., & Salmen, M. (2019). *Resilient technology strategy in volatile environments: Derivation of requirements to enable long-term strategic positioning in times of VUCA*. In *PICMET '19: 2019 Portland International Conference on Management of Engineering and Technology* (pp. 1–8). IEEE. https://doi.org/DOI: 10.23919/PICMET.2019.8893840

Schultheis, A., Alt, B., Bast, S., Guldner, A., Jilg, D., Katic, D., Mundorf, J., Schlagenhauf, T., Weber, S., Bergmann, R., Bergweiler, S., Creutz, L., Dartmann, G., Malburg, L., Naumann, S., Rezapour, M., & Ruskowski, M. (2024). EASY: Energy-Efficient Analysis and Control Processes in the Dynamic Edge-Cloud Continuum for Industrial Manufacturing. *KI -. Kunstliche Intelligenz*. Advance online publication. DOI: 10.1007/s13218-024-00868-3

Schwab, K. (2016). *The fourth industrial revolution*. World Economic Forum. https://www.weforum.org/about/the-fourth-industrial-revolution-by-klaus-schwab

Schwark, F., Dawel, L., & Pehlken, A. (2023). Sustainability Digital Twin: A tool for the manufacturing industry. *Procedia CIRP, 116*, 143–148.

Schwarz, L. (2024, May 22). *AI in manufacturing: Benefits and 15 use cases*. NetSuite. https://www.netsuite.com/portal/resource/articles/erp/ai-in-manufacturing.shtml

Scott, W. R. (2008). *Institutions and organizations: Ideas and interests* (3rd ed.). Sage.

Sekti, B. A. (2025a). Adversarial machine learning in industrial cybersecurity: Challenges and solutions. In *AI-Enhanced Cybersecurity for Industrial Automation*. DOI: 10.4018/979-8-3373-3241-3.ch018

Sekti, B. A. (2025b). Machine learning and artificial intelligence (AI) in manufacturing. In *Integrating Digital Innovation and Integrated Frameworks in Manufacturing*. DOI: 10.4018/979-8-3373-1082-4.ch011

Selvapriya, R., Gopinath, M., Sampathkumar, P., Velmurugan, D., & Suresh, S. (2025). AI reduces environmental toxins in civil engineering. In *AI Methods for Environmental Protection and Resource Conservation* (pp. 83–106). DOI: 10.4018/979-8-3373-3246-8.ch004

Serac, C. A.SERAC. (2023). DIGITAL TRANSFORMATION VULNERABILITIES: ASSESSING THE RISKS AND STRENGTHENING CYBER SECURITY. *The Annals of the University of Oradea Economic Sciences*, *3*(1), 771–781. DOI: 10.47535/1991AUOES32(1)059

Setyadi, A., Soekotjo, S., Lestari, S. D., Pawirosumarto, S., & Damaris, A. (2025). Trends and Opportunities in Sustainable Manufacturing: A Systematic Review of Key Dimensions from 2019 to 2024 [Review of *Trends and Opportunities in Sustainable Manufacturing: A Systematic Review of Key Dimensions from 2019 to 2024*]. *Sustainability, 17*(2), 789. Multidisciplinary Digital Publishing Institute. DOI: 10.3390/su17020789

Shafi, I., Mazhar, M., Fatima, A., Álvarez, R. M., Miró, Y., Espinosa, J. C. M., & Ashraf, I. (2023). Deep Learning-Based Real Time Defect Detection for Optimization of Aircraft Manufacturing and Control Performance. *Drones (Basel)*, *7*(1), 31. DOI: 10.3390/drones7010031

Shahin, M., Chen, F. F., Hosseinzadeh, A., & Zand, N. (2023). Using machine learning and deep learning algorithms for downtime minimization in manufacturing systems: An early failure detection diagnostic service. *International Journal of Advanced Manufacturing Technology*, *128*(9-10), 3857–3883. DOI: 10.1007/s00170-023-12020-w

Shankar, A., Behl, A., Pereira, V., Chavan, M., & Chirico, F. (2024). Exploring enablers and inhibitors of AI-enabled drones for manufacturing process audits: A mixed-method approach. *Business Strategy and the Environment, 33*(5), 3749–3768. DOI: 10.1002/bse.3679

Shanthi, D., Ashok, G., Biswal, C., Udharika, S., Varshini, S., & Sindhu, G. (2025). AI-driven adaptive IT training: A personalized learning framework for enhanced knowledge retention and engagement. *Metallurgical and Materials Engineering, 136-145.*

Sharma, V., Tripathi, A. K., & Mittal, H. (2022). *Technological revolutions in smart farming: Current trends, challenges & future directions.*

Sharma, A., Priya, G. S. K., & Bandyopadhyay, S. (2023). Industrial decarbonization: A revolution ahead. *Clean Technologies and Environmental Policy, 25*(8), 2467–2468. DOI: 10.1007/s10098-023-02602-3

Shelton, B. E., & Hedley, N. R. (2004). Exploring a cognitive basis for learning spatial relationships with augmented reality. *Technology, Instruction. Cognition and Learning, 1*(4), 323.

Shih, W. C. (2020, September–October). Global supply chains in a post-pandemic world. *Harvard Business Review, 98*(5), 82–89. https://hbr.org/2020/09/global-supply-chains-in-a-post-pandemic-world

Shishodia, A., Sharma, R., Rajesh, R., & Munim, Z. H. (2023). Supply chain resilience: A review, conceptual framework and future research. *International Journal of Logistics Management, 34*(4), 879–908. DOI: 10.1108/IJLM-03-2021-0169

Shook, J. (2009, February 3). *The remarkable Chief Engineer*. Lean Enterprise Institute. https://www.lean.org/the-lean-post/articles/the-remarkable-chief-engineer/

Siddique, I. (2024). Sustainable Manufacturing Practices: Reducing Waste and Energy Consumption in Production Systems. SSRN *Electronic Journal*. DOI: 10.2139/ssrn.4919617

Siemens, A. G. Digital Factory. (2017). *Leading as the digital enterprise: Electronics Works Amberg — Digitalization practices* [Brochure]. https://assets.new.siemens.com/siemens/assets/api/uuid%3A73a13464-b0d2-4a40-acd1-a79919ac98e2/brochure-electronics-works-amberg-digitalization-practices-e.pdf

Sihag, V., Choudhary, G., Choudhary, P. K., & Dragoni, N. (2023). Cyber4Drone: A Systematic Review of Cyber Security and Forensics in Next-Generation Drones [Review of *Cyber4Drone: A Systematic Review of Cyber Security and Forensics in Next-Generation Drones*]. *Drones, 7*(7), 430. Multidisciplinary Digital Publishing Institute. DOI: 10.3390/drones7070430

Silva, A. S. S., Machado, J., & Sampaio, P. (2024). Predictive quality model for customer defects. *The TQM Journal, 36*(9), 155–174. DOI: 10.1108/TQM-09-2023-0302

Simeth, A., Kumar, A. A., & Plapper, P. (2025). Flexible and robust detection for assembly automation with YOLOv5: A case study on HMLV manufacturing line. *Journal of Intelligent Manufacturing, 36*(5), 3447–3463. DOI: 10.1007/s10845-024-02411-5

Singh, A. R., Kumar, R. S., Bajaj, M., Khadse, C. B., & Зайцев, Є. (2024). Machine learning-based energy management and power forecasting in grid-connected microgrids with multiple distributed energy sources. *Scientific Reports, 14*(1). Advance online publication. DOI: 10.1038/s41598-024-70336-3 PMID: 39160194

Singh, A., Kaur, P., & Kumar, R. (2023). Artificial intelligence, machine learning and deep learning in advanced robotics: A review. *Journal of Artificial Intelligence and Technology, 3*(1), 22–36. DOI: 10.1016/j.jaite.2023.03.001

Singhaphandu, R., & Pannakkong, W. (2024). A review on enabling technologies of industrial virtual training systems. [IJKSS]. *International Journal of Knowledge and Systems Science, 15*(1), 1–33. DOI: 10.4018/IJKSS.352515

Singh, B., Kumar, N., Kaur, G., Akhai, S., Abbass, M., & Kaur, P. (2025). Emerging trends in electric automotive instrumentation and control systems: A review of current trends and future technologies. *AIP Conference Proceedings, 3343*, 020012. DOI: 10.1063/5.0292567

Singh, S., & Akhai, S. (2024). Electronics in precision control for mechanical engineering: A brief overview. *AIP Conference Proceedings, 3121*, 020007. DOI: 10.1063/5.0221464

Sinha, B. B., & Dhanalakshmi, R. (2021). Recent advancements and challenges of Internet of Things in smart agriculture: A survey. *Future Generation Computer Systems, 126*, 169–184. DOI: 10.1016/j.future.2021.08.006

Sledz, A., Unger, J., & Heipke, C. (2020). UAV-BASED THERMAL ANOMALY DETECTION FOR DISTRIBUTED HEATING NETWORKS. *˜The œinternational Archives of the Photogrammetry, Remote Sensing and Spatial Information Sciences. The International Archives of the Photogrammetry, Remote Sensing and Spatial Information Sciences, 499.* Advance online publication. DOI: 10.5194/isprs-archives-XLIII-B1-2020-499-2020

Smartsheet. (2023). *Guide to gap analysis with examples*. Smartsheet. Retrieved September 16, 2025, from https://www.smartsheet.com/gap-analysis-method-examples

Solomon, B. D. (2023). Organisation for Economic Co-operation and Development (OECD). In *Dictionary of Ecological Economics: Terms for the New Millennium*. https://doi.org/DOI: 10.4337/9781788974912.O.20

Song, C., Shin, S. Y., & Shin, K. S. (2024). Implementing the dynamic feedback-driven learning optimization framework: A machine learning approach to personalize educational pathways. *Applied Sciences (Basel, Switzerland)*, *14*(2), 916. DOI: 10.3390/app14020916

Sophos. (2024, May 28). *The State of Ransomware in Manufacturing and Production 2024*. Sophos. https://www.sophos.com/en-us/whitepaper/state-of-ransomware-in-manufacturing-and-production

Soufi, Z., David, P., & Yahouni, Z. (2021). A methodology for the selection of Material Handling Equipment in manufacturing systems. *IFAC-PapersOnLine*, *54*(1), 122–127. DOI: 10.1016/j.ifacol.2021.08.193

Spaltini, M., Acerbi, F., Pinzone, M., Gusmeroli, S., & Taisch, M. (2022). Defining the roadmap towards Industry 4.0: The 6Ps maturity model for manufacturing SMEs. *Procedia CIRP*, *105*, 631–636. https://www.sciencedirect.com/science/article/pii/S2212827122001068. DOI: 10.1016/j.procir.2022.02.105

Srinivasan, B., Iqbal, M. U., Shahab, M. A., & Srinivasan, R. (2022). Review of virtual reality (VR) applications to enhance chemical safety: From students to plant operators. *Journal of Chemical Health and Safety*, *29*(3), 246–262. DOI: 10.1021/acs.chas.2c00006

Stanton, M. C., Kalonde, P. K., Zembere, K., Spaans, R. H., & Jones, C. M. (2021). The application of drones for mosquito larval habitat identification in rural environments: A practical approach for malaria control? *Malaria Journal*, *20*(1), 244. Advance online publication. DOI: 10.1186/s12936-021-03759-2 PMID: 34059053

Stephani, H., Weibel, T., Rösch, R., & Moghiseh, A. (2023). Challenges and approaches when realizing online surface inspection systems with deep learning algorithms. *Discover Data, 1*(1), 3. Advance online publication. DOI: 10.1007/s44248-023-00004-w

Steuer, J. (1992). Defining virtual reality: Dimensions determining telepresence. *Journal of Communication, 42*(4), 73–93. DOI: 10.1111/j.1460-2466.1992.tb00812.x

Stroud, D., & Weinel, M. (2020). A safer, faster, leaner workplace? Technical-maintenance worker perspectives on digital drone technology 'effects' in the European steel industry. *New Technology, Work and Employment, 35*(3), 297–313. DOI: 10.1111/ntwe.12174

Strubell, E., Ganesh, A., & McCallum, A. (2020). Energy and policy considerations for modern deep learning research. AAAI 2020 - 34th AAAI Conference on Artificial Intelligence. DOI: 10.1609/aaai.v34i09.7123

Sukathong, S., Suksawang, P., & Naenna, T. (2021). Analyzing the importance of critical success factors for the adoption of advanced manufacturing technologies. *International Journal of Engineering Business Management, 13*, 18479790211055057. https://journals.sagepub.com/doi/full/10.1177/18479790211055057. DOI: 10.1177/18479790211055057

Sullivan, G. R., & Harper, M. V. (1996). *Hope is not a method: What business leaders can learn from America's Army*. Times Business. https://www.penguinrandomhouse.com/books/175183/hope-is-not-a-method-by-gordon-r-sullivan/

Sundaram, S., & Zeid, A. (2023). Artificial Intelligence-Based Smart Quality Inspection for Manufacturing. *Micromachines, 14*(3), 570. DOI: 10.3390/mi14030570 PMID: 36984977

Sundaram, S., & Zeid, A. (2024). Technical language processing for Prognostics and Health Management: Applying text similarity and topic modeling to maintenance work orders. *Journal of Intelligent Manufacturing*. Advance online publication. DOI: 10.1007/s10845-024-02323-4

Sun, Y., Fesenko, H., Kharchenko, V., Luo, Z., Kliushnikov, I., Illiashenko, O., Morozova, O., & Sachenko, A. (2022). UAV and IoT-Based Systems for the Monitoring of Industrial Facilities Using Digital Twins: Methodology, Reliability Models, and Application. *Sensors (Basel), 22*(17), 6444. DOI: 10.3390/s22176444 PMID: 36080903

Sun, Y., Zhang, M., Chang, J., & Wang, C. (2025). Unpacking Boundary-Spanning Search and Green Innovation for Sustainability: The Role of AI Capabilities in the Chinese Manufacturing Industry. *Sustainability (Basel), 17*(14), 6439. Advance online publication. DOI: 10.3390/su17146439

Suresh, S. (2024). Virtual reality in engineering: Innovations and applications. In *Creating Immersive Learning Experiences Through Virtual Reality (VR)* (pp. 287–310). DOI: 10.4018/979-8-3693-6407-9.ch013

Suresh, S., Natarajan, E., Sudhagar, S., Boopathi, S., & Elayaraja, R. (2024). Additive manufacturing for a sustainable production: Materials, processes, and impact. In *Futuristic Technology for Sustainable Manufacturing* (pp. 1–16). DOI: 10.4018/979-8-3693-1862-1.ch001

Suresh, S., Velmurugan, D., Priya, E. L. D., & Selvapriya, R. (2024). AI for enhanced drone capabilities in industry 5.0. In *Drone Applications for Industry 5.0* (pp. 75–91). DOI: 10.4018/979-8-3693-2093-8.ch006

Suresh, S., Velmurugan, D., Vasudevan, N., Surekha Rathi Samundi, D., & Elayaraja, R. (2025). Autonomous UAV swarms: Innovations and applications. In *Innovations and developments in unmanned aerial vehicles* (pp. 95–128). IGI Global., DOI: 10.4018/979-8-3693-8462-6.ch005

Su, Y., & Xu, G. (2025). The Influencing Mechanism of Intelligent Production on Carbon Emission Efficiency of Equipment Manufacturing Industry in China. *Processes (Basel, Switzerland), 13*(4), 1102. DOI: 10.3390/pr13041102

Syberfeldt, A., Danielsson, O., & Gustavsson, P. (2017). Augmented reality smart glasses in the smart factory: Product evaluation guidelines and review of available products. *IEEE Access : Practical Innovations, Open Solutions, 5*, 9118–9130. DOI: 10.1109/ACCESS.2017.2703952

Szira, Z., Varga, E., Csegődi, T. L., & Milics, G. (2023). The Development of Drone Techology and its Regulation in the European Union. *EU Agrarian Law, 12*(1), 35–41. DOI: 10.2478/eual-2023-0005

Tabbakh, A., Amin, A., Islam, M., Mahmud, G., Chowdhury, I. K., & Mukta, M. S. H. (2024). Towards sustainable AI: A comprehensive framework for Green AI. *Discover Sustainability, 5*(1), 408. Advance online publication. DOI: 10.1007/s43621-024-00641-4

Taneja, T., Bhatia, B. S., & Akhai, S. (2024). *Assessing groundwater quality in Rampura-Bathinda for physicochemical parameters & heavy metal concentrations using electronic equipment*. In *2024 1st International Conference on Sustainability and Technological Advancements in Engineering Domain (SUSTAINED)* (pp. 791–796). IEEE. DOI: 10.1109/SUSTAINED63638.2024.11073880

Taneja, T., Bhatia, B. S., & Akhai, S. (2025). Agricultural resilience: Water quality and human well-being. In *Emerging smart agricultural practices using artificial intelligence* (pp. 1–23). Wiley., DOI: 10.1002/9781394274277.ch1

Tang, L., Jing, K., He, J., & Stanley, H. E. (2016). Complex interdependent supply chain networks: Cascading failure and robustness. *Physica A*, *443*, 58–69. DOI: 10.1016/j.physa.2015.09.082

Tao, F., Cheng, J., Qi, Q., Zhang, M., Zhang, H., & Sui, F. (2018). Digital twin-driven product design, manufacturing and service with big data. *International Journal of Advanced Manufacturing Technology*, *94*(9-12), 3563–3576. DOI: 10.1007/s00170-017-0233-1

Tappia, E., Marchet, G., Melacini, M., & Perotti, S. (2015). Incorporating the environmental dimension in the assessment of automated warehouses. *Production Planning and Control*, *26*(10), 824–838. DOI: 10.1080/09537287.2014.990945

Tatsi, K., Hutchinson, T. H., Ehrlich, N., Correia, M., & Handy, R. D. (2016). Toxicity of Pristine and Aged Coated Copper Oxide Engineered Nanomaterials (CuO ENMs) to the Earthworm E. fetida. *Research Portal Denmark*, 120. https://local.forskningsportal.dk/local/dki-cgi/ws/cris-link?src=dtu&id=dtu-fba0fc31-7b81-4f4d-96f0-421cc3b994bc&ti=Toxicity%20of%20Pristine%20and%20Aged%20Coated%20Copper%20Oxide%20Engineered%20Nanomaterials%20(CuO%20ENMs)%20to%20the%20Earthworm%20E.%20fetida

TechTarget. (2015, October). *VUCA (volatility, uncertainty, complexity and ambiguity)*. WhatIs.com (TechTarget). https://www.techtarget.com/whatis/definition/VUCA-volatility-uncertainty-complexity-and-ambiguity

Teece, D. J. (2007). Explicating dynamic capabilities: The nature and microfoundations of (sustainable) enterprise performance. *Strategic Management Journal*, *28*(13), 1319–1350. DOI: 10.1002/smj.640

Teece, D. J., Pisano, G., & Shuen, A. (1997). Dynamic capabilities and strategic management. *Strategic Management Journal*, *18*(7), 509–533. DOI: 10.1002/(SICI)1097-0266(199708)18:7<509::AID-SMJ882>3.0.CO;2-Z

Tercan, H., & Meisen, T. (2022). Machine learning and deep learning based predictive quality in manufacturing: a systematic review [Review of *Machine learning and deep learning based predictive quality in manufacturing: a systematic review*]. *Journal of Intelligent Manufacturing, 33*(7), 1879. Springer Science+Business Media. DOI: 10.1007/s10845-022-01963-8

Thareja, P., & Akhai, S. (2016). Processing Aluminum Fly Ash Composites via Parametric Analysis of Stir Casting. *Journal of Advanced Research in Manufacturing. Material Science & Metallurgical Engineering, 3*(3&4), 21–28.

Thimm, H. (2023). Data modeling and NLP-based scoring method to assess the relevance of environmental regulatory announcements. *Environment Systems & Decisions, 43*(3), 416–432. DOI: 10.1007/s10669-023-09900-7

Thomas, G., & Sule, M. (2022). A service lens on cybersecurity continuity and management for organizations' subsistence and growth. *Organizational Cybersecurity Journal Practice Process and People, 3*(1), 18–40. DOI: 10.1108/OCJ-09-2021-0025

Thorén, K., & Vendel, M. (2019). Backcasting as a strategic management tool for meeting VUCA challenges. *Journal of Strategy and Management, 12*(2), 298–312. DOI: 10.1108/JSMA-10-2017-0072

Thornton, P. H., Ocasio, W., & Lounsbury, M. (2012). *The institutional logics perspective: A new approach to culture, structure, and process*. Oxford University Press., https://global.oup.com/ukhe/product/the-institutional-logics-perspective-9780199601943 DOI: 10.1093/acprof:oso/9780199601936.001.0001

Tianyuan, L., Guo, H., Danilov, C., & Nahrstedt, K. (2020). *A Privacy-Preserving Data Collection and Processing Framework for Third-Party UAV Services*. 683. DOI: 10.1109/TrustCom50675.2020.00095

Tmušić, G., Manfreda, S., Aasen, H., James, M. R., Gonçalves, G., Dor, E. B., Brook, A., Polinova, M., Justel, J. J. A., Mészáros, J., Zhuang, R., Johansen, K., Malbéteau, Y., de Lima, M. I. P., Davids, C., Herban, S., & McCabe, M. F. (2020). Current Practices in UAS-based Environmental Monitoring. *Remote Sensing (Basel), 12*(6), 1001. DOI: 10.3390/rs12061001

To, A., Liu, M., Hairul, M. H. B. M., Davis, J. G., Lee, J. S. A., Hesse, H., & Nguyen, H. D. (2021). Drone-Based AI and 3D Reconstruction for Digital Twin Augmentation. In *Lecture notes in computer science* (p. 511). Springer Science+Business Media. DOI: 10.1007/978-3-030-77626-8_35

Todescato, M., Braholli, O., Chaltsev, D., Blasio, I. D., Don, D., Egger, G., Emig, J., Monizza, G. P., Sacco, P., Siegele, D., Steiner, D., Terzer, M., Riedl, M., Giusti, A., & Matt, D. T. (2023). Sustainable manufacturing through application of reconfigurable and intelligent systems in production processes: A system perspective. *Scientific Reports, 13*(1), 22374. Advance online publication. DOI: 10.1038/s41598-023-49727-5 PMID: 38104211

Tøndel, I. A., Line, M. B., & Jaatun, M. G. (2014). Information security incident management: Current practice as reported in the literature. *Computers & Security, 45*, 42–57. DOI: 10.1016/j.cose.2014.05.003

Torab-Miandoab, A., Samad-Soltani, T., Jodati, A., & Rezaei-Hachesu, P. (2023). Interoperability of heterogeneous health information systems: a systematic literature review [Review of *Interoperability of heterogeneous health information systems: a systematic literature review*]. *BMC Medical Informatics and Decision Making, 23*(1). BioMed Central. DOI: 10.1186/s12911-023-02115-5

Toussaint, M., Krima, S., & Panetto, H. (2024). Industry 4.0 data security: A cybersecurity frameworks review. *Journal of Industrial Information Integration, 39*, 100604. DOI: 10.1016/j.jii.2024.100604

Tran, T.-H., & Nguyen, D. D. (2022). Management and Regulation of Drone Operation in Urban Environment: A Case Study. *Social Sciences (Basel, Switzerland), 11*(10), 474. DOI: 10.3390/socsci11100474

Triantafyllou, A., Sarigiannidis, P., Bibi, S., Vakouftsi, F., & Pantzios, V. (2020). *Modelling deployment costs of Precision Agriculture Monitoring Systems. 5*, 252. DOI: 10.1109/DCOSS49796.2020.00048

Triebe, M. J., Deng, S., Pérez-Cardona, J. R., Joung, B. G., Wu, H., Shakelly, N., Pieper, J. P., Zhou, X. K., Maani, T., Zhao, F., & Sutherland, J. W. (2023). Perspectives on future research directions in green manufacturing for discrete products. *Green Manufacturing Open, 1*(1), 10. DOI: 10.20517/gmo.2022.11

Troise, C., Corvello, V., Ghobadian, A., & O'Regan, N. (2022). How can SMEs successfully navigate VUCA environment: The role of agility in the digital transformation era. *Technological Forecasting and Social Change, 174*, 121227. DOI: 10.1016/j.techfore.2021.121227

Trstenjak, M., Mustapić, M., Gregurić, P., & Opetuk, T. (2023). Use of Green Industry 5.0 Technologies in Logistics Activities. *TehničkiGlasnik, 17*(3), 471–477. DOI: 10.31803/tg-20230518185836

Tsai, W.-H., & Lin, W.-H. (2024). Production Decision Model for the Cement Industry in Pursuit of Carbon Neutrality: Analysis of the Impact of Carbon Tax and Carbon Credit Costs. *Sustainability (Basel), 16*(6), 2251. DOI: 10.3390/su16062251

Tsay, C. H. H., Kofinas, A., & Luo, J. (2018). Enhancing student learning experience with technology-mediated gamification: An empirical study. *Computers & Education, 121*, 1–17. DOI: 10.1016/j.compedu.2018.01.009

Tukamuhabwa, B. R., Stevenson, M., Busby, J., & Zorzini, M. (2015). Supply chain resilience: Definition, review and theoretical foundations for further study. *International Journal of Production Research, 53*(18), 5592–5623. DOI: 10.1080/00207543.2015.1037934

Tupa, J., Simota, J., & Steiner, F. (2017). Aspects of risk management implementation for Industry 4.0. *Procedia Manufacturing, 11*, 1223–1230. DOI: 10.1016/j.promfg.2017.07.248

Tuptuk, N., & Hailes, S. (2018). Security of smart manufacturing systems. *Journal of Manufacturing Systems, 47*, 93–106. DOI: 10.1016/j.jmsy.2018.04.007

Turner, C., Oyekan, J., Garn, W., Duggan, C., & Abdou, K. (2022). Industry 5.0 and the Circular Economy: Utilizing LCA with Intelligent Products. *Sustainability (Basel), 14*(22), 14847. DOI: 10.3390/su142214847

Tushman, M. L., & O'Reilly, C. A. (2002). *Winning through innovation: A practical guide to leading organizational change and renewal* (Updated ed.). Harvard Business Press. https://books.google.com/books?id=CvEplLBTN3QC

Tyagi, A. K., Bhatt, P., Chidambaram, N., & Kumari, S. (2024). Artificial intelligence empowered smart manufacturing for modern society: a review. *Artificial Intelligence-Enabled Digital Twin for Smart Manufacturing*, 55-83.

U.S. Department of Justice. (2021, June 7). *Department of Justice seizes $2.3 million in cryptocurrency paid to the ransomware extortionists DarkSide* [Press release]. https://www.justice.gov/archives/opa/pr/department-justice-seizes-23-million-cryptocurrency-paid-ransomware-extortionists-darkside

U.S. Department of Labor. (2020). *Workforce Innovation and Opportunity Act.* https://www.dol.gov/agencies/eta/wioa

U.S. Government Accountability Office. (2024, December 2). *Export controls: Commerce implemented advanced semiconductor rules and took steps to address compliance challenges* (GAO-25-107386). U.S. Government Accountability Office. https://www.gao.gov/products/gao-25-107386

Uçar, A., Karaköse, M., & Kırımça, N. (2024). Artificial intelligence for predictive maintenance applications: Key components, trustworthiness, and future trends. *Applied Sciences (Basel, Switzerland), 14*(2), 898. DOI: 10.3390/app14020898

UNESCO. (2021). *Education and digital transformation.* United Nations Educational, Scientific and Cultural Organization. https://unesdoc.unesco.org/ark:/48223/pf0000374719

UNIDO. (2019). *Industrial development report 2020: Industrialization, trade, and development in the digital era.* United Nations Industrial Development Organization. https://www.unido.org/resources-publications-flagship-publications-industrial-development-report-series/industrial-development-report-2020

Unilever, P. L. C. (2024). *Annual report and accounts 2023.* https://www.unilever.com/files/66bc4aea-608f-46ee-8da3-cde0ec8ebe90/unilever-annual-report-and-accounts-2023.pdf

Unilever, P. L. C. (2024). *Climate Transition Action Plan (updated 2024).* https://www.unilever.com/files/8b5df5f6-cb90-40fd-9691-38d06905d81d/unilever-climate-transition-action-plan-updated-2024.pdf

Urama, J., Wiren, R., Galinina, O., Kauppi, J., Hiltunen, K., Erkkilä, J., Chernogorov, F., Eteläaho, P., Heikkilä, M., Torsner, J., Andreev, S., & Valkama, M. (2020). UAV-Aided Interference Assessment for Private 5G NR Deployments: Challenges and Solutions. *IEEE Communications Magazine, 58*(8), 89–95. DOI: 10.1109/MCOM.001.2000042

Valdivia, S., Backes, J. G., Traverso, M., Sonnemann, G., Cucurachi, S., Guinée, J. B., Schaubroeck, T., Finkbeiner, M., Leroy-Parmentier, N., Ugaya, C. M. L., Peña, C., Zamagni, A., Inaba, A., Amaral, M., Berger, M., Dvarionienė, J., Vakhitova, T., Benoît-Norris, C., Prox, M., & Goedkoop, M. (2021). Principles for the application of life cycle sustainability assessment. *The International Journal of Life Cycle Assessment, 26*(9), 1900–1905. DOI: 10.1007/s11367-021-01958-2

Valtonen, A., & Holopainen, M. (2025). Mitigating employee resistance and achieving well-being in digital transformation. *Information Technology & People, 38*(8), 42–72. DOI: 10.1108/ITP-05-2024-0701

van Bos, B., Mallion, A., Wilson, C., Zwicker, E., Schler, A., & Black, T. (2015, November 3). Robotic Inspection Solutions for Petrochemical Pressure Vessels. *Abu Dhabi International Petroleum Exhibition and Conference.* DOI: 10.2118/177913-MS

Van der Meer, N., van der Werf, V., Brinkman, W. P., & Specht, M. (2023). Virtual reality and collaborative learning: A systematic literature review. *Frontiers in Virtual Reality*, *4*, 1159905. DOI: 10.3389/frvir.2023.1159905

Van Erp, T., Carvalho, N. G. P., Gerolamo, M. C., Gonçalves, R., Rytter, N. G. M., & Gladysz, B. (2024). Industry 5.0: A new strategy framework for sustainability management and beyond. *Journal of Cleaner Production*, *461*, 142271. DOI: 10.1016/j.jclepro.2024.142271

Van Rossem, C. (2008). *Individual producer responsibility in the WEEE Directive – From theory to practice?* [Doctoral thesis, International Institute for Industrial Environmental Economics, Lund University]. LUCRIS/Lund University. https://lucris.lub.lu.se/ws/portalfiles/portal/5603015/1266800.pdf

Vance, D., Jin, M., Price, C., Nimbalkar, S. U., & Wenning, T. (2023). Smart manufacturing maturity models and their applicability: A review. *Journal of Manufacturing Technology Management*, *34*(5), 735–770. DOI: 10.1108/JMTM-03-2022-0103

Vance, D., Jin, M., Wenning, T., Nimbalkar, S., & Price, C. (2025). Next-Level Energy Management in Manufacturing: Facility-Level Energy Digital Twin Framework Based on Machine Learning and Automated Data Collection. *Energies*, *18*(13), 3242. DOI: 10.3390/en18133242

Vater, J., Harscheidt, L., & Knoll, A. (2019).. . *Smart Manufacturing with Prescriptive Analytics*, *224*, 224–228. Advance online publication. DOI: 10.1109/ICITM.2019.8710673

Ventura, S. M., Bellagente, P., Rinaldi, S., Flammini, A., & Ciribini, A. L. C. (2023). Enhancing Safety on Construction Sites: A UWB-Based Proximity Warning System Ensuring GDPR Compliance to Prevent Collision Hazards. *Sensors (Basel)*, *23*(24), 9770. DOI: 10.3390/s23249770 PMID: 38139616

Verhoef, P. C., Broekhuizen, T., Bart, Y., Bhattacharya, A., Dong, J. Q., Fabian, N., & Haenlein, M. (2021). Digital transformation: A multidisciplinary reflection and research agenda. *Journal of Business Research*, *122*, 889–901. DOI: 10.1016/j.jbusres.2019.09.022

Verma, D. K. (2018). Analysis of Smart Manufacturing Technologies for Industry Using AI Methods. *Türk BilgisayarveMatematikEğitimi Dergisi*, *9*(2), 529–540. DOI: 10.17762/turcomat.v9i2.13857

Verpoort, P. C., Gast, L., Hofmann, A., & Ueckerdt, F. (2024). Impact of global heterogeneity of renewable energy supply on heavy industrial production and green value chains. *Nature Energy*, *9*(4), 491–503. DOI: 10.1038/s41560-024-01492-z

Verschuur, J., Koks, E. E., & Hall, J. W. (2021). Global economic impacts of the Suez Canal blockage. *Nature Sustainability*, *4*(9), 749–757. DOI: 10.1038/s41893-021-00738-9

Vespoli, S., Mattera, G., Marchesano, M. G., Nele, L., & Guizzi, G. (2025),Adaptive manufacturing control with Deep Reinforcement Learning for dynamic WIP management in industry 4.0, Computers & Industrial Engineering, Volume 202, 2025,110966, ISSN 0360-8352, doi:DOI: 10.1016/j.cie.2025.110966

Vial, G. (2019). Understanding digital transformation: A review and a research agenda. *The Journal of Strategic Information Systems*, *28*(2), 118–144. DOI: 10.1016/j.jsis.2019.01.003

Vidyalakshmi, G. V., Gopikrishnan, S., Boulila, W., Koubâa, A., & Srivastava, G. (2025). Digital Twins and Cyber-Physical Systems: A New Frontier in Computer Modeling. *Computer Modeling in Engineering & Sciences*, *143*(1), 51–113. DOI: 10.32604/cmes.2025.057788

Villalonga, A., Negri, E., Biscardo, G., Castaño, F., Haber, R. E., Fumagalli, L., & Macchi, M. (2021). A decision-making framework for dynamic scheduling of cyber-physical production systems based on digital twins. *Annual Reviews in Control*, *51*, 357–373. DOI: 10.1016/j.arcontrol.2021.04.008

Vysocky, A., & Novak, P. (2016). Human-robot collaboration in industry. *MM Science Journal*, *9*(2), 903–906. DOI: 10.17973/MMSJ.2016_06_201611

Wadhwa, A. S., Abbass, M., Akhai, S., Kumar, D., & Kumar, P. (2025). Integrating Taguchi optimization for multi-criteria decision making in engineering applications. In *Recent theories and applications for multi-criteria decision-making* (pp. 125–150). IGI Global., DOI: 10.4018/979-8-3693-6502-1.ch005

Wadhwa, A. S., Akhai, S., Abbass, M., Chouksey, A., Tiwari, S., & Taneja, T. (2025). Computational intelligence-driven design and optimization of polyurethane belt-type oil skimmer for sustainable manufacturing using SolidWorks 3D CAD. In *Computational intelligence for sustainable manufacturing of advanced materials* (pp. 445–464). IGI Global., DOI: 10.4018/979-8-3693-7974-5.ch019

Wagire, A. A., Joshi, R., Rathore, A. P. S., & Jain, R. (2021). Development of maturity model for assessing the implementation of Industry 4.0: Learning from theory and practice. *Production Planning and Control*, *32*(8), 603–622. DOI: 10.1080/09537287.2020.1744763

Wahdaniah, S., Sucianti, R., Ambalele, E., & Tellu, A. H. (2023). Human Resource Management Transformation in the Digital Age: Recent Trends and Implications. *International Journal of Applied Research and Sustainable Sciences, 1*(3), 239–258. DOI: 10.59890/ijarss.v1i3.902

Walsh, G., Andersen, N. S., Stoianov, N., & Jänicke, S. (2023). Towards a New Symbology and Visual Interaction Design for Sub-Sea Military Operations. *Research Portal Denmark*, 24. https://local.forskningsportal.dk/local/dki-cgi/ws/cris-link?src=sdu&id=sdu-145ca22b-f48d-48cd-8ffd-bb523392d677&ti=Towards%20a%20New%20Symbology%20and%20Visual%20Interaction%20Design%20for%20Sub-Sea%20Military%20Operations

Walter, S. (2023). AI impacts on supply chain performance : A manufacturing use case study. *Discover Artificial Intelligence, 3*(1), 18. Advance online publication. DOI: 10.1007/s44163-023-00061-9 PMID: 40476886

Waltersmann, L., Kiemel, S., Stuhlsatz, J., Sauer, A., & Miehe, R. (2021). Artificial Intelligence Applications for Increasing Resource Efficiency in Manufacturing Companies—A Comprehensive Review. *Sustainability (Basel), 13*(12), 6689. DOI: 10.3390/su13126689

Wamba-Taguimdje, S. L., Fosso Wamba, S., Kala Kamdjoug, J. R., & Tchatchouang Wanko, C. E. (2020). Influence of artificial intelligence (AI) on firm performance: The business value of AI-based transformation projects. *Business Process Management Journal, 26*(7), 1893–1924. Advance online publication. DOI: 10.1108/BPMJ-10-2019-0411

Wang, C., Pan, R., Wan, X., Tan, Y., Xu, L., McIntyre, R. S., Choo, F. N., Tran, B., Ho, R., Sharma, V. K., & Ho, C. (2020a). A longitudinal study on the mental health of general population during the COVID-19 epidemic in China. *Brain, Behavior, and Immunity, 87*, 40–48. Advance online publication. DOI: 10.1016/j.bbi.2020.04.028 PMID: 32298802

Wang, F., Zou, Y., Zhang, C., Buzzatto, J., Liarokapis, M., Castillo, E. del R., & Lim, J. B. P. (2023). UAV navigation in large-scale GPS-denied bridge environments using fiducial marker-corrected stereo visual-inertial localisation. *Automation in Construction, 156*, 105139. DOI: 10.1016/j.autcon.2023.105139

Wang, L., Wang, X. V., & Bi, Z. M. (2020b). Smart manufacturing and real-time monitoring: AI-enabled adaptive production control. *Journal of Manufacturing Systems, 57*, 457–467. DOI: 10.1016/j.jmsy.2020.10.006

Wang, S., Li, J., Song, J., Li, Y., & Sherk, M. (2018). Institutional pressures and product modularity: Do supply chain coordination and functional coordination matter? *International Journal of Production Research, 56*(20), 6644–6661. DOI: 10.1080/00207543.2018.1444807

Wang, S., Smith, M., & Gershenfeld, N. (2023). Voxelcopter. *Modular Autonomous Aerial Systems, 1*, 1–2. Advance online publication. DOI: 10.1145/3623263.3629155

Wang, S., Wan, J., Zhang, D., Li, D., & Zhang, C. (2020c). Towards smart factory for Industry 4.0: A self-organized multi-agent system with big data-based feedback and coordination. *Computer Networks, 101*, 158–168. DOI: 10.1016/j.comnet.2015.12.017

Wang, W., Wang, P., Zhang, H., Chen, X., Wang, G., Lu, Y., Chen, M., Liu, H., & Li, J. (2023). A Real-Time Defect Detection Strategy for Additive Manufacturing Processes Based on Deep Learning and Machine Vision Technologies. *Micromachines, 15*(1), 28. DOI: 10.3390/mi15010028 PMID: 38258148

Wang, X., Zhu, J., Su, C., & Zhen, X. (2024). Cascade failure modeling and resilience analysis of mine cyber physical systems under deliberate attacks. *Journal of Safety Science and Resilience = An Quan Ke Xue Yu Ren Xing (Ying Wen), 5*(3), 266–280. DOI: 10.1016/j.jnlssr.2024.03.007

Wan, J., Li, X., Dai, H.-N., Kusiak, A., Martinez-Garcia, M., Li, D., Wan, J., Li, X., Dai, H.-N., Kusiak, A., Martinez-Garcia, M., & Li, D. (2020). Artificial-Intelligence-Driven Customized Manufacturing Factory: Key Technologies, Applications, and Challenges. *Proceedings of the IEEE, 109*(4), 377–398. DOI: 10.1109/JPROC.2020.3034808

Waqas, A., Kang, D., & Cha, Y. (2023). Deep learning-based obstacle-avoiding autonomous UAVs with fiducial marker-based localization for structural health monitoring. *Structural Health Monitoring, 23*(2), 971–990. DOI: 10.1177/14759217231177314 PMID: 38405115

Watson, R., Kamel, M., Zhang, D., Dobie, G., MacLeod, C., Pierce, G., & Nieto, J. (2021). Dry Coupled Ultrasonic Non-Destructive Evaluation Using an Over-Actuated Unmanned Aerial Vehicle. *IEEE Transactions on Automation Science and Engineering, 19*(4), 2874–2889. DOI: 10.1109/TASE.2021.3094966

Weick, K. E., & Sutcliffe, K. M. (2007). *Managing the unexpected: Resilient performance in an age of uncertainty* (2nd ed.). Jossey-Bass., https://www.wiley.com/en-gb/Managing%2Bthe%2BUnexpected%3A%2BResilient%2BPerformance%2Bin%2Ban%2BAge%2Bof%2BUncertainty%2C%2B2nd%2BEdition-p-9780787996499

Westkämper, E., & Warnecke, H. J. (1994). Zero-Defect Manufacturing by Means of a Learning Supervision of Process Chains. *CIRP Annals*, *43*(1), 405–408. DOI: 10.1016/S0007-8506(07)62240-9

West, S. M., Whittaker, M., & Crawford, K. (2019). Discriminating systems. *AI Now*, *2019*, 1–33.

Whiteman, W. E. (1998). *Training and educating Army officers for the 21st century: Implications for the United States Military Academy*. U.S. Army War College. https://archive.org/details/DTIC_ADA345812

Wicaksono, H., Trat, M., Bashyal, A., Boroukhian, T., Felder, M., Ahrens, M., Bender, J., Groß, S., Steiner, D., July, C., Dorus, C., & Zoerner, T. (2024). Artificial-intelligence-enabled dynamic demand response system for maximizing the use of renewable electricity in production processes. *International Journal of Advanced Manufacturing Technology*, *138*(1), 247–271. DOI: 10.1007/s00170-024-13372-7

Wieland, A., & Wallenburg, C. M. (2013). The influence of relational competencies on supply chain resilience: A relational view. *International Journal of Physical Distribution & Logistics Management*, *43*(4), 300–320. DOI: 10.1108/IJPDLM-08-2012-0243

Wilson, H. J., & Daugherty, P. R. (2018). Collaborative intelligence: Humans and AI are joining forces. *Harvard Business Review*, *96*(4), 114–123. https://hometownhealthonline.com/wp-content/uploads/2019/02/ai2-R1804J-PDF-ENG.pdf

Womack, J. P., Jones, D. T., & Roos, D. (1990). *The machine that changed the world*. Rawson Associates. https://openlibrary.org/books/OL22145160M/The_machine_that_changed_the_world

World Bank. (2024). *World Development Indicators* [Database]. World Bank. https://datatopics.worldbank.org/world-development-indicators/

World Economic Forum. (2019, January). *Fourth Industrial Revolution: Beacons of Technology and Innovation in Manufacturing*. World Economic Forum. https://www3.weforum.org/docs/WEF_4IR_Beacons_of_Technology_and_Innovation_in_Manufacturing_report_2019.pdf

World Economic Forum. (2020). The future of jobs report 2020. https://www.weforum.org/reports/the-future-of-jobs-report-2020

World, V. U. C. A. (2025). *VUCA – Leaders with vision, understanding, clarity, agility!* https://www.vuca-world.org/

Wright, S., Pickton, D. W., & Callow, J. (2002). Competitive intelligence in UK firms: A typology. *Marketing Intelligence & Planning*, *20*(6), 349–360. DOI: 10.1108/02634500210445400

Wu, D., Ren, A., Zhang, W., Fan, F., Liu, P., Fu, X., & Terpenny, J. (2018). Cybersecurity for digital manufacturing. *Journal of Manufacturing Systems*, *48*, 3–12. DOI: 10.1016/j.jmsy.2018.03.006

Wu, D., Rosen, D. W., Wang, L., & Schaefer, D. (2013). Cloud-based design and manufacturing: A new paradigm in digital manufacturing and design innovation. *Computer Aided Design*, *45*(2), 104–118. DOI: 10.1016/j.cad.2012.08.006

Wuest, T., Weimer, D., Irgens, C., & Thoben, K. D. (2016). Machine learning in manufacturing: Advantages, challenges, and applications. *Production & Manufacturing Research*, *4*(1), 23–45. Advance online publication. DOI: 10.1080/21693277.2016.1192517

Xavier, B. M., Thollander, P., Hilletofth, P., & Johansson, M. (2024). Exploring energy management integration into upstream supply chains: A systematic literature review. *Frontiers in Energy Research*, *12*, 1425795. Advance online publication. DOI: 10.3389/fenrg.2024.1425795

Xiao, J., Zhang, R., Zhang, Y., & Feroskhan, M. (2025). Vision-Based Learning for Drones: A Survey. *IEEE Transactions on Neural Networks and Learning Systems*, *36*(9), 15601–15621. DOI: 10.1109/TNNLS.2025.3564184 PMID: 40327484

Xia, T., Sun, H., Ding, Y., Han, D., Qin, W., Seidelmann, J., & Xi, L. (2023). Digital twin-based real-time energy optimization method for production line considering fault disturbances. *Journal of Intelligent Manufacturing*, *36*(1), 569–593. DOI: 10.1007/s10845-023-02219-9

Xia, Y., & Ahmad, M. (2025). Industrial Robotics, Resource Efficiency, Energy Transition, and Environmental Quality: Designing a Sustainable Development Goals Framework for G7 Countries in the Presence of Geopolitical Risk. *Sustainability (Basel)*, *17*(5), 1960. DOI: 10.3390/su17051960

Xu, D., & Jaggars, S. S. (2014). Performance gaps between online and face-to-face courses: Differences across types of students and academic subject areas. *The Journal of Higher Education*, *85*(5), 633–659. DOI: 10.1080/00221546.2014.11777343

Xu, L. D., Xu, E. L., & Li, L. (2018). Industry 4.0: State of the art and future trends. *International Journal of Production Research*, *56*(8), 2941–2962. DOI: 10.1080/00207543.2018.1444806

Xu, X., Lu, Y., Vogel-Heuser, B., & Wang, L. (2021). Industry 4.0 and Industry 5.0—Inception, conception and perception. *Journal of Manufacturing Systems, 61*, 530–535. DOI: 10.1016/j.jmsy.2021.10.006

YahiaMarzouk, Y., & Jin, J.YahiaMarzouk. (2022). The relationship between environmental scanning and organizational resilience: Roles of process innovation and environmental uncertainty. *Frontiers in Environmental Science, 10*, 966474. DOI: 10.3389/fenvs.2022.966474

Yakovenko, Y., & Shaptala, R. (2024). Study of digital twins as the driving force of digital transformation and achieving the goals of sustainable development. *Technology Audit and Production Reserves*, *2*(4(76)), 11–20. DOI: 10.15587/2706-5448.2024.301423

Yan, W., Wang, J., Lu, S., Zhou, M., & Peng, X. (2023). A Review of Real-Time Fault Diagnosis Methods for Industrial Smart Manufacturing [Review of *A Review of Real-Time Fault Diagnosis Methods for Industrial Smart Manufacturing*]. *Processes, 11*(2), 369. Multidisciplinary Digital Publishing Institute. DOI: 10.3390/pr11020369

Yang, H., Rao, P., Simpson, T. W., Lu, Y., Witherell, P., Nassar, A. R., Reutzel, E. W., & Kumara, S. (2020). Six-Sigma Quality Management of Additive Manufacturing. *Proceedings of the IEEE, 109*(4), 347–376. DOI: 10.1109/JPROC.2020.3034519 PMID: 34248180

Yang, X., Smith, T. M., Prado, A., & Yang, Y. (2025). Net-zero greenhouse gas mitigation potential across multi-tier supply chains. *Communications Earth & Environment, 6*(1), 230. Advance online publication. DOI: 10.1038/s43247-025-02173-9

Yan, P., Abdulkadir, A., Luley, P.-P., Rosenthal, M., Schatte, G. A., Grewe, B. F., & Stadelmann, T. (2024). A Comprehensive Survey of Deep Transfer Learning for Anomaly Detection in Industrial Time Series: Methods, Applications, and Directions. *IEEE Access : Practical Innovations, Open Solutions, 12*, 3768–3789. DOI: 10.1109/ACCESS.2023.3349132

Yao, K., Lin, C.-L., & Pan, C.-H. (2024). Industrial Sustainable Development: The Development Trend of Programmable Logic Controller Technology. *Sustainability (Basel), 16*(14), 6230. DOI: 10.3390/su16146230

Yarger, H. R. (2006). *Strategic theory for the 21st century: The little book on big strategy*. U.S. Army War College Press., https://press.armywarcollege.edu/monographs/723/

Yazdi, M. (2024). Augmented reality (AR) and virtual reality (VR) in maintenance training. In *Advances in Computational Mathematics for Industrial System Reliability and Maintainability* (pp. 169–183). Springer Nature Switzerland. DOI: 10.1007/978-3-031-53514-7_10

Yitmen, İ., & Almusaed, A. (2024). Synopsis of Industry 5.0 Paradigm for Human-Robot Collaboration. In *Artificial intelligence*. IntechOpen., DOI: 10.5772/intechopen.1005583

Yi, W., & Sutrisna, M. (2020). Drone scheduling for construction site surveillance. *Computer-Aided Civil and Infrastructure Engineering*, *36*(1), 3–13. DOI: 10.1111/mice.12593

Yoo, S., Reza, S., Tarashiyoun, H., Ajikumar, A., & Moghaddam, M. (2024). AI-Integrated AR as an Intelligent Companion for Industrial Workers: A Systematic Review. *IEEE Access : Practical Innovations, Open Solutions*, *12*, 191808–191827. DOI: 10.1109/ACCESS.2024.3516536

Yuan, M. L., Ong, S. K., & Nee, A. Y. C. (2008). Augmented reality for assembly guidance using a virtual interactive tool. *International Journal of Production Research*, *46*(7), 1745–1764. DOI: 10.1080/00207540600972935

Yuan, Y., Ma, G., Cheng, C., Zhou, B., Zhao, H., Zhang, H., & Ding, H. (2019). *Artificial Intelligent Diagnosis and Monitoring in Manufacturing. arXiv*. Cornell University., DOI: 10.1093/nsr/nwz190

Yu, J., Welford, R., & Hills, P. (2006). Industry responses to EU WEEE and ROHS Directives: Perspectives from China. *Corporate Social Responsibility and Environmental Management*, *13*(5), 286–299. DOI: 10.1002/csr.131

Zahid, A., Ferraro, A., Petrillo, A., & De Felice, F. (2025). Exploring the Role of Digital Twin and Industrial Metaverse Technologies in Enhancing Occupational Health and Safety in Manufacturing. *Applied Sciences (Basel, Switzerland)*, *15*(15), 8268. DOI: 10.3390/app15158268

Zainol, Z., Hassan, K., Wan Hussein, W. M. H., & Phuoc, J. C. (2019). *Adaptive regulation for Industry 4.0.* In *Proceedings of the International Conference on Social Science 2019 (ICSS 2019)* (pp. 1180–1183). Atlantis Press (Advances in Social Science, Education and Humanities Research). https://doi.org/DOI: 10.2991/icss-19.2019.8

Zaphiris, P., & Ioannou, A. (2015). *Learning and Collaboration Technologies* (Vol. 9192). Springer. DOI: 10.1007/978-3-319-20609-7

Zavala-Alcívar, A., Verdecho, M. J., & Alfaro-Saiz, J. J. (2020). A conceptual framework to manage resilience and increase sustainability in the supply chain. *Sustainability (Basel)*, *12*(16), 6300. DOI: 10.3390/su12166300

Zhang, C., Juraschek, M., & Herrmann, C. (2024). Deep reinforcement learning-based dynamic scheduling for resilient and sustainable manufacturing: A systematic review [Review of *Deep reinforcement learning-based dynamic scheduling for resilient and sustainable manufacturing: A systematic review*]. *Journal of Manufacturing Systems, 77*, 962. Elsevier BV. DOI: 10.1016/j.jmsy.2024.10.026

Zhang, K., Chermprayong, P., Xiao, F., Tzoumanikas, D., Dams, B., Kay, S., Koçer, B. B., Burns, A., Orr, L., Alhinai, T., Choi, C. Y., Darekar, D. D., Li, W., Hirschmann, S., Soana, V., Ngah, S. A., Grillot, C., Sareh, S., Choubey, A., & Kovač, M. (2022). Aerial additive manufacturing with multiple autonomous robots. *Nature*, *609*(7928), 709–717. DOI: 10.1038/s41586-022-04988-4 PMID: 36131037

Zhang, Y., Xu, C., Yu, S., & Xie, S. (2019). Machine vision-based inspection for industrial quality control: A review. *Journal of Manufacturing Systems*, *51*, 1–11. DOI: 10.1016/j.jmsy.2019.08.001

Zhao, G., Ruan, D., Wang, Q., Zhang, X., & Wang, Y. (2018). Systemic boundaries in industrial systems: A new concept defined to improve LCA for metallurgical and manufacturing systems. *Journal of Cleaner Production*, *187*, 717–729. DOI: 10.1016/j.jclepro.2018.03.263

Zhong, R. Y., Xu, C., Chen, C., & Huang, G. Q. (2017). Big Data Analytics for Physical Internet-based intelligent manufacturing shop floors. *International Journal of Production Research*, *55*(9), 2610–2621. Advance online publication. DOI: 10.1080/00207543.2015.1086037

Zhong, S., & Pearce, J. M. (2017). Tightening the loop on the circular economy: Coupled distributed recycling and manufacturing with recyclebot and RepRap 3-D printing. *Resources, Conservation and Recycling*, *128*, 48–58. DOI: 10.1016/j.resconrec.2017.09.023

Zhong, W., Rasouli, S., Singh, A. K., Mohandes, S. R., Antwi-Afari, M. F., Cheung, C., Manu, P., & Agrawal, U. S. (2025). The Adoption of UAVs for Enhancing Safety in Construction Industry: A Systematic Literature Review. *Intelligent Infrastructure and Construction*, *1*(1), 3. DOI: 10.3390/iic1010003

Zhong, Y., Wang, Z., Yalamanchili, A., Yadav, A., Srivatsa, B. N. R., Saripalli, S., & Bukkapatnam, S. (2020). Image-based flight control of unmanned aerial vehicles (UAVs) for material handling in custom manufacturing. *Journal of Manufacturing Systems*, *56*, 615–621. DOI: 10.1016/j.jmsy.2020.04.004

Zhou, T., Dallasanta, K., Nazarenko, L., Schmidt, G. A., & Jin, Z. (2021). The impact of increasing stratospheric radiative damping on the quasi-biennial oscillation period. *Atmospheric Chemistry and Physics, 21*(9), 7395–7407. Advance online publication. DOI: 10.5194/acp-21-7395-2021

Zhou, Z., & Zhang, J. (2025). Manufacturing enterprise digital transformation, financial flexibility, and financial risk—Evidence from China. *International Review of Financial Analysis, 104*, 104279. Advance online publication. DOI: 10.1016/j.irfa.2025.104279

Zhu, Q. (2016). *Institutional pressures and support from industrial zones for motivating sustainable production among Chinese manufacturers. International Journal of Production Economics, 181*(Part B), 402–409. https://doi.org/DOI: 10.1016/j.ijpe.2016.06.027

Zhukabayeva, T., Zholshiyeva, L., Karabayev, N., Khan, S., & Alnazzawi, N. (2025). Cybersecurity Solutions for Industrial Internet of Things–Edge Computing Integration: Challenges, Threats, and Future Directions. *Sensors (Basel), 25*(1), 213. DOI: 10.3390/s25010213 PMID: 39797003

Zhu, S., Hu, H., Yang, H., Qu, Y., & Li, Y. (2023). Mini-Review of Best Practices for Greenhouse Gas Reduction in Singapore's Semiconductor Industry. *Processes (Basel, Switzerland), 11*(7), 2120. DOI: 10.3390/pr11072120

Zhu, X., Lai, P. L., Li, X., Wang, Y., & Pei, X. (2025). A Systematic Review of Literature on the Use of Extended Reality in Logistics and Supply Chain Management Education: Evolution of Research Themes and System-Level Trends. *Systems, 13*(7), 514. DOI: 10.3390/systems13070514

Zoubaidi, Z., Herrou, B., & Sekkat, S. (2025). Towards Intelligent Manufacturing: AI Based Defect Prediction and MES Integration. *Research Square (Research Square)*. DOI: 10.21203/rs.3.rs-7521361/v1

Zou, J., Chang, Q., Arinez, J., & Xiao, G. (2017). Data-driven modeling and real-time distributed control for energy efficient manufacturing systems. *Energy, 127*, 247–257. DOI: 10.1016/j.energy.2017.03.123

Zschech, P., Weinzierl, S., & Kraus, M. (2025). Inherently Interpretable Machine Learning: A Contrasting Paradigm to Post-hoc Explainable AI. *Business & Information Systems Engineering*. Advance online publication. DOI: 10.1007/s12599-025-00964-0

Попова, Л. (2024). Immersive Technologies as a Modern Educational Strategy for Training Future Specialists. *Педагогічна освіта: Теорія і практика. Психологія. Педагогіка*, (42(2)), 51–58.

About the Contributors

Shalin Hai-Jew works as a case manager for Sedgwick County. Prior, she has worked for many years in higher education, as a grant writer, instructional designer, instructor, and other roles. She has B.A.s in English and psychology from the University of Washington (Seattle), a master's in English (UW), and an Ed.D. in Educational Leadership from Seattle University, where she was a Morford Scholar. She tested into the University of Washington at age 14. She has lived and worked abroad for four years. She has edited several dozen nonfiction texts around the topics of education and technologies, among others. She has researched and written a dozen edited texts. She was born in Huntsville, Alabama, in the U.S.

***

Mahapara Abbass is a distinguished academician and researcher in structural and sustainable engineering, with expertise in low-carbon materials, geopolymer concrete, and AI-enabled applications in urban and infrastructure systems. Her professional contributions include teaching at UG and PG levels, supervising M.Tech and Ph.D. research scholars, and mentoring innovation-driven projects. Recognized among Emerging Global Researchers (2025) in sustainable materials and AI-driven engineering, Dr. Abbass has published numerous research papers in SCI, SCIE, and Scopus-indexed journals, authored several academic books, and secured multiple design and utility patents. A respected reviewer, technical expert, and research mentor, she continues to advance sustainable construction, smart materials, and technology-driven engineering solutions.

Shalom Akhai is a distinguished engineering academician, researcher, and innovator with more than eighteen years of professional experience in academia, industry, and consultancy. He is the Founder of IGNIS Technical Solutions, an MSME-registered consultancy specializing in sustainable systems, energy-efficient design, and AI-enabled smart technologies. Recognized among the World's Top 2% Scientists (Stanford–Elsevier Global Ranking 2025), Dr. Akhai has authored over 100 publications in SCI, Scopus, IEEE, CRC, Springer, and IGI Global journals and books, along with more than 70 granted or published patents. He has also written nine academic books and presented his work at numerous national and international conferences. A respected reviewer, mentor, and innovation ambassador, Dr. Akhai continues to contribute significantly to sustainable engineering, advanced technologies, and research-driven development.

Ing. Otmane Azeroual is a world-renowned senior researcher, lecturer, and thought leader in digital transformation, artificial intelligence, and advanced data-driven research infrastructures. His groundbreaking interdisciplinary work at the nexus of computer science, higher education research, innovation management, and industry has set new standards in how data and AI reshape science and society. His research on the design, deployment, and critical evaluation of Current Research Information Systems (CRIS) — combined with cutting-edge AI-powered analytics — is internationally acclaimed and among the most cited in the field. Dr. Azeroual's pioneering contributions drive the evolution of intelligent data ecosystems that empower Open Science, elevate transparency, and revolutionize strategic decision-making in science policy worldwide. Holding a Ph.D. in Engineering Informatics and a robust foundation in business information systems and software engineering, he has authored numerous high-impact publications in top-tier journals and has been a sought-after keynote and plenary speaker at prestigious global conferences. His expertise is recognized by editorial boards and scientific committees internationally, where he shapes the future agenda of research data science. Bridging academia and private industry, Dr. Azeroual uniquely fuses rigorous theoretical insight with practical innovation. His leadership in both sectors fuels transformative digital strategies and sustainable technological solutions. As an inspiring educator, he equips emerging researchers and professionals with cutting-edge knowledge in AI, data science, research analytics, and project management — preparing them to excel in complex, dynamic environments. Passionate about harnessing AI for evidence-based governance, Dr. Azeroual actively collaborates across disciplines to build resilient, transparent, and data-driven research infrastructures that set global benchmarks. His visionary work not only advances scientific understanding but also catalyzes paradigm shifts that benefit institutions, governments, and society at large.

K. Arthi, working as Associate Professor in SRM University, has more than 13 years of teaching and research experience. She has earned her M.E. and Ph.D. in Computer Science Engineering from College of Engineering, Anna University, Chennai in 2005 and 2015 respectively. Specializing in IoT and Machine Learning, she has published more than 25 articles in reputed journals and holds two patents in her area of specialization. She has professional membership in ACM and ISTE, through which she has organised multiple events.

M. Vasudevan is a seasoned Divisional Manager with over 21 years of experience in the automotive industry. Currently, he serves as a PMO Lead at Switch Mobility Limited, part of the Hinduja Group, specializing in electric vehicle manufacturing and based in Chennai. His extensive expertise spans Plant Engineering, Manufacturing, Product Development, Process Management and Development (NPI), Service, and Project Management. Recognized for his passion in training and consulting, Vasudevan holds multiple certifications, including a Premiership in Business Management from the ICTRD, Project Management certification from NPTEL, MS Project Management from Microsoft, and Zone Trainer certification from JCI. His academic qualifications are equally distinguished, with degrees including a B.E. in Industrial Engineering, an M.S. in Engineering Management, an M.A. in Business Economics, an MBA in Operations Management, and diplomas in Safety and Machine Tool Maintenance and Repairs. Currently, Vasudevan is pursuing a Ph.D. in Management from Hindustan Institute of Technology and Science (HITS)

Priyanka N. is currently working as an assistant professor at the Vellore Institute of Technology, Vellore. She did her Ph.D. at Anna University, Chennai. She did her Master's degree in software engineering at Anna University, Chennai. Her areas of interest include software-defined networks, next-generation wireless networks, the Internet of Things, and network security.

R. Renugadevi completed her PhD in 2022 from Anna University, Chennai. She is currently working as Associate professor in Saveetha Engineering College. She has more than 20 years of teaching experience in engineering colleges.She published more than 40 papers in International journals, conferences and holds 10 patents. Her research interest includes Machine learning, Internet of Things, Wireless Sensor Networks and Cloud computing. She is a life member of ISTE, CSI and IAENG.

Ardra Rianisa is an Information Technology professional with a strong interest in digital technology and information systems. Ardra actively engages in writing academic papers and popular articles that discuss current trends, innovations, and developments in the field of information technology. With a solid understanding of technology and its practical applications, Ardra is committed to continuously expanding knowledge and contributing insights to the digital and academic community.

Göran Roos is a globally recognised expert in intellectual capital management, manufacturing competitiveness, industry policy, and innovation systems. With around 19,000 Google citations and more than 300 publications, he has established himself as a leading authority on strategic management, economic complexity, and sustainable industrial development. Academic Credentials and Appointments Professor Roos holds a PhD from the University of South Australia (2019), an MBA from INSEAD (1989), and an MSc in Engineering Physics from Chalmers University of Technology (1982). His current academic positions and fellowships include: • Adjunct Professor at the Institute for Sustainability, Energy and Resources (ISER), The University of Adelaide (2023-2025) • Fellow of the Australian Academy of Technological Sciences and Engineering (ATSE) since 2014 • Fellow of the Royal Swedish Academy of Engineering Sciences (IVA) since 2015 Areas of Expertise Intellectual Capital and Knowledge Management Professor Roos is a world-renowned expert in intellectual capital, having developed theories and implemented practical solutions that have shaped the

Binastya Anggara Sekti has been a lecturer at the Faculty of Computer Science, Esa Unggul University, Jakarta, since 2016. He is actively involved in the Artificial Intelligence and Information Security Study Group. He holds a Bachelor's degree in Informatics Engineering from the Bandung Institute of Technology and a Master's degree in Information Systems Management from Trisakti University, Jakarta. Prior to entering academia, he accumulated over 25 years of professional experience as an IT practitioner in various companies across Jakarta. His work focused on developing data centers for internet service providers and delivering IT solutions for the manufacturing and oil & gas sectors. His research interests include artificial intelligence, cybersecurity, and the development of IT infrastructure.

J.Shobana received Ph.D. in Computer Science and Engineering. Working as an Assistant Professor in the Department of Computer Science and Engineering, SRM Institute of Science and Technology has been serving the Education Profession for the past 20 years. Her area of interest is Text Mining, Natural Language Processing, Artificial Intelligence and Machine Learning.

Index

A

C

D

E

G

H

I

M

O

P

Q

R

S

V

W

www.ingramcontent.com/pod-product-compliance
Lightning Source LLC
LaVergne TN
LVHW062334300926
849161LV00007B/1689
9798337357560